METHODS IN ENZYMOLOGY

Editors-in-Chief

ANNA MARIE PYLE
Departments of Molecular, Cellular and Developmental Biology and Department of Chemistry
Investigator, Howard Hughes Medical Institute
Yale University

DAVID W. CHRISTIANSON
Roy and Diana Vagelos Laboratories
Department of Chemistry
University of Pennsylvania
Philadelphia, PA

Founding Editors

SIDNEY P. COLOWICK and NATHAN O. KAPLAN

VOLUME SIX HUNDRED AND SEVENTY SEVEN

METHODS IN ENZYMOLOGY

Small Angle Scattering Part A: Methods for Structural Investigation

Edited by

JOHN A. TAINER
Robert A. Welch Chair in Chemistry
Departments of Molecular & Cellular Oncology
and of Cancer Biology Director Structural Biology
University of Texas MD Anderson Cancer Center
Houston, TX

ELSEVIER

ACADEMIC PRESS
An imprint of Elsevier

Academic Press is an imprint of Elsevier
50 Hampshire Street, 5th Floor, Cambridge, MA 02139, United States
525 B Street, Suite 1650, San Diego, CA 92101, United States
The Boulevard, Langford Lane, Kidlington, Oxford OX5 1GB, United Kingdom
125 London Wall, London, EC2Y 5AS, United Kingdom

First edition 2022

ISBN: 978-0-323-99179-7
ISSN: 0076-6879

For information on all Academic Press publications
visit our website at https://www.elsevier.com/books-and-journals

Publisher: Zoe Kruze
Developmental Editor: Federico Paulo S. Mendoza
Production Project Manager: Sudharshini Renganathan
Cover Designer: Greg Harris

Typeset by STRAIVE, India

Contents

Contributors

Pau Bernadó
Centre de Biologie Structurale (CBS), Université de Montpellier, INSERM, CNRS, Montpellier, France

Marina Cagnes
National Deuteration Facility, ANSTO, Lucas Heights, NSW, Australia

Leonie Chatzimagas
Theoretical Physics and Center for Biophysics, Saarland University, Saarbrücken, Germany

Tamim A. Darwish
National Deuteration Facility, ANSTO, Lucas Heights, NSW, Australia

Joshua T. Del Mundo
Department of Chemical Engineering, The Pennsylvania State University, University Park, PA, United States

Anthony P. Duff
National Deuteration Facility, ANSTO, Lucas Heights, NSW, Australia

Xianyang Fang
Beijing Advanced Innovation Center for Structural Biology, School of Life Sciences, Tsinghua University, Beijing, China

Frank Gabel
Univ. Grenoble Alpes, CEA, CNRS, IBS, Grenoble, France

José Gallego
CITSAM, Universidad Católica de Valencia, C/Quevedo 2, Valencia, Spain

Richard E. Gillilan
Cornell High Energy Synchrotron Source, Cornell University, Ithaca, NY, United States

Enrique D. Gomez
Department of Chemical Engineering; Department of Materials Science and Engineering and Materials Research Institute, The Pennsylvania State University, University Park, PA, United States

Esther W. Gomez
Department of Chemical Engineering; Department of Biomedical Engineering, The Pennsylvania State University, University Park, PA, United States

Melissa A. Graewert
European Molecular Biology Laboratory–Hamburg Unit, Hamburg, Germany

Michal Hammel
Molecular Biophysics and Integrated Bioimaging, Lawrence Berkeley National Laboratory, Berkeley, CA, United States

Jochen S. Hub
Theoretical Physics and Center for Biophysics, Saarland University, Saarbrücken, Germany

Greg L. Hura
Molecular Biophysics and Integrated Bioimaging, Lawrence Berkeley National Laboratory, Berkeley; Chemistry and Biochemistry Department, University of California Santa Cruz, Santa Cruz, CA, United States

Cy M. Jeffries
European Molecular Biology Laboratory (EMBL), Hamburg, Germany

Inokentijs Josts
The Hamburg Advanced Research Center for Bioorganic Chemistry (HARBOR); Department of Chemistry, Institute for Biochemistry and Molecular Biology, University of Hamburg, Hamburg, Germany

Dominique-Maurice Kehlenbeck
The Hamburg Advanced Research Center for Bioorganic Chemistry (HARBOR); Department of Chemistry, Institute for Biochemistry and Molecular Biology, University of Hamburg, Hamburg, Germany; Life Sciences Group, Institut Laue–Langevin; Partnership for Structural Biology (PSB), Grenoble, France

Anwen M. Krause-Heuer
National Deuteration Facility, ANSTO, Lucas Heights, NSW, Australia

Susan Krueger
NIST Center for Neutron Research, NIST, Gaithersburg, MD, United States

Annette Eva Langkilde
Department of Drug Design and Pharmacology, University of Copenhagen, Universitetsparken 2, Copenhagen, Denmark

Samuel Lenton
Drug Delivery and Biophysics of Biopharmaceuticals, Department of Pharmacy, University of Copenhagen, Copenhagen, Denmark; Division of Theoretical Chemistry, Lund University, Lund, Sweden

Anne Martel
Large Scale Structures Group, Institut Laue-Langevin, Grenoble, France

Michael Moir
National Deuteration Facility, ANSTO, Lucas Heights, NSW, Australia

Julius Nitsche
Department of Chemistry, Institute for Biochemistry and Molecular Biology, University of Hamburg, Hamburg, Germany

Suzette A. Pabit
School of Applied and Engineering Physics, Cornell University, Ithaca, NY, United States

Javier Pérez
Orme des Merisiers, Saint-Aubin, France

Zuzanna Pietras
Department of Physics, Chemistry, and Biology, Linköping University, Linköping, Sweden

Lois Pollack
School of Applied and Engineering Physics, Cornell University, Ithaca, NY, United States

Carl Recsei
National Deuteration Facility, ANSTO, Lucas Heights, NSW, Australia

Agata Rekas
National Deuteration Facility, ANSTO, Lucas Heights, NSW, Australia

Sintu Rongpipi
Advanced Light Source and Materials Science Division, Lawrence Berkeley National Laboratory, Berkeley, CA, United States

Daniel J. Rosenberg
Molecular Biophysics and Integrated Bioimaging, Lawrence Berkeley National Laboratory; Graduate Group in Biophysics, University of California Berkeley, Berkeley, CA, United States

Robert A. Russell
National Deuteration Facility, ANSTO, Lucas Heights, NSW, Australia

Amin Sagar
Centre de Biologie Structurale (CBS), Université de Montpellier, INSERM, CNRS, Montpellier, France

Josue San Emeterio
School of Applied and Engineering Physics, Cornell University, Ithaca, NY, United States

Marie Skepö
Division of Theoretical Chemistry, Lund University; LINXS—Institute of Advanced Neutron and X-ray Science, Lund, Sweden

Dmitri I. Svergun
European Molecular Biology Laboratory–Hamburg Unit, Hamburg, Germany

Aurélien Thureau
Orme des Merisiers, Saint-Aubin, France

Henning Tidow
The Hamburg Advanced Research Center for Bioorganic Chemistry (HARBOR); Department of Chemistry, Institute for Biochemistry and Molecular Biology, University of Hamburg, Hamburg, Germany

Mark D. Tully
BioSAXS beamline, BM29, European Synchrotron Radiation Facility, ESRF, Grenoble, France

Patrice Vachette
Université Paris-Saclay, CEA, CNRS Institute for Integrative Biology of the Cell, Gif-sur-Yvette, France

Bente Vestergaard
Department of Drug Design and Pharmacology, University of Copenhagen, Universitetsparken 2, Copenhagen, Denmark

Yun-Xing Wang
Protein-Nucleic Acid Interaction Section, Center for Structural Biology, National Cancer Institute, National Institutes of Health, Frederick, MD, United States

Andrew E. Whitten
Australian Nuclear Science and Technology Organisation, New Illawarra Rd, Lucas Heights, NSW, Australia

Karyn L. Wilde
National Deuteration Facility, ANSTO, Lucas Heights, NSW, Australia

Kathleen Wood
Australian Nuclear Science and Technology Organisation, New Illawarra Rd, Lucas Heights, NSW, Australia

Lin Yang
National Synchrotron Light Source II, Brookhaven National Laboratory, Upton, NY, United States

Nageshwar R. Yepuri
National Deuteration Facility, ANSTO, Lucas Heights, NSW, Australia

Preface

"*And yet it moves*"—This phrase, attributed to Galileo Galilei, emphasizes the critical importance of defining matter with its movement over time for understanding a system.

Advances in experimental and computational structural methods are increasingly making transformative improvements in defining and predicting biological mechanisms. At the heart of these improvements is the critical ability to visualize both structural detail at the level of their molecular building blocks along with their changing conformation and assemblies as a function of time. Prior *Methods in Enzymology* volumes on diffraction methods and cryo-electron microscopy have presented detailed techniques for accurate and often high-resolution structures for crystallized and cryo-cooled biological macromolecules and assemblies. Here the contributing authors for *Scattering Methods for Biological Macromolecules* focus on procedures employing the scattering of X-rays (SAXS) and neutrons (SANS) by biological macromolecules, as divided into Volume 677, *Part A, Methods for Structural Investigation* and Volume 678, *Part B, Methods for Structural Interpretation.* These scattering methods can comprehensively measure and model molecular structures in solution, including their flexibility and dynamic functional conformations and assemblies. The analyses not only can approach near physiological conditions but also can broadly evaluate solution states and conditions so as to better understand the relationships between macromolecular structures, their solution environments, and their functionally important flexibility and dynamics. In essence, defining the time-dependent conformation and assembly state of macromolecules in their biological environment is foundational and predictive knowledge for their activities and regulation. In this context, these volumes encompass the capabilities that make scattering methods a premier approach for measuring molecular conformations and assemblies in both space and time, as well as for efficiently leveraging information on atomic models from all other techniques.

Each volume is organized into sections and topics pertaining to the spectrum of scattering methods and the logical sequence of steps involved in first investigating (PART A) and then interpreting structure (PART B). Part A is primarily devoted to the experimental aspects of scattering experiments and data, starting from sample preparation and handling, followed by methods of data collection and data assessment. Part B primarily discusses data and structural interpretation, covering various aspects of scattering measurements and methods for their analysis as well as the resulting structural insights. Armed by the information in these articles, the reader can proceed knowledgeably and productively toward their structural goals with an appreciation of the practical and fundamental aspects of each technique, including its potential to substantially augment combined and integrated structural approaches.

In providing these two volumes on scattering methods, the authors and I collectively strived to attain a threefold overall goal: (1) to give biochemists and biologists an introduction to the field of macromolecular structure analysis by scattering, offering them practical guidance to available techniques and pathways to gain functionally important insights into macromolecular structures; (2) to give scattering practitioners a comprehensive spectrum of current techniques available to them from robust to state-of-the-art and into pushing the technology's frontiers; and (3) to provide sufficient knowledge and guidance for interested researchers to effectively apply scattering approaches to combined methods and integrated structural biology. The authors of these chapters include both those upon whose shoulders we stand in developing new methods as well as established leaders and new pioneers in extending the reach and impact of scattering methods for structural biology.

In Volume 677 Graewert and Svergun provide the strong foundation for *Part A, Methods for Structural Investigation* in Chapter 1, by comprehensively describing sample environments and requirements for biological SAXS. In Chapters 2–5, Pollack, Duff, Krueger, and Jeffries and colleagues provide expert information on equipment and procedures to harness the power of neutrons for contrast and other experiments. In Chapters 6–7, Rosenberg and Vachette and their colleagues detail considerations and methods for coupling size-exclusion chromatography with scattering (SEC-SAXS) including informative videos. Tidow and Jost along with Martel and Gabel then describe powerful approaches to time-resolved SAXS (Chapter 8) and SANS (Chapter 9). Vestergaard and Langkilde (Chapter 10) provide an essential guide for the application of SAXS to the biologically and medically important process of protein fibrillation. Pioneering methods for high-pressure SAXS experiments and for soft X-ray scattering are described by Gillilan (Chapter 11) and Gomez and colleagues (Chapter 12), respectively. In Chapter 13, Yang discusses SAXS on lipid membranes and in Chapter 14 Todow and colleagues present SANS on membrane protein by stealth constitution systems. Predicting scattering with explicit solvent is detailed by Hub and Chatzimagas, who describe the WAXSiS web server for calculation of SAXS/WAXS curves based on explicit-solvent molecular dynamics, as a useful tool to extract maximum information from the SAXS profile (Chapter 15). In Chapter 16, Skepö and colleagues describe sample preparation and data collection considerations for applying SAXS to highly flexible protein samples. Wang and colleagues (Chapter 17) consider SAXS analyses of RNA, including an approach to simplifying exploration of conformational space and tying this in with SAXS constraints. In Chapter 18, Bernadó and Sagar provide unique information for enhanced biological insights from SAS data, including conformational rearrangements and transient biomolecular complexes, to anchor this volume.

Trewhella provides the foundation for Volume 678, *Part B, Methods for Structural Interpretation* in Chapter 1, with insights into the method and its development into a mainstream technique at the frontier of integrative structural biology, including key standards for data quality, model validation, publication, and data sharing. In Chapter 2, Hub and Chatzimagas describe a rigorous protocol for Bayesian interpretation of SAXS-driven MD simulations, models, and methods that will aid understanding of biological function by enabling understanding of conformational dynamics. Whitten and Jeffries describe SANS data analysis and modeling (Chapter 3), including a clear theoretical overview plus a step-by-step procedure for available software. Gabel describes time-resolved small-angle neutron scattering (Chapter 4) with an exemplary macromolecular system including the practical use of temperature and thermophilic systems to control kinetics. In Chapter 5, Gomez and colleagues describe informative procedures for predicting scattering contrast, identifying energies plus methods for interpretation suitable for measurements using resonant soft X-ray scattering (RSoXS) and near-edge X-ray absorption fine structure (NEXAFS) spectroscopy to gain structural insights. Grant (Chapter 6) describes his *ab initio* modeling algorithm DENSS, which allows examination of low-resolution density fluctuations inside the particle envelope, which would seem to have many biological applications. In Chapter 7, Mertens considers the components and developments in the ATSAS package for its user-driven applications informing data reduction, analysis, and modeling strategies. Schneidman and colleagues describe a protocol for accurate integrative modeling of antibody-antigen complexes in Chapter 8, including SAXS profiles, protein-protein docking, deep-learning models, and statistical potentials. In Chapter 9, Sattler and colleagues present the integration of SAS and

NMR as complementary solution structural techniques to better describe structure with conformational dynamics with application guidelines depending upon the nature of the system. Skepö and coauthors present SAXS spectra analyses for highly flexible protein sample solutions in Chapter 10 with the definable parameters, model-free analysis methods, plus advanced methods to examine ensembles and crowded solutions. In Chapter 11, Brosey and colleagues present an efficient protocol for applying high-throughput SAXS to screening the binding of chemical ligands. In Chapter 12, Tsutakawa and colleagues consider SAXS applications to improved structure prediction. Vestergaard and Langkilde (Chapter 13) describe analysis of protein fiber structures by SAXS. In Chapter 14, Hura and Murray reveal elegant and practical web-accessible approaches to analyses of similarity and correlated SAXS data sets to complete the volume.

In practice it is nearly impossible to navigate through scattering experiments, controls, data validation and analysis, and data interpretation, which can be conceptually simple but are usually technically sophisticated, without more methods detail than is available in published papers. Journal publications typically are space limited and focused upon new results. In their sharing of techniques and methods that they have painstakingly developed, validated, and optimized by years of tests and efforts, the authors herein made a remarkable contribution and investment toward the reproducibility and advanced impact of scattering for structural biology. In this regard, I acknowledge the generosity of all the contributing authors who strived to reveal their insights and explicit instructions on how to utilize their valuable methods to our readers. I furthermore thank our many reviewers, whose time and attention contributed valuable suggestions to improve the chapters. I thank Jill Trewhella, Greg Hura, and Michal Hammel for insightful thoughts on chapters and reviewers. I am grateful to Anna Pyle and David Christianson (Co-Editors-in-Chief of *Methods in Enzymology*) for the invitation and strong support. Most credit for the quality of the two volumes should go to the authors and reviewers, along with David Christianson, who was a force of nature in implementing reviewing and editing. I thank Paulo Mendoza and the Elsevier staff for their patient and expert efforts at all stages to strengthen the volumes for readers. I acknowledge my National Cancer Institute R35 CA220430 support for my efforts on the project "*Mesocale and Nanoscale Technologies Integrated by Structures for DNA Repair Complexes* (MANTIS-DRC)," and my Welch Chemistry Chair, which together enabled me to work on SAXS methods development and applications. I admit that I am responsible for any mistakes and omissions. Yet, due to the combined efforts of authors, reviewers, and editors, the chapters in these two volumes provide a depth of coverage and clarity of presentation suitable for readers spanning from students to experts.

Together the combined chapters show that scattering methods have advanced to the point that these articles are likely to become classic "go to" references that stand the test of time. So we expect that many laboratories will find these collected methods helpful as a starting point to adapt for their own systems and to extend these methods for in-depth understanding of mechanism, regulation, and functions. Reflecting on these chapters, it is an honor to be the editor of both volumes, where leaders and innovators I have admired, along with early investigators who are pushing methods forward into new frontiers, provide inspiration and insights to me and others working on hybrid methods to unveil biology at the nanoscale. In my view, the collected articles in these two volumes provide a critical enabling path to defining macromolecular structures and mechanisms in solution, including biologically relevant flexibility in both functionally important complexes and conformations.

John A. Tainer
University of Texas MD Anderson Cancer Center

CHAPTER ONE

Advanced sample environments and sample requirements for biological SAXS

Melissa A. Graewert* and Dmitri I. Svergun
European Molecular Biology Laboratory–Hamburg Unit, Hamburg, Germany
*Corresponding author: e-mail address: m.graewert@embl-hamburg.de

Contents

Abstract

In this chapter, we discuss the various sample delivery systems that are available at most biological SAXS beamlines. The focus is laid on the EMBL Biosaxs beamline P12 at the Petra 3 storage ring on the DESY site in Hamburg, Germany.

The minimal requirements necessary to prepare samples are described specifically for macromolecular samples in solution and the background is given on how the physical properties of the scattering process itself determine the sample requirements. We provide a number of exemplary applications as well as guidelines for selecting and performing the right data collection strategy depending on the scientific question at hand. This chapter is aimed at novices to the SAXS technique with biochemical background as well as more experienced users setting out to employ more advanced SAXS set-ups.

Methods in Enzymology, Volume 677
ISSN 0076-6879
https://doi.org/10.1016/bs.mie.2022.08.009

Abbreviations

SAXS small angle X-ray scattering
SEC size exclusion chromatography

1. Introduction

The notion that a biological small angle scattering (SAXS) measurement is a conceptually simple experiment is widely accepted. This argument is often used to advertise for the addition of this technique to integrated research projects (Da Vela & Svergun, 2020; Trewhella, 2022). In general, the sample is "simply" placed in an X-ray beam and the scattered photons are recorded on a (typically, two-dimensional) detector. However, due to the underlining physics of the scattering process itself, there are a number of issues that have to be considered when designing how the sample is actually placed into the beam (Fig. 1).

These include—but are not restricted to

I. the per se very low scattering signal of the macromolecules—this requires the reduction of parasitic scattering from the instrument

II. the subtractive nature of the scattering experiment (to account for background scattering) requires a two-step measurement (sample and pure solvent) in a reproducible manner

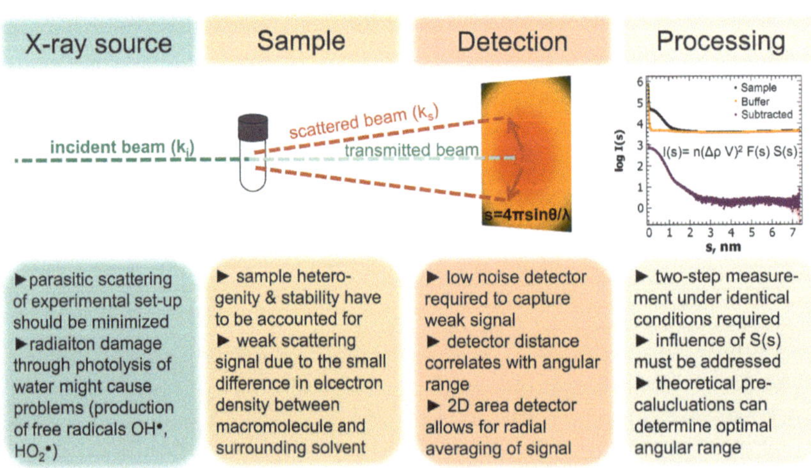

Fig. 1 Schematic representation of a SAXS experiment and pinpointing of various considerations for instrumental set-up. Note, scattering angle between the transmitted and scattered beam is depicted here as 2θ.

III. the potential alteration of the molecules through radiation induced effects (radiation damage)
IV. the generally limited amount of sample to conduct the experiments due to the difficulties often encountered when expressing and purifying biological macromolecules
V. the competitive access for beam time requiring optimization of individual steps such as sample reloading, cleaning of sample chamber between the samples

To address these issues and enable biological solution scattering experiments for a broader community, essential amount of effort has been placed into the automation of sample loading at synchrotron radiation (SR) facilities offering SAXS technology as well as by developers of laboratory-based SAXS instruments. Specifications of such sample loading devices include measures to not only automate loading and cleaning processes, but can also allow for some sample moderation on-the-fly. Examples and applications of modern sample loading robots (for automated "batch" analysis) are discussed in Section 3.

A more recent milestone in sample delivery for biological solution SAXS has been the introduction of online purification systems. Here, the incorporation of liquid chromatography (LC) based separation techniques such as size exclusion chromatography (SEC) is used to improve the sample quality of the SAXS data by ensuring sample monodispersity. This data collection mode, referred to as SEC-SAXS, is in high demand by the user community. Examples, applications and further developments are highlighted in Section 4.

Based on our experience at the EMBL beamline P12 (Petra3 storage ring at DESY, Hamburg, Germany) a large majority of measurements are performed with the automated sample loading robot (batch mode) and/or SEC-SAXS mode (Fig. 2). About 10% of the experiments performed at P12 employ specialized sample environments. These environments are used for the analysis of (i) tricky systems (toxic, sticky, highly viscose, etc.), (ii) time-resolved studies, (iii) screening methods and (iv) for other more unique questions. Due to the current upgrades of synchrotron radiation (SR) facilities to 4th generation SR, we expect the demands of screening options to gain in popularity. The anticipated increase in coherence and flux, as well as decrease in expected sample consumption will open new venues for SAXS experiments, which we could call 4D-SAXS. Here, the notion is to probe the changes of the 3-dimensional composition of the sample by adding a fourth dimension. This concept may include screening

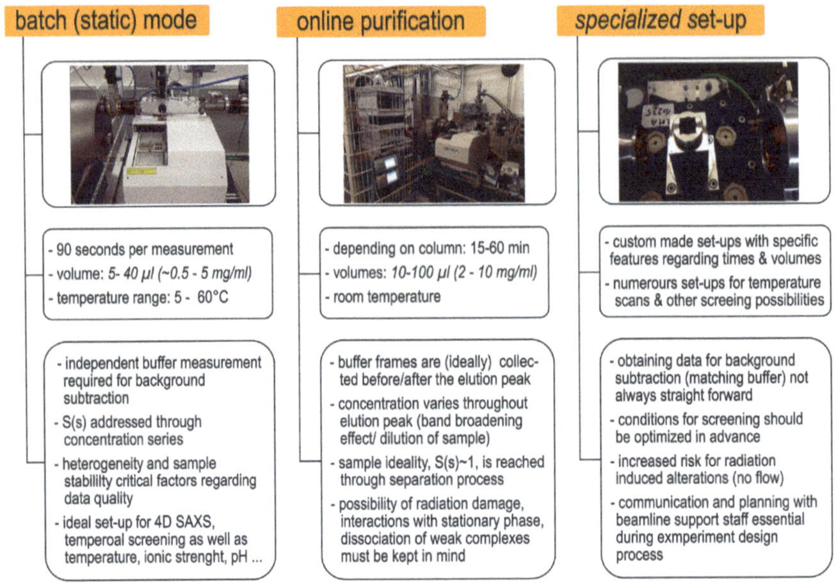

Fig. 2 Comparison of various biological SAXS data collection modes. Indicated values refer to the reference parameters for P12.

for temporal changes or reactions dependent on other parameters such as ligand binding, complex formation, ionic strength, pH, etc. Section 5 gives examples of various sample environments used at P12 and allowing for 4D-SAXS studies.

Once the mode of data collection is selected, the next question arises regarding sample requirements. To understand the essential steps to generate good data an overall understanding to the scattering process is key, to define the best strategy for collecting reasonable data.

2. Correlation between the physical process of the scattering event and sample requirements

Over the past two decades, the power of SAXS to complement and support a number of structural as well as bioanalytical studies has been well accepted by the research community and scattering data has been incorporated into a long list of integrated research projects (Brosey & Tainer, 2019; Trewhella, 2022). The dissemination of these success stories and description of the unique niche that SAXS occupies has naturally inspired other scientists to seek out the possibility to add SAXS to their own research.

Thereby the first questions asked by the novices about this technique refer to the sample requirements: how much material is needed and are there any restrictions to the solvent?

Fig. 3 demonstrates that the answers to these questions are connected to a number of factors. Ultimately, the sample properties and the posed scientific questions will determine the data collection strategy, which is best suited for the individual projects. Depending on this, the sample requirements can then be defined.

In the following chapters, the different sample modes are described in more detail along with guidelines on how to prepare for data collection. Examples of recent studies are given.

For all collection strategies, a simplified understanding of the physical properties of the scattering process can be beneficial as these directly translate into a better understanding of the sample requirements. This may guide a SAXS user to optimize the sample preparation to obtain good scattering data and assess if the quality is sufficient to draw the desired conclusions.

As we demonstrate later, automation at the SAXS beamline—regarding not only measurement itself but also in terms of automated on-the-fly analysis—has transferred SAXS to an almost "black box operation," allowing one to obtain a wrath of useful information such as ab initio models along with a number of valuable SAXS derived parameters including molecular weight (MW) estimates, radius of gyration, R_g, and maximum dimension, D_{max}. At P12, these results are automatically reported to the user within minutes of the experiment.

However, to prepare for the measurements and to perform on-site trouble shooting a short list of key principles should be taken into account and their influence on the data quality understood.

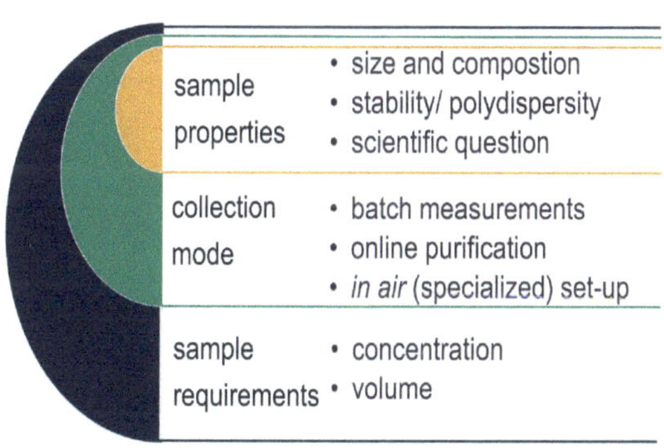

Fig. 3 The sample requirements depend on a number of different factors.

2.1 The scattering process and increasing the intensity of the signal

The scattering experiment and process itself may be broken down into these conceptually simple terms:

X-rays are directed at a volume comprising a number of molecules tumbling freely in solution (Fig. 1). If the electron density of these molecules (ρ_{sample}) is different from the surrounding solvent ($\rho_{solvent}$), then the path (of a small subset) of the X-rays will be altered. Due to interactions of the X-rays with the electrons of the atomic shell of the particles, the electrons start to oscillate and act as a dipole, thereby sending out spherical waves, with the same energy (elastic scattering). Within a particle, the scattering amplitudes of all atoms are summed up, and are measured as intensity, typically recorded on a two-dimensional (2D) area detector. Thus, the probability of a scattering event happening lies in the difference of the electron density between the molecules of interest and the surrounding solvent ($\Delta\rho = \rho_{sample} - \rho_{solvent}$). This difference, however, is indeed very small. For example, the electron density of an average protein is $\rho_{protein} \sim 0.43$ electrons/Å^3, and thus only slightly higher than the average electron density of water, $\rho_{water} = 0.33$ electrons/Å^3.

As once calculated by H.B. Stuhrmann: a 1 mg/mL solution of a globular protein of 15 kDa molecular mass, such as lysozyme or myoglobin will scatter in the order of 1 photon in 10^6 incident photons (Stuhrmann, 1980). Thus, during a SAXS measurement the signal of interest is very weak, and in order to properly extract this signal, low-noise detectors are required as well as an optimized beam set-up to minimize the contribution of the instrumental parasitic background scattering. Ideally the solvent, i.e., the buffer in which the sample is measured should be kept minimalistic to not further reduce this very small difference. For example, the addition of either ~35% *v*/v glycerol, ~3MNaCl, or ~1.2 M sucrose to a buffer will result in an approximate 50% reduction in the net scattering intensity measured from proteins in solution (Graewert & Jeffries, 2017). The influence of additives to the buffers later also becomes an issue when we look at the necessity of measuring a blank measurement for background subtraction.

Commonly the output of the SAXS experiment is visualized in a one-dimensional scattering profile, where the detected intensity is plotted as I(s) vs s (Fig. 1). Here, s is referred to as momentum transfer and defined as $s = 4\pi \sin\Theta/\lambda$, where λ is the X-ray wavelength, and 2Θ is the angle between the incident and scattered beam. Note, the momentum transfer is sometimes denoted as q rather than s.

Fig. 4 depicts some examples of such scattering curves collected from a diverse set of macromolecules. At other parts in this book, the wrath of information derived from these plotted SAXS curves is described in more detail.

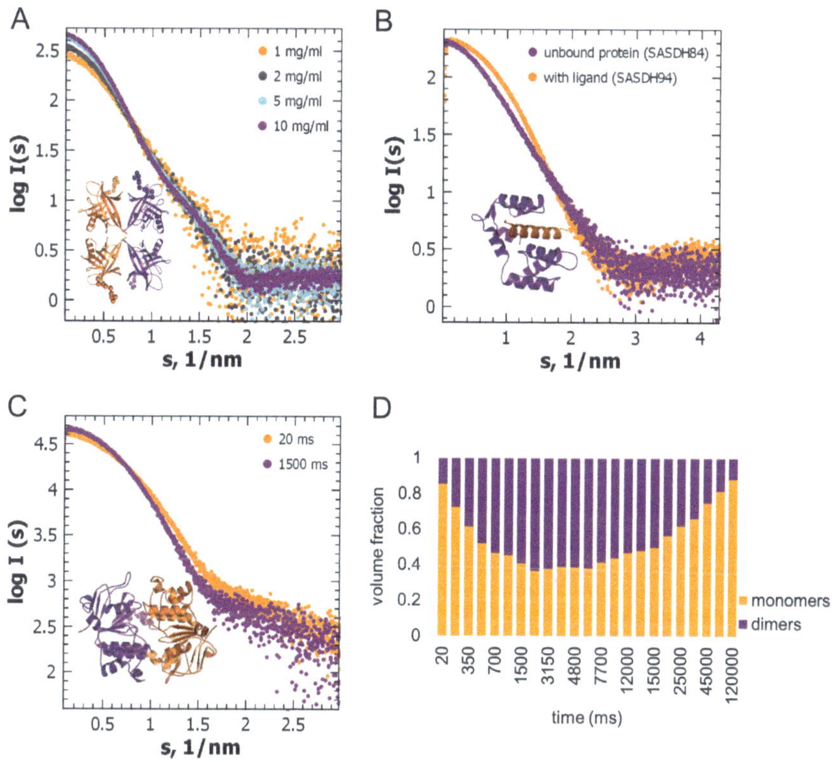

Fig. 4 Examples of SAXS studies conducted with different collection strategies (A) batch mode: dimer/tetramer equilibrium of PfLCN. measured at different concentrations as indicated. Inlay shows the organization of the tetrameric organization as cartoon representation. SASBDB entry: SASDH92, (B) SEC-SAXS data collected from TgELC2: TgMyoA compared to SAXS data of TgELC2 by itself. Inlay shows a cartoon representation of the fragment (orange) bound to TgMyoA (purple). SASBDB entry: SASDH94, (C) mechanistic understanding of the gliding motility in Apicomplexa. (D) TR-SAXS: comparison of two SAXS profiles collected after 20 ms (*orange*) and 1500 ms (*purple*) after addition of ATP to MsbA-NBD. The inlay shows the cartoon representation of the dimeric state of the transporter. (E) SAXS analysis of 12 time points collected during the TR-SAXS experiment indicating the volume fraction of dimers and monomers. Note, until 1500 ms the system shifts toward dimeric fraction and after ATP hydrolysis shifts back to the monomeric state. SASBDB entry: SASDGV5.

Overall, the scattering intensity, I, for chaotically tumbling systems, depends on the number, n, of the density particles, as well as the volume, V, of a single particle. Thus, the forward scattering I(0), i.e., the hypothetical[a] scattering at the angle s = 0, can be described as.

$$I(0) = n\,\Delta\rho^2\,V^2 \tag{1a}$$

When taking the angular dependency of s, into account (Eq. **1a**) is extended by the form factor of the particle, F(s), resulting in

$$I(s) \approx n\,\Delta\rho^2\,V^2\,F(s) \tag{1b}$$

The ultimate goal of many SAXS experiments is the interpretation of F(s).

Note that, (Eq. **1b**) only holds true for dilute ideal solutions, where there is no interaction between the individual molecules. For completeness, in the general case of solutions accounting for interactions between multiple particles the scattering curve is described with

$$I(s) = n\,\Delta\rho^2\,V^2\,F(s)\,S(s) \tag{1c}$$

whereby, S(s) describes the interference of scattering waves between different neighboring particles in the sample.[b] For dilute solutions, the structure factor (S(s)) approaches 1, and may be neglected.

In a solution scattering experiment, the low s regime is often of special importance. Here, the influence of the larger distances are reflected. At these low angles, the Guinier approximation ($I(s) \sim I(0)e^{-s^2\,R_g^2/3}$) is valid (Guinier, 1939). Because of this exponential relation, a linear fit to the transformed plot of ln(I) vs s^2, allows for the derivation of R_g and the forward scattering I(0).

Here, we want to first concentrate on the practical implementations of (Eq. 1a) and how this translates into requirements for sample preparation and data collection.

As an experimentalist, one can obviously "squeeze" the strongest signal out by aiming for high concentrations, and thus increasing n, the number of molecules in the illuminated area. Furthermore, keeping the buffer of the sample as simple as possible leads to higher $\Delta\rho$ and consequently to higher scattering intensities. In conclusion, one could argue that the answer to the

[a] Hypothetical as this cannot be measured do to the overlap with the majority portion of the X-rays, which directly passes through the illuminated volume without being scattered.

[b] Should not be confused with structure factors in crystallography that refer to diffraction amplitudes arising from preserved atomic positions between crystal lattice planes.

question on sample requirements for an ideal SAXS experiment would be: concentrate the sample as high as possible and transfer into water. However, anybody with biochemical experience will know that that is not a feasible approach. Not all biological samples are easily brought to high concentrations, and many systems require certain substances in the buffer to stabilize the protein and prevent aggregation. Thus, it becomes necessary to carefully weigh the best conditions for the sample itself against the aimed resolution of the SAXS experiment.

Thus, it is important to highlight, that before blindly starting the SAXS measurement, it can be beneficial to take the time to look at the expected results of the experiments. This can be done by taking advantage of programs that allow one to calculate theoretical scattering curves from existing models obtained experimentally with high resolution techniques such as macromolecule crystallography, nuclear magnetic resonance spectroscopy, cryo electron microscopy or obtained computationally with prediction modeling programs. The program Crysol from the ATSAS package, for example, can easily and rapidly produce such a theoretical curve (Svergun, Barberato, & Koch, 1995). Other available programs are listed in (Gräwert & Svergun, 2020; Trewhella, 2022).

Fig. 4 emphasizes different SAXS studies that were performed to analyze conformational changes. Fig. 4A shows the concentration series (1–10 mg/mL) of the lipocalin protein from malaria parasite *Plasmodium falciparum* (*Pf*LCN) collected in SAXS batch mode at P12 (Burda et al., 2020). From this, the tetrameric structure of typical lipocalin monomers could be derived and its concentration-dependent organization. The findings support the understanding of the essential role that *Pf*LCN plays in in counteracting oxidative stress-induced cell damage during multiplication of parasites within erythrocytes. The data measured at different concentrations can be described with an equilibrium between dimeric and tetrameric species. At higher concentrations the sample is shifted toward the tetrameric state, which becomes evident with the shift to higher I(0) values, the faster descend of the curve at low s, and the more pronounced bump at $s \sim 1.2\,nm^{-1}$.

In Fig. 4B an example is shown in which SEC-SAXS was employed to study the binding of *Toxoplasma gondii* myosin essential light chain 2, TgELC2, bound to myosin A C-terminus, TgMyoA, (Pazicky et al., 2020). The SAXS curve of the 15 kDa TgELC2 protein by itself (purple) points to a rather flexible structure (by its almost featureless SAXS profile). After binding of the small 2.6 kDa fragment, a more compact globular complex is formed. This can be detected from the SEC-SAXS curve of the complex (orange) which shows a flatter descend of the scattering intensity at low s. Further, a decrease in R_g from 2.1 for TgELC2 to 1.7 for the complex is observed. These data (in the context of the larger structural and biophysical study) shed more light on our understanding of the mechanism of gliding in the *Apicomplexa*, a phylum of intracellular, parasitic, single cell eukaryotes with high medical and agricultural relevance.

In the final example, Fig. 4C and D, a time-resolved SAXS study was performed (TR-SAXS) to examine conformational changes during the reaction cycle that involves transient dimerization of the nucleotide-binding domains (NBDs) of an ATP-binding cassette (ABC) transporter, MsbA (Josts et al., 2020). Here, the stopped-flow device described in more detail in subchapter 5 was used to trigger the dimerization event upon addition of ATP. Fig. 4C displays the SAXS profiles collected at the beginning of the experiment ($t = 20$ ms; orange) and at a later time point ($t = 1500$ ms; purple). The difference in the two curves can be explained with a shift from monomeric MsbA to dimeric MsbA, manifested by the increase in I(0) and the faster descend (larger R_G) for the latter. This dimerization state is a transient state, as the complete time series (up to $t = 12$ s) showed, that after hydrolysis of ATP the protein dissociates again into the monomeric fractions. In Fig. 4D, the volume fractions for monomeric (orange) and dimeric (purple) state are shown that were derived from the individual SAXS curves collected at 12 time points during the experiment. This example demonstrates the potential of TR-SAXS in its current state to detect kinetic parameters associated with conformational changes in the low-millisecond regime.

With these examples, we compare different states and demonstrate how do the conformational and assembly states manifest themselves in the SAXS data. Later, we will look at the different parameters for the data collection of the examples, but here we want to highlight, that different scientific questions will require exploration of a different ranges of spatial and temporal resolution. Obviously, if the aim of the experiment is to not simply differentiate between different oligomeric states (example a), but to detect ligand introduced conformational changes (example b), then a higher resolution (i.e., a stronger signal at higher angles) will be required to draw meaningful conclusions.

As mentioned, one way to increase the overall intensity of the SAXS signal is to utilize higher concentrations. However, forcing samples to such crowded conditions is not always the ultimate solution. It is important to remember, that sample quality especially in terms of monodispersity plays an important role in the successful outcome of the SAXS experiment (Jacques & Trewhella, 2010; Jeffries et al., 2016).

The scattering of the impurities are summed up with the measured scattering of the sample. As derived from Eq. (1c), the scattering probability increases quadratically with size of the analyte so that large contaminations such as aggregates and higher oligomeric species will contribute noticeably to the scattering curve—even if they are only present at low concentrations. Thus, when aiming for higher concentrations and a minimalistic buffer system, one must prevent the formation of such aggregates. Advanced sample characterization and detection of potential impurities is, therefore, key for a successful data collection.

Besides the parameters discussed here that contribute to the intensity of the scattering signal, there are two other important factors from the scattering process that have direct implementations on the design of the experimental delivery system (Fig. 1).

- background subtraction, i.e., the requirement of second measurement of solely the solvent under precisely the same experimental conditions to account for the portion of scattering signal that is derived from the instrumental setup as well as the solvent itself
- addressing the presence of structure factor, typically by measuring a concentration series, i.e., the sample with varying concentration of the solute followed by the extrapolation to infinite dilution

2.2 Accounting for background subtraction with a second measurement

A very important step of the SAXS experiment is to record a second measurement of the identical solution lacking the sample of interest. The recorded scattering image contains the scattering contribution from the buffer, the sample cell/capillary as well as other instrumental (parasitic) scattering. By measuring the buffer alone, one obtains a curve that can be subtracted to reduce the sample scattering signal to that of the molecule of interest (Fig. 1).

One major pitfall of many first-time SAXS users is the lack of appreciation for the sensitivity of the method to even very small discrepancies between correctly- and incorrectly-matched buffers. Thus, when designing the experiment and deciding the collection strategy, the reproducibility of obtaining the solvent background must be carefully considered.

As described in more detail in subchapter 5, the buffer measurement is especially tricky for many of the specialized sample chambers. Here, proper alignment and normalization play a crucial role. For static (batch) measurements with an automated sample loading robot, the buffer measurements can easily be conducted in the same capillary so that the problem is reduced to providing a fully matching buffer, ideally by the means of dialysis. Other possibilities for buffer matching are discussed in subchapter 3.

One strong advantage of SEC-SAXS collection mode is the ease at which suitable frames for background subtraction can be obtained by simply collecting data frames before and/or after the elution peak of the sample. In most cases such frames are easily identified—provided that radiation induced aggregation with subsequent capillary fouling is kept to a minimum. Recent developments in software processing tools as implemented for example in Scatter IV (Tully, Tarbouriech, Rambo, & Hutin, 2021), Chromixs (Panjkovich & Svergun, 2018) and BioXTAS RAW (Hopkins, Gillilan, & Skou, 2017) facilitate the step identifying the best frames for buffer subtraction and taking possible buffer drifts across the elution profile into account.

2.3 Addressing the influence of structure factor with a concentration series

On a final note regarding the physical properties of the scattering event and sample requirement, we want to highlight the importance of measuring SAXS over an available range of concentrations.

In most cases, the aim of a SAXS study is to derive the form factor, F(s), according to 1c. This is best achieved when the experimental conditions are selected in such a manner that the sample approaches ideality, i.e., that the structure factor, S(s), approaches 1 and may be neglected. A simple approach commonly performed for most biological solution SAXS measurements is to collect data over a wider range of concentrations. Normally a four point concentration series including, for example 10, 5, 2.5, and 1 mg/mL is sufficient (similar to the example shown in Fig. 4A). Simple Guinier analysis is performed were the data is plotted as ln(I) vs s^2 and from the linear fit I(0) is retrieved from the y-axis intercept and R_g from the slope. This allows one to then quickly assess and compare the behavior of the sample across the measured concentration range (for example, by plotting R_g vs concentration). With the Guinier analysis, the intensities at low angles are emphasized and thus the information on the large distances due to the reciprocal nature of the scattering profile. Thereby, attractive forces which lead to aggregation and the presence of larger species becomes visible as an upward turn in the Guinier analysis. Repulsive interactions, on the other hand manifest themselves as downward turns.

Again, the approach of measuring over a concentration range is common practice when performing batch measurements with the automated sample loading robot. For screening "4D-SAXS" experiments with specialized delivery systems, the sample is normally characterized well in advance so that the sample is often kept at a steady concentration while other parameters such as ligand concentration, temperature, ionic strength, etc. are varied.

SEC-SAXS measurements are also typically performed at a single concentration—here the separation across the column is usually sufficient to ensure sample ideality so that the S(s) may be neglected.

2.4 Summary of general sample requirements

In general, the sample preparation for SAXS experiments is straight forward compared to macromolecular crystallography or cryo-electron microscopy and the sample requirements are less stringent. However, there are some potential pitfalls that the experimentalist should take into account when designing the SAXS experiment. Thereby the three main challenges are to

I. Deliver pure samples at high enough concentration (keeping in mind that the probability that a photon will be scattered by the solute is very low). Sample quality must be ensured so that meaningful data is collected

II. Ensure that the buffer does indeed match the composition of the sample solvent. Then the background subtraction can be performed properly and the subtracted profile reflects solely the scattering contribution of the molecules of interest

III. Make sure that the proper concentration range has been covered so the information of interest can be derived. The final amount of material required varies with the collection strategy and sample environment

Fig. 2 summarizes the different collection strategies with their advantages and disadvantages. As a general guideline the typical sample ranges measured with the different sample environments at the beamline P12 are provided. Typical sample volumes that are currently needed are 40–200 μL at concentrations of 5–10 mg/mL. These requirements are comparable with those for other structural studies.

3. Automated sample loading robotics

With the increased interest of SAXS as a tool to study the shape and size of biological macromolecules in solution, the necessity for improved sample handling at the SR facilities grew significantly. Thus, the implementation of various automated sample loading robotics was rewarded with the highly efficient use of biological SAXS beamlines allowing to facilitate and speed up measurements (examples from various beamlines include: David & Perez, 2009; Hura et al., 2009; Martel, Liu, Weiss, Niebuhr, & Tsuruta, 2012, Pernot et al., 2013; Round et al., 2015, Osaka et al., 2016; Cowieson et al., 2020). These developments are one of the major contributors to the grown success of biological solution SAXS over the last 15 years.

Originally, the beam time use at synchrotrons was accompanied by tedious manual filling of sample cells—a tricky act when having to prevent the introduction of air bubbles—spending valuable time washing said sample cells as well as constant re-setting of the protective interlock system. Besides the convenience of reducing the required human input (and thus human errors) in these various steps, the automated sample loading robots also contribute to increased data quality. As they ensure precise sample loading, the measurement become more reproducible. This increases the confidence in the required blank (buffer) measurements, which allows for optimal background subtraction. Especially when performing comparative (screening) studies the reproducibility of the measurements is a key aspect.

The gained robustness and consistency in data collection allows the experimentalists to shift the focus on sample preparation as well as trouble shooting during the beamtime. In most cases, these robotic systems are coupled to an automated software pipeline for data reduction and analysis providing real-time data analysis during the measurement (Franke, Kikhney, & Svergun, 2012). Thus, while the measurements of the queued samples continue automatically, the experimentalist can investigate the results and if necessary improve the data with on-site sample handling. The implemented solutions developed of the various beamlines differ slightly and thereby take the requests for the individual user communities into account.

At P12, the 2nd generation sample robot from ARINAX (Grenoble, France) is in place. This device originated from a collaboration between the SAXS development team in Hamburg, the Frauenhofer institute as well as the ESRF. In 2007, the first version of the prototype was installed and the successful use by the first 50 user groups was reported in 2008 (Round et al., 2008). This prototype was installed at the X33 beamline of the EMBL, at the storage ring DORIS-III (DESY, Hamburg). The device was specifically designed for biological solution experiments. Low availability of sample amounts and potential stickiness of the samples were factored in when tailoring the controlled delivery of biological samples into the exposure cell in a high-throughput manner.

A full integration with the beamline control software was essential for remote and unsupervised performance of the measurements. A detailed description of the updated version of the automated sample loading robot is provided here (Round et al., 2015) (Fig. 5).

Features of the current robotic system include

I. Sample exposure unit, SEU

The temperature-controlled SEU maintains the sample capillary in a vacuum environment which reduces the background signal through the prevention of air scattering. The sample exposure pod itself is a modular component that can be readily exchanged. Calculations of an optimal sample capillary size depending on X-ray energy/absorption and radiation issues are discussed in more detail (Schroer et al., 2018). At P12 a 1 mm diameter quartz glass capillaries are presently in use. For sample loading 5 μL are already sufficient, however, ideally, more is loaded (>20 μL) to allow for "flow exposure mode" leading to a reduction in radiation induced damage (see below).

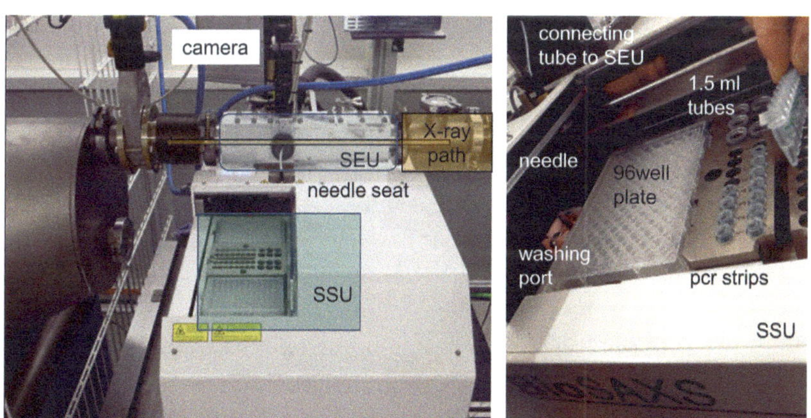

Fig. 5 Overview of the automated sample loading robot at the beamline P12. Special features are indicated. *SEU*, sample exposure unit; *SSU*, sample storage unit.

II. Liquid handling features and optics

Precise sample loading is achieved through a number of special features including a sample viewing camera perpendicular to the X-ray beam. This allows the detection of liquid/air interface and is used to place the sample at the correct position in the beam.

A robust needle is used to suck up the sample. Besides simply loading the sample, a number of extended liquid handling features such as mixing, recuperating or time delays can be included into the measurement routine. The most prominent option is "flow mode" which is used by default and allows the sample to move through the capillary during the exposure, which significantly minimizes radiation damage (Jeffries, Graewert, Svergun, & Blanchet, 2015). Thus at P12, usually 20–40 μL of sample are loaded. During the overall exposure of 1–5 s 20–40 individual frames are collected that are then assessed regarding any signs of radiation induced changes. This is done automatically by the pipeline software and the information of how many suitable frames were used for the final averaged curve is provided. According to this, the collection protocol may be adapted for the specific sample (reduce/increase number of frames, loading volume, etc.). Alternatively, the sample may be chemically altered (addition of so-called scavengers) to enable longer exposure periods resulting in better signal-to-noise ratios.

The needle position is calibrated in such a manner that the dead volume inside the sample vessel is <4 μL.

III. Sample storage unit, SSU

The thermostatically controlled SSU has room for a number of different vessels that may be queued for measurements. This reduces the number of times that the safety interlock system has to be set, while ensuring sample stability, for example, by keeping the samples chilled. Besides temporal screens temperature scans can, therefore, also elegantly be performed by step-wise heating up the sample environment. If designed in the proper manner, these measurements can then be submitted as "unattended overnight" measurements to ensure an optimal use of the allocated beamtime.

At P12, the SSU can presently house two 96-well plates as well as PCR strips (32 vessels) and 1.5 mL microfuge tubes (12 vessels). The temperature range available in the SSU is from 5 °C to 60 °C.

IV. Software solution: versatile graphical user interface, Becquerel

For convenient control of the measurements by the sample robot system, a very intuitive submission and queuing software solution is provided. The program called Becquerel uses a graphical user interface, which allows for simple selection from a visual representation of the sample storage unit arrangement. At the beamline, the users require only a short introduction to this system before they can autonomously continue their measurements (or even submit through remote access). Only a few parameters are changed for each measurement such as sample name, pairing to matching buffer, and concentration. A wide variety of parameters are set to default values for standard measurements, but may be adapted easily for specialized experiments. The auxiliary system BMS, beamline meta server, initiates the execution of the actual command execution (see Franke et al., 2012 for details). The system is versatile so that new hardware components and work routines can be added. In the next chapter, we demonstrate an easy conversion of the interface between the batch and SEC modes (Hajizadeh, Franke, & Svergun, 2018). Once a job is submitted, the measuring protocol includes a controlled loading of the sample, followed by cleaning and drying of the cell between measurements. Currently, roughly 12 samples together with matching buffers can be measured per hour with this collection scheme. The system is reliable and the failure rate of sample loading is $<0.5\%$; it offers a client-server-based network protocol, locally and remotely.

3.1 Examples

As already emphasized, the advancements in sample handling have revolutionized the way SAXS measurements are performed allowing SAXS to increasingly contribute to various scientific studies. This holds true for not only protein based samples but also for a broad range of various types of macromolecular solutions. Here, we want to point out examples of high scientific out put that strongly benefitted from the versatile and automated sample delivery system.

During the Covid pandemic a number of studies were performed that benefitted from the complementary information offered with SAXS. In the study described here (Custódio et al., 2020) SAXS was used as a complementary method to characterize synthetic nanobodies (sybodies) that were generated against SARS-Cov2 spike glycoprotein. In the first instance,

SAXS was used to screen biophysical properties of a subset of sybodies that were preselected as potential binders. Assessment of D_{max} and derived MW estimates was possible within a few minutes after the measurements. With this a qualitative estimate regarding the potential of the individual binders to from dimers was made. With the current settings at P12, the screening of dozens sybodies could be achieved within <2 h. In the next step, the capacity of most promising candidates from SAXS and biochemical studies was assessed to bind to the receptor binding domain, RBD, of SARS-Cov2 spike. For the best candidate, an advanced SAXS modeling using the data from the individual sybody, RBD and their complex was employed to assess the binding interface. This analyses indicated that the selected sybody effectively blocked the binding site for the cell receptor ACE2. The SAXS modeling was complemented by a detailed cryo-EM analysis of the whole spike protein revealing the binding modes of the sybodies.

In another study, other components such as the coronaviral non-structural proteins form the replication-transcription complex Nsp7 and Nsp8 were examined with SAXS at P12 in combination with native mass-spectrometry (Krichel et al., 2021).

A further important SAXS contribution in fight against SARS-Cov2 were the characterization of mRNA-based vaccines. These studies benefitted from automated sample loading robot and the high-throughput operation which allowed the screening of a wide range of parameters regarding mRNA-filled lipid nanoparticles. In continuation of a long lasting collaboration with University of Mainz and BioNTech company, staff from P12 were involved in the studies of lipid nanoparticles with SAXS, which allowed the determination of the overall size and shape of the protective lipid capsules. On the other hand, the inner packing of the mRNA molecules can be analyzed as well. Here, the automated sample loading robot allowed researchers to screen various formulations and determine the sample stability under quasi physiological conditions. Changes in pH were also analyzed to study the effect on the fragile mRNA molecules upon uptake into the various host cells. A number of publications (Nogueira et al., 2020; Siewert et al., 2020; Uebbing et al., 2020) demonstrate the efficient use of the P12 set-up for the analysis of lipid base nano carriers.

To further illustrate the versatile application range of the automatic sample handling robot, we point out a number of the studies performed *in cellulo*. In one study (Lahey-Rudolph et al., 2020), SAXS was used to screen for a formation of protein microcrystals grown within the host insect cells. Due to the high-brilliance and low-background of P12, rapid and reliable detection of the crystals was possible through the presence of Bragg peaks in the recorded SAXS profiles. Another set of experiments included the assessment of the effect of antibiotics on whole bacterial cells. With these studies the aim was to establish the pattern detection system which would allow for a classification of different antibiotic working mechanisms.

3.2 Designing/performing and assessing a SAXS experiment in batch mode: Automated sample loading robot

Access to SR facilities dedicated to biological solution SAXS is still limited. After competitive application processes, and successful submission of beamtime proposals, beamshifts are allocated. To make the best use of the beamtime, we have listed a few aspects to consider in the following check list.

In the end, it is up to the user to understand their sample properties and make the right decisions on how to use the available set-ups for the best characterization. Local beamline support staff are available for consultation. The earlier communication is initiated the more productive the use of the allocated beamtime will be.

(1) Planning

- Contact local support/check webpages and information sent out by the user office on:
  - o minimum volume to load sample cell (taking dead volumes of tube into account
  - o suggested concentration for optimal measurement
  - o estimate of number of measurements per hour
  - o any other suggestions on how to prepare samples (i.e., what tubes/ vessels to use)
  - o confirm availability of any equipment that might be needed on site (such as nanodrop, incubators, centrifuge, dialysis stations, etc.)
- Calculate amount of sample that will be needed; plan for additional material so that back up measurements may be performed in the case of unexpected events
- Determine the most suitable way to obtain matching buffer (by default we suggest dialysis, other means or SEC, the use of diafiltration devices, desalting columns.)
- Request shipment information and plan enough time for transportation

- *For screening studies*: optimize the conditions that will be tested (regarding sample concentration); ideally perform in-depth characterization of the sample in advance

(2) Sample preparation

- User visit on-site:
  - **o** highly concentrated samples:

    Thaw, spin/filter, re-measure concentration/prepare dilutions series in vessels that may be placed into the robot, make sure no air-bubbles are in the sample.
  - **o** unstable samples stored at low concentrations:

    Perform step-wise concentration of the solute, thereby, after different intervals, a small aliquot is removed for analysis (UV absorption and SAXS analysis) and the concentration procedure continued.
- Advice for mail-in/remote user operation:
  - **o** Ideally samples are prepared in final vessels, so that they may be directly loaded into the automated sample loading robot
  - **o** Quick spinning of the sample, perhaps filtering or some dilution steps may be performed by the local support, depending on available capacity and time constraints
  - **o** At the beamline P12: best use 96-well plate or PCR strips; catalog numbers are normally provided on the user dedicated webpages

(3) Data assessment and trouble shooting

- Many beamlines offer on-the-fly automated analysis pipelines that allow a quick assessment of data quality:
  - **o** Check for the number of frames used in the automated averaging process: this will allow you to assess potential radiation-induced damage
  - **o** Check the quality of background subtraction. The behavior at high angles (over subtraction or featureless curves) are indications for a potentially miss-matched buffer
  - **o** Check for consistency over concentration range (e.g., decrease of R_G and D_{max} values at lower concentrations indicate removal of larger oligomeric species/contaminations/aggregates, increase of these values are indications for the present of repulsive forces). Perhaps further concentration points have to be measured. In some case re-measuring of the concentration series is advised if these values are inconsistent

- o Check for consistency regarding the MW estimates
- o If, an independent model is at hand it can be used to quick check whether the overall shape, scattering profile is according to the expectations. Check if the achieved resolution is sufficient for retrieving the information needed for the scientific question at hand

4. Coupling of online purification systems

The lab-based experimentalist with experience in the preparation of biological macromolecules will be all too familiar with the often encountered issues regarding sample stability over time. Thus, there is an understanding that the notion to deliver a monodisperse sample for SAXS analysis is not always that straight forward. Intrinsic equilibria may lead to the formation of the larger oligomeric species and the formation of aggregates over time. This is, unfortunately, more often the default state than the exception. As highlighted in Section 2, even small amounts of such aggregates can influence the SAXS data in such a way that it is no longer suitable for further analysis. Due to these considerations, the success of SEC-SAXS over the past decade can easily be comprehended. Here, the possibility is provided to generate the "almost perfect" sample directly at the beamline through an on-line sample polishing step with the removal of contaminating species. As the name implies, size exclusion chromatography is based on the separation of the different species by their size. With the means of liquid chromatography (LC) the sample is "pushed" through a stationary phase which in many cases is made up of spherical silica or sugar-based beads. The beads contain small pores in which the samples diffuse. The depth in which the various components penetrate into the cavities, and thereby their retention time in the chromatographic resin, is dependent on the hydrodynamic radius of the molecules (Stoke's radius, Siegel & Monty, 1965). To some extent, the chemical interaction with the beads themselves will also influence the elution behavior of the molecules. Naturally, larger molecules that may only experience a short migration delay during their passage through the column will elute before smaller molecules that are held back for a longer period. The resolution that determines if two different species may be separated depends on the column material, length of the column and of the chemical and physical properties of the molecules themselves.

These parameters, especially the choice of column, may (and should) be tested offline before the beamtime. The more effort that is put into optimizing the separation conditions before the SAXS measurement, the better the outcome of the scattering experiment. As we will see below, using light scattering techniques, in this case static light scattering (LS) methods is a good approach to prepare for the experiment. In the same way, the X-ray scattering is more sensitive to larger molecules, LS will also pick up on larger contaminations as compared to UV light absorption for example.

In 2004, the first successful use of online SEC was demonstrated at biological SAXS beamline BioCAT in Chicago (APS, USA; Mathew, Mirza, & Menhart, 2004). This was followed by a similar experiment at the Photon Factory BL10C (Ibaraki, Japan) in 2008 (Watanabe & Inoko, 2009).

With their dual modality system, the bioSAXS SWING beam line at the Soleil synchrotron (Saint-Aubin, France) for the first time offered users to easily opt for the additional online purification (David & Perez, 2009). Here, the LC system is used for both static single sample measurements as well as for the online purification mode.

Fueled by this success and the overall acceptance of the SAXS community, all major biological SAXS beam lines offer SEC-SAXS collection mode in some kind of way (Pérez & Vachette, 2017).

In the meantime, it has even been successfully adapted for the less-intense X-ray lab sources utilizing the powerful flux delivered by the metal-jet technology (Bucciarelli et al., 2018) or rotating-anode laboratory instrument (Inoue, Nakagawa, & Morishima, 2019).

Another adaptation of the approach has been the implemented for small-angle neutron scattering (SANS), for example, on the D22 SANS instrument at the ILL (Grenoble, France; Jordan et al., 2016; Johansen, Pedersen, Porcar, Martel, & Arleth, 2020). More recently, SEC was coupled to inverse contrast matching SANS to study the spatial arrangements of the N-terminal domains of components of cyanobacterial circadian clock system (Yunoki et al., 2022),

In 2012, the usefulness of adding additional detectors to the set-up, such as the downstream installment of a refractometer at SWING beamline was demonstrated (Berthaud, Manzi, Pérez, & Mangenot, 2012). At P12 in Hamburg, a split valve at the outlet of the size exclusion column is used to create two parallel and identical paths. This allows one to collect SAXS data in parallel to light scattering data for the independent assessment of the molecular weight of the eluting components (Graewert et al., 2020, 2015). The set-up also includes UV–vis, multiangle laser light scattering

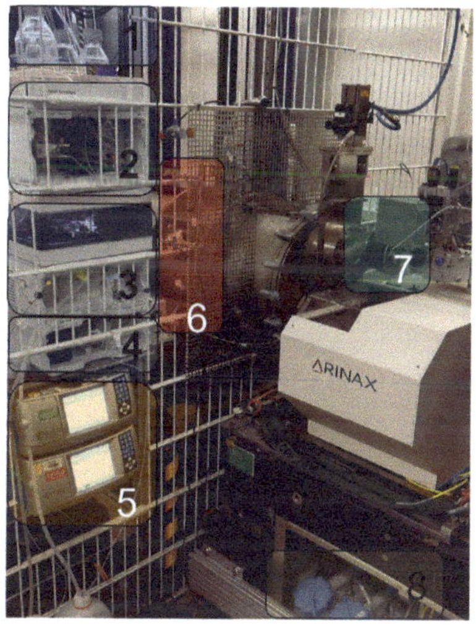

Fig. 6 SEC/MALLS set-up at the beamline P12. Specific features are highlighted.

(with a quasi-elastic light-scattering (QELS) option) and refractive index detectors. The hydrodynamic radius, R_H, that can be derived from the QELS detectors gives further information on the shape of the sample and may be usefully compared with R_g derived from SAXS. The current set-up is depicted in Fig. 6.

For the full implementation of SEC systems (and also for adapting to other online separation tools), the following considerations were useful (references to the features depicted in Fig. 6 are given):

I. Column connection to the sample exposure unit, SEU

For the most convenient integration of the LC based pre-purification possibilities an adapter allows the direct connection of the column or detector to the SAXS capillary embedded in the vacuum chamber (7). The setup profits from the vacuum environment significantly reducing the parasitic scattering. When determining the optimal diameter and length of the connecting tubes, the overall pressure build-up as well as spatial rearrangement of the LC system within the experimental hutch have to be considered.

A valve port for easy switching between SEC and batch data collection mode does not only facilitate the switching between the different collection modes but also profits from the robotic cleaning procedures that are in place (8). After each SEC run, the capillary may be rinsed off to remove any coating that might have built up during the run (see radiation damage below). The automated process, which uses a detergent emulsion followed by water rinsing is especially convenient as it prevents the need for entering the experimental hutch and having to break and reset the interlock safety system.

II. Synchronization of LC and SAXS, software integration

For automated and convenient data collection during the SEC-SAXS runs communication tools between beamline control software and chromatography instrument are essential. At P12, this communication has been incorporated into the data collection software system Becquerel (Graewert et al., 2020; Hajizadeh et al., 2018). Here, the injection of sample onto the column is queued into the process, so that the user can easily start the run and not worry about missing any important time frames during the run. The temperature controlled auto sampler connected to the LC pumps allows for prolonged storing/queuing of the samples (3). UV absorption data, is automatically collected as complementary data in parallel and provided with the data output files. Note, currently additional data such as MALLS is collected separately with respective software programs.

III. Optimized SEC parameters for sufficient resolution

Ideally the SEC parameters are optimized before the beamtime to provide sufficient separation resolution. The SEC run at the beamline itself should be considered more like a final polishing step. Peaks of interest should be ideally baseline-separated allowing to obtain scattering data arising solely from the molecule of interest.

Most importantly the choice of column is the essential component for a good separation (6). In addition, mobile phase and loading also contribute to obtaining the desired resolution.

Regarding the sample concentration that is loaded, ideally the aim should be to reach approximately a concentration of 0.5–1 mg/mL at the elution peak. (Note for small columns such as GE columns 5/150, 2.4 mL, the dilution effect is ~2 and for the larger 10/300 dimension columns, 24 mL, the dilution effect is ~8). As mentioned in subchapter 2, in most cases SEC-SAXS can be performed at a single concentration. Here, the notion that the S(s) approaches 1 is given by the separation and rather dilute state of the sample. The injection volume should also be considered. Often the tendency to overload the column can be observed. As the separation is also strongly dependent on the width migration front of the sample passing through the stationary phase, loading too much sample may worsen the separation. As a rule of thumb, one should never load more than 50 µL onto a 2.4 mL column and nor more than 100 µL onto the larger 24 mL column.

Regarding the right choice of column consultation of the tables and selection guides provided by the column manufactures is a good way to start. Thereby, the expected molecular weight of interest is normally the most important selection criteria. Buffer capability should naturally also be assessed.

At most SR facilities a number of standard columns are provided for the users. However, ensuring correct separation resolution lies in the responsibility of the user themselves.

IV. Radiation damage

As explained above, radiation induced alteration to biological macromolecules is often observed in solution SAXS experiments. Due to the constant flow during a SEC-SAXS run, one would expect that the continuous illumination of fresh sample would prevent such a reaction. Still, due to the laminar flow and the decreased speed at which the sample is replaced along the wall of the cylindrical capillary, a build-up can often be observed, at least at flow rates below 0.2 mL/min. Addition of scavengers such as 3–5% glycerol, or 1 mM TCEP may reduce the potential damage.

Other beamlines have implemented other successful approaches in dealing with radiation damage. For example, the co-flow (sheath-flow) approach applied at the Australian synchrotron (Ryan et al., 2018). Here, a flow of matched buffer is applied to focus the sample through the center of the cell.

V. Data volume and processing

Different strategies for data collection are in place when it comes to deciding what part of the chromatography run should be captured by the 2D area detector. At P12, the default setting is to collect all the scattering frames throughout the run. This includes an excess of buffer frames before the void volume of the column, then the separated peaks with some more buffer frames at the end of the run. The excess in buffer frames allows for the assessment of the beam and elution stability, and are ideally used for background subtraction.

In cases where problematic aggregates (in terms of stickiness to the capillary) are expected, one could consider to first start data collection after the elution peak, as the illumination of potential aggregates increases the risk of capillary fouling. Data after the elution peak are often resourceful when addressing said fouling effects, but also may have slight miss matches in buffer composition.

Naturally, the "collect all" approach can easily lead to large data sets, which might take some time to process. However, with the development of dedicated software for the processing of SEC-SAXS data, such as the ATSAS program Chromixs (Panjkovich & Svergun, 2018) or BioXTAS RAW (Hopkins et al., 2017) have facilitated the subsequent processing of the data. With Chromixs, a folder containing up to 10,000 frames can easily be opened. With the automated as well as manual selection modes, frames of interest—either sample or buffer—can be selected and a final scattering profile obtained. This is automatically displayed in the program Primus which allows for a quick assessment of the essential parameters such as R_g, D_{max} as well as MW estimates (Konarev, Volkov, Sokolova, Koch, & Svergun, 2003).

With these tools, the success of a SEC-SAXS run is assessed within minutes after the final frames are collected and potential improvements to the sample and experiment itself can be initiated while still at the beamline. Some of these steps are now also performed automatically. For example, an automated pipeline is available to provide preliminary SAXS analysis and the first-step reconstruction of membrane proteins in detergent solutions (Molodenskiy, Mertens, & Svergun, 2020).

As a side note, currently there are more programs being developed to assess the information within an elution peak by apply evolving factor analysis or Gaussian decomposition (such as EFAMIX from Konarev, Graewert, & Jeffries, 2022; in BioXTAS RAW from Hopkins et al., 2017; in US-SOMO from Brookes, Vachette, Rocco, & Pérez, 2016). These approached give a possibility to decompose data collected by SEC from partially overlapping components. However, despite the availability of such software solutions to handle polydisperse samples, a major effort should be placed into achieving monodispersity with the use of the right purification and storage protocol.

As for samples measured with the automated sample loading robot, before investing the beamtime for a prolonged SEC-SAXS run, sample quality should be controlled independently of SAXS through complementary techniques, such as LS techniques, native electrophoresis gels and/or offline SEC runs.

Despite all the advances laid out here for the use of SEC mode, there are some considerations to be made that make this collection not the best strategy for all cases. There are a number of issues to consider such as the dilution effect, potential radiation damage, interaction with the column material, which might lead to dissociation of the complex of interest, and the overall restriction to the mobile phase and screening options.

To address some of these limitation, different modes of online purification have thus been tested and implemented. Other separation techniques, such ion exchange based techniques have been reported (Brennich, Round, & Hutin, 2017, Hutin, Brennich, Maillot, & Round, 2016; Konarev et al., 2022). Here, the challenge is to address the ever varying buffer during the elution process, (normally achieved in changes in ionic strength or pH). Further implementations of other online purification tools are to be expected in the future.

In summary, over the past decade SEC-SAXS has established itself within the user community and has evolved into a highly appreciated main stream tool at most biological SAXS beamlines.

Especially, during the unavoidable switch to increased mail-in operation during the current pandemic, SEC-SAXS data collection has played a major role in this smooth transition. The integrated polishing of the sample on-the-fly nicely substitutes for the otherwise required on-site sample handling to remove contaminations.

4.1 Examples

Currently, 60% of the users at the beamline P12 opt for the SEC data collection mode (either exclusively or in combination with other batch experiments).

Some of the recent highlights of scientific outcome resulting from the experiments include the characterization of the elongate rod-like form of a ^{11}C-aptamer with application was applied for in vivo imaging of Ehrlich ascites carcinoma (Ozerskaya et al., 2021). This example emphasizes the versatile application of SEC-SAXS for the studies of oligonucleotides as well.

An example showing the potential of using SAXS to retrieve structural information on the formation of higher oligomeric species is given with the study on low temperature states of GTP- and GDP-tubulin (Shemesh et al., 2021). Corroborated with SEC-LS analysis, their data show SEC-SAXS chromatogram is consistent with an isodesmic thermodynamic model of longitudinal tubulin self-association into 1D oligomers, terminated by the formation of tubulin single rings.

Obviously, SEC-SAXS is a good option for studying complex formation (see also Fig. 4B). In another study (Macpherson et al., 2021) so-called knob domains, which are small peptides derived from bovine antibodies with ultra-long heavy chain, were studied regarding their potential in modulating the complement system. With a hybrid biophysical approach including SEC-LS and SEC-SAXS, the binding was confirmed and allosteric effects propagating >50 Å from the binding sites could be described. This study further reports the use of these antibody fragments as potential therapeutic.

In another study that elegantly combined SEC-SAXS with SEC-LS, a conformational ensemble was described for muscular multidomain protein myotillin in solution and allowed the detection of small traces of higher oligomeric species such as dimers. The subsequent SAXS-based flexibility study of the protein suggests that it is restrictive in its flexibility and explores a manifold of rather compact conformations (Kostan et al., 2021). The derived structural model gives insights into how myotilin may interact with other muscular proteins that are important for the contraction of striated muscle.

4.2 Designing/performing and assessing a SAXS experiment in SEC-SAXS mode

As a reminder, access to SR facilities dedicated for biological solution SAXS is still limited, thus the use of SEC-SAXS as opposed to batch measurement should be critically considered due to the longer acquisition times. As mentioned, potential column interactions, dissociation of weak complexes, and risk of radiation damage should be taken into account before opting for this collection mode. Ideally, if enough sample is available a mixed approach of SEC as well as batch measurements might bring advantages. As a prerequisite for a successful SEC-SAXS data collection we strongly recommend an offline separation (ideally with LS detection) in advance. Below are important points for the SEC-SAXS operation based on the experience gained at P12 operation.

(1) Planning

- Contact local support/check webpages and information sent out by the user office on:

- o compatibility/availability of the selected column and buffer system with the LC and beam set-up
- o minimum volume to load (taking dead volumes into account);
- o suggested concentration for optimal measurement
- o estimate of duration of a measurement, taking connection of the column, washing steps as well as equilibration times into account
- o ideally, if a number of runs are planned, column and buffer should be selected in such a manner that time consuming switching of columns and buffers is kept to a minimum

- Calculate amount of sample that will be needed; plan for additional material so that back up measurements may be performed in the case of unexpected events. Request shipment information and plan enough time for transportation

(2) Sample and set-up preparation

- SEC-SAXS operations can be performed in mail-in operation. Here, the on-site sample handling is normally reduced to a minimum (mostly including a filtration or long centrifugation to remove any aggregates that might have formed during storage and/or shipment). In this case, clear instructions should be shared on how to thaw and treat the sample thereafter
- It is important that the buffer is filtered and sufficiently degassed. On site vacuum filtration devices are usually available for this. In case of mail-in operation, ideally a stock solution in sufficient amount is provided, and the local support staff will prepare the buffer accordingly
- The column should be washed and equilibrated extensively. To maximize the use of the beamtime this could be done in parallel to batch measurements
- Importantly the pressure of the LC system should be monitored constantly to detect any possible issues in advance
- The detectors required for the runs should be sufficiently equilibrated as well. If the splitter is being used for the parallel collection of LS data, then the even flow of the two outlets should be confirmed

(3) Sample submission

- Due to potential risk of radiation induced damaging with subsequent capillary fouling submission of numerous samples for queuing is currently not supported at P12

- It is suggested to assess state of the capillary before each submission and to make use of automated cleaning procedure that is in place for cleaning the capillary between individual runs
- Safety guards and cross checking of the individual components are in place to make sure that the beam and all other hardware components are available and idle before the sample is injected onto the column
- When using additional detectors, their idle state should be confirmed and input for data collection be prepared
- Communication between beamline software and LC equipment itself is in place so that a "one button" submission is sufficient to start the next run
- During the run, the collection of 2D images may be monitored, along with beam stability and constant flow of the LC pumps (no pressure build up, no introduction of air bubbles, etc.).

(4) Data assessment and trouble shooting

- Here too, on-the-fly automated analysis pipelines are in place, so that the outcome of the measurement can be assessed in a timely manner (normally within 3–5 min after the run; during this time the capillary can ideally be cleaned and prepared for the next run).
- For best assessment of the run, the radially averaged frames are examined with the program Chromixs, which will reveal the issues that might have arisen during the run including: drift in beam, buffer issues, capillary fouling. The elution profile will also provide information if the resolution of the separation was sufficient. The program allows for production of a final curve that may be analyzed regarding the derived MW estimates, for example, or a quick cross check with available models (using Crysol).
- Trouble shooting may include the addition of scavengers to the running buffer in case of radiation induced damage, extensive cleaning of the capillary might be necessary to remove the strong build up on the capillary walls, readjustment of the chromatography parameters, measuring the sample in batch mode, repeating the chromatographic separation with additional detectors

5. Specialized sample environments

Currently at the beamline P12, a large majority of the users (>90%) come to perform standard measurements with the automated sample loading robot and LC set-up.

The other 10% perform more specialized experiments that sometimes require their own set-ups. One thing that all these experiments have in common is the necessity to reach out to the beamline support staff and discuss the set-up in detail before the actual measurements. A number of issues that have to be addressed include: vacuum or in-air measurements, alignment possibilities (for example, with the available x-y stage), required network connections, sample volumes, temperatures, etc. Many more issues also depend on the sample type itself. As these preparations are very individual and in most cases require a project-to-project consultation and development of experimental protocols, going into exact protocols for each sample environment would be beyond the scope of this overview. Here, we are listing a number of examples of such experiments as inspiration on the vast possibilities given a scattering endstation. With the experience over the past years, we cannot emphasize strongly enough to commence communication with the beamline support staff as soon as possible during the planning stage. Ideally some pre-testing may be performed to make sure that the experiment is not hindered, e.g., by unaccepted firewall issues or missing passwords and administrative access rights. Accommodating various experimental set-ups and processes presents a significant challenge regarding the associated requirements on hardware synchronization for data collection. The aim is to be flexible on the one hand without compromising robustness and beam quality on the other hand. Importantly, the software management system at P12 also allows for flexible adaptation for new set-ups, but still, advanced planning and communications are extremely important.

We divide our specialized data collection modes at P12 into two types: the vacuum measurements and the in-air set-ups. For the latter the automated sample loading robot is removed from the beam path, allowing for an air gap in which the new sample holder in placed. This gap is defined by two sealed windows. The gap should be kept as small as possible to prevent unnecessary loss of beam due to parasitic scattering (increases the X-ray absorption as well as increase in higher background signal). The windows are often made of kapton or mica material. A number of other materials may be used for the "specialized" SAXS set-ups that should be tested well in advance. Fig. 7 shows some of our frequently used in-air set-ups.

Here, the first two examples are specialized holders with the capability for temperature controlled measurements, whereby the choice lays in the nature of the sample. The mica holder (Fig. 7A) is suitable for non-liquid solutions (e.g., gels) and the sample is placed (sandwiched) between two mica windows.

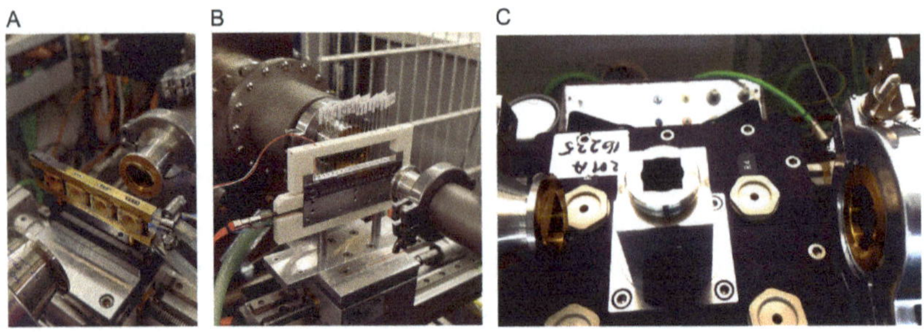

Fig. 7 Different customized set-ups for biological SAXS experiments currently used at the beamline P12. (A) Temperature controlled capillary holder. (B) Temperature controlled mica windows. (C) Stopped-flow device for TR-SAXS.

The next set-up depicted in Fig. 7B, is a temperature controlled (from 4 °C to 100 °C) capillary holder consisting of a Peltier element and the possibility to place quartz capillaries in the intended slots. It is designed in such a manner that different temperature ranges are easily reached with relative fast temperature jumps. This set-up is ideal for highly viscous samples or other tricky samples with toxic or other corrosive properties. Also samples that are known to be "sticky" and often result in extensive capillary fouling should be measured in such a system, as the exchange of the individual quartz capillaries is more straightforward. Here, too a wider range of apolar solvents may be used compared to the standard automated sample loading robot. Examples for the use of this system include the study of fibrillation processes and it might also be more suitable for samples where evaporation could be an issue in particular if applying incubation at high temperature or over long time periods.

The biggest pool of "non-standard experiments" that are performed at the beamline P12, and which is the most advanced regarding autonomous user operation is the stopped-flow set-up (Bio-logic, Seyssinet-Pariset, France) for time-resolved studies (Fig. 7C, Blanchet et al., 2015). Here, a rapid mixing device for fast time-kinetic is integrated and has been used, for example, to study the effects of ATP binding on protein conformation and association state (Josts et al., 2020). The results from this study are shown in Fig. 4C and D and discussed above. Regarding the detailed sample requirements for this study: 1 mM MsbA NBD was mixed in a 1:1 (v/v) ratio with Mg^{2+}-ATP (0.9 mM). Thereby, 60 µL of each component were mixed for each timeseries.

In another study (Lizarrondo et al., 2021) TR-SAXS shed light on the interaction of two adaptor domains involved in clathrin-mediated endocytosis, which is important for cellular functions such as nutrient uptake, cell signaling, and cell adhesion. In this context, the two binding domains, ANTH and ENTH, form a stable complex in the presence of phosphatidylinositol 4,5-bisphosphate (PIP2) on a fast subsecond timescale. Complex formation was triggered with mixing of ANTH and ENTH at equimolar ratios. After mixing, SAXS curves were collected over the first 400 ms and the increase in R_G over time is consistent with the formation of larger species. Singular value decomposition of the curves further revealed the presence of three components: the monomers, the final complex and an intermediate state. The concentration used was 2 mg/mL and in the presence of 200 μM PIP2 and the dead time of the study was 5 ms.

Currently, the trigger options of TR-SAXS reactions are being extended by a tunable Nd:YAG–laser system (Ekspla, Lithuania) covering the wavelengths 335–500 nm & 1065–2500 nm. Via a fiber the laser ray is directed to the sample position in the experimental hutch. The laser, reaching a repetition rate of 10 Hz can trigger temperature jumps through heating and may also be used for uncaging experiments or direct activation of photosensitive proteins.

For the future, TR-SAXS is expected to play an even more prominent role. The expected increase in flux and coherence with anticipated upgrades to 4th generation synchrotrons, will afford the possibilities to follow yet faster biological process in action and in solution.

Besides these more frequently used set-ups shown in Fig. 7, P12 offers other environments including various microfluidic chips for small-sample volume mixing and analysis. With the so-called SAXS-CD, screening with as little as 2 μL of sample, 3 μL of mixing buffer and 3 μL of screening buffer can be performed (Schwemmer et al., 2016). In the flat compact-disc-sized device, small reservoirs and liquid channels are incorporated for easy and reproducible mixing. The challenge in this set-up is the proper alignment in the beam, the currently high cost of the chip itself and protection against radiation damage. However, the possibility to gain valuable SAXS information on smallest volumes is the basis for further developments in this direction. Another microfluidic set-up was more recently developed for the use at P12—to assess to possibility to use SAXS to study the effect the Tetraherz illumination (Schroer et al., 2021).

For more advanced studies in terms of soft matter, a scanning stage is available at P12 that allows, for example, the analysis of tissues or bones. Here, the idea is to use a small X-ray beam and probe the specimen in two dimensions which results in a real space map of scattering intensity.

This can be used, for example, to study the organization within cells or the orientation of hydroxyapatite crystals within bones. Recently, a study was performed to generate maps of the mineral particle thickness measured along the different zones of the sword of a swordfish (Schmidt et al., 2019). This project revealed a higher mineral particle thickness at the tip compared to the base of the sword. This information completes the in-depth understanding of the base-to-tip-wise aging of bone.

As mentioned, these specialized sample environments are placed in the air-gap and thus are all referred to the in air set-ups. On a positive side, for many in-air setups (e.g., solid, powders or gel-like samples) no matching buffer subtraction is needed and the data may be used as they are measured. This significantly eases the demands on the data quality and allows one to reliably detect the necessary features (e.g., peak positions, if observed). For the in-air configurations, 3–4 blocks of approximately 1–2 weeks time are set aside each beam year at P12, the rest is in-vacuum operation.

There is another non-standard measurement performed in vacuum, with the automated sample loading robot: an anomalous SAXS (ASAXS) mode. Here, the energy of the incident X-rays is selected to be close to the binding energy of the electrons of certain atoms. This allows one to specifically exploit small changes of the scattering properties of these atoms. ASAXS on biological samples is potentially extremely powerful but also extremely challenging because of the very small effect. With recently published pilot ASAXS experiments from dilute solutions of ferritin and cobalt-loaded apoferritin (Wieland et al., 2021), the high brilliance of P12 beamline was shown to be useful for biological ASAXS. Extraction of the Fe- or Co-specific anomalous dispersion terms from the ASAXS data was possible to determine the spatial distribution of either iron or cobalt atoms incorporated into the ferritin/apoferritin protein cages. These studies were followed by a more versatile set of experiments (Gruzinov et al., 2021) on silica-covered gold nanoparticles as well as micelles with bromine counter ions. With this it is concluded that the ASAXS option is demonstrated to be a useful tool for further characterization of functionalized nanomaterials for medical, pharmaceutical, and biological applications, offering also advanced possibilities to study biological samples. ASAXS is technically demanding regarding the data sets needed at different energies (and ensuring sample and beam stability along with required calibrations), however it does benefit from the possibility to use the robotic sample delivery. The analysis tools at P12 have been extended so they may be used for the processing of ASAXS data to retrieve the element-specific information.

5.1 General thoughts on designing "non-standard" experiments

- Sample holder material:
  - o What material will be placed in the X-ray beam? How does this contribute to parasitic scattering? If applicable, how will the "air-gap" be kept to a minimum?
  - o How will the temperature be controlled through-out the measurement?
  - o How will the samples be exchanged?
  - o Was the fact, that the experiment takes place inside the specific chamber of the experimental hutch taken into account?
- Sample alignment:
  - o How will the sample be aligned in the beam? How will reproducibility of sample position be ensured?
- Required measurements:
  - o How will the 2D detector images be calibrated, can the standard calibration with silver behenate ($C_{22}H_{43}AgO_2$) be performed with the proposed sample holder?
  - o Can water, empty cell, buffer measurements for background subtraction be easily performed?
- Data processing and quality control:
  - o Can standard pipeline be used to process the data or are customized adaptations required?
  - o What means are in place for quick assessment of data quality control at the beamline to allow for the detection and introduction of adaptation to ensure the meaningful collection of the data at the beamline?
- Other considerations:
  - o Ensure that all electrical equipment has current testing and tagging prior to bringing it on site
  - o Are there any other take care with any other safety and reliability factors, such as pressure ratings/certification of components and systems, leak and spill containment, preventing of mechanical collision with beamline equipment required?

References

Berthaud, A., Manzi, J., Pérez, J., & Mangenot, S. (2012). Modeling detergent organization around aquaporin-0 using small-angle X-ray scattering. *Journal of the American Chemical Society, 134*(24), 10080–10088. https://doi.org/10.1021/ja301667n.

Blanchet, C. E., Spilotros, A., Schwemmer, F., Graewert, M. A., Kikhney, A., Jeffries, C. M., et al. (2015). Versatile sample environments and automation for biological solution X-ray scattering experiments at the P12 beamline (PETRA III, DESY). *Journal of Applied Crystallography*, *48*(Pt. 2), 431–443.

Brennich, M. E., Round, A. R., & Hutin, S. (2017). Online size-exclusion and ion-exchange chromatography on a SAXS beamline. *Journal of Visualized Experiments*, *119*, e54861.

Brookes, E., Vachette, P., Rocco, M., & Pérez, J. (2016). *US-SOMO* HPLC-SAXS module: Dealing with capillary fouling and extraction of pure component patterns from poorly resolved SEC-SAXS data. Journal of Applied Crystallography, 49(Pt. 5), 1827–1841.

Brosey, C. A., & Tainer, J. A. (2019). Evolving SAXS versatility: Solution X-ray scattering for macromolecular architecture, functional landscapes, and integrative structural biology. *Current Opinion in Structural Biology*, *58*, 197–213.

Bucciarelli, S., Midtgaard, S. R., Nors Pedersen, M., Skou, S., Arleth, L., & Vestergaard, B. (2018). Size-exclusion chromatography small-angle X-ray scattering of water soluble proteins on a laboratory instrument. *Journal of Applied Crystallography*, *51*(Pt. 6), 1623–1632.

Burda, P., Crosskey, T., Lauk, K., Zurborg, A., Söhnchen, C., Liffner, B., et al. (2020). Structure-based identification and functional characterization of a lipocalin in the malaria parasite plasmodium falciparum. *Cell Reports*, *31*(12), 107817.

Cowieson, N. P., Edwards-Gayle, C., Inoue, K., Khunti, N. S., Doutch, J., Williams, E., et al. (2020). Beamline B21: High-throughput small-angle X-ray scattering at diamond light source. *Journal of Synchrotron Radiation*, *27*(Pt. 5), 1438–1446.

Custódio, T. F., Das, H., Sheward, D. J., Hanke, L., Pazicky, S., Pieprzyk, J., et al. (2020). Selection, biophysical and structural analysis of synthetic nanobodies that effectively neutralize SARS-CoV-2. *Nature Communications*, *11*(1), 5588.

Da Vela, S., & Svergun, D. I. (2020). Methods, development and applications of small-angle X-ray scattering to characterize biological macromolecules in solution. *Current Research in Structural Biology*, *2*, 164–170.

David, G., & Perez, J. (2009). Combined sampler robot and high-performance liquid chromatography: A fully automated system for biological small-angle X-ray scattering experiments at the synchrotron SOLEIL SWING beamline. *Journal of Applied Crystallography*, *42*, 892–900.

Franke, D., Kikhney, A. G., & Svergun, D. I. (2012). Automated acquisition and analysis of small angle X-ray scattering data. *Nuclear Instruments and Methods in Physics Research Section A: Accelerators, Spectrometers, Detectors and Associated Equipment*, *689*, 52–59.

Graewert, M. A., & Jeffries, C. M. (2017). Sample and buffer preparation for SAXS. *Advances in Experimental Medicine and Biology*, *1009*, 11–30.

Graewert, M. A., Franke, D., Jeffries, C. M., Blanchet, C. E., Ruskule, D., Kuhle, K., et al. (2015). Automated pipeline for purification, biophysical and x-ray analysis of biomacromolecular solutions. *Scientific Reports*, *5*, 10734.

Graewert, M. A., Da Vela, S., Gräwert, T. W., Molodenskiy, D. S., Blanchet, C. E., Svergun, D. I., et al. (2020). Adding size exclusion chromatography (SEC) and light scattering (LS) devices to obtain high-quality small angle X-ray scattering (SAXS) data. *Crystals*, *10*(11), 975 (1-18).

Gräwert, T. W., & Svergun, D. I. (2020). Structural modeling using solution small-angle X-ray scattering (SAXS). *Journal of Molecular Biology*, *432*(9), 3078–3092.

Gruzinov, A. Y., Schroer, M. A., Manalastas-Cantos, K., Kikhney, A. G., Hajizadeh, N. R., Schulz, F., et al. (2021). Anomalous SAXS at P12 beamline EMBL Hamburg: Instrumentation and applications. *Journal of Synchrotron Radiation*, *28*(Pt. 3), 812–823.

Guinier, A. (1939). La diffraction des rayons X aux très petits angles: Application à l'étude de phénomènes ultramicroscopiques. *Annals of Physics*, *11*(12), 161–237. https://doi.org/10.1051/anphys/193911120161.

Hajizadeh, N. R., Franke, D., & Svergun, D. I. (2018). Integrated beamline control and data acquisition for small-angle X-ray scattering at the P12 BioSAXS beamline at PETRAIII storage ring DESY. *Journal of Synchrotron Radiation*, *25*, 906–914.

Hopkins, J. B., Gillilan, R. E., & Skou, S. (2017). BioXTAS RAW: Improvements to a free open-source program for smallangle X-ray scattering data reduction and analysis. *Journal of Applied Crystallography*, *50*, 1545e1553.

Hura, G. L., Menon, A. L., Hammel, M., Rambo, R. P., Poole, F. L., 2nd, Tsutakawa, S. E., et al. (2009). Robust, high-throughput solution structural analyses by small angle X-ray scattering (SAXS). *Nature Methods*, *6*(8), 606–612.

Hutin, S., Brennich, M., Maillot, B., & Round, A. (2016). Online ion-exchange chromatography for small-angle X-ray scattering. *Acta Crystallographica Section D*, *72*, 1090–1099.

Inoue, R., Nakagawa, T., Morishima, K., et al. (2019). Newly developed Laboratory-based Size exclusion chromatography Small-angle x-ray scattering System (La-SSS). *Scientific Reports*, *9*(1), 12610. Published 2019 Aug 30.

Jacques, D. A., & Trewhella, J. (2010). Small-angle scattering for structural biology—Expanding the frontier while avoiding the pitfalls. *Protein Science: A Publication of the Protein Society*, *19*(4), 642–657.

Jeffries, C. M., Graewert, M. A., Blanchet, C. E., Langley, D. B., Whitten, A. E., & Svergun, D. I. (2016). Preparing monodisperse macromolecular samples for successful biological small-angle X-ray and neutron-scattering experiments. *Nature Protocols*, *11*(11), 2122–2153.

Jeffries, C. M., Graewert, M. A., Svergun, D. I., & Blanchet, C. E. (2015). Limiting radiation damage for high-brilliance biological solution scattering: Practical experience at the EMBL P12 beamline PETRAIII. *Journal of Synchrotron Radiation*, *22*, 273–279.

Johansen, N. T., Pedersen, M. C., Porcar, L., Martel, A., & Arleth, L. (2020). Introducing SEC–SANS for studies of complex self-organized biological systems. *Acta Crystallographica Section D*, *74*, 1178–1191.

Jordan, A., Jacques, M., Merrick, C., Devos, J., Forsyth, V. T., Porcar, L., et al. (2016). SEC-SANS: Size exclusion chromatography combined *in situ* with small-angle neutron scattering. *Journal of Applied Crystallography*, *49*(Pt. 6), 2015–2020.

Josts, I., Gao, Y., Monteiro, D. C. F., Niebling, S., Nitsche, J., Veith, K., et al. (2020). Structural kinetics of MsbA investigated by stopped-flow time-resolved small-angle X-ray scattering. *Structure*, *28*, 348–354.

Konarev, PV, Graewert, MA, Jeffries, CM, et al. (2022). EFAMIX, a tool to decompose inline chromatography SAXS data from partially overlapping components. *Protein Science*, *31*(1), 269–282.

Konarev, P., Volkov, V. V., Sokolova, A. V., Koch, M. H., & Svergun, D. I. (2003). PRIMUS: A Windows PC-based system for small-angle scattering data analysis. *Journal of Applied Crystallography*, *36*(5), 1277–1282. Web.

Kostan, J., Pavšič, M., Puž, V., Schwarz, T. C., Drepper, F., Molt, S., et al. (2021). Molecular basis of F-actin regulation and sarcomere assembly via myotilin. *PLoS Biology*, *19*(4), e3001148.

Krichel, B., Bylapudi, G., Schmidt, C., Blanchet, C., Schubert, R., Brings, L., et al. (2021). Hallmarks of Alpha- and Betacoronavirus non-structural protein 7 + 8 complexes. *Science Advances*, 7(10), eabf1004.

Lahey-Rudolph, J. M., Schönherr, R., Jeffries, C. M., Blanchet, C. E., Boger, J., Ferreira Ramos, A. S., et al. (2020). Rapid screening of in cellulo grown protein crystals via a small-angle X-ray scattering/X-ray powder diffraction synergistic approach. *Journal of Applied Crystallography*, *53*(5), 1169–1180.

Lizarrondo, J., Klebl, D. P., Niebling, S., Abella, M., Schroer, M. A., Mertens, H., et al. (2021). Structure of the endocytic adaptor complex reveals the basis for efficient membrane anchoring during clathrin-mediated endocytosis. *Nature Communications, 12*(1), 2889.

Macpherson, A., Laabei, M., Ahdash, Z., Graewert, M. A., Birtley, J. R., Schulze, M. E., et al. (2021). The allosteric modulation of complement C5 by knob domain peptides. *eLife, 10*, e63586.

Martel, A., Liu, P., Weiss, T. M., Niebuhr, M., & Tsuruta, H. (2012). An integrated high-throughput data acquisition system for biological solution X-ray scattering studies. *Journal of Synchrotron Radiation, 19*, 431–434.

Mathew, E., Mirza, A., & Menhart, N. (2004). Liquid-chromatography-coupled SAXS for accurate sizing of aggregating proteins. *Journal of Synchrotron Radiation, 11*(Pt. 4), 314–318.

Molodenskiy, D. S., Mertens, H., & Svergun, D. I. (2020). An automated data processing and analysis pipeline for transmembrane proteins in detergent solutions. *Scientific Reports, 10*(1), 8081.

Nogueira, S. S., Schlegel, A., Maxeiner, K., Weber, B., Barz, M., Schroer, M., et al. (2020). Polysarcosine-functionalized lipid nanoparticles for therapeutic mRNA delivery. *ACS Applied Nano Materials, 3*(11), 10634–10645.

Osaka, K., et al. (2016). High-throughput and automated SAXS/USAXS experiment for industrial use at BL19B2 in SPring-8. In *AIP Conf. Proc. 1741*, 030003.

Ozerskaya, A. V., Zamay, T. N., Kolovskaya, O. S., Tokarev, N. A., Belugin, K. V., Chanchikova, N. G., et al. (2021). ^{11}C-radiolabeled aptamer for imaging of tumors and metastases using positron emission tomography—Computed tomography. *Molecular Therapy- -Nucleic Acids, 26*, 1159–1172.

Panjkovich, A., & Svergun, D. I. (2018). CHROMIXS: Automatic and interactive analysis of chromatography-coupled small-angle X-ray scattering data. *Bioinformatics, 34*, 1944–1946.

Pazicky, S., Dhamotharan, K., Kaszuba, K., Mertens, H. D. T., Gilberger, T., Svergun, D., et al. (2020). Structural role of essential light chains in the apicomplexan glideosome. *Communications Biology, 3*(1), 568.

Pérez, J., & Vachette, P. (2017). A successful combination: Coupling SE-HPLC with SAXS. *Advances in Experimental Medicine and Biology, 1009*, 183–199. https://doi.org/10.1007/978-981-10-6038-0_11.

Pernot, P., Round, A., Barrett, R., De Maria Antolinos, A., Gobbo, A., Gordon, E., et al. (2013). Upgraded ESRF BM29 beamline for SAXS on macromolecules in solution. *Journal of Synchrotron Radiation, 20*(Pt. 4), 660–664.

Round, A. R., Franke, D., Moritz, S., Huchler, R., Fritsche, M., Malthan, D., et al. (2008). Automated sample-changing robot for solution scattering experiments at the EMBL Hamburg SAXS station X33. *Journal of Applied Crystallography, 41*(Pt. 5), 913–917.

Round, A., et al. (2015). BioSAXS Sample Changer: A robotic sample changer for rapid and reliable high-throughput X-ray solution scattering experiments. *Acta Crystallographica, Section D: Biological Crystallography, 71*, 67–75.

Ryan, T. M., Trewhella, J., Murphy, J. M., Keown, J., Casey, L. W., Pearce, F. G., et al. (2018). An optimized SEC-SAXS system enabling high X-ray dose for rapid SAXS assessment with correlated UV measurements for biomolecular structure analysis. *Journal of Applied Crystallography, 51*, 97–111.

Schmidt, F. N., Zimmermann, E. A., Walsh, F., Plumeyer, C., Schaible, E., Fiedler, I., et al. (2019). On the origins of fracture toughness in advanced Teleosts: How the swordfish Sword's bone structure and composition allow for slashing under water to kill or stun prey. *Advanced Science (Weinheim, Baden-Wurttemberg, Germany), 6*(12), 1900287.

Schroer, M. A., Blanchet, C. E., Gruzinov, A. Y., Gräwert, M. A., Brennich, M. E., Hajizadeh, N. R., et al. (2018). Smaller capillaries improve the small-angle X-ray scattering signal and sample consumption for biomacromolecular solutions. *Journal of Synchrotron Radiation*, *25*(Pt. 4), 1113–1122.

Schroer, M. A., Schewa, S., Gruzinov, A. Y., Rönnau, C., Lahey-Rudolph, J. M., Blanchet, C. E., et al. (2021). Probing the existence of non-thermal terahertz radiation induced changes of the protein solution structure. *Scientific Reports*, *11*(1), 22311.

Schwemmer, F., Blanchet, C. E., Spilotros, A., Kosse, D., Zehnle, S., Mertens, H. D., et al. (2016). LabDisk for SAXS: A centrifugal microfluidic sample preparation platform for small-angle X-ray scattering. *Lab on a Chip*, *16*(7), 1161–1170.

Shemesh, A., Ginsburg, A., Dharan, R., Levi-Kalisman, Y., Ringel, I., & Raviv, U. (2021). Structure and energetics of GTP- and GDP-tubulin isodesmic self-association. *ACS Chemical Biology*, *16*(11), 2212–2227.

Siegel, L. M., & Monty, K. J. (1965). Determination of molecular weights and frictional ratios of proteins in impure systems: Aggregation of urease. *Journal of Chromatography*, *19*(4), 494–499. https://doi.org/10.1016/0006-291x(65)90152-x.

Siewert, C. D., Haas, H., Cornet, V., Nogueira, S. S., Nawroth, T., Uebbing, L., et al. (2020). Hybrid biopolymer and lipid nanoparticles with improved transfection efficacy for mRNA. *Cell*, *9*(9), 2034.

Stuhrmann, H. (1980). In H. Winick, & S. Doniach (Eds.), *Synchrotron radiation research* (pp. 513–531). New York: Plenum Press.

Svergun, D. I., Barberato, C., & Koch, M. H. (1995). CRYSOL a program to evaluate X-ray solution scattering of biological macromolecules from atomic coordinates. *Journal of Applied Crystallography*, *28*, 768e773.

Trewhella, J. (2022). Recent advances in small-angle scattering and its expanding impact in structural biology. *Structure*, *30*(1), 15–23.

Tully, M. D., Tarbouriech, N., Rambo, R. P., & Hutin, S. (2021). Analysis of SEC-SAXS data via EFA deconvolution and scatter. *Journal of Visualized Experiments: JoVE*, *2021*(167). https://doi.org/10.3791/61578. Published 2021 Jan 28.

Uebbing, L., Ziller, A., Siewert, C., Schroer, M. A., Blanchet, C. E., Svergun, D. I., et al. (2020). Investigation of pH-responsiveness inside lipid nanoparticles for parenteral mRNA application using small-angle X-ray scattering. *Langmuir: The ACS Journal of Surfaces and Colloids*, *36*(44), 13331–13341.

Watanabe, Y., & Inoko, Y. (2009). Size-exclusion chromatography combined with small-angle X-ray scattering optics. *Journal of Chromatography A*, *1216*(44), 7461–7465.

Wieland, D., Schroer, M. A., Gruzinov, A. Y., Blanchet, C. E., Jeffries, C. M., & Svergun, D. I. (2021). ASAXS measurements on ferritin and apoferritin at the bioSAXS beamline P12 (PETRA III, DESY). *Journal of Applied Crystallography*, *54*(Pt. 3), 830–838.

Yunoki, Y., Matsumoto, A., Morishima, K., Martel, A., Porcar, L., Sato, N., et al. (2022). Overall structure of fully assembled cyanobacterial KaiABC circadian clock complex by an integrated experimental-computational approach. *Communications Biology*, *5*(1), 184.

CHAPTER TWO

Contrast variation SAXS: Sample preparation protocols, experimental procedures, and data analysis

Josue San Emeterio, Suzette A. Pabit, and Lois Pollack*
School of Applied and Engineering Physics, Cornell University, Ithaca, NY, United States
*Corresponding author: e-mail address: lp26@cornell.edu

Contents

Methods in Enzymology, Volume 677
ISSN 0076-6879
https://doi.org/10.1016/bs.mie.2022.08.007

Abstract

Proteins and nucleic acids, alone and in complex are among the essential building blocks of living organisms. Obtaining a molecular level understanding of their structures, and the changes that occur as they interact, is critical for expanding our knowledge of life processes or disease progression. Here, we motivate and describe an application of solution small angle X-ray scattering (SAXS) which provides valuable information about the structures, ensembles, compositions and dynamics of protein-nucleic acid complexes in solution, in equilibrium and time-resolved studies. Contrast variation (CV-) SAXS permits the visualization of the distinct molecular constituents (protein and/or nucleic acid) within a complex. CV-SAXS can be implemented in two modes. In the simplest, the protein within the complex is effectively rendered invisible by the addition of an inert contrast agent at an appropriate concentration. Under these conditions, the structure, or structural changes of only the nucleic acid component of the complex can be studied in detail. The second mode permits observation of both components of the complex: the protein and the nucleic acid. This approach requires the acquisition of SAXS profiles on the complex at different concentrations of a contrast agent. Here, we review CV-SAXS as applied to protein-nucleic acid complexes in both modes. We provide some theoretical framework for CV-SAXS but focus primarily on providing the necessary information required to implement a successful experiment including experimental design, sample quality assessment, and data analysis.

1. Introduction

Protein-nucleic acid complexes play essential roles in biology. Protein interactions with DNA contribute to processes as fundamental as: packaging or remodeling chromatin, transcription, replication, recombination, damage repair, or DNA modification (Brenner & Miller, 2014). RNA-protein complexes exhibit biological activities that extend far beyond gene expression. A myriad of newly-identified RNA binding proteins have been reported; yet the structural interactions that facilitate biological activity have not yet been established for many of these complexes (Beckmann, Castello, & Medenbach, 2016; Van Nostrand et al., 2020). The exploding interest in RNA underscores the need for structural characterization, alone and with partners. Finally, either DNA or RNA can be packaged by proteins into viruses, a key concern at the present time.

Traditional high-resolution structural methods such as X-ray crystallography or cryo-electron microscopy (cryo-EM) have provided exceptional insight into the interactions that stabilize complexes (Galej et al., 2016; Luger, Mäder, Richmond, Sargent, & Richmond, 1997; Yan et al., 2015; Yusupov et al., 2001); however, the molecules are often immobilized, either through crystal contacts or by plunge freezing. Though extensive computation has been used to sequence a series of static snapshots (Frank & Ourmazd, 2016), the opportunity to measure complexes, in solution, at ambient (or higher) temperature, or in real time, during a reaction offers unique and valuable information.

Because SAXS provides structural information about biomolecules in solution at practically any temperature or buffer condition, it has often been used to complement higher resolution structural techniques (Putnam, Hammel, Hura, & Tainer, 2007; Stuhrmann, 2008; Tsutakawa, Hura, Frankel, Cooper, & Tainer, 2007). Expansion into the wide angle (or WAXS) regime sharpens the spatial resolution of solution scattering to a few angstroms, and is being applied to study proteins, DNA and RNA (e.g., He, Chen, Pollack, & Kirmizialtin, 2021; Makowski, Rodi, Mandava, Devarapalli, & Fischetti, 2008; Zuo et al., 2006). Solution X-ray scattering is a versatile and efficient tool to study biological systems; if the molecules are soluble, they can be measured, and their structures can be postulated, even if the biomolecular sample is not monodisperse. Specifically, for this latter case, SAXS with in-line size-exclusion chromatography (SEC-SAXS) can separate heterogeneous biomolecules by size immediately before measurement (David & Pérez, 2009; Pérez & Nishino, 2012). This technique is covered in other chapters in this volume. A true strength of SAXS is the study (measurement) of the conformational dynamics of biomolecules, either as an ensemble in equilibrium studies (Tria, Mertens, Kachala, & Svergun, 2015), or sequentially, through time-resolved studies which can be triggered using rapid mixing (e.g., Plumridge et al., 2018; Pollack et al., 1999) (with small or large partners) or through light (e.g., Lamb, Zoltowski, Pabit, Crane, & Pollack, 2008) or heat activation (e.g., Thompson et al., 2019).

SAXS has been most effectively applied to reveal the conformations of either protein or nucleic acid components in solution (Svergun & Koch, 2003). The signal strength scales with the electron density difference between the biomolecule and the solvent that surrounds it. The electron densities of both proteins and nucleic acids exceed that of the surrounding solvent, under typical conditions. Of particular interest to this chapter, we note that proteins and nucleic acids have different electron densities, hence they scatter with different strengths. We will show that the measured signal of a complex contains contributions that reflect the structure of each component, as well as an additional term that contains information about the relative positions of the two. It can therefore be difficult to disentangle the signals arising from individual components using conventional SAXS methods. The contrast variation (CV) method we describe circumvents this challenge. It exploits the density difference referenced above and permits the visualization of each component within a complex. Like traditional SAXS, CV-SAXS can then be used to monitor the changing structures of these distinct components as they interact.

This chapter provides essential guidelines for designing and performing CV-SAXS experiments on protein-nucleic acid complexes. We introduce the basic theory of CV-SAXS, discuss sample considerations, sample quality assessment, experimental practices and design, as well as data analysis approaches. Finally, we illustrate the types of results that can be extracted, using both DNA-protein and RNA-protein complexes as examples. Throughout the chapter, techniques used to overcome the unique challenges of CV-SAXS are extensively addressed.

2. Background: SAXS and CV-SAXS

2.1 The origin of any SAXS signal: Electron density differences (or contrast) between molecule and background solvent

Before providing details about contrast variation SAXS, we first briefly review regular SAXS. More comprehensive reviews can be found in Ref. (Blanchet & Svergun, 2013; Jacques & Trewhella, 2010; Koch, Vachette, & Svergun, 2003; Svergun & Koch, 2003) as well as other chapters in this book. It is relatively straightforward to compute the magnitude of scattering, the so-called scattering amplitude, from a single molecule in solution. This amplitude A depends on the molecule's electron density, ρ_M, its volume, V_M, and the arrangement of electrons within the molecule, e.g., its shape. This shape is reflected by an angle dependent form factor; the scattering decreases relative to its value in the forward or zero-angle ($\theta = 0$) direction in a way that depends on the structure of the scatterer. This form factor F is typically written as a function of scattering vector $q = 4\pi \sin\theta/\lambda$, where 2θ is the scattering angle and λ is the X-ray wavelength. The geometry of a typical scattering experiment is shown in Fig. 1. Experiments measure the intensity of the scattered radiation, $I(q)$, which is the product of amplitude A and its complex conjugate. Because most biomolecules are dissolved in solvent with electron density ρ_S, A is proportional to the electron density difference between the molecule and the background solvent, $\Delta\rho_M = \rho_M - \rho_S$. This difference represents the excess electron density of the solute relative to the solvent and is called the contrast of the system. The scattering amplitude and intensity for a molecule of electron density ρ_M and form factor $F(q)$ in a solvent with electron density ρ_S can be written as:

$$A(q) = (\Delta\rho_M)\ V_M\ F(q)$$
$$I(q) = \|A(q)\|^2 = (\Delta\rho_M)^2\ V_M^2\ \langle|\ F(q)|^2\rangle \quad (1)$$

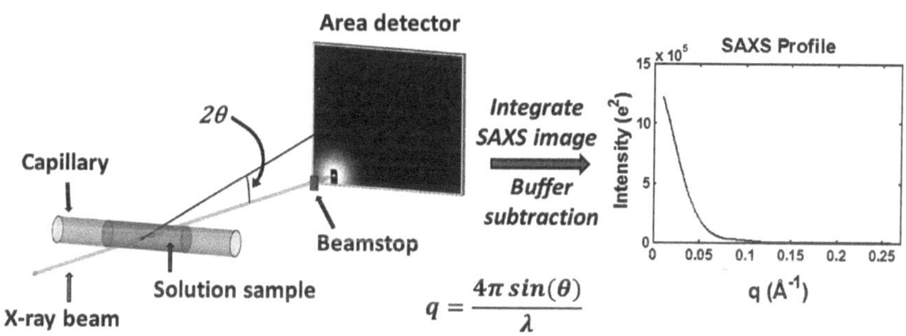

Fig. 1 Schematic of a typical small-angle x-ray scattering (SAXS) experiment. An X-ray beam from either a synchrotron or home/lab source is incident on a capillary containing a solution of macromolecules. The primary sample is typically a buffered solution containing protein, nucleic acid, or protein-nucleic acid complex at dilute concentrations. To avoid radiation damage from the X-ray beam, the sample is typically oscillated or flowed continuously. The scattered X-rays are imaged onto an area detector while the primary beam is either blocked or greatly attenuated by a beamstop. These images are then pooled, averaged, and azimuthally integrated to obtain SAXS profiles of intensity as a function of scattering vector, $I(q)$. For each biomolecule containing sample, a corresponding measurement of the buffer is made, and the resulting buffer profile is subtracted from the biomolecule sample profile to obtain the SAXS scattering profile from the macromolecule. SAXS intensities can be calibrated on an absolute scale. *From Tokuda, J. M., Pabit, S. A., & Pollack, L. (2016). Protein–DNA and ion–DNA interactions revealed through contrast variation SAXS.* Biophysical Reviews, *8(2), 139–149. https://doi.org/10.1007/s12551-016-0196-8.*

$\langle |F(q)|^2 \rangle$ represents the rotational average over many molecules in solution and is the product of $F(q)$ and its complex conjugate. It is often written as $P_M(q)$. The scattering intensity for this molecule in solution is written:

$$I(q) = (\Delta\rho_M)^2 \; V_M^2 \; P_M(q) \tag{2}$$

2.2 Contrast variation SAXS

We begin our discussion of contrast variation SAXS by modifying Eq. (1) to account for scattering from a system containing two species of molecules with different electron densities, on a uniform solvent background. We are most interested in protein-nucleic acid complexes. Nucleic acids are more electron dense than proteins, because of their phosphate backbone. If component 1 represents the nucleic acid, and component 2 is the protein, the scattering amplitude of one unit of the complex can be expressed as the sum of the amplitude of each component: $A(q) = \Delta\rho_1 V_1 F_1(q) + \Delta\rho_2 V_2 F_2(q)$.

Here, ρ_i is the electron density of macromolecule i (= 1 or 2) and ρ_s is the electron density of the solvent. The excess electron density of component 1 is written as $\Delta\rho_1 = \rho_1 - \rho_s$. When the amplitudes are summed, and the intensity computed following the procedure outlined above, the following expression results:

$$I(q) = \Delta\rho_1^2 V_1^2 P_1(q) + \Delta\rho_2^2 V_2^2 P_2(q) + 2\Delta\rho_1 \Delta\rho_2 V_1 V_2 P_{12}(q) \tag{3}$$

The first two terms of the above equation represent the scattering from the first and second molecular components, while the third term, the so-called cross-term, contains information about the relative positions of the two components. Although we discuss only the two-component case, this argument can be generalized to any number of molecular constituents.

To perform a contrast variation measurement of the type described herein, the density of the solvent, ρ_S, is increased following the addition of an inert contrast agent. The increase leads to a reduction of $\Delta\rho_M$ for each component. Experiments taken under multiple contrast conditions, where the contribution from the components is varied in a controlled way, can be used to separate the scattering from the components.

An extreme, but useful case arises when the electron density of the solvent, ρ_s is increased to equal ρ_2. Under this condition, the second and third terms of Eq. (3) become zero. At this so-called 'match point', one component of our complex has the same electron density as the solvent, and

$\Delta\rho_2$ vanishes. Eq. (3) reduces to Eq. (2) where only a single species is present and its scattering detected, although at decreased contrast relative to the condition where no contrast agent is added.

At this point, it is useful to introduce some numbers. For a DNA-protein complex in buffer, we need the electron densities of the solvent, the protein and the DNA. These densities can vary, depending on the details of the system (packing of a protein, for example); however, we quote generally accepted values (Svergun & Koch, 2003) and illustrate the method using a cartoon in Fig. 2. The illustration represents a higher electron density nucleic acid surrounded by a lower electron density protein in a standard buffer (top panel) and a contrast matched buffer (bottom panel). The electron density for most buffers is close to that of water, $\rho_s = 330\,\mathrm{e/nm^3}$. The protein electron density is typically $420\,\mathrm{e/nm^3}$, and the ubiquitous phosphate atoms along each nucleic

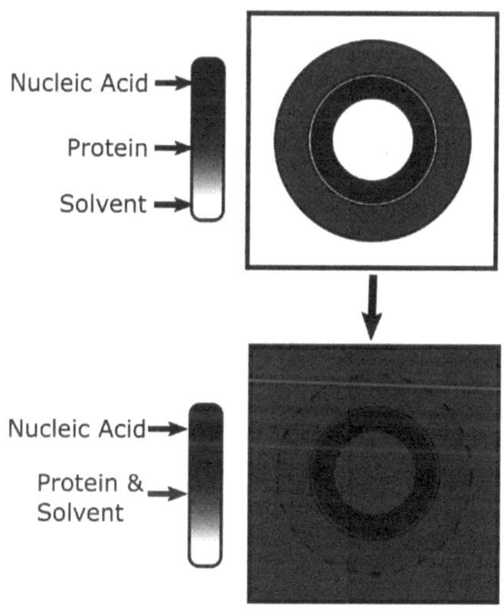

Fig. 2 Illustration of the principle of contrast variation SAXS. In CV-SAXS, the electron density of the solvent is increased until it matches one of the components of a multicomponent system. In this figure, the electron density is represented as color. In the top panel (no contrast agent), we model a protein-nucleic acid complex in regular aqueous solvent. In the bottom panel, the electron density of the solvent is increased (schematically shown as color change from white to red to black) by the addition of sucrose until it matches the electron density of the protein (red). Under this contrast matched condition any scattering signal that contains a contribution from the protein component blends into the background, e.g., disappears. Only the scattering from the more electron dense nucleic acid is detected. *From San Emeterio, J., & Pollack, L. (2020). Visualizing a viral genome with contrast variation small angle X-ray scattering.* Journal of Biological Chemistry, 295*(47), 15923–15932. https:/doi.org/10.1074/jbc.RA120.013961.*

acid strand enhance the electron density of DNA to 550 e/nm^3. We note that these values are quoted in the literature (Svergun & Koch, 2003). Our experience suggests that the actual contrast of nucleic acids is slightly lower. In a standard buffer (Fig. 2, top panel), the excess scattering density of the protein or DNA component is easily approximated by subtracting the density of the solvent from the value quoted above. Of course, this subtraction assumes that the electron density is constant and uniform, which is not always strictly true, but it is a reasonable approximation for most proteins and nucleic acids. As the biomolecule is measured with increasing electron density in the buffer, this difference decreases. As explained in more detail later, sucrose is our contrast agent of choice to increase the electron density of the buffer. It is highly soluble in aqueous solutions and appears to be inert. When enough sucrose is added to the buffer, its electron density can reach the value quoted for the protein. The electron density difference between protein and background vanishes under this condition. In the language of Eq. (3), assigning component 1 to the DNA and 2 to the protein, $\Delta\rho_2 = 0$ and any SAXS measurement tracks only the DNA conformation (Fig. 2, bottom panel). CV-SAXS, in this mode, can be used to mask or 'hide' the protein in a protein-nucleic acid complex (Tokuda, Pabit, & Pollack, 2016). Because of the solubility limit of sucrose, it is impossible to 'match' the density of the nucleic acid component. Thus, using sucrose, measurements can be performed up to, but not far in excess of the match point for proteins.

Contrast variation (CV) for biological systems is well established for small angle neutron scattering (SANS). Comprehensive reviews can be found in Ref. (Gabel, 2015; Krueger, 2017; Whitten & Trewhella, 2009) as well as in other chapters in this volume. Although CV-SAXS and CV-SANS share the same underlying principle ('matching' hence nulling out one component of the scattering, while leaving the other, albeit reduced), the different scattering physics for neutrons vs. X-rays creates profound differences in how the method is applied. Neutrons scatter through interactions with atomic nuclei, via nuclear forces, while X-rays scatter through interactions with electrons. Because the scattering length of the biomolecules lies between the values for H_2O or D_2O, scattering lengths that match those of biomolecules can be reached by mixing the two at the proper ratio, and a selected component can be hidden by the solvent (Krueger, 2017; Whitten & Trewhella, 2009). SANS, therefore, has the advantage that either the protein or nucleic acid can be targeted for 'blanking'. In some cases, it has been noted that added D_2O can lead to aggregation (Krueger, 2017).

CV-SAXS derives its power from the ready availability of high intensity X-ray beams. With high intensity, synchrotron X-rays, it takes seconds to minutes to collect CV-SAXS data, in contrast to minutes to hours for CV-SANS (Krueger, 2017). The short collection times and higher signal-to-noise of CV-SAXS enables time resolved contrast variation experiments. These measurements generally require short exposure times because measurements are made in real time, or a limited amount of biomolecules are available. In a time-resolved mode, the dynamic interaction of proteins and nucleic acids can be monitored in real time (Chen et al., 2017). As an important note, equilibrium CV-SAXS experiments can be effectively performed on the latest generation of lab sources.

Finally, although the schematic of Fig. 2 focuses on measuring at the protein contrast match point, measurements can be performed at varying sucrose concentrations in the buffer, ranging from 0% to 85% w/v (weight of sucrose in grams per volume of solution in milliliters) sucrose, or 65% w/w (weight of sucrose per weight of solution) sucrose. We note the increased difficulty of working with highly viscous solutions, above 50–60% w/v sucrose. At sucrose concentrations below the match point of the protein, both protein and nucleic acid components scatter, and their contribution(s) to the overall signal vary in a readily quantifiable way. Multiple copies of Eq. (3) can be written to describe measurements at specific contrast levels. These simultaneous equations can be used to decompose the scattering of a complex under multiple conditions into the scattering of its constituents and of course, the cross term.

3. What information can be extracted from CV-SAXS data?

3.1 Special case: Analysis at the match point

For CV-SAXS data acquired at the protein match point ($\Delta\rho_2 = 0$) analysis is identical to that performed for a standard SAXS experiment. At this condition, nucleic acid conformations are revealed directly from CV-SAXS data. The analysis and modeling of SAXS data are reviewed in these Refs.: (Da Vela & Svergun, 2020; Kikhney & Svergun, 2015; Schneidman-Duhovny, Kim, & Sali, 2012; Weiel, Reinartz, & Schug, 2019). Robust analysis packages are readily available, and provide structural information ranging from global structural parameters, such as molecular weight, largest spatial extent of the molecule or radius of gyration, or through molecular shape reconstructions. Complex stoichiometry can be determined in some conditions for both standard and CV-SAXS (see examples). Finally, dynamic structural ensembles can be created that yield scattering profiles consistent with the SAXS data (Tria et al., 2015).

3.2 Measurements at different contrast values: Zero-angle scattering and radius of gyration

Some of the best-known SAXS-derived structural parameters, radius of gyration, R_g, and the forward or zero-angle ($q = 0$) scattering, $I(0)$, also provide insightful information when used in CV-SAXS studies. The radius of gyration is the second moment of the electron distribution in the molecule, it provides a measure of its size. The forward scattering is proportional to the number of excess electrons in the molecule squared (see Eq. (3), and note that $P_M(q)$ is defined to be 1 at $q = 0$). Both can be easily obtained using the Guinier approximation (Guinier, 1939), which models the scattering at the lowest angle as a simple Gaussian. The zero-angle scattering must be extrapolated from the fit because it is typically occluded by the large intensity of the unscattered beam (or beamstop, when used). We can compute $I(q = 0)$ using the Guinier approximation (Guinier, 1939), and note that this analysis is 'hardwired' into many standard SAXS analysis programs, including ATSAS (Franke et al., 2017) or BioXTAS RAW (Hopkins, Gillilan, & Skou, 2017). If the low q data are of poor quality (for example, if there are concerns about effects of interparticle interactions (Skou, Gillilan, & Ando, 2014) or access to low q data is limited for technical reasons), both $I(0)$ and R_g can be computed through a pair distance distribution function (defined below) with software such as GNOM (Svergun, 1992) or BIFT (Hansen, 2000). Here, we discuss the structural information that can be readily extracted from these parameters alone and from their dependence on contrast.

The forward intensity $I(0)$ can be used to measure the molecular weight of a macromolecule if the detector is 'absolutely calibrated', meaning that a standard sample like water has been used to convert the measured intensity into absolute units (Orthaber, Bergmann, & Glatter, 2000). The value of $I(0)$ (in absolute units of cm^{-1}) can be expressed in terms of the mass of the macromolecule, M (in units of g/mol), the partial specific volume, v (in units of cm^3/g), the excess scattering density, $\Delta\rho_M$ (in units of e/cm^3), the concentration of macromolecules, c (in units of g/mL), Avogadro's number, N_A (6.02×10^{23} mol^{-1}), and the classical radius of the electron, r_o (2.8179×10^{-13} cm) (Feigin & Svergun, 1987; Mylonas & Svergun, 2007; Orthaber et al., 2000):

$$I(0) = \frac{cMv^2}{N_A} r_o^2 \Delta\rho_M{}^2 = \frac{cMv^2}{N_A} r_o^2 (\rho_M - \rho_S)^2 \tag{4}$$

The above expression clearly shows the quadratic dependence of $I(0)$ on electron density difference and can be exploited to find the match point, the point where the molecule 'disappears' into the solvent background (Ibel & Stuhrmann, 1975; Inoko, Yamamoto, Fujiwara, & Ueki, 1992). Measurements of $I(0)$ can be performed at several different contrast agent concentrations. A linear fit of the square root of $I(0)$ vs. solvent scattering density ρ_s or contrast agent concentration in % sucrose can be computed. In principle, the x-intercept of the best fit line reveals the match point of the molecule (Ibel & Stuhrmann, 1975; Inoko et al., 1992; Sardet, Tardieu, & Luzzati, 1976; Stuhrmann & Miller, 1978).

In addition, the absolute scattering at $q = 0$ informs about the stoichiometry of a molecular complex, useful for determining whether the complex forms in a 1:1 ratio, for example. This information can be quite helpful in solving structures of complexes acquired by CV-SAXS; it is important to know how many copies of each molecule are present. As an example, if the protein and RNA form a complex in a 1:1 ratio, the extrapolated $I(0)$ values derived from an RNA-alone and a protein-alone can be used to estimate the $I(0)$ corresponding to the protein-RNA complex using:

$$I(0) = I_{RNA}(0) + 2\sqrt{I(0)_{RNA}}\sqrt{I(0)_{protein}} + I_{protein}(0). \tag{5}$$

The radius of gyration, R_g informs about the overall size of a molecule. It is easy to measure and provides ready information not only about size but more importantly, changes in molecular size that accompany dynamic molecular motions or binding. The dependence of R_g on contrast provides guidance about molecular composition, as described by Stuhrmann and coworkers (Ibel & Stuhrmann, 1975). The measured radius of gyration follows an inverse square relationship with contrast:

$$\begin{aligned} R_g{}^2 &= R_c{}^2 + \frac{\alpha}{\Delta\rho} - \frac{\beta}{\Delta\rho^2} \\ \alpha &= \frac{1}{V}\int \rho(r)\, r^2 dr \\ \beta &= \frac{1}{V^2}\left(\int \rho(r) r dr\right)^2 \end{aligned} \tag{6}$$

A plot of the measured R_g vs. the inverse of average contrast of the molecule is referred to as a Sturhmann plot. Several relevant structural parameters can be obtained by fitting the quadratic dependence on contrast in the above equation. R_c is the radius of gyration of the same molecule, but with homogeneous density. The term α is the second moment of the density variations; it characterizes the distribution of phases within the particle (Stuhrmann, 2008). This term is particularly useful for complexes with a radial mass distribution, i.e., viruses or apoferritin. A positive α indicates a higher density component on outside of the complex, whereas a negative α indicates higher density towards the core of the complex. The term β relates to the square of the first moment of the density fluctuations. A non-zero β indicates that the two phases of the complex do not share the same center of mass. For a molecule of homogeneous density, there should not be any dependence on contrast, as α and β are both zero.

3.3 Structural information can be obtained from the entire SAXS profile: Pair distance distribution function P(R), reconstructions, and ensemble modeling

Below we provide additional useful approaches for analyzing CV-SAXS data. Examples illustrating their application are provided in the last sections of this chapter.

Structural changes can be readily inferred by computing the pair distance distributions, P(R) from the scattering curves. P(R) reflects the sets of distances between all pairs of electrons in the macromolecule and can be derived from an inverse Fourier Transform of the scattering profile. Typically, P(R) is computed using the program GNOM (Svergun, 1992) from the ATSAS package (Franke et al., 2017) or from BIFT (Hansen, 2000). A peak in the P(R) curve often reflects a repeated length scale in a molecule, such as the radius of a sphere, or the diameter of a cylinder. This information provides an immediate assessment of repeating structures in real space (Svergun & Koch, 2003).

A single SAXS profile is often interpreted using the assumption that the system is homogeneous and monodisperse. The resulting parameters correspond to average molecular features. Under contrast matched conditions, where the entire structure has a reasonably uniform excess electron density, bead model reconstruction programs can be applied to suggest this average structure. These bead model SAXS reconstructions are typically obtained through DAMMIF (Svergun, 1999). We note that this program does not account for the differing electron densities of protein-nucleic acid complexes. Programs based on electron density reconstructions such as DENSS (Grant, 2018) might be able to overcome these limitations and assign the proper electron densities.

In many cases, structural variation is critically important for biological activity, so the data benefit from an interpretation that invites more advanced ensemble modeling approaches. It can be particularly valuable to compare data with profiles computed from simulations or models, when ensembles of conformations are present. CV-SAXS data on nucleic acids, acquired at the protein match point, are amenable to this more sophisticated approach. Typically, these approaches require an initial atomic structure which can be refined to fit the data. For instance, in the case of molecules that may have more than one conformation in solution, ensemble optimization methods (EOM) can be used to propose ensembles whose summed scattering profiles look like the data (Tria et al., 2015). Here, it is essential to computationally create a large pool of as many molecular conformations as possible. This deep structural pool is selectively refined over many cycles of a genetic algorithm. After convergence, a minimal set of structures is identified whose scattering profiles best recapitulate the data.

Reconstructions illustrate the real space positions of the components in a complex; however, obtaining them by 'inverting' the data is more involved for complexes than proposing structures for single component objects. Fortunately, there are packages available that exploit measurements of a contrast series to simultaneously fit two different molecular components. An algorithm available for this computation is MONSA (Svergun, 1999; Svergun & Nierhaus, 2000). The advantage of this method is that we can obtain reconstructions from both components in the assembly as well as their relative spacing.

4. Experimental considerations

4.1 General beamline details

A CV-SAXS experiment can be carried out using equipment available at most standard SAXS beamlines. Powerful lab sources can also be used for equilibrium measurements. However, the changes to the sample (either biomolecules in buffer or buffer background), especially the increased viscosity of samples prepared at high sucrose concentrations, require that attention be paid to both sample preparation and handling during the measurement. In this section, we discuss experimental requirements specific to CV-SAXS.

4.2 Buffer subtraction

As explained in Ref. (Pollack, 2011) measuring the SAXS profile of a biomolecule requires that two sample solutions be prepared, measured, and properly subtracted. The first sample contains the molecule of interest in the selected buffer. The second sample is identical to the first but does not contain the biomolecule of interest. The scattering of this 'buffer background' includes the contributions from the buffer, sample cell, and any background scattering from the beamline, which should of course be reduced, but can rarely be eliminated. Assuming the sample cell thickness is identical, the scattering intensity from the two samples must be normalized to account for beam intensity variations. We use either a semi-transparent beamstop or a pin diode to record the beam intensity at the beamstop. After normalization, the biomolecule absent (buffer background) measurement is subtracted from the biomolecule present sample; the resulting profile reflects only the scattering of the biomolecular sample. This buffer subtraction is absolutely critical. It is essential to prepare a buffer background solution for each biomolecular sample, with components that as closely as possible match that of the biomolecular solution conditions. As discussed below, this simple buffer subtraction is often one of the biggest challenges of CV-SAXS experiments. Accurate buffer matching is an important step in acquiring useful data.

4.3 Contrast agent

In an X-ray scattering contrast variation experiment, we require a simple method for increasing the electron density of the solvent relative to its intrinsic value. Historically, this has been accomplished by adding salts (Fernandez, Riske, Amaral, Itri, & Lamy, 2008; Schneidman-Duhovny, Hammel, Tainer, & Sali, 2013), glycerol (Bolze, Hörner, & Ballauff, 1996; Hickl, Ballauff, & Jada, 1996; Hirai et al., 2018; Ibel & Stuhrmann, 1975) and/or sucrose (Bolze, Ballauff, Kijlstra, & Rudhardt, 2003; Ibel & Stuhrmann, 1975; Pabit et al., 2020; San Emeterio & Pollack, 2020). Although any small molecule could be used (in theory), there are important considerations when selecting a contrast agent.

Ideally, the solubility of the contrast agent must be such that it can be added in sufficient amounts to reach the match point of at least the lower density component in the complex. Its addition should not alter the shape or properties of the molecule and it must have relatively high (reasonable) transmission of X-rays. The first requirement is obvious if we want to reach contrast levels that allow a match. The second requirement cannot be understated and is often dependent on the biological system being studied. In some cases, added contrast agents can induce aggregation (Krueger, 2017), can alter the ionic strength of the solution or induce osmotic or crowding effects in the molecule. These considerations underscore the need for control experiments, to ensure that any structural changes are not the result of the added agent. Finally, the last requirement will ensure that the remaining signal is not degraded by excessive absorption of X-rays. For water, for example, the optimal sample cell or sample capillary thickness that balances sample size against absorption is between 0.9 and 2.0 mm for typical X-ray energies (Schroer et al., 2018). If the contrast agent significantly increases the sample absorption, signal transmission at these 'optimal' values, computed for lower density water, may be reduced. For these reasons, a limited selection of contrast agents is available to experiments on biological molecules.

Sucrose is the most commonly used contrast agent for CV-SAXS experiments, and our contrast agent of choice. It is soluble in water up to a limit of 67% w/w sucrose (89% w/v sucrose) at room temperature. When added to buffers at a concentration of 50–65% w/v sucrose, the electron density increase is sufficient to match that of most proteins. Sucrose does not alter the conformation of many proteins (Kim et al., 2003; Lee & Timasheff, 1981), has no effect on the ion atmosphere around nucleic acids (Blose et al., 2011) and does not significantly absorb X-rays (Gabel, Engilberge, Pérez, & Girard, 2019). As an illustration, Fig. 3 shows how increasing sucrose concentrations decrease and eventually mask the scattering of a protein sample. With added sucrose, the overall signal intensity decreases, reflecting the decrease in contrast between protein and solvent, yet the shape of the curve ideally remains unchanged, showing that the sucrose does not dramatically alter the macromolecule's structure. When using highly concentrated sucrose containing solutions, it is important to consider their increased viscosities. For the range of concentrations needed to reach the match point of proteins, the viscosity can reach ~150 times that of water (Mathlouthi & Génotelle, 1995). The high viscosity creates several challenges in sample preparation for contrast variation experiments. We provide strategies that overcome these challenges in later sections.

Several other molecules have been suggested as contrast agents for CV-SAXS studies. Due to the limitations of sucrose, Gabel et al. (2019) have proposed the use of medical contrast media, iohexol and Gd-HPDO3A as contrast agents. These molecules are more electron dense than sucrose, and can therefore more easily reach the match point of proteins, at a lower

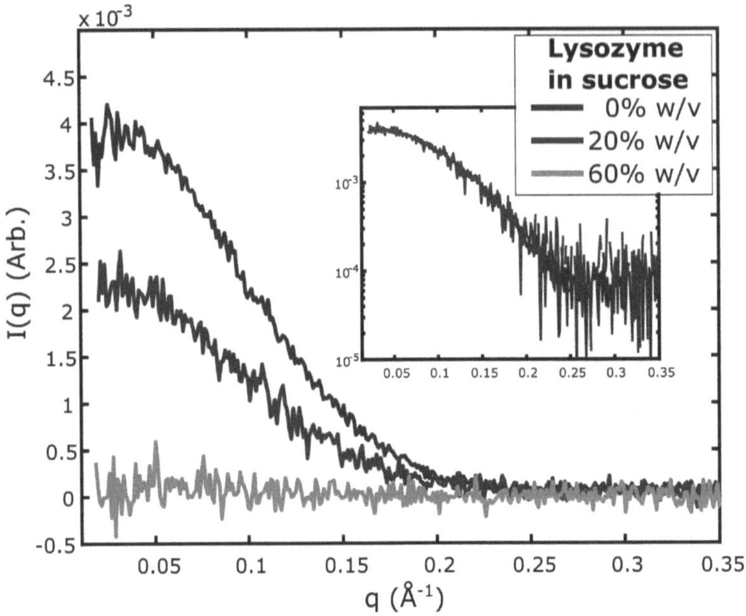

Fig. 3 Scattering from lysozyme (4 mg/mL) in buffers containing different sucrose concentrations. Measurements were made on a lab source (BioXolver, Xenocs) using 10 exposures of 120 s each. As the contrast is varied through the addition of sucrose, the scattering intensity of lysozyme is reduced. At 20% *w/v* sucrose, the signal is reduced but maintains the same shape as 0% w/v sucrose. At 60% w/v sucrose the signal from lysozyme disappears below the noise threshold. Inset shows the scattering at 0% and 20% w/v sucrose scaled and superimposed to demonstrate that there are no changes in the shape of the scattering profile upon the addition of sucrose.

viscosity (Rickwood, Ford, & Graham, 1982). However, they have much lower X-ray transmission than sucrose, so would require changes to the pathlength to optimize the signal. Furthermore, their compatibility with biomolecules would have to be assessed. Other possible agents include salt, which can increase electron density depending on the identity of ions chosen. However, salt can also affect the conformations, or aggregation states of highly charged biomolecules, such as RNA (Schlatterer et al., 2008). Salt also has a limited ability to increase contrast. Even when used in high (~Molar) concentrations, the solution density increases to ~370 e/nm^3 (Stuhrmann & Miller, 1978), which is less than half of the density change required to reach the protein match point. Glycerol is also commonly used as a contrast agent. Its role as a stabilizer and cryoprotectant in protein preparations

(Vagenende, Yap, & Trout, 2009) makes it an ideal candidate for these experiments. However, pure glycerol has an electron density of $410\,e/nm^3$ (Stuhrmann & Miller, 1978), which is still lower than that of the average value for a protein. In addition, high concentrations of glycerol can alter the hydration shell of proteins (Hirai et al., 2018; Vagenende et al., 2009). Finally, as with sucrose, glycerol containing solutions can have very high viscosity (Glycerine Producers' Association, 1963). For all of these reasons, we recommend the use of sucrose as a contrast agent for CV-SAXS experiments.

4.4 Biomolecule considerations: Molecular weight, purity, concentration

For CV-SAXS, as for most regular SAXS experiments, biomolecules must be of high purity, and monodisperse (Jacques, Guss, & Trewhella, 2012; Jacques & Trewhella, 2010). In addition, the relative masses of protein and nucleic acid in a given complex are critical considerations in establishing its suitability for CV-SAXS measurements. As a general guideline, the complex should derive at least half of its overall molecular weight from nucleic acid, relative to protein. At the match point, the signal from the remaining component is reduced relative to its value in buffer alone. The noise is also increased relative to the same standard SAXS measurement due to the increased background from the added sucrose (Sedlak, Bruetzel, & Lipfert, 2017). Although possible, it would be challenging to perform a CV-SAXS experiment in which the nucleic acid component is much smaller than the protein component. Even close to the match point, the contribution of the cross term of Eq. (3), could contribute significantly to, or even dominate the measured signal.

A final important concern is the biomolecule concentration. While higher biomolecule concentrations enhance the signal strength, and are especially valuable for measurements at higher sucrose concentrations where the signal is significantly reduced, interparticle interference can occur and will impact the shape of the profile at the lowest angles (Skou et al., 2014). The degree of complex formation is also dependent on the concentration of the components and must be taken into consideration, for example, one component may be present in excess or a different stoichiometry might be created. If possible, it is ideal to characterize and/or purify the complex ahead of the SAXS measurement. Note that excess protein will not contribute to the scatter if measurements are made at the contrast matched condition.

4.5 Preparing the samples: Biomolecules in buffer and its buffer background

One of the main challenges of CV-SAXS stems from the need to ensure proper buffer matching in the presence of contrast agents. We consider buffer subtraction to be acceptable for CV-SAXS experiments when the high angle baselines match within a fraction of a percent (at most 0.5% scattering intensity difference at high q when it is known that the sample does not scatter at these values). Some biomolecules, including duplex DNA and RNA, have significant scattering that extends into the wide angle regime (Zuo et al., 2006). Typically, a sucrose mismatch would appear at the largest scattering angles, e.g., $q > 0.3\,\text{Å}^{-1}$ or so, where the signal from the macromolecule is low and much of the scattering is derived from the buffer. One major source of measurement disagreement between the biomolecular sample and buffer sample arises if the sucrose concentration(s) in their preparation is not the same. At high q, even slight deviations in sucrose concentration can lead to high angle scattering of similar magnitude to that from the biomolecule. Thus, it is not possible to know if any high q scattering originates from the biomolecule itself (and is interesting) or from a buffer mismatch (and is an experimental error). For this reason, special care must be taken to ensure that the sucrose concentration is consistent between biomolecule samples and their buffer background solutions. We use two strategies to prepare the sucrose samples: accurate pipetting and dialysis.

Pipetting is perhaps the most straightforward method for preparing sucrose solutions. For accurate preparation of biomolecular samples containing sucrose and their matching buffers, also in sucrose, a highly concentrated sucrose solution is prepared and mixed with each to obtain the desired final sucrose concentration. Note that mixing of buffered biomolecules and buffered sucrose solutions leads to dilution, so the biomolecules must initially be at high concentration to ensure that the proper final concentration can be reached. We present a practical method to prepare this stock sucrose solution. First, make the desired stock sucrose concentration in weight percentage (w/w). Weighing is the most precise way of combining sucrose and solvent. Then, using sucrose density tables (Anderson, 1966; Asadi, 2006; Darros-Barbosa, Balaban, & Teixeira, 2003), convert the weight percentage (% w/w) to volume percentage (% w/v). This allows preparation of different concentrations of biomolecules and buffers in sucrose by direct pipetting (mixing volumes). To illustrate, we describe the preparation of a 60% w/w sucrose solution at room temperature. Sucrose density tables tell us that

60% w/w sucrose is equivalent to 77.2% w/v sucrose, achieved by multiplying the weight percentage with the density, $\%(\text{w/w}) \times \rho = \%(w/v)$. In this example, $60\%(\text{w/w}) \times 1.286\ (\text{g/mL}) = 77.2\%\ (\text{w/v})$. To prepare a 20 mL stock solution of 60% w/w sucrose in a centrifuge tube, first weigh the centrifuge tube and add the necessary amount of sucrose: $20\,\text{mL} \times 77.2\%\ (\text{w/v}) = 15.44\,\text{g}$ of sucrose. Separately prepare a $10\times$ buffering solution containing all the necessary ions and buffering agents needed in the buffer but with no sucrose. Add the necessary amount of the $10\times$ buffer solution (2 mL) to the sucrose powder. Then, add the necessary amount of water to reach the final weight needed to maintain the weight percentage, $15.44\,\text{g} \div 60\%(\text{w/w}) = 25.73\,\text{g}$ final weight. For most buffers, we can approximate the amount of water that needs to be added by subtracting the sucrose and buffer weight from the final volume ($25.73\,\text{g} - 15.44\,\text{g} - 2\,\text{g} = 8.3\,\text{g} \sim 8.3\,\text{mL}$ water), however careful weighing will result in the most accurate results. Lastly, to solubilize the sucrose, place the solution on a nutator or similar device for several hours until all sucrose is fully solubilized. Gentle shaking is recommended as rapid shaking can create bubbles in the sucrose solution. If not removed (through centrifugation) these bubbles can affect the sample preparation and SAXS measurement. To enhance solubility, the solution can be heated in a water bath at temperatures near, but no higher than 60 °C. Higher temperatures can break sucrose down into fructose and glucose and induce caramelization. These reactions are irreversible. The sucrose concentration can be verified with a refractometer. This stock sucrose solution can be used to prepare additional samples, by judicious mixing. We recommend mixing samples by thoroughly pipetting up and down while avoiding introduction of bubbles.

Care should be taken in pipetting, as the stock sucrose solution is highly viscous. The use of positive displacement pipettes allows more accurate pipetting when dealing with high viscosity solutions. We recommend wiping any excess sucrose solution from outside of the displacement tips to prevent sucrose concentration mismatches. Practice is recommended when learning to reproducibly pipet viscous solutions.

The pipetting method has the advantage of real time sample preparation. Samples can be adjusted, changed and improved based on examination of the acquired experimental data. However, due to the difficulty of pipetting sucrose reproducibly, buffer mismatches are common and repeated measurements, performed on independently prepared samples, are often needed to ensure consistent results.

For the pipetting method, samples containing biomolecules at the highest sucrose concentrations are made from biomolecule stock solutions (in no-sucrose buffer) that are 3–5 times more concentrated than in the final states. Unfortunately, some proteins are prone to aggregation at such high concentrations. For these biomolecules another method can be utilized: dialysis.

Using this method, samples are dialyzed into the desired sucrose-containing buffer in advance of the experiment using standard biochemical methods. This method is slower than pipetting: proper dialysis into high sucrose buffers can require 1–2 days and must be assayed beforehand to ensure that equilibrium has been reached. The dialysis time can be reduced by pre-mixing the biomolecule with sucrose to a concentration close to the target concentration. In this way the dialysis serves to ensure an exact match as opposed to a full exchange. To assay the necessary time of dialysis, several buffers can be prepared and dialyzed in the same way as the biomolecule would be. The sucrose concentration of these buffers can then be measured with a refractometer to confirm it matches the buffer it is being dialyzed into. Despite these lengthy steps, this preparation method guarantees that biomolecule containing samples will match (have the same sucrose concentration as) the buffer. Although dialysis can be difficult for small volumes (such as those used for SAXS), there are commercially available products that can facilitate this process. In our lab, we have used Slide-A-Lyzer™ MINI (Thermo Scientific™) and Dialysis Buttons™ (Hampton Research). Although more reliable in the long run, this method has the disadvantage of using more biomolecules and being time intensive. It also cannot be adjusted "on the fly," so conditions must be identified well in advance of the experiment. The biomolecules must also remain viable and stable in solution for days, as opposed to the minutes that elapse when prepared by pipetting.

4.6 Loading the sample at the beamline

Contrast variation experiments can be performed using any standard SAXS beamline configuration and equipment; although the vast majority of our work has employed synchrotrons, CV-SAXS experiments on static samples are possible even on commercially available lab sources (see Fig. 3 for an example). In most cases, and for most setups, sample volumes between 20 and 60 μL are used. As discussed above, each biomolecule measurement requires a matching buffer for background subtraction. We normally acquire

SAXS profiles this order: buffer, biomolecule, buffer. A "good" data point (from the beamline perspective) has indistinguishable pre and post buffers.

Even when all samples are carefully prepared (as described above) challenges still remain. The largest challenge to CV-SAXS data acquisition comes from the sucrose itself. Three of the major difficulties of handling sucrose at the beamline (and their resolution) are discussed here: compatibility of highly viscous sucrose with pipetting 'robots', radiation damage of sucrose, and proper cleaning of the sample cell between measurements to ensure a reproducible background.

Given the high viscosity of sucrose solutions, special care should be taken when loading samples. This is particularly important when beamlines employ automatic pipetting robots for sample loading (Acerbo, Cook, & Gillilan, 2015; Yang et al., 2020), where a reduction in pipetting speed is recommended to ensure that the sample is withdrawn correctly. As an alternative, we frequently employ manual loading of samples. Finally, it is essential to (optically) check for the presence of bubbles in the sample plug (the sample loaded into the optically transparent sample cell that the X-rays pass through, see Fig. 1, 'solution sample'). Most beamlines have cameras that permit visualization of the liquid sample for exactly this reason.

Care must be taken to avoid radiation damage of the sucrose. Although sucrose can remediate X-ray damage of biological samples (Kuwamoto, Akiyama, & Fujisawa, 2004), a high radiation dose can induce a change in the sucrose solution itself (Wolfrom, Binkley, & McCabe, 1959). This damage manifests as a change in the color and/or the appearance of bubbles in the irradiated volume. To mitigate X-ray exposure, sample plugs can be oscillated, or the sample can simply flow through the cell during the measurement (Acerbo et al., 2015; Schroer et al., 2018; Yang et al., 2020). If possible, the use of a co-flow sample cell can help by ensuring that the sample is not resting on (stuck to) the capillary wall (Kirby et al., 2016).

Finally, enhanced cleaning protocols must be implemented when dealing with sucrose solutions. We suggest starting any CV-SAXS measurement series by taking the SAXS profile of pure water in the sample cell, to get a good, 'clean' measure of the background. It is important to repeat this measurement at different points throughout a measurement series by loading a pure water sample, to ensure that the profiles have not deviated from the first one.

Due to the viscosity of solutions and the stickiness of sucrose, cleaning times that work for standard SAXS samples might be too short to clear the sucrose between measurements. We almost always increase the length of the 'detergent' step in the cleaning cycle. Failure to properly clean the sample cell can result in sample contamination and improper buffer matching. In the event that the capillary or cell gets too dirty, or sucrose has been accidentally seared on the capillary, it is important to move the X-ray beam to sample a clean part of the cell, and to restart the sequence by measuring a scattering profile of pure water.

5. Doing the experiment

5.1 Finding and exploiting the match point

We have employed two different CV-SAXS approaches to reveal the structures of the macromolecular constituents of a protein-nucleic acid complex. In the first, extensive experimentation is employed to identify, for each system, the sucrose concentration where the protein signal vanishes below the noise floor (the "match point"). For these measurements, data are initially acquired on a sample that contains ONLY the protein component of the complex in solution, at various sucrose concentrations, to identify the match point: the optimal sucrose concentration for subsequent measurement of the complex. Note that it is always most favorable to use the lowest possible sucrose concentration, both for ease in sample preparation, and to maximize the signal from the nucleic acid component. Subsequent measurements are taken on the complex at the sucrose concentration where the protein signal is zero.

Eq. (4) guides experiment to determine the match point for a component with a given electron density. In practice, we use two different methods to find it. In the first, the protein scattering is measured in at least three distinct sucrose concentrations. For these measurements, it is important to maintain a constant protein concentration, as the signal amplitude is used to extrapolate to the match point. First, the measured scattering intensity for each profile is extrapolated to its zero angle value, $I(0)$ as described above. A plot of the square root of $I(0)$ vs. sucrose concentration is made, and fit with a line. The x-intercept of this curve yields the sucrose concentration where the intensity goes to zero. Although this method works in principle, it is important to confirm the derived value, by preparing a protein sample at the indicated sucrose concentration. In some cases, the noise floor (protein signal is smaller than the noise) appears before the predicted zero intensity.

Once this "match" condition is identified, numerous measurements should be performed (though some may have been performed in search of the match point and could be re-used, if taken using the same sample).

Measurements of the protein alone, nucleic acid alone, and the complex (at appropriate concentrations) should be performed in the preferred buffer for the system, at 0% sucrose. Note, it would be ideal to have a stoichiometrically correct complex for this measurement (all the protein and nucleic acids molecules are bound together), so we recommend that additional characterization and/or purification of the complex is performed. If the concentrations are known, measurement of the molecular weight (through $I(0)$) can be used to assess the composition, though some uncertainly remains and should be considered (e.g., how to unambiguously distinguish a population of pure dimers from a population comprised of a mixture of tetramers and unbound molecules). In addition to the three contrast measurements described above, the protein, nucleic acid and complex should all be measured at the pre-determined match point. Finally, two measurements should be made of the protein and nucleic acid components alone, at an intermediate value of sucrose, to test for sucrose dependent effects on the components. This protocol corresponds to a set of at least eight measurements.

The initial three, regular-contrast measurements are an essential part of a SAXS analysis and provide a benchmark for the contrast variation measurements to be performed. The second set of three verifies that the match point is indeed properly set (protein scatter should vanish) and serves as a comparison of the nucleic acid in the unbound and bound state, at the same contrast level. The additional measurement of the protein and nucleic acid at the intermediate value of contrast ensures that the sucrose does not, on its own, modify the structure(s) of the molecule.

Forgoing a comprehensive series, the match point may also be determined by trial and error. A sample is prepared and measured at a sucrose concentration that contrast matches other protein systems (based on experience or literature values). If scattering from the protein is still visible, the sucrose concentration is varied over a small range until the scattering can no longer be measured. From our experience, we find that this method can take less time than the more robust method listed above; however, samples must be prepared by pipetting and not by dialysis to ensure rapid turnaround.

5.2 Performing a contrast series to reveal the structures of all components of a complex

In a second approach, systematic measurements are performed at varying sucrose concentrations at several, preselected contrast points. For a full contrast series, scattering profiles should be acquired at around five distinct contrast values. This value is largely empirical and agrees with recommendations for contrast variation neutron experiments (Whitten, Cai, & Trewhella, 2008). Ideally, measurements should be well spaced according to contrast (not grouped around the highest contrast, for example), so that significant variations can be observed and fit. All of the above described considerations should apply here (measurements should be performed on the independent components to ensure that there is no sucrose-related conformational change).

We note that these two approaches are not mutually exclusive; the contrast variation series can be used to determine the match point and provide useful information on the effect of the contrast agent on the molecules. However, we can also use the contrast series of the complex to obtain additional information of its structure. Although CV-SAXS data analysis is simplest when measuring at the match point because the signal arises exclusively from the nucleic acid component of the complex, it does not provide the full extent of information that can be derived from the series (the protein conformation in addition to the nucleic acid conformation).

5.3 Benchmarks to assess data quality and effectiveness of the method

From the above discussion it should be clear that acquisition of high quality, CV-SAXS relies on two important considerations. First, buffer subtraction must be accurate, meaning that the sucrose concentration in the buffer background sample must be as close as possible to that of the biomolecule sample, and the sample cell must remain uncontaminated between measurements. Second, as mentioned above, the contrast agent must not, by itself, alter the structure of the biomolecules.

The first consideration, accurate background subtraction can be readily assessed by a careful series of controls and some basic understanding of scattering. If, for example, the buffer background contains more sucrose than the biomolecule sample, the subtraction may yield curves that drop below zero (have negative values as an average) at high angle. This situation is unphysical and the data must be retaken. The simplest way to assess the quality of the match between buffer background and biomolecule sample is to compare the relative high to low angle data acquired from a single component sample (such as protein only) when sucrose is added, to its value taken under standard conditions (if known). For the second consideration, it is important that for each single component in the complex no change occurs in the shape of the scattering as the contrast is reduced (see Fig. 3). Any differences in the shapes of the profile could indicate either a change in the biomolecule structure, a background mismatch, or possibly a large variation in the electron density of the molecule across its volume. In any case, it is important to determine the origin of any sucrose-dependent changes. These controls ensure that the buffer has the same sucrose concentration as the biomolecule samples, and that the data acquired on the complex (prepared at the same time as these samples), should be reliable. If the buffer has a different sucrose concentration, the subtractions may become unreliable. Thus, sample preparation (as described above) is critical to the success of CV-SAXS experiments.

6. Examples of information extracted

In the remaining sections, we demonstrate the power of CV-SAXS through examples taken from selected studies. We first consider studies conducted on protein-RNA and protein-DNA complexes at the protein match point. The goal of the first two studies is to measure the nucleic acid (RNA or DNA) conformation(s) in a protein-nucleic acid complex. The final study illustrates the reconstruction of nucleic acid and protein complex using SAXS data acquired through a contrast series.

6.1 At the match point: First example shows primary microRNA binding to a microprocessor protein

We first illustrate the application of CV-SAXS in studies of a protein-RNA complex consisting of a primary microRNA (pri-miRNA) and the DGCR8 protein (Pabit et al., 2020). The protein is part of the microprocessor complex that cleaves the primary microRNA as it matures to a functional microRNA. The goal of this work is to understand DGCR8's role in identifying a cleavage site on the RNA. Data acquired at the 0% sucrose condition (traditional SAXS) and at the protein match point elucidate the binding stoichiometry of the complex and suggest that the protein bends the RNA once it is bound, potentially exposing the cleavage site (Pabit et al., 2020).

6.1.1 Complex stoichiometry can be inferred from I(0) and validated with CV-SAXS

Fig. 4 shows scattering profiles from standard and CV-SAXS studies of the primary microRNA-DGCR8 complex (Pabit et al., 2020). We first focus on the data of Fig. 4A, acquired under standard SAXS conditions (no sucrose). We extrapolate our measured curves to $I(q=0)$ using GNOM (Svergun, 1992) for each of the three samples of interest: protein alone,

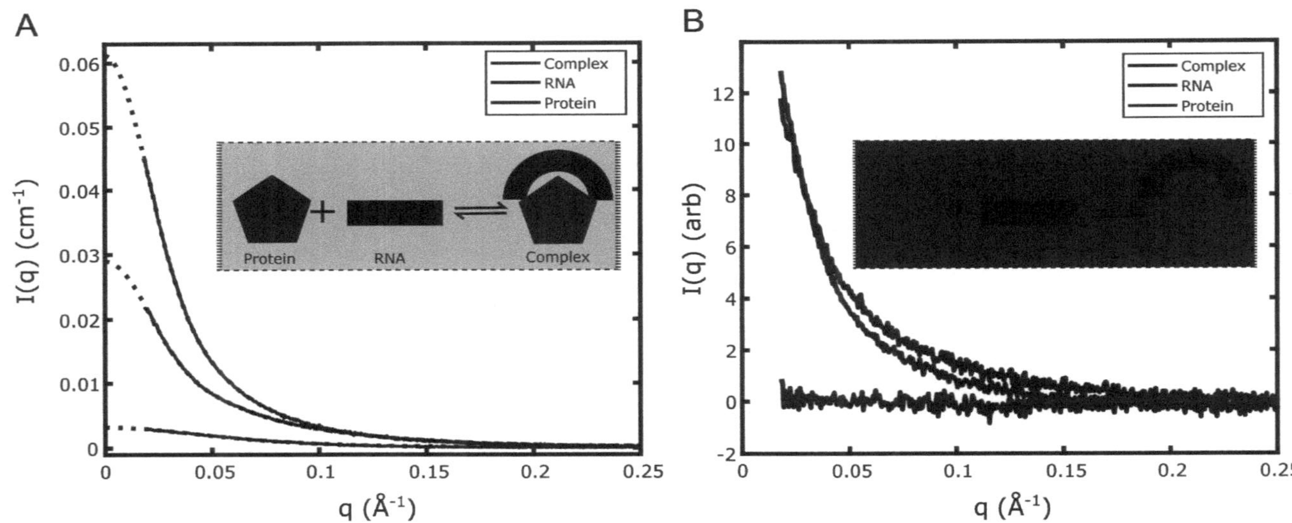

Fig. 4 CV-SAXS of primary microRNA and the protein DGCR8 core demonstrates that a protein can bend a rigid RNA molecule. (A) SAXS profiles, in absolute units, of DGCR8-core protein alone (red), primary microRNA (pri-miR-16) alone (black), and the complex formed (blue) by binding of pri-miR-16 RNA to the DGCR8 core are shown in buffer containing 0% sucrose. The dotted lines in the figure denote the GNOM extrapolation to extract $I(q = 0)$. Comparison of the $I(0)$ value of the complex to the monomer units in absolute scale show the protein and the RNA form a 1:1 complex in the bound state. (B) SAXS profiles of protein alone (red), RNA alone (black), and complex (blue) in buffer containing 50% w/v sucrose are shown. Insets illustrate the contrast-matching method and the formation of the protein-nucleic acid complex. Under contrast matching conditions (inset of B) the molecules are in a buffer that has the same electron density as the protein (*red*), so the signal from the protein vanishes. Only the signal from the nucleic acid contributes to the measured SAXS curve of the complex (*blue*). *From Pabit, S. A., Chen, Y.-L., Usher, E.T., Cook, E.C., Pollack, L., & Showalter, S.A. (2020). Elucidating the role of microprocessor protein DGCR8 in bending RNA structures.* Biophysical Journal, 119*(12), 2524–2536. https://doi.org/10.1016/j.bpj.2020.10.038.*

RNA alone and protein and RNA in complex. We then apply absolute calibration to convert these values into absolute scattering units. Plugging these values into Eq. (5) (which assumes equal concentrations of the two components), we find that the stoichiometry of the protein and RNA is consistent with 1:1 binding, meaning that the zero angle scattering, $I(0)$ of the complex equals that of the sum of the contributions from the protein alone, the RNA alone and the cross term. Fig. 4B shows data acquired at the sucrose match point for the DGCR8 protein. Here, the zero-angle scattering of the RNA alone is very similar to the scattering of the complex in solution at the "match" condition. The similar values of $I(0)$ obtained for the two measurements allows us to conclude that one RNA molecule is present per complex, supporting our hypothesis of 1:1 binding at 0% sucrose. For complexes where the 2 components bind in different ratios, the zero-angle scattering provides a clue to the makeup of the complex.

6.1.2 Radius of gyration can indicate nucleic acid conformational changes upon protein binding

In addition to stoichiometry, structural information about the components of the complex can be extracted via measurements of R_g. Specifically, when contrast matching of the protein component can be achieved, the R_g of the contrast matched sample can be compared to that of the unbound sample, acquired at 0% sucrose. Standard analysis, performed using BioXTAS RAW (Hopkins et al., 2017; Nielsen et al., 2009), easily yields these values. Any difference between them could suggest that protein binding results in RNA structural changes.

6.1.3 Interpreting structural changes through distance distribution analysis and shape reconstructions

SAXS profiles provide much more information than the two parameters discussed above, R_g and $I(0)$. A more thorough analysis involves comparing information derived from the full scattering profiles of the nucleic acid alone and in complex with the protein at the match point. For this primary microRNA and DGCR8 complex, we computed the pair distance distribution functions, P(R), from the profiles of the RNA shown in Fig. 4 (in the free and bound states). These curves are shown in Fig. 5A.

The P(R) of the RNA alone (no sucrose, no protein partner) resembles that of a cylindrical rod with a 20 Å diameter. This length scale can simply be read off the plot, by finding the R value of the peak in the P(R) curve.

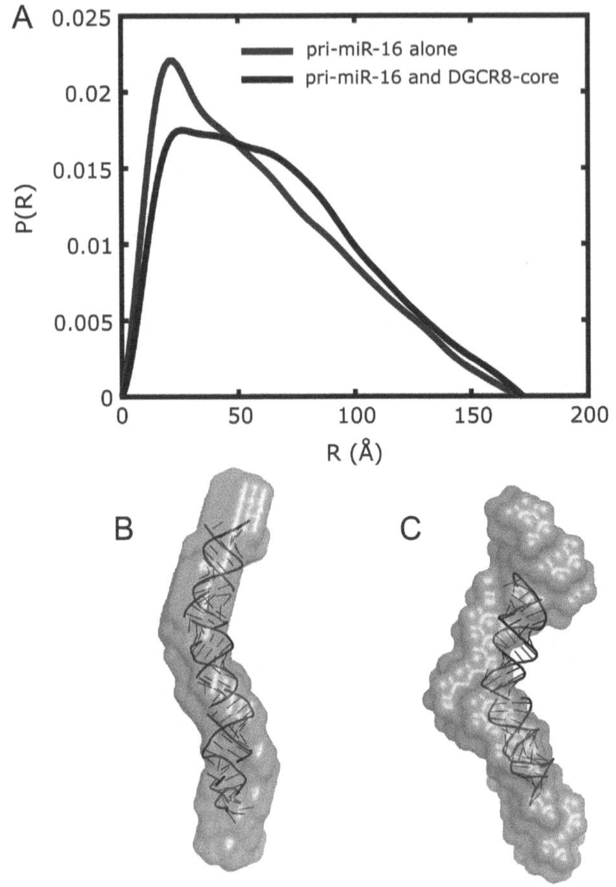

Fig. 5 Comparisons of pair-distance distribution functions and solution SAXS reconstructions suggest a notable bend of the pri-miR-16 upon binding the DGCR8 core. (A) P(R) of pri-miR-16 alone (*blue*) and pri-miR-16 bound to the DGCR8 core (*red*) with the protein signal blanked out by the contrast-matching agent (50% w/v sucrose) is shown. For ease of comparison, P(R) curves are normalized by dividing each curve by the area under the curve. By itself, the RNA shows a P(R) distribution characteristic of a cylindrical rod, while the P(R) of the RNA in complex with the protein implies a more bent structure. This is evident in the averaged shape envelopes from SAXS reconstructions shown in panel (B) for pri-miR-16 alone (*blue*) and (C) pri-miR-16 bound to DGCR8 core (*red*). The atomic model in panel (B) was determined from MD simulations. The atomic model in panel (C) was determined by imposing a bend in the model from (B) that matches the shape reconstruction. *From Pabit, S. A., Chen, Y.-L., Usher, E. T., Cook, E.C., Pollack, L., & Showalter, S.A. (2020). Elucidating the role of microprocessor protein DGCR8 in bending RNA structures.* Biophysical Journal, 119*(12), 2524–2536. https://doi.org/10.1016/j.bpj.2020.10.038.*

The curve exhibits a featureless decay with increasing distance, R. In contrast, the P(R) of the RNA in the complex is consistent with a bent structure, suggesting that the bending is facilitated by the protein.

To convert these plots into simpler-to-visualize forms, the P(R) curves can be used to generate ab-initio models of the molecules using shape reconstruction software such as DAMMIN/DAMMIF (Franke et al., 2017; Svergun, 1999) from the ATSAS suite. These low-resolution shape reconstructions display the real space changes in the nucleic acid structure that result from protein binding. Although the reconstructions are neither unique nor high resolution, the information they convey is consistent with the changing features displayed in the P(R) plots. Fig. 5B and C present the dummy atom reconstructions. Displaying the data in this format is often the most straightforward way to convey changes. As additional validation of these structural changes, we applied Molecular Dynamics (MD) simulations and a bending algorithm to the primary microRNA and obtained results consistent with the protein bending the RNA. Full details are reported in ref. (Pabit et al., 2020).

6.2 At the match point with changing conformations, equilibrium and time resolved studies: Second example reports studies of DNA dynamics in nucleosome core particles

For some systems, interactions with multiple partners is required for proper biological function. For example, many different proteins interact with DNA to facilitate gene expression. Within chromatin, DNA is stored in fundamental units known as nucleosome core particles (NCP). In each NCP, $\sim$146 base pairs of genomic DNA is tightly wound for storage, but must also be available for ready release (Cutter & Hayes, 2015). Numerous protein partners are involved in packaging the DNA and in effecting its selective release for future processing, including histone proteins and their variants as well as other chromatin binding proteins. Fundamental biophysical studies can be used to reveal the interactions that facilitate tight storage of DNA in NCPs, but also identify 'loose points' that may be targeted by other protein factors for easy release. In many cases, the release of DNA from the histone core depends on both the DNA sequence as well as the specific proteins (or variants) present in the histone core. We have used CV-SAXS to study the patterns of release of different DNA sequences from the native histone core (Chen et al., 2014). Studies of DNA conformation in the presence of chromatin remodelers, such as Chd1 are also facilitated by CV-SAXS (Tokuda et al., 2018).

For some of these experiments, monovalent salt is used to weaken the electrostatic interactions that stabilize the complex. At physiological levels of salt, the DNA is fully packaged (wound around) the histone core. Once these interactions are reduced by the addition of salt, the DNA is selectively released in a sequence dependent way. When the added salt reaches 1 M concentration, the complex is fully dissociated (Chen et al., 2017). Of interest are the intermediate structures that populate this unwinding landscape.

The use of CV-SAXS as a viable method for studying NCPs was first reported in 1992 by ref. (Inoko et al., 1992) using NCP from rat thymus nucleosomes. That work reported values of R_g, $I(0)$ and α and inspired our recent studies of the salt dependence of DNA release from the histone core (Chen et al., 2014) shown in Fig. 6. This later work benefits from the use of the artificial 601 DNA (Lowary & Widom, 1998) which is tightly positioned, in conjunction with a solved NCP crystal structure (Luger et al., 1997). Knowledge of NCP structure enables application of more advanced modeling efforts (Mauney, Tokuda, Gloss, Gonzalez, & Pollack, 2018; Tokuda et al., 2018).

6.2.1 P(R) analysis provides in depth information about the changing DNA structures

Here, because the stoichiometry of the NCP is known and large shape changes are expected upon DNA release, it is difficult to interpret changes using R_g as the sole structural metric. Instead, we use the information encoded in the entire scattering curve to facilitate analysis. Fig. 6A compares the structure of 147-base pair DNA fully wrapped in an NCP (PDB ID: 1AOI) (Luger et al., 1997) to a fully extended DNA molecule. The full power of the contrast variation method is on display when comparing the evolution of the P(R) as a function of increasing salt concentration in 0% sucrose (Fig. 6B) and at the match point of 50% w/v sucrose (Fig. 6C). In the absence of sucrose, the peaks broaden, and the maximum extent of the complex increases with salt, but the data cannot be readily interpreted because they contain information about the proteins as well. Once the DNA dissociates, the conformations and association states of the proteins in the histone core becomes unknown. Analysis by SAXS would be very difficult under these circumstances. However, at the match point, the signal from all proteins becomes indistinguishable from the background, and the clear signal from only the DNA structure(s) is revealed. Shifting peaks can be readily interpreted using models, and from these data, the changing conformations of the DNA are readily observed. As the DNA unwraps, the pair distance distribution changes to match that of a fully extended DNA molecule as depicted in P(R) from model systems (Fig. 6D).

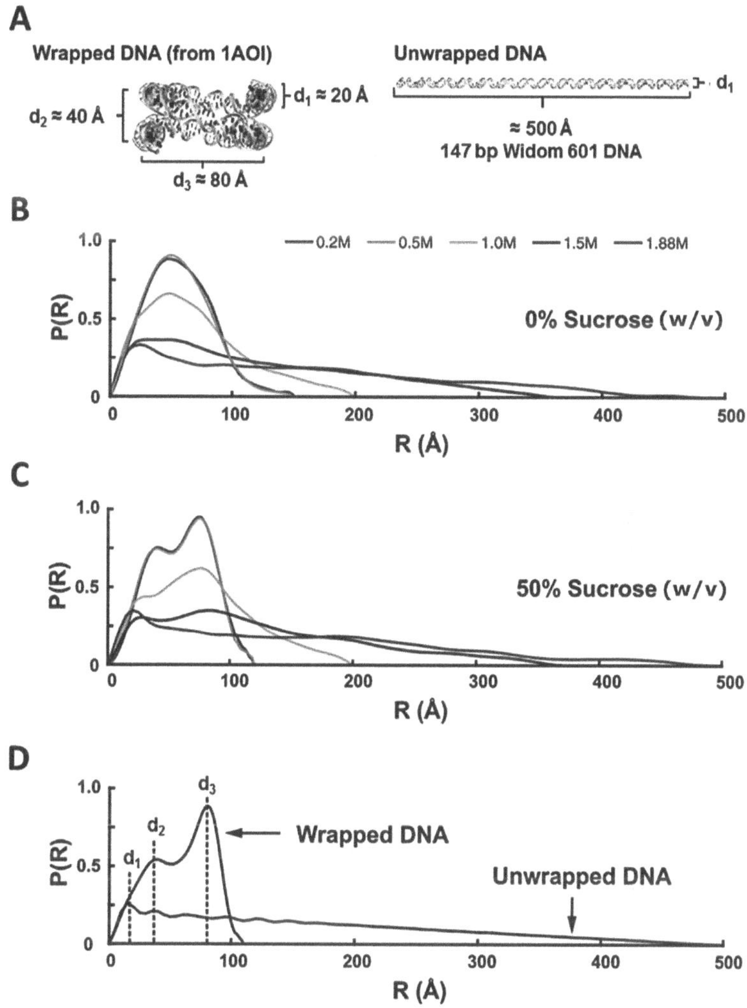

Fig. 6 Application of solution contrast variation to monitor DNA unwrapping during the salt-induced disassembly of nucleosome core particles (NCP). (A) DNA models for the expected end states of the NCP at low NaCl concentration (completely wrapped) and high NaCl concentration (completely unwrapped). (B) P(R) curves for the NCP measured in 0% sucrose and various NaCl concentrations. (C) P(R) curves for the NCP measured in 50% w/v sucrose and various NaCl concentrations. At this point, the protein is masked, and the prominent features appear in the P(R). (D) P(R) curves determined for the models in panel (A). Peaks in the P(R) curves can be associated with structural features as follows: d_1, diameter of the duplex DNA; d_2, distance between overlapping DNA ends; d_3, diameter of the overall wrapped structure. *From Tokuda, J. M., Pabit, S. A., & Pollack, L. (2016). Protein–DNA and ion–DNA interactions revealed through contrast variation SAXS.* Biophysical Reviews, *8(2), 139–149. https://doi.org/10.1007/s12551-016-0196-8.*

6.2.2 Structural ensemble modeling provides essential information about the conformations present in equilibrium salt titrations, or as a function of added proteins (tri-nucleosomes, remodelers)

Because of the wide range of conformations populated as a function of changing salt, this system is ideally modeled using ensemble methods. At the match point, only the nucleic acid structures need to be considered. Programs and strategies used for modeling were discussed in previous sections. For these studies, we made use of ensemble optimization method, EOM (Tria et al., 2015). One of the more challenging aspects of using EOM is the need to create a pool of structures that contains all or as many as possible accessible conformations. This can be especially challenging for large molecules, such as long DNA with hundreds of base pairs. In early studies (Chen et al., 2014, 2017), we simply extended the crystal structure by releasing bits of DNA from each end. This approach, while naïve, did provide insight into conformations assumed by the DNA. More recently, we have extended coarse grain DNA models, such as cgDNA (Petkevičiūtė, Pasi, Gonzalez, & Maddocks, 2014) which accounts for the mechanical properties of DNA sequences. We performed CV-SAXS experiments on long DNAs wrapped around histone cores to focus on the role of DNA sequence in release from NCPs (Mauney et al., 2018) as well as the conformations of multiple nucleosomes in tri-nucleosome chains (Mauney, Muthurajan, Luger, & Pollack, 2021). For these cases, the CV-SAXS measurement, performed at the match point, is relatively straightforward. Because proteins are matched, we can also see the effect of added chromatin remodelers (Chd1) without observing the protein remodelers themselves, which would complicate the scattering profile (Tokuda et al., 2018).

6.2.3 CV-SAXS experiments can be performed in a time-resolved mode, to watch real time sequence of events

Chen et al. (2014, 2017) demonstrates the coupling of CV-SAXS, time-resolved experiments using a stopped-flow mixer and EOM to investigate the real time release of DNA from the nucleosome core particles. In these studies, data acquired during a mixing experiment was described by EOM generated models. The power of this type of analysis is illustrated in Fig. 7 where we see the conformational dynamics and the different structural changes DNA undergoes as it dissociates from the histone core in a time dependent manner. Of particular interest is the asymmetry of release, discussed in Ref. (Chen et al., 2017).

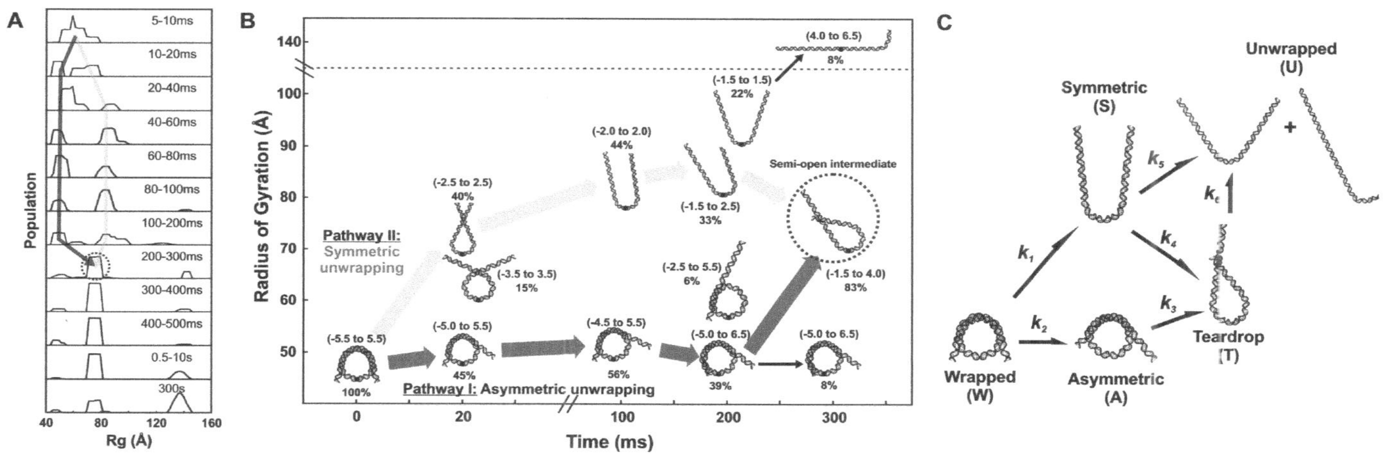

Fig. 7 CV-SAXS applied to the time-resolved disassembly of DNA from nucleosome core particles in 50% w/v sucrose. Disassembly is triggered by the addition of 1.2 M NaCl, also in 50% w/v sucrose. Time resolved CV-SAXS data were analyzed with the ensemble optimization method (EOM) to select the DNA structures whose computed scattering profiles best recapitulate the data at different times. (A) R_g histograms from DNA models selected by EOM that best recapitulate the SAXS data. Red and green arrows highlight two pathways through which DNA structures change before settling into a prominent peak after 300 ms (circled in red). (B) DNA models selected by EOM before ($t = 0$) and after mixing into 1.2 M NaCl (at $t = 20$ ms, 100 ms, 200 ms, and 300 ms). Green and red arrows highlight two major pathways through which DNA unwraps to form the teardrop DNA structure. Black arrows show minor pathways. Under moderate salt conditions that favor partial disassembly, the majority of structures unwrap symmetrically and asymmetrically before converging into the teardrop structure. (C) Kinetic scheme for complete disassembly with pathways inferred from prominent DNA structures selected by EOM. *From Chen, Y., Tokuda, J.M., Topping, T., Meisburger, S.P., Pabit, S.A., & Gloss, L.M., et al. (2017). Asymmetric unwrapping of nucleosomal DNA propagates asymmetric opening and dissociation of the histone core.* Proceedings of the National Academy of Sciences, 114*(2), 334–339. https://doi.org/10.1073/pnas.1611118114.*

6.2.4 *Additional controls: Compare with other methods (validate)*

Finally, as with many biophysical experiments, CV-SAXS benefits from combination with other experimental methods. Although information about the protein composition of the histone core cannot be easily obtained from CV-SAXS due to the sheer number of components involved, the coupling to other techniques, such as single molecule Fluorescence Resonance Energy Transfer (smFRET), which selectively reports on specific protein constituents, can add significant biological understanding (Chen et al., 2017), can aid in the interpretation of data, and can validate the selection of models using the algorithms described above.

6.3 Analysis of contrast series: Third example to extract protein as well as RNA structure

6.3.1 *Structure of a non-enveloped virus (or virus like particle)*

For more complex systems, containing multiple molecules, specifically many copies of a given protein, a full understanding of the complex may require a determination of the protein structure in addition to the nucleic acid structure. The structure(s) of both can be obtained from a contrast series on the complex.

We describe recent studies on bacteriophage MS2, a non-enveloped virus or a virus that lacks a lipid membrane (San Emeterio & Pollack, 2020). Bacteriophages are large complexes, consisting of an outer protein shell (or capsid) that surrounds (encapsidates) a nucleic acid genome. For MS2, the capsid contains 180 proteins, and the single-stranded genome consists of 3569 RNA nucleotides. Scattering profiles acquired at different contrast levels (Fig. 8A) can be deconvoluted to reflect the distinct contribution of protein vs. RNA constituents.

6.3.2 *Analysis via P(R)—Gateway for modeling*

Following data acquisition at 5 different contrast values (see Fig. 8), we first examine a P(R) series. As the contrast increases, and the weight of the scattering becomes dominated by the RNA core, the peak of the P(R) shifts to a lower radius. However, as is the case with all the above-described experiments, the most useful application of P(R) is the comparison between no sucrose (highest contrast) and the protein match point, where information about the relative spatial distributions of the protein and genome can be derived. These curves are shown in Fig. 8B (San Emeterio & Pollack, 2020).

Here, data acquired at a series of contrasts show dramatic changes in features of the scattering profiles, such as shifting positions of minima,

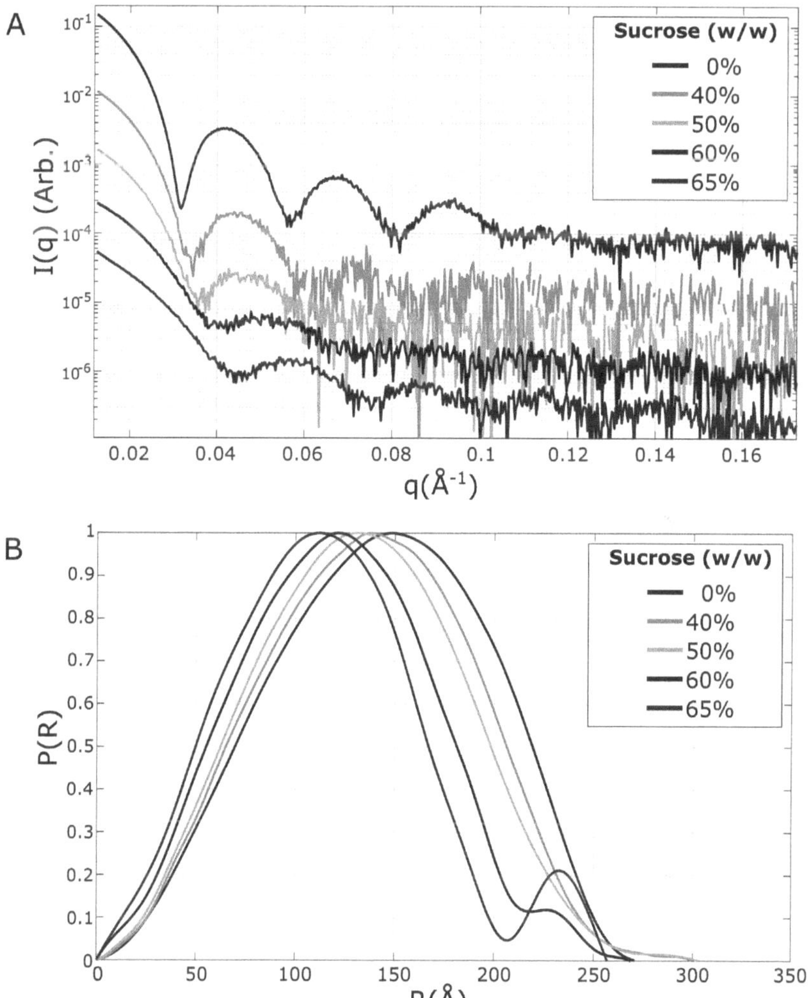

Fig. 8 CV-SAXS measurements on bacteriophage MS2 at various sucrose concentrations given in % w/w. *SAXS* profiles (A) are shown with an offset to aid in visualization. As the solution contrast increases, the scattering changes to reflect a higher contribution from the RNA core (relative to the protein contribution). The computed pair distance distribution functions P(R) (B) are normalized to enable comparison. As the solution contrast increases, the peak of the P(R) shifts to a lower distance. Beyond the match point (near but likely just below 60% w/w sucrose) the contribution from the protein shell reappears as the second peak in the P(R). *From San Emeterio, J., & Pollack, L. (2020). Visualizing a viral genome with contrast variation small angle X-ray scattering.* Journal of Biological Chemistry, 295*(47), 15923–15932. https://doi.org/10.1074/jbc.RA120.013961.*

reflecting changing dimensions. This information can be used as input to model-building programs that specifically account for components with differing electron densities, at different contrast values. Programs such as MulCh (Whitten et al., 2008) and MONSA (Svergun, 1999; Svergun & Nierhaus, 2000) are straightforward to use.

We applied MONSA to create reconstructions of the CV-SAXS series of bacteriophage MS2 (Fig. 9). These model structures are pictured along with asymmetric cryo-EM reconstructions of the MS2 phage (Koning et al., 2016). The reconstruction on the left recapitulates the general location and distribution of the RNA (blue) and the protein (red) of the higher resolution cryo-EM reconstruction on the right. As with any SAXS reconstruction, the results are neither unique nor high resolution, but they provide insight into the structure of the distinct components that contribute to the overall complex.

As an alternative, it is possible to decompose the scattering of a complex into the scattering of the constituents, using a mathematical algorithm. An application for performing this calculation is contained in the Mulch (Whitten et al., 2008) software, which is available through a web server.

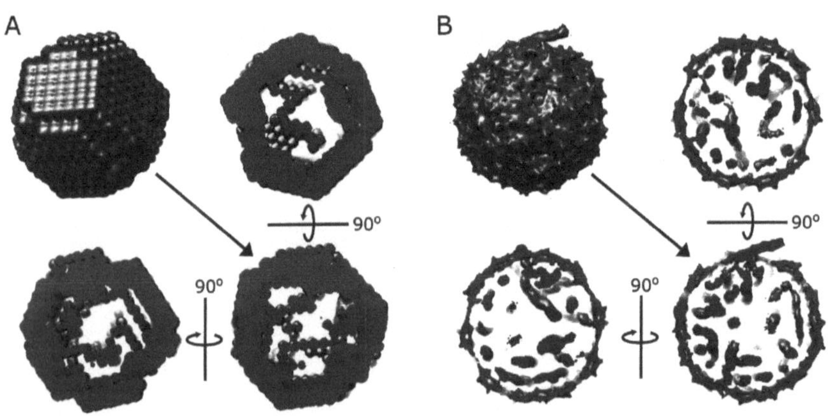

Fig. 9 MONSA reconstructions from CV-SAXS data on bacteriophage MS2 (A) compared to the models derived from cryo-EM (EMD-3404/3403) (B). Protein is shown in *red*, and RNA is shown in *blue*. The full reconstruction and three orthogonal cross-sections are shown for each case. Although the spatial resolution obtained through SAXS is lower and the reconstruction is not unique, SAXS data are much simpler to acquire than asymmetric cryo-EM reconstructions. Similar structural features are captured by both methods, including a small, protruding piece of RNA that may reflect the position of the maturation protein. *From San Emeterio, J., & Pollack, L. (2020). Visualizing a viral genome with contrast variation small angle X-ray scattering.* Journal of Biological Chemistry, 295*(47), 15923–15932. https://doi.org/10.1074/jbc.RA120.013961.*

7. Conclusions and outlook

Contrast variation SAXS is a powerful tool that can provide information about the structure(s) and dynamics of protein-nucleic acid complexes. In its simplest implementation, CV-SAXS can effectively render the protein component of the complex invisible. In this mode, the structure of the nucleic acid component can be easily determined. In another mode, a series of CV-SAXS measurements can provide information from both the protein and nucleic acid components of a complex. Perhaps the greatest strength of CV-SAXS is its ability to provide structural information on flexible systems and its utility in time-resolved experiments.

Largely motivated by the recognized importance of complexes in biological processes, measurement of the structure(s) of protein-nucleic acid complexes is now becoming routine. New tools, such as Cryo-EM expand the range of experimental tools and offer new opportunities. CV-SAXS, in conjunction with modeling, contributes by enabling measurements on complexes over a wide variety of sizes, and elucidating the dynamics of flexible molecules as they respond in real time to stimuli. CV-SAXS provides unique insight into the structure and dynamics of protein-nucleic acid complexes.

In this chapter, we have briefly provided the theoretical motivation for CV-SAXS, discussed practical considerations, experimental practices, as well as examples illustrating its application. We have outlined the procedures and pitfalls associated with CV-SAXS measurements with the aim of providing the necessary guidelines for a successful CV-SAXS experiment. Of interest is the successful demonstration of CV-SAXS measurements on laboratory sources. Future applications, especially those exploiting time-resolved measurements are anticipated to provide unique perspectives on the assembly of macromolecular complexes of biological importance. Potential targets range from complexes involved in gene expression and regulation, all the way through large multimeric complexes and viruses.

Acknowledgments

We thank current and former Pollack Lab members for helping to make CV-SAXS a routine experimental technique. We are especially grateful to Brady Sites, Tong (George) Wang, Kara Zielinski and Joshua Tokuda for help with this manuscript. This work is/was supported by grants from the National Institutes of Health grants EUREKA R01-GM088645, R01-GM085062, R35-GM122514, T32GM008267 and the National Science Foundation under award DBI-1930046. SAXS data were acquired on a lab source acquired through National Institutes of Health grant S10OD028617. SAXS data were also acquired at the G-1 beamline at CHESS, supported by National Science Foundation (NSF) & National Institutes of Health/National Institute of General Medical Sciences (NIH/NIGMS) via NSF award (DMR-0936384 to CHESS)), at the BioCAT beamline of the Advanced Photon Source, a US Department of Energy (DOE) Office of Science User Facility operated for the DOE Office of Science by Argonne National Laboratory under Contract No. DE-AC02-06CH11357. BioCAT is supported by Grant 9 P41 GM103622 from the National Institute of General Medical Sciences (NIGMS) of the NIH.

References

Acerbo, A. S., Cook, M. J., & Gillilan, R. E. (2015). Upgrade of MacCHESS facility for X-ray scattering of biological macromolecules in solution. *Journal of Synchrotron Radiation*, *22*(1), 180–186. https://doi.org/10.1107/S1600577514020360.

Anderson, N. G. (1966). *The development of zonal centrifuges and ancillary Systems for Tissue Fractionation and Analysis*. U.S. Department of Health, Education, and Welfare, Public Health Service, National Cancer Institute.

Asadi, M. (2006). Tables. In *Beet-sugar handbook* (pp. 779–801). John Wiley & Sons, Ltd. https://doi.org/10.1002/9780471790990.oth1.

Beckmann, B. M., Castello, A., & Medenbach, J. (2016). The expanding universe of ribonucleoproteins: Of novel RNA-binding proteins and unconventional interactions. *Pflügers Archiv - European Journal of Physiology*, *468*(6), 1029–1040. https://doi.org/10.1007/s00424-016-1819-4.

Blanchet, C. E., & Svergun, D. I. (2013). Small-angle X-ray scattering on biological macromolecules and nanocomposites in solution. *Annual Review of Physical Chemistry*, *64*(1), 37–54. https://doi.org/10.1146/annurev-physchem-040412-110132.

Blose, J. M., Pabit, S. A., Meisburger, S. P., Li, L., Jones, C. D., & Pollack, L. (2011). Effects of a protecting Osmolyte on the ion atmosphere surrounding DNA duplexes. *Biochemistry*, *50*(40), 8540–8547. https://doi.org/10.1021/bi200710m.

Bolze, J., Ballauff, M., Kijlstra, J., & Rudhardt, D. (2003). Application of small-angle X-ray scattering as a tool for the structural analysis of industrial polymer dispersions. *Macromolecular Materials and Engineering*, *288*(6), 495–502. https://doi.org/10.1002/mame.200390046.

Bolze, J., Hörner, K. D., & Ballauff, M. (1996). Adsorption of the nonionic surfactant Triton X-405 on polystyrene latex particles as monitored by small-angle X-ray scattering. *Langmuir*, *12*(12), 2906–2912. https://doi.org/10.1021/la951073c.

Brenner, S., & Miller, J. H. (2014). *Brenner's encyclopedia of genetics*. Elsevier Science.

Chen, Y., Tokuda, J. M., Topping, T., Meisburger, S. P., Pabit, S. A., Gloss, L. M., et al. (2017). Asymmetric unwrapping of nucleosomal DNA propagates asymmetric opening and dissociation of the histone core. *Proceedings of the National Academy of Sciences*, *114*(2), 334–339. https://doi.org/10.1073/pnas.1611118114.

Chen, Y., Tokuda, J. M., Topping, T., Sutton, J. L., Meisburger, S. P., Pabit, S. A., et al. (2014). Revealing transient structures of nucleosomes as DNA unwinds. *Nucleic Acids Research*, *42*(13), 8767–8776. https://doi.org/10.1093/nar/gku562.

Cutter, A. R., & Hayes, J. J. (2015). A brief review of nucleosome structure. *FEBS Letters*, *589*(20 Part A), 2914–2922. https://doi.org/10.1016/j.febslet.2015.05.016.

Da Vela, S., & Svergun, D. I. (2020). Methods, development and applications of small-angle X-ray scattering to characterize biological macromolecules in solution. *Current Research in Structural Biology*, *2*, 164–170. https://doi.org/10.1016/j.crstbi.2020.08.004.

Darros-Barbosa, R., Balaban, M. O., & Teixeira, A. A. (2003). Temperature and concentration dependence of density of model liquid foods. *International Journal of Food Properties*, *6*(2), 195–214. https://doi.org/10.1081/JFP-120017815.

David, G., & Pérez, J. (2009). Combined sampler robot and high-performance liquid chromatography: A fully automated system for biological small-angle X-ray scattering experiments at the synchrotron SOLEIL SWING beamline. *Journal of Applied Crystallography*, *42*(5), 892–900. https://doi.org/10.1107/S0021889809029288.

Feigin, L. A., & Svergun, D. I. (1987). In G. W. Taylor (Ed.), *Structure analysis by small-angle X-ray and Neutron scattering* Springer US. https://doi.org/10.1007/978-1-4757-6624-0.

Fernandez, R. M., Riske, K. A., Amaral, L. Q., Itri, R., & Lamy, M. T. (2008). Influence of salt on the structure of DMPG studied by SAXS and optical microscopy. *Biochimica et Biophysica Acta (BBA) - Biomembranes*, *1778*(4), 907–916. https://doi.org/10.1016/j.bbamem.2007.12.005.

Frank, J., & Ourmazd, A. (2016). Continuous changes in structure mapped by manifold embedding of single-particle data in cryo-EM. *Methods*, *100*, 61–67. https://doi.org/10.1016/j.ymeth.2016.02.007.

Franke, D., Petoukhov, M. V., Konarev, P. V., Panjkovich, A., Tuukkanen, A., Mertens, H. D. T., et al. (2017). ATSAS 2.8: A comprehensive data analysis suite for small-angle scattering from macromolecular solutions. *Journal of Applied Crystallography*, *50*(4), 1212–1225. https://doi.org/10.1107/S1600576717007786.

Gabel, F. (2015). Chapter thirteen—Small-angle neutron scattering for structural biology of protein–RNA complexes. In S. A. Woodson, & F. H. T. Allain (Eds.), *Vol. 558. Methods in enzymology* (pp. 391–415). Academic Press. https://doi.org/10.1016/bs.mie.2015.02.003.

Gabel, F., Engilberge, S., Pérez, J., & Girard, E. (2019). Medical contrast media as possible tools for SAXS contrast variation. *IUCrJ*, *6*(4), 521–525. https://doi.org/10.1107/S2052252519005943.

Galej, W. P., Wilkinson, M. E., Fica, S. M., Oubridge, C., Newman, A. J., & Nagai, K. (2016). Cryo-EM structure of the spliceosome immediately after branching. *Nature*, *537*(7619), 197–201. https://doi.org/10.1038/nature19316.

Glycerine Producers' Association. (1963). *Physical properties of glycerine and its solutions*. https://www.aciscience.org/docs/Physical_properties_of_glycerine_and_its_solutions.pdf.

Grant, T. D. (2018). Ab initio electron density determination directly from solution scattering data. *Nature Methods*, *15*(3), 191–193. https://doi.org/10.1038/nmeth.4581.

Guinier, A. (1939). La diffraction des rayons X aux très petits angles: Application à l'étude de phénomènes ultramicroscopiques. *Annales de Physique*, *11*(12), 161–237. https://doi.org/10.1051/anphys/193911120161.

Hansen, S. (2000). Bayesian estimation of hyperparameters for indirect Fourier transformation in small-angle scattering. *Journal of Applied Crystallography*, *33*(6), 1415–1421. https://doi.org/10.1107/S0021889800012930.

He, W., Chen, Y.-L., Pollack, L., & Kirmizialtin, S. (2021). The structural plasticity of nucleic acid duplexes revealed by WAXS and MD. *Science Advances*, 7(17), eabf6106. https://doi.org/10.1126/sciadv.abf6106.

Hickl, P., Ballauff, M., & Jada, A. (1996). Small-angle X-ray contrast-variation study of micelles formed by poly(styrene)−poly(ethylene oxide) block copolymers in aqueous solution. *Macromolecules*, *29*(11), 4006–4014. https://doi.org/10.1021/ma951480v.

Hirai, M., Ajito, S., Sugiyama, M., Iwase, H., Takata, S., Shimizu, N., et al. (2018). Direct evidence for the effect of glycerol on protein hydration and thermal structural transition. *Biophysical Journal*, *115*(2), 313–327. https://doi.org/10.1016/j.bpj.2018.06.005.

Hopkins, J. B., Gillilan, R. E., & Skou, S. (2017). BioXTAS RAW: Improvements to a free open-source program for small-angle X-ray scattering data reduction and analysis. *Journal of Applied Crystallography*, *50*(5), 1545–1553. https://doi.org/10.1107/S1600576717011438.

Ibel, K., & Stuhrmann, H. B. (1975). Comparison of neutron and X-ray scattering of dilute myoglobin solutions. *Journal of Molecular Biology*, *93*(2), 255–265. https://doi.org/10.1016/0022-2836(75)90131-X.

Inoko, Y., Yamamoto, M., Fujiwara, S., & Ueki, T. (1992). X-ray scattering study of the shape of the DNA region in nucleosome Core particle with synchrotron radiation. *The Journal of Biochemistry*, *111*(3), 310–316. https://doi.org/10.1093/oxfordjournals.jbchem.a123755.

Jacques, D. A., Guss, J. M., & Trewhella, J. (2012). Reliable structural interpretation of small-angle scattering data from bio-molecules in solution—The importance of quality control and a standard reporting framework. *BMC Structural Biology*, *12*, 9. https://doi.org/10.1186/1472-6807-12-9.

Jacques, D. A., & Trewhella, J. (2010). Small-angle scattering for structural biology-expanding the frontier while avoiding the pitfalls: Small-angle scattering for structural biology. *Protein Science*, *19*(4), 642–657. https://doi.org/10.1002/pro.351.

Kikhney, A. G., & Svergun, D. I. (2015). A practical guide to small angle X-ray scattering (SAXS) of flexible and intrinsically disordered proteins. *FEBS Letters*, *589*(19, part A), 2570–2577. https://doi.org/10.1016/j.febslet.2015.08.027.

Kim, Y.-S., Jones, L. S., Dong, A., Kendrick, B. S., Chang, B. S., Manning, M. C., et al. (2003). Effects of sucrose on conformational equilibria and fluctuations within the native-state ensemble of proteins. *Protein Science*, *12*(6), 1252–1261. https://doi.org/10.1110/ps.0242603.

Kirby, N., Cowieson, N., Hawley, A. M., Mudie, S. T., McGillivray, D. J., Kusel, M., et al. (2016). Improved radiation dose efficiency in solution SAXS using a sheath flow sample environment. *Acta Crystallographica. Section D, Structural Biology*, *72*(Pt 12), 1254–1266. https://doi.org/10.1107/S2059798316017174.

Koch, M. H. J., Vachette, P., & Svergun, D. I. (2003). Small-angle scattering: A view on the properties, structures and structural changes of biological macromolecules in solution. *Quarterly Reviews of Biophysics*, *36*(2), 147–227. https://doi.org/10.1017/S0033583503003871.

Koning, R. I., Gomez-Blanco, J., Akopjana, I., Vargas, J., Kazaks, A., Tars, K., et al. (2016). Asymmetric cryo-EM reconstruction of phage MS2 reveals genome structure in situ. *Nature Communications*, 7(1), 12524. https://doi.org/10.1038/ncomms12524.

Krueger, S. (2017). Designing and performing biological solution small-angle neutron scattering contrast variation experiments on multi-component assemblies. In B. Chaudhuri, I. G. Muñoz, S. Qian, & V. S. Urban (Eds.), *Vol. 1009. Biological small angle scattering: Techniques, strategies and tips* (pp. 65–85). Singapore: Springer. https://doi.org/10.1007/978-981-10-6038-0_5.

Kuwamoto, S., Akiyama, S., & Fujisawa, T. (2004). Radiation damage to a protein solution, detected by synchrotron X-ray small-angle scattering: Dose-related considerations and suppression by cryoprotectants. *Journal of Synchrotron Radiation*, *11*(6), 462–468. https://doi.org/10.1107/S0909049504019272.

Lamb, J. S., Zoltowski, B. D., Pabit, S. A., Crane, B. R., & Pollack, L. (2008). Time-resolved dimerization of a PAS-LOV protein measured with Photocoupled small angle X-ray scattering. *Journal of the American Chemical Society*, *130*(37), 12226–12227. https://doi.org/10.1021/ja804236f.

Lee, J. C., & Timasheff, S. N. (1981). The stabilization of proteins by sucrose. *Journal of Biological Chemistry*, *256*(14), 7193–7201. https://doi.org/10.1016/S0021-9258(19)68947-7.

Lowary, P. T., & Widom, J. (1998). New DNA sequence rules for high affinity binding to histone octamer and sequence-directed nucleosome positioning11Edited by T. Richmond. *Journal of Molecular Biology*, *276*(1), 19–42. https://doi.org/10.1006/jmbi.1997.1494.

Luger, K., Mäder, A. W., Richmond, R. K., Sargent, D. F., & Richmond, T. J. (1997). Crystal structure of the nucleosome core particle at 2.8 Å resolution. *Nature*, *389*(6648), 251–260. https://doi.org/10.1038/38444.

Makowski, L., Rodi, D. J., Mandava, S., Devarapalli, S., & Fischetti, R. F. (2008). Characterization of protein fold by wide-angle X-ray solution scattering. *Journal of Molecular Biology*, *383*(3), 731–744. https://doi.org/10.1016/j.jmb.2008.08.038.

Mathlouthi, M., & Génotelle, J. (1995). Rheological properties of sucrose solutions and suspensions. In M. Mathlouthi, & P. Reiser (Eds.), *Sucrose: Properties and applications* (pp. 126–154). US: Springer. https://doi.org/10.1007/978-1-4615-2676-6_6.

Mauney, A. W., Muthurajan, U. M., Luger, K., & Pollack, L. (2021). Solution structure(s) of trinucleosomes from contrast variation SAXS. *Nucleic Acids Research*, *49*(9), 5028–5037. https://doi.org/10.1093/nar/gkab290.

Mauney, A. W., Tokuda, J. M., Gloss, L. M., Gonzalez, O., & Pollack, L. (2018). Local DNA sequence controls asymmetry of DNA unwrapping from nucleosome Core particles. *Biophysical Journal*, *115*(5), 773–781. https://doi.org/10.1016/j.bpj.2018.07.009.

Mylonas, E., & Svergun, D. I. (2007). Accuracy of molecular mass determination of proteins in solution by small-angle X-ray scattering. *Journal of Applied Crystallography*, *40*(s1), s245–s249. https://doi.org/10.1107/S002188980700252X.

Nielsen, S. S., Toft, K. N., Snakenborg, D., Jeppesen, M. G., Jacobsen, J. K., Vestergaard, B., et al. (2009). BioXTAS RAW, a software program for high-throughput automated small-angle X-ray scattering data reduction and preliminary analysis. *Journal of Applied Crystallography*, *42*(5), 959–964. https://doi.org/10.1107/S0021889809023863.

Orthaber, D., Bergmann, A., & Glatter, O. (2000). SAXS experiments on absolute scale with Kratky systems using water as a secondary standard. *Journal of Applied Crystallography*, *33*(2), 218–225. https://doi.org/10.1107/S0021889899015216.

Pabit, S. A., Chen, Y.-L., Usher, E. T., Cook, E. C., Pollack, L., & Showalter, S. A. (2020). Elucidating the role of microprocessor protein DGCR8 in bending RNA structures. *Biophysical Journal*, *119*(12), 2524–2536. https://doi.org/10.1016/j.bpj.2020.10.038.

Pérez, J., & Nishino, Y. (2012). Advances in X-ray scattering: From solution SAXS to achievements with coherent beams. *Current Opinion in Structural Biology*, *22*(5), 670–678. https://doi.org/10.1016/j.sbi.2012.07.014.

Petkevičiūtė, D., Pasi, M., Gonzalez, O., & Maddocks, J. H. (2014). cgDNA: A software package for the prediction of sequence-dependent coarse-grain free energies of B-form DNA. *Nucleic Acids Research*, *42*(20), e153. https://doi.org/10.1093/nar/gku825.

Plumridge, A., Katz, A. M., Calvey, G. D., Elber, R., Kirmizialtin, S., & Pollack, L. (2018). Revealing the distinct folding phases of an RNA three-helix junction. *Nucleic Acids Research*, *46*(14), 7354–7365. https://doi.org/10.1093/nar/gky363.

Pollack, L. (2011). SAXS Studies of Ion–Nucleic Acid Interactions. *Annual Review of Biophysics*, *40*(1), 225–242. https://doi.org/10.1146/annurev-biophys-042910-155349.

Pollack, L., Tate, M. W., Darnton, N. C., Knight, J. B., Gruner, S. M., Eaton, W. A., et al. (1999). Compactness of the denatured state of a fast-folding protein measured by submillisecond small-angle x-ray scattering. *Proceedings of the National Academy of Sciences of the United States of America*, *96*(18), 10115–10117.

Putnam, C. D., Hammel, M., Hura, G. L., & Tainer, J. A. (2007). X-ray solution scattering (SAXS) combined with crystallography and computation: Defining accurate macromolecular structures, conformations and assemblies in solution. *Quarterly Reviews of Biophysics*, *40*(3), 191–285. https://doi.org/10.1017/S0033583507004635.

Rickwood, D., Ford, T., & Graham, J. (1982). Nycodenz: A new nonionic iodinated gradient medium. *Analytical Biochemistry*, *123*(1), 23–31. https://doi.org/10.1016/0003-2697(82)90618-2.

San Emeterio, J., & Pollack, L. (2020). Visualizing a viral genome with contrast variation small angle X-ray scattering. *Journal of Biological Chemistry*, *295*(47), 15923–15932. https://doi.org/10.1074/jbc.RA120.013961.

Sardet, C., Tardieu, A., & Luzzati, V. (1976). Shape and size of bovine rhodopsin: A small-angle X-ray scattering study of a rhodopsin-detergent complex. *Journal of Molecular Biology*, *105*(3), 383–407. https://doi.org/10.1016/0022-2836(76)90100-5.

Schlatterer, J. C., Kwok, L. W., Lamb, J. S., Park, H. Y., Andresen, K., Brenowitz, M., et al. (2008). Hinge stiffness is a barrier to RNA folding. *Journal of Molecular Biology*, *379*(4), 859–870. https://doi.org/10.1016/j.jmb.2008.04.013.

Schneidman-Duhovny, D., Hammel, M., Tainer, J. A., & Sali, A. (2013). Accurate SAXS profile computation and its assessment by contrast variation experiments. *Biophysical Journal*, *105*(4), 962–974. https://doi.org/10.1016/j.bpj.2013.07.020.

Schneidman-Duhovny, D., Kim, S. J., & Sali, A. (2012). Integrative structural modeling with small angle X-ray scattering profiles. *BMC Structural Biology*, *12*(1), 17. https://doi.org/10.1186/1472-6807-12-17.

Schroer, M. A., Blanchet, C. E., Gruzinov, A. Y., Gräwert, M. A., Brennich, M. E., Hajizadeh, N. R., et al. (2018). Smaller capillaries improve the small-angle X-ray scattering signal and sample consumption for biomacromolecular solutions. *Journal of Synchrotron Radiation*, *25*(Pt 4), 1113–1122. https://doi.org/10.1107/S1600577518007907.

Sedlak, S. M., Bruetzel, L. K., & Lipfert, J. (2017). Quantitative evaluation of statistical errors in small-angle X-ray scattering measurements. *Journal of Applied Crystallography*, *50*(Pt 2), 621–630. https://doi.org/10.1107/S1600576717003077.

Skou, S., Gillilan, R. E., & Ando, N. (2014). Synchrotron-based small-angle X-ray scattering (SAXS) of proteins in solution. *Nature Protocols*, *9*(7), 1727–1739. https://doi.org/10.1038/nprot.2014.116.

Stuhrmann, H. B. (2008). Small-angle scattering and its interplay with crystallography, contrast variation in SAXS and SANS. *Acta Crystallographica Section A Foundations of Crystallography*, *64*(1), 181–191. https://doi.org/10.1107/S0108767307046569.

Stuhrmann, H. B., & Miller, A. (1978). Small-angle scattering of biological structures. *Journal of Applied Crystallography*, *11*(5), 325–345. https://doi.org/10.1107/S0021889878013473.

Svergun, D. I. (1992). Determination of the regularization parameter in indirect-transform methods using perceptual criteria. *Journal of Applied Crystallography*, *25*(4), 495–503. https://doi.org/10.1107/S0021889892001663.

Svergun, D. I. (1999). Restoring low resolution structure of biological macromolecules from solution scattering using simulated annealing. *Biophysical Journal*, *76*(6), 2879–2886.

Svergun, D. I., & Koch, M. H. J. (2003). Small-angle scattering studies of biological macromolecules in solution. *Reports on Progress in Physics*, *66*(10), 1735–1782. https://doi.org/10.1088/0034-4885/66/10/R05.

Svergun, D. I., & Nierhaus, K. H. (2000). A map of protein-rRNA distribution in the 70 SEscherichia coli ribosome. *Journal of Biological Chemistry*, *275*(19), 14432–14439. https://doi.org/10.1074/jbc.275.19.14432.

Thompson, M. C., Barad, B. A., Wolff, A. M., Sun Cho, H., Schotte, F., Schwarz, D. M. C., et al. (2019). Temperature-jump solution X-ray scattering reveals distinct motions in a dynamic enzyme. *Nature Chemistry*, *11*(11), 1058–1066. https://doi.org/10.1038/s41557-019-0329-3.

Tokuda, J. M., Pabit, S. A., & Pollack, L. (2016). Protein–DNA and ion–DNA interactions revealed through contrast variation SAXS. *Biophysical Reviews*, *8*(2), 139–149. https://doi.org/10.1007/s12551-016-0196-8.

Tokuda, J. M., Ren, R., Levendosky, R. F., Tay, R. J., Yan, M., Pollack, L., et al. (2018). The ATPase motor of the Chd1 chromatin remodeler stimulates DNA unwrapping from the nucleosome. *Nucleic Acids Research*, *46*(10), 4978–4990. https://doi.org/10.1093/nar/gky206.

Tria, G., Mertens, H. D. T., Kachala, M., & Svergun, D. I. (2015). Advanced ensemble modelling of flexible macromolecules using X-ray solution scattering. *IUCrJ*, *2*(2), 207–217. https://doi.org/10.1107/S205225251500202X.

Tsutakawa, S. E., Hura, G. L., Frankel, K. A., Cooper, P. K., & Tainer, J. A. (2007). Structural analysis of flexible proteins in solution by small angle X-ray scattering combined with crystallography. *Journal of Structural Biology*, *158*(2), 214–223. https://doi.org/10.1016/j.jsb.2006.09.008.

Vagenende, V., Yap, M. G. S., & Trout, B. L. (2009). Mechanisms of protein stabilization and prevention of protein aggregation by glycerol. *Biochemistry*, *48*(46), 11084–11096. https://doi.org/10.1021/bi900649t.

Van Nostrand, E. L., Freese, P., Pratt, G. A., Wang, X., Wei, X., Xiao, R., et al. (2020). A large-scale binding and functional map of human RNA-binding proteins. *Nature*, *583*(7818), 711–719. https://doi.org/10.1038/s41586-020-2077-3.

Weiel, M., Reinartz, I., & Schug, A. (2019). Rapid interpretation of small-angle X-ray scattering data. *PLoS Computational Biology*, *15*(3), e1006900. https://doi.org/10.1371/journal.pcbi.1006900.

Whitten, A. E., Cai, S., & Trewhella, J. (2008). MULCh: Modules for the analysis of small-angle neutron contrast variation data from biomolecular assemblies. *Journal of Applied Crystallography*, *41*(1), 222–226. https://doi.org/10.1107/S0021889807055136.

Whitten, A. E., & Trewhella, J. (2009). Small-angle scattering and neutron contrast variation for studying bio-molecular complexes. In R. S. Foote, & J. W. Lee (Eds.), *Micro and nano technologies in bioanalysis: Methods and protocols* (pp. 307–323). Humana Press. https://doi.org/10.1007/978-1-59745-483-4_20.

Wolfrom, M. L., Binkley, W. W., & McCabe, L. J. (1959). The effect of ionizing radiation on carbohydrates. The irradiation of sucrose and methyl α-D-Glucopyranoside1. *Journal of the American Chemical Society*, *81*(6), 1442–1446. https://doi.org/10.1021/ja01515a039.

Yan, C., Hang, J., Wan, R., Huang, M., Wong, C. C. L., & Shi, Y. (2015). Structure of a yeast spliceosome at 3.6-angstrom resolution. *Science*, *349*(6253), 1182–1191. https://doi.org/10.1126/science.aac7629.

Yang, L., Antonelli, S., Chodankar, S., Byrnes, J., Lazo, E., & Qian, K. (2020). Solution scattering at the life science X-ray scattering (LiX) beamline. *Journal of Synchrotron Radiation*, *27*(3), 804–812. https://doi.org/10.1107/S1600577520002362.

Yusupov, M. M., Yusupova, G. Z., Baucom, A., Lieberman, K., Earnest, T. N., Cate, J. H. D., et al. (2001). Crystal structure of the ribosome at 5.5 Å resolution. *Science*, *292*(5518), 883–896. https://doi.org/10.1126/science.1060089.

Zuo, X., Cui, G., Merz, K. M., Zhang, L., Lewis, F. D., & Tiede, D. M. (2006). X-ray diffraction “fingerprinting” of DNA structure in solution for quantitative evaluation of molecular dynamics simulation. *Proceedings of the National Academy of Sciences*, *103*(10), 3534–3539. https://doi.org/10.1073/pnas.0600022103.

CHAPTER THREE

Deuteration for biological SANS: Case studies, success and challenges in chemistry and biology

Anthony P. Duff*, Marina Cagnes, Tamim A. Darwish, Anwen M. Krause-Heuer, Michael Moir, Carl Recsei, Agata Rekas, Robert A. Russell, Karyn L. Wilde, and Nageshwar R. Yepuri
National Deuteration Facility, ANSTO, Lucas Heights, NSW, Australia
*Corresponding author: e-mail address: Anthony.Duff@ansto.gov.au

Contents

Methods in Enzymology, Volume 677
ISSN 0076-6879
https://doi.org/10.1016/bs.mie.2022.08.030

Abstract

Small angle neutron scattering is a powerful complementary technique in structural biology. It generally requires, or benefits from, deuteration to achieve its unique potentials. Molecular deuteration has become a mature expertise, with deuteration facilities located worldwide to support access to the technique for a wide breadth of structural biology and life sciences. The sorts of problems well answered by small angle scattering and deuteration involve large (>10 Å) scale flexible movements, and this approach is best used where high-resolution methods (crystallography, NMR, cryo-EM) leave questions unanswered. This chapter introduces deuteration, reviewing biological deuteration of proteins, lipids and sterols, and then steps through the ever-expanding range of deuterated molecules being produced by chemical synthesis and enabling sophisticated experiments using physiologically relevant lipids. Case studies of recent successful use of deuteration may provide illustrative examples for strategies for future experiments. We discuss issues of nomenclature for synthesised molecules of novel labeling and make recommendations for their naming. We reflect on our experiences, with cost associated with achieving an arbitrary deuteration level, and on the benefits of experimental co-design by user scientist, deuteration scientist, and neutron scattering scientist working together. Although methods for biological and chemical deuteration are published in the public domain, we recommend that the best method to deuterate is to engage with a deuteration facility.

Abbreviations

B. subtilis	*Bacillus subtilis*
DDM	n-dodecyl-β-D-maltopyranoside
DMPC	1,2-dimyristoyl-d54-*sn*-glycero-3-phosphocholine
DMPC	1,2-Dimyristoyl-d54-sn-glycero-3-phosphocholine
DOPC	1,2-oleoyl-*sn*-glycero-3-phosphocholine
DOTMA	1,2-di-O-oleyl-3-trimethylammonium propane
DPPC	1,2-dipalmitoyl-*sn*-glycero-3-phosphocholine
DSPC	1,2-distearoyl-*sn*-glycero-3-phosphocholine
E. coli	*Escherichia coli*
Monoolein	1-(cis-9-Octadecenoyl)-*rac*-glycerol, elsewhere glycerol monoolein
NR	neutron reflectometry
nSLD	neutron scattering length density
OG	octyl β-D-glucopyranoside
P. pastoris	*Pichia pastoris*
PC	phosphatidylcholine
POPC	1-palmitoyl, 2-oleoyl-sn-glycero-3-phosphocholine
S. cerevisiae	*Saccharomyces cerevisiae*
SANS	small angle neutron scattering
SAXS	small angle X-ray scattering
SEC	size-exclusion chromatography

Facilities

NDF National Deuteration Facility, ANSTO
D-Lab, ILL. The Deuteration Laboratory (D-Lab) of the Life Sciences group at the Institut Laue-Langevin, Grenoble (ILL).

Definitions and terminology

Deuteration level: Number of deuterium atoms in non-exchangeable hydrogen atom sites, divided by the number of non-exchangeable hydrogen atom sites, expressed as a percentage.
Exchangeability of hydrogen atom sites is taken to be for aqueous near-neutral pH. Buried exchangeable hydrogen atom sites may not exchange on relevant timescales, but are considered exchangeable.
Non-exchangeable: Hydrogen atoms bound to carbon atoms.
Exchangeable: hydrogen atoms bound atoms other than carbon.
Deuterated: prepared with deuterium atoms in the place of non-exchangeable hydrogen atoms.
Unlabeled: Not deuterated. Having natural abundance hydrogen in all hydrogen atom sites. Elsewhere the synonyms "non-deuterated", "protiated" or "hydrogenated" may be used.

1. Deuteration for small angle neutron scattering

Molecular deuteration, for contrast in small angle neutron scattering (SANS), has become a mature field. Its importance in biological SANS, and other neutron scattering, has been well reviewed (Ashkar et al., 2018; Dunne et al., 2017; Fisher, Jackson, Kovalevsky, Oksanen, & Wacklin, 2017; Haertlein et al. , 2016; Le Brun, Clifton, Holt, Holden, & Lakey, 2016; Okuda et al., 2021).

A consistent theme in these reviews is that, in structural biology, deuteration and SANS is a powerful complementary method to the dominant, high-resolution, data-rich, methods of macromolecular crystallography, nuclear magnetic resonance (NMR) and cryo-electron microscopy.

Biological small angle scattering provides extra information for questions of large-scale flexibility or disorder. SANS with deuteration improves on small angle X-ray scattering (SAXS) for questions of internal flexibility, or where the system requires contrast-matching to observe part of the sample.

1.1 Manipulating contrast

Contrast in X-ray scattering of biological samples is important, but it is usually not modified by altering the nature of the biomolecules themselves. Proteins have a small natural X-ray scattering contrast against water, but with high flux X-ray sources, the collection of statistically strong data is very good. Nucleic acids have strong X-ray contrast against protein, and solvent, which can be helpful. Oils and fatty acids have low X-ray contrast with water. Contrast can be manipulated for SAXS experiments (Kawaguchi, Hamanaka, Kito, & Machida, 1991), but the contrast-altering molecules introduce variable effects on the protein being studied.

In SANS, for hydrogen-rich materials, as are common in biology, deuteration offers an opportunity to manipulate contrast with minimal physiochemical effects. Hydrogen-1 (^{1}H) scatters neutrons very differently to deuterium ($^{2}H{=}D$) (refer to Table 1 and Fig. 1). Substitution of hydrogen atoms with deuterium atoms, in a hydrogen-rich molecule, dramatically alters its neutron scattering properties.

In practice, it is the neutron scattering length density (nSLD) that is considered in the design of a neutron scattering experiment. The nSLD is the sum of coherent neutron scattering lengths of all of the isotopes of the molecule, divided by the volume of the molecule:

$$\mathrm{nSLD} = \sum \mathrm{b}_{\mathrm{coherent}}/\mathrm{volume}$$

Table 1 Coherent neutron scattering lengths for selected atoms. The atoms H, C, N and O dominate the composition of many biological molecules.

Isotope	Barns, coherent (fm, 10^{-15} m)
H	−3.739
^{1}H	−3.741
$^{2}H{=}D$	6.671
C	6.646
N	9.360
O	5.803

The symbol "H" is used for natural abundance hydrogen, which is 99.985% ^{1}H and 0.015% ^{2}H. The neutron scattering from ^{1}H has a negative amplitude, over 7.4 units difference from ^{2}H and C, which creates great potential for changing neutron scattering length by deuteration.

This data is taken from https://www.ncnr.nist.gov/resources/n-lengths/elements/h.html.

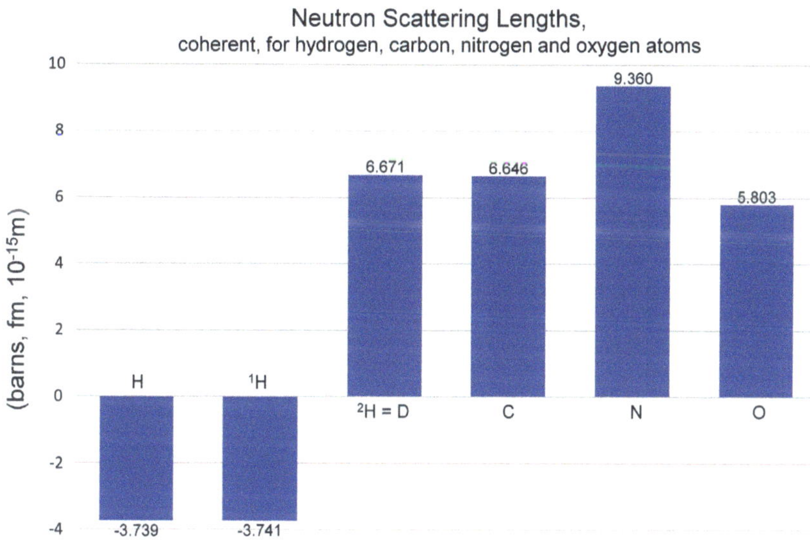

Fig. 1 Neutron scattering lengths (coherent) of common biological atoms. A graphical presentation of the date from Table 1 illustrates the very different scattering length of ^1H from ^2H, C, N and O.

which can be written as.

$$nSLD = \sum b_{coherent} \text{ x } N_A / \left(MW \text{ x } V_p\right)$$

where N_A is avagadro's number; MW is molecular weight; and V_p is the molecule's partial specific volume (ml/g), inverse density. In practice, these calculations, including an estimation of the partial specific volume, are done by an online nSLD calculator, such as MULCh, at, https://smb-research.smb.usyd.edu.au/NCVWeb/input.jsp (Whitten, Cai, & Trewhella, 2008).

While the sum of coherent scattering lengths, and measurement of the bulk density of the material, can be accurate, the partial specific volume of the molecule, dissolved, or in interaction with other molecules, may differ due to temperature and molecular packing. Nevertheless, an estimate is a good starting point.

Table 2 shows MULCh output for example biological molecules, for each molecule unlabeled, and 100% deuterated. A selection of these are graphed in Fig. 2. Table 2 and Fig. 2 show estimate nSLD values for selected

Table 2 Neutron scattering length density (nSLD) for selected biological molecules. Estimates for partial specific volume are for the unlabeled version.

Molecule	Partial specific volume (ml/g)	nSLD (10^{-6} Å^{-2}) Molecule unlabeled in H_2O	in D_2O	Molecule 100% deuterated in H_2O	in D_2O
Water	1.003	−0.560			6.388
T4 Lysozyme. Uniprot P00720.	0.740	1.794	3.290	6.381	7.876
MBP From *E. coli*. Uniprot P0AEX9	0.741	1.818	3.092	6.531	7.804
RNA. $(ACGU)_{100}$	0.572	3.409	4.444	6.066	7.101
DNA. $(ACGT)_{100}$	0.589	3.249	3.946	5.689	6.389
Palmitic acid. $C_{16}H_{32}O_2$	1.048	−0.040	0.194	7.196	7.429
Monoolein. $C_{21}H_{40}O_2$	0.993	0.224	0.578	6.955	7.309
Cholesterol. $C_{27}H_{46}O$	1.075	0.191**	0.343	6.982	7.133[a]
POPC. $C_{42}H_{82}NO_8P$	0.944	0.279	7.444	0.279	7.444
Cellulose. $(C_6H_{10}O_5)_n$	0.727	1.608	3.203	5.333	6.926

[a](Moulin et al., 2018) determined different nSLD values, 0.21 (not 0.19) for h-cholesterol, and 7.65 (not 7.13) for d-cholesterol. These differences are likely due to inaccuracy in the estimation of the partial specific volume. The higher nSLD may reflect that cholesterol packs more tightly in lipid than to itself. taken from MULCh (Whitten et al., 2008).

molecules, unlabeled or 100% deuterated. nSLD values can be readily calculated for any input deuteration level. Partial deuteration levels that are predicted to give nSLD matchout in 100% D_2O, for the proteins and nucleic acids, are given in Table 3.

Fig. 2 can be used to assist in the design of a neutron scattering experiment. An nSLD calculator should be used to calculate nSLD values for the actual molecules to be used. Fig. 2 shows molecules as undeuterated or 100% deuterated. Partial deuteration will give a line intermediate between the two. If an nSLD match-out is desired, then a deuteration level should be chosen so that the nSLD line intersects with the water line. The choice of 100% D_2O for the matchout condition is commonly made on the basis of minimising incoherent scattering noise, which arises almost entirely from the ^1H atoms.

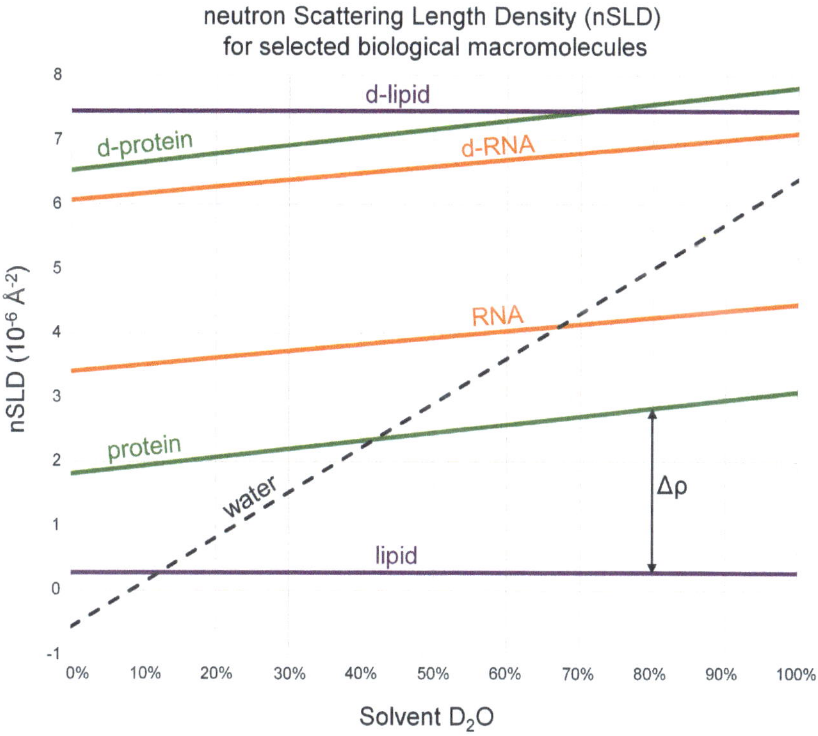

Fig. 2 nSLDs for selected biological molecules. The estimated nSLDs for selected biological molecules are graphed versus the deuteration level of the solvent, from 0% to 100% D_2O. Molecules may have an increase in nSLD with increasing deuteration of the solvent due to solvent-exchangeable hydrogen atoms. Deuteration of the macromolecules is shown at 100% deuteration, indicating the maximum increase in nSLD that can be achieved by deuteration. In principle, any level of deuteration may be chosen. nSLD matchout occurs at solvent D_2O concentrations where two lines cross. A commonly desired matchout is for the dashed water line to cross the molecules of lesser-interest, at 100% D_2O. Partial deuteration is often chosen such that the nSLD at 100% D_2O matches that of water, 6.39×10^{-6} Å^{-2}, which achieves contrast matchout in 100% D_2O, which is the solvent choice of minimum incoherent scattering noise. $\Delta\rho$, the symbol for nSLD contrast, is indicated, between lipid and protein in 80% D_2O. For this illustration, the molecular choices made are: Lipid – POPC; protein – MBP; RNA $(ACGU)_{100}$.

Table 3 Matchout deuteration levels.

Molecule	Deuteration level for nSLD matchout in 100% D_2O
Lysozyme	76.6%
MBP	70.0%
RNA	73.2%
DNA	76.2%

The nSLD calculator can be used to predict the deuteration level required to achieve an nSLD match-out in 100% D_2O. Moderate protein to protein variation occurs due to different amino acid compositions.

2. Protein SANS

2.1 Protein deuteration for SANS with contrast variation

Small angle neutron scattering with solvent contrast variation is often an extension of SAXS, used on multi-protein assemblies. Further characterisation of the internal disposition of protein subunits involves: (1) deuterating one or more of the protein chains; (2) measuring SANS across a range of solvents varying from 0% to 100% D_2O, leaving solutes and other solution components the same across the samples. This technique is particularly applicable to multi-protein assemblies where there are large scale internal dynamics, relevant to function. Crystallography and cryo-EM can give snapshots, and many snapshots give the picture, but solution scattering is better for dynamics in the solution state. The SANS with contrast variation method, with reports from 1975 (Engelman, Moore, & Schoenborn, 1975) is well developed, see Table 4 Rows 3–7 for examples. Protein deuteration via recombinant bacterial expression has mostly achieved a robust reliability (Duff, Wilde, Rekas, Lake, & Holden, 2015).

More challenging protein deuteration includes segmental labeling of a single polypeptide protein. This has been successfully achieved, e.g., by (Sonntag et al., 2017), see Table 4 Row 2. The methods were developed for NMR (Ashkar et al., 2018). However, our experience is that the ligation of separately purified protein fragments can be unreliable and inefficient. We speculate that the mechanism of protein ligation, whether with intein chemical ligation, sortase A, or butelase 1, the structural rigidity of the protein fragments can limit the ligation, and that successful results are rare and case-specific.

Other advanced strategies for protein deuteration are yeast, insect cell, mammalian and cell-free expression (Ashkar et al., 2018). Cell-free expression with labeling by exogenous supply of labeled amino acids is an established method for advanced labeling for NMR, which in theory would work for SANS for proteins that require cell-free expression. However, the costs for the higher yields required for SANS are large.

Related to protein deuteration by recombinant expression is the controlled deuteration of nucleic acids. Chromosomal DNA is naturally abundant and readily deuterated (Shotton et al., 1997). High-yield custom sequences are obtained using high copy number plasmids (Zhu et al., 2010). High yield deuterated RNA has been obtained by over-expression of tRNAs (Jünemann et al., 1996).

Table 4 Examples used with deuteration for neutron scattering, and description of the deuteration used.

	Protein: Protein complexes requiring protein deuteration, for SANS with contrast match-out or contrast variation	**Deuterated molecules and source**
1	Ibrahim et al. (2017) conducted time-resolved SANS experiments on the fluorescent protein (GFP) while it was processed by the archaeal AAA + PAN (proteasome activating nucleotidase) unfoldase. They used perdeuterated GFP and unlabeled PAN, which was matched-out in 42% D_2O. This strategy provided maximum possible nSLD contrast for the GFP from the background solvent and PAN, which provided stronger signal	Green fluorescent protein (GFP) was perdeuterated produced at the D-Lab, ILL
2	Sonntag et al. (2017) used segmental labeling within the single-chain protein, RNA binding protein (RBP) TIA-1. They perdeuterated selected domains of and then used the protein-ligase Sortase A to stitch back together the unlabeled and labeled domains. The same protein ligation method had been previously used for C-13 and N-15 labeling, for NMR studies. Perdeuteration gave strong nSLD contrast against the unlabeled protein, matched-out in 42% D_2O.	The domains of RBP TIA-1 were perdeuterated used ^{2}H-glucose and 100% D_2O M9 media.
3	Dunne et al. (2017) present a case study deuteration and SANS with contrast variation experiment featuring the use of both *E. coli* and *P. pastoris* for the deuteraton. They report methods for the production of 100% D_2O matched-out deuteration. They additionally refer to nine similar reports of deuteration and SANS with contrast variation (page 426, references 4–12), dating from 2007 to 2014.	The proteins were deuterated using 85% D_2O media, MBP using *E. coli* with ^{1}H-glycerol and SCR-19/20 with *P. pastoris*, using 0.5% ^{1}H methanol, at the D-Lab, ILL.
4	Okuda et al. (2021) discuss protein deuteration for SANS with contrast variation. Three examples are given, their references 8–10. They discuss the difficulties faced without having experience in protein deuteration. They also discuss the advantage of "inverse contrast matching", where the protein of interest is unlabeled, and other proteins are labeled to match out to 100% D_2O, with the advantage of minimized incoherent neutron scattering noise from ^{1}H atoms.	Proteins are perdeuterated with 100% D_2O and ^{2}H-glucose. Proteins are partially deuterated using 75% D_2O and 75% ^{2}H-glucose (25% ^{1}H-glucose)
5	Grosas, Rekas, Mata, Thorn, & Carver (2020) investigated deuterated aB-crystallin crowded with unlabeled crowding agent Ficoll 400 and unlabeled gamma-crystallin and aA-crystallin. Perdeuteration was used to provide maximum neutron contrast for the aB-crystallin relative to the other components that would be matched-out, at 28% D_2O for the ficoll, and 40% D_2O for the unlabeled proteins.	aB-crystallin was deuterated to 99.8%D using 100% D_2O media containing 40 g/L ^{2}H-glycerol, as the carbon source, at the NDF

Continued

Table 4 Examples used with deuteration for neutron scattering, and description of the deuteration used.—cont'd

	Protein: Protein complexes requiring protein deuteration, for SANS with contrast match-out or contrast variation	Deuterated molecules and source
6	Investigating protein-pectin nutrient delivery systems, (Xu et al., 2018) used deuterated b-lactoglobulin A (blgA) with an array of different pectins of variable size and charge. As the nSLD of pectin in 100% D_2O (2.2×10^{-6} Å^{-2}) is too close to that of unlabeled blgA (2.9×10^{-6} Å^{-2}), the blgA was deuterated, as its deuteration was much more readily feasible. blgA was deuterated to a ~ 100% D_2O matchout (nSLD = 6.1×10^{-6} Å^{-2}), which provided sufficient nSLD contrast to visualise assembly of pectin around protein clusters in D_2O and behaviour of the protein in H_2O buffer.	β-Lactoglobulin A (blgA) was partially deuterated, to 74%D by using 90% D_2O and 40 g/L ^{1}H-glycerol in the growth media, at the NDF
7	(Furlong et al., 2018) used SANS with deuteration to probe the internal domain structure of the B and C subunits of *Proteus mirabilis* Suppressor of copper-sensitivity (SCS) protein. The SANS extended on the prior X-ray crystallography and SAXS by confirming that the system internal asymmetry was not dynamic, but fixed. Using mutant proteins that become fixed in their interaction, the B subunit was deuterated, and the C subunit was used unlabeled.	Protein ScsB was deuterated to approximately a 100% D_2O nSLD matchpoint, using 90% D_2O and 40 g/L ^{1}H-glycerol, at the NDF
Matching out (making invisible) the solubilising components of the system		
8	Midtgaard et al. (2018) used SANS to analyse the five membrane proteins: bacteriorhodopsin, maltoporin, photosystem I, ionotropic glutamate receptor A2, and sarco/endoplasmatic reticulum Ca2+ ATPase. Each was solubilised by solvent nSLD contrast matched detergents: OG, and DDM. The heads and tails of the detergents were deuterated such that they had the same neutron scattering length density as 100% D_2O. This way, the obtained SANS data only consists of the single contrast signal stemming from the purified membrane protein.	The tail of octyl β-D-glucopyranoside (OG) was deuterated to 94% D, and the head group to 52% D; The tail of n-dodecyl-β-D-maltopyranoside (DDM) was deuterated to 89% D, and the head group to 57% D, at the NDF.
9	Golub et al. (2022) performed SEC-SANS on the membrane proteins, Photosystems I (PSI) and II (PSII), to verify the crystallography and cryo-EM in the non-frozen solution state. They used unlabeled DDM and deuterated matched-out DDM, and showed that results using unlabeled DDM systematically overestimated the size of the protein.	Deuterated DDM, with tail deuteration of 89%, and head group deuteration of 57%, both corresponding to an nSLD of 6.36×10^{-6} Å^{-2}, almost perfectly matching 100% D_2O, was supplied by the NDF

10	Using deuteration and SANS, (Nickels et al., 2017) were able to characterise single membranes of living *Bacillus subtilis* and provide the first experimental evidence for hypothesised lipid rafts. They grew the bacteria in partially deuterating conditions (90% D_2O, ^{1}H-glucose), and fed the culture key protiated and deuterated fatty acids, having blocked the cellular pathways for fatty acid catabolism and anabolism, to produce deuterated cells except for the protiated fatty acid lipids. In SANS measurements, the partially deuterated cells matched out in 85% D_2O, and the signal obtained came from the membranes. Analysis of the living lamellar membrane, compared to similar analysis of synthetic vesicles, indicated lateral de-mixing on a length scale consistent with the lipid raft hypothesis.	*B. subtilis* were deuterated by culturing in 90% D_2O and ^{1}H-glucose. The first key deuterated fatty acid, palmitic acid-d_{31}, was from Sigma-Aldrich. The second, perdeuterated anteiso-pentadecanoic acid (D-a15:0) was produced according to the method developed and published by the NDF, (Yepuri et al., 2013).
Membrane-characterising experiments that required deuterated naturally produced lipids and sterols		
11	D38-cholesterol was used as a Raman active probe for imaging intracellular cholesterol. Their use of selective vibrational imaging is analogous to matchout-methods (Alfonso García, Pfisterer, Riezman, Ikonen, & Potma, 2015). Commercial D7-cholesterol was used and found inadequate. The team produced D38-cholesterol, determined it to have a deuteration level of 86.2%, and found it sufficient for their work.	D7-cholesterol was obtained from Sigma-Aldrich. D38-cholesterol from *S. cerevisiae* (strain RH6829) grown per (Souza et al., 2011), but in 100% D_2O with 1.25% ^{1}H-glucose.
12	Moulin et al. (2018) demonstrated the capability of using deuterated cholesterol in neutron reflectometry. Cholesterol was mixed into dipalmitoylphosphatidylcholine (DPPC), each deuterated or unlabeled, in different combinations. The location of the cholesterol was able to be modelled against the NR data.	D-Cholesterol was perdeuterated using *P. pastoris* with ^{1}H-glycerol in 100% D_2O, at the D-Lab ILL. DPPC-d62 was obtained from Avanti Polar Lipids
13	In neutron reflectometry experiments, (Waldie et al., 2018) used perdeuterated cholesterol for maximum nSLD contrast of the cholesterol against the lipid, to highlight the location of the cholesterol.	Perdeuterated cholesterol was produced using *P. pastoris*, and phosphatidylcholine (PC) lipids were produced using deuterated *E. coli*, by the method of (Maric et al., 2015), at the D-Lab, ILL
14	Using SANS and neutron reflectometry, (Waldie et al., 2019) produced and used match-out cholesterol so that it could be matched against the D_2O solvent, to better observe the unlabeled lipids in the bilayers.	Partially deuterated, matchout cholesterol was produced using *P. pastoris*, using D_2O and ^{1}H-glycerol, at the D-Lab, ILL.

Continued

Table 4 Examples used with deuteration for neutron scattering, and description of the deuteration used.—cont'd

	Protein: Protein complexes requiring protein deuteration, for SANS with contrast match-out or contrast variation	**Deuterated molecules and source**
15	Sebastiani et al. (2021) used deuterated cholesterol and lipids match-out in SANS to study the apolipoprotein ApoE interactions in liquid nano particles (LNPs). They were able to discover the distribution of the lipid components of the LNPs. Deuterated and unlabeled cholesterol were mixed to get the desired average nSLD.	d83-DSPC was obtained from Avanti Polar Lipids d-MC3 (dilinoleylmethyl-4-dimethylaminobutyrate, DLin-MC3-DMA) from AstraZeneca D-linoleic acid D-ethyl ester from Cambridge Isotope Laboratories. d-cholesterol was supplied from both D-Lab ILL using *P. pastoris*, and NDF using *S. cerevisiae*.
16	Luchini et al. (2021) used neutron reflectometry with deuterated lipid extracts and deuterated cholesterol to study the interaction of proteins with membrane lipids. Due to lack of access to the physiologically relevant lipids in deuterated form, they produced a natural mix of lipids from deuterated yeast. Separately supplied d-cholesterol meant that the cholesterol composition of the lipid could be readily controlled. The membrane was deuterated to give good contrast for the unlabeled protein.	Deuterated natural lipids were extracted from *P. pastoris* at the D-Lab, ILL. D-Cholesterol (85%) *S. cerevisiae* was supplied from the NDF.
17	Santamaria et al. (2022) used deuterated cholesterol and lipids for SANS, in a highly integrative set of experiments involving all of crystallography, NMR, cryo-EM and neutron scattering. Deuteration of the model membrane was needed to clearly observe the SARS-CoV-2 Spike protein fusion domain, and this was critical in being able to reveal different roles of the Spike protein membrane fusion domain.	d45-Cholesterol 85% deuterated from *S. cerevisiae*, NDF Lipids were extracted from *Pichia pastoris*, D-Lab, -ILL
Cases that used deuterated oleic acid or its derivatives		
18	McCoy et al. (2019) selectively used deuterated oleyl amidopropyl betaine, a model zwitterionic surfactant, and deuterated additive p-xylene, to study micellular fluids. Contrast variation was used to determine precise partitioning within the micelle.	Oleyl amidopropyl betaine-d_{33} from the NDF. p-Xylene-D10 was obtained from Sigma

19	In a study of lipid cubic phase systems (Cleveland IV et al., 2021) used deuteration for contrast matching of the lipid cubic phase to high D_2O. This removed complexity of the cubic phase and allowed direct study of the distribution and behaviour of the membrane protein bacteriorhodopsin (bR), and solubilising detergents, in the cubic phase.	Deuterated monoolein was supplied the NDF. Octyl glucoside (n-octyl-β-D-glucoside, OG), tail-deuterated, and fully-deuterated were purchased from Anatrace.
20	To study the conformation of model peptides in the lipid bilayer of the bicontinuous cubic phase, (Conn et al., 2021) used deuterated monoolein to match-out the cubic phase to high D_2O. Unlabeled and deuterated monoolein were mixed to achieve the desired nSLD.	Deuterated monoolein was supplied by the NDF.
21	In a SANS study, (van 't Hag et al., 2019) demonstrated that deuterated monoolein-based lipid bilayers used for membrane protein crystallisation can be matched-out, and they observed the isolated scattering from encapsulated peptides during the crystal growth process. A D_2O nSLD matchpoint achieved by mixing deuterated and unlabeled monoolein.	Deuterated monoolein was obtained from the NDF.
22	Johansen et al. (2022) investigated the pentameric Mg^{2+} channel protein, CorA, using SEC-SANS, to complement the atomic resolution magic-angle spinning solid-state NMR (MAS-NMR), and compared with cryo-EM results. For SEC-SANS, CorA was solubilised with stealth lipid nanodiscs made from match-out POPC and match-out deuterated membrane scaffold protein MSP1E3D1. For additional datasets, SEC-SANS was also conducted using matched-out DDM. In both approaches, the membrane protein solubilising components are rendered invisible, and the SANS data refects solely the undeuterated membrane protein. Their results revealed the structure of the pentameric membrane protein was the same in the nanodisc as when solubilised by DDM.	Matchout deuterated MSP1E3D1 was produced at the D-Lab, ILL Matchout deuterated POPC, with 94% D in the tail, and 71% D in the head group, was supplied by the NDF. Matchout deuterated DDM, as per (Midtgaard et al., 2018), was supplied by the NDF
23	Delhom et al. (2020) used deuterated POPC and the perdeuterated sterol ergosterol to study antifungal drugs in model membranes. Using multidimensional neutron contrast variation, they revealed different drug behaviours in model lipid bilayers and yeast phospholipid extract bilayers.	Deuterated POPC was supplied by the NDF. Per-deuterated ergosterol d44 lipid extracts of *P. pastoris* per (de Ghellinck et al., 2014) at the D-Lab, ILL

Continued

Table 4 Examples used with deuteration for neutron scattering, and description of the deuteration used.—cont'd

	Protein: Protein complexes requiring protein deuteration, for SANS with contrast match-out or contrast variation	**Deuterated molecules and source**
24	Bremer et al. (2017) used neutron membrane diffraction to study the interaction between the plant cold-regulation protein COR15A and membranes. Deuterated POPC was used for its differing nSLD to probe the membrane-protein distance. The average nSLD of the lipid was adjusted by mixing unlabeled and deuterated POPC.	Deuterated POPC was supplied by the NDF. POPC from Avanti Polar Lipid 10:90 POPC-d to POPC ratio
25	Using NR and SANS with contrast variation, (Waldie et al., 2021) used deuterated lipid-cholesterol model membranes and unlabeled apolipoprotein variants to study protein interactions with differing lipid saturation levels.	Deuterated cholesterol has obtained from the D-Lab, ILL Deuterated POPC was supplied by the NDF. Deuterated DMPC and tail-deuterated DMPC were obtained from Avanti Polar Lipids
26	In neutron reflectometry experiments, (Waldie et al., 2020) examined lipid exchange involving lipid tail saturation and "good" and "bad" cholesterol (high- and low-density lipoproteins, HDL & LDL). Lipids, deuterated and unlabeled, were chosen noting the importance of correct membrane compositions. Deuteration was essential the experiment, for creating nSLD contrast, and deuteration choices were made to achieve better contrast for the experiments.	Deuterated POPC was supplied by the NDF. Deuterated phosphatidylcholine (PC) was synthesised D-Lab ILL Deuterated cholesterol was obtained *P. pastoris* at D-Lab, ILL Deuterated DMPC was obtained from Avanti Polar Lipids
27	Bryant et al. (2019) published a validation of deuterated model membranes. Using methods including neutron membrane diffraction and SAXS, they measured characterised properties including melting temperature, against deuteration, and discussed implications to the general assumption that the properties of deuterated lipids are identical to those of protiated lipids.	Deuterated DOPC was obtained from the NDF. Tail-deuterated DSPC, DPPC, DMPC were obtained from Avanti lipids.

It may be desired for the contrast variation to include contrast match-out measurement of the deuterated protein, and this requires that the deuteration level is limited. Typically, for a globular protein, a deuteration level of ~75% is chosen, so that it will be matched out in 100% D_2O. Alternatively, it may be considered less important to have a match-out measurement of the deuterated protein, and more important to have greater contrast between the components. In these cases, a high deuteration level is sought. A decision is whether to use a deuterated carbon source, incurring more consumable expense to achieve great contrast, and thus greater signal for determining the intra-disposition of the labeled components. An example is provided by (Ibrahim et al., 2017), see Table 4 Row 1.

2.2 Matching out all but the protein: Invisible detergents and nanodiscs

SANS with contrast matching is a powerful and popular technique for solution scattering of membrane proteins. Membrane proteins require solubilisation, such as by detergents, and these detergents may be contrast "matched-out", to be effectively invisible in the experiment.

Examples include (Sebastiani et al., 2021) who used deuteration to match out lipids to observe the protein in interaction with the lipids (Table 4, Row 15), and (Midtgaard et al., 2018) who demonstrated, using deuterated detergents octyl β-D-glucopyranoside (OG) and n-dodecyl-β-D-maltopyranoside (DDM), that five membrane proteins could be solubilised, purified by size exclusion chromatography, and characterised by SANS in high D_2O, with the detergents matched-out (Table 4 Row 8). (Golub et al., 2022) similarly used match-out DDM to characterise the solution state of photosystems I and II (Table 4 Row 9). (Nickels et al., 2017) used deuteration for matching out entire cells, except for lipids, to obtain the first experimental evidence for lipid rafts in living *Bacillus subtilis* (Table 4, Row 10).

Noting the importance of the lipids constituting a membrane to the structure and function of membrane-bound proteins (Bagatolli, Ipsen, Simonsen, & Mouritsen, 2010), detergent-solubilised membrane proteins may not be a physiologically relevant model system. To achieve a better experimental membrane, nanodiscs were developed. Nanodiscs have a prior history of use in NMR, and have been demonstrated to work in SANS, e.g., Maric et al. (2014) and Johansen et al. (2022). They are composed of a lipid bilayer inside a protein belt. The protein belt is typically a modified apolipoprotein, such as MSP1D1, that produces nanodiscs of defined size.

The protein belt and lipids are all deuterated to the one contrast match-out condition, commonly 100% D_2O. The membrane proteins are used unlabeled, which means the proteins are produced by their standard methods, and have good nSLD contrast against the solvent.

3. Membranes: Sterols and physiologically relevant lipids

Deuterated lipids and sterols can be produced and purified from bacterial and yeast membranes. For example, phosphatidylcholine (PC) production is reported with controlled deuteration using *E. coli* (Maric et al., 2015), and natural unsaturated perdeuterated lipids using *Pichia pastoris* (de Ghellinck et al., 2014), and sterols are now routinely produced using transgenic yeasts (Fisher et al., 2017).

In addition to physiologically relevant lipids, cholesterol is of particular interest as it is an important component in eukaryotic plasma membranes (Mouritsen & Zuckermann, 2004). The sterols have fluidity effects on and in the membrane, and are similarly considered important or critical components of model membranes (Patel et al., 2020). Several publications making use of deuterated cholesterol from yeasts, and natural lipids from bacteria have been reported (See Table 4, Rows 11–17).

Cholesterol with the deuterium labeling of the C17 alkyl chain is possible to access via chemical synthesis. Deuteration of the whole molecule would be very challenging to achieve synthetically, and so biosynthetic methods have been employed to access highly deuterated cholesterol (>85%D). Biodeuteration provides a simple, cost-effective alternative, and opens up a wider range of deuteration for researchers. The expression of cholesterol from genetically-modified yeast produces cholesterol that can be easily purified from the cells.

4. Deuteration by chemical synthesis

The synthesis of deuterated molecules is unique in the sense that the broad field of deuteration science can be applicable to any class of organic molecule. This can provide both challenges and opportunities for deuteration scientists, as they require a breadth of experience in working with a wide variety of types of molecules and also reaction scales.

There are several categories of molecules that have demonstrated utility for SANS, which are summarised below. Other categories of deuterated molecules have application for other neutron techniques.

4.1 Carboxylic acids, alcohols, amines, alkyl halides, thiols, alkanes

Long chain carboxylic acids are one of the most common precursors utilised in deuteration chemistry. The direct deuteration of long chain carboxylic acids via hydrothermal hydrogen-deuterium exchange facilitates access to a wide variety of other deuterated molecules through these versatile precursors. Deuterated carboxylic acids can undergo numerous functional group interconversions to access alcohols, alkyl halides thiols, amides, esters, amines and other relevant functional groups through standard organic chemistry techniques. For instance, deuterated fatty acids serve as the precursors for synthesis of phospholipids, and the deuteration of these individual components allows for tailored deuteration labeling strategies based on user requirements.

The branched acid phytanic acid has been deuterated as a precursor to phytantriol. Phytantriol is of interest as it forms particles with internal bicontinuous cubic phase structures when dispersed with non-ionic surfactants at ambient and physiological temperatures. These structures are interesting for potential bioactive payload release. Cubosomes assembled with deuterated phytantriol have been studied using neutron scattering techniques with contrast variation (Conn et al., 2021; Yepuri et al. , 2019).

4.2 Unsaturated fatty acids

Saturated fatty acids serve as precursors for the synthesis of unsaturated fatty acids (such as oleic acid and linoleic acid). Unsaturated fatty acids are unable to be deuterated directly under hydrothermal conditions due to the presence of sensitive double bonds. The synthetic route to deuterated oleic acid was reported by (Darwish et al., 2013). The unsaturated fatty acid can be produced on a multi-gram scale in an eight-step convergent synthesis from deuterated azelaic acid-d_{14} and deuterated nonanoic acid-d_{17}. These unsaturated fatty acids are often desirable molecules in their own right, and are used as precursors for the synthesis of phospholipids and glycerides, which are extremely important to biological model membranes.

4.3 DDM/OG surfactants

Surfactants have found application in drug delivery, formulations and cosmetic products. Of note, deuterated surfactants have been successfully utilised to study biological systems through neutron experiments (Hiruma-Shimizu, Shimizu, Thompson, Kalverda, & Patching, 2015; Midtgaard et al., 2018; Osborne & Chabre, 1982). A variety of deuterated surfactants have been reported in the literature, including but not limited to, non-ionic surfactants such as n-octyl β-D-glucopyranoside (OG), n-dodecyl β-D-maltoside (DDM), lauryl maltose neopentyl glycol (LMNG) and ionic surfactants such as lauryldimethylamine oxide (LDAO) and sodium dodecyl sulfate (SDS). Generally, specific deuteration levels are required in the polar and non-polar portions of the surfactant to achieve contrast matching with an aqueous environment. In most cases, fatty acids with specific levels of deuteration are utilised, along with commercially available deuterated reagents and deuteration techniques to access the desired detergents. For the synthesis of deuterated glycosides a particularly useful method for hydrogen-deuterium exchange of the sugar group is isotopic exchange over Raney nickel. C—H bonds adjacent to hydroxy groups readily undergo exchange in heavy water under mild conditions without racemisation of the sugarmoiety.

4.4 Ionic liquids

Ionic liquids have numerous applications as powerful solvents and electrolytes, although few applications to biological studies. It was recently found that common imidazolium-based ionic liquids can be reliably deuterated under hydrothermal conditions (Akutsu, Cagnes, Niizeki, Hasegawa, & Darwish, 2018; Akutsu-Suyama, Cagnes, Tamura, Kanaya, & Darwish, 2019). Based on the reaction conditions used the level of deuteration can be carefully controlled.

4.5 Phospholipids—many examples

Phospholipids are key components of all cell membranes. Deuterated phospholipids are critical to enabling the use of physiologically relevant model membranes. Deuterated analogues of several phospholipid classes have been reported, with varying levels of deuteration. Glycerophospholipids can be selectively deuterated at one or both of the acyl chains, the glycerol backbone and the polar head group, depending on the application. 1-Palmitoyl-2-oleoyl-sn-glycero-3-phosphocholine (POPC) is the most abundant phospholipid in most eukaryotic cells and tissues. Analogues with specific parts of the molecule deuterated have been successfully synthesised, including perdeuterated POPC-d_{82} (Yepuri et al., 2016).

4.6 Glycerides—Monoolein and Triolein

Deuterated glycerides can be synthesised from deuterated saturated and unsaturated fatty acids. The monoglyceride of oleic acid, 1-oleoyl-rac-glycerol (monoolein), has been utilised in the in the fields of drug delivery, emulsion stabilisation and protein crystallisation Deuterated monoolein has been used to create lipid bilayers in the form of contrast matched bicontinuous cubic phases to study the structure and function of important membrane-bound proteins (Cleveland IV et al., 2021; Conn et al., 2021; van 't Hag et al., 2019).

In a similar manner, deuterated triglycerides such as tripalmitin and triolein can be synthesised by the esterification of glycerol with three equivalents of fatty acid.

4.7 Oleic acid and its derivatives

Oleic acid is the precursor to monoolein, 1-monoolein (1-(cis-9-Octadecenoyl)-*rac*-glycerol), a major constituent lipid in biology. (Kulkarni, Wachter, Iglesias-Salto, Engelskirchen, & Ahualli, 2011) It is a centrally important component in the fields of drug delivery, emulsion stabilisation and protein crystallisation. Consequently, it, or its derivatives, is highly desired in many experiments using model membrane systems. The first demonstration of high yield chemical synthesis of deuteration monoolein. While oleic acid can be produced biosynthetically, its chemical synthesis is a reliable way to produce multi gram quantities of this important material.

Synthesis of deuterated [D32]oleic acid(Darwish et al., 2013) has enabled neutron scattering experiments to be markedly improved in terms of physiologically relevant model membranes. Subsequently, the yields and efficiency of production have improved, to multi-gram scale, at ~$3000 per gram. Several commonly desired physiological lipids derived from deuterated oleic acid are routinely produced. One is POPC, which is often the predominant lipid in eukaryotic cell membranes, where it is crucial for the fluidity of membranes under physiological conditions.(Yepuri et al., 2016) See Fig. 3 for the structures of deuterated monoolein and derivates already supplied and published.

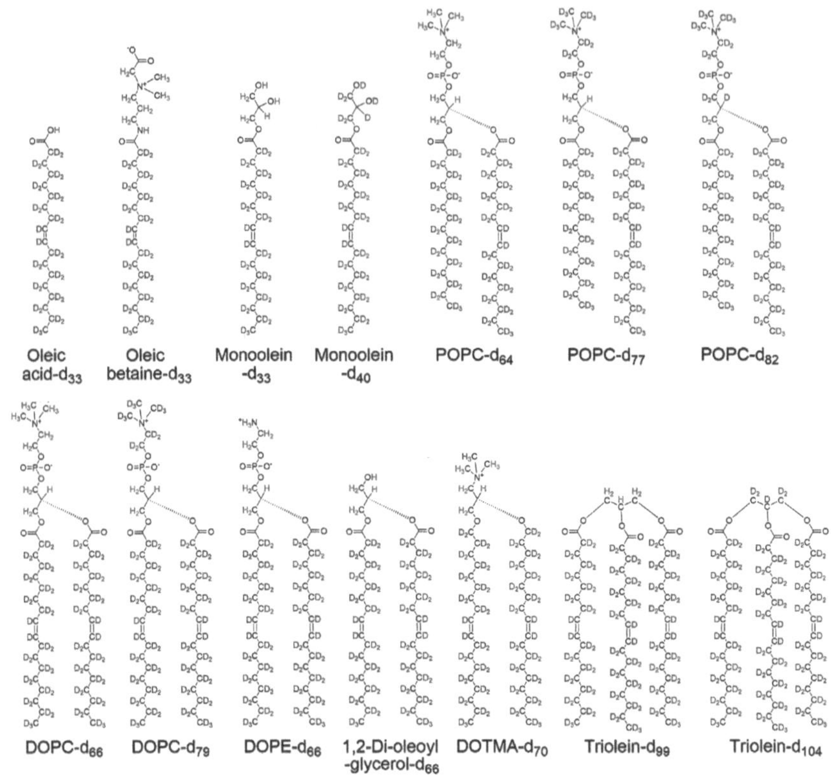

Fig. 3 Deuterated oleic acids and derivates. All of the molecules shown have been requested and supplied by the NDF, for increasingly sophisticated model membranes for neutron scattering. For published experiments using these, see Table 4: oleic betaine, Row 18; monoolein, Rows 19–21; POPC, Rows 22–26; DOPC Row 27.

Monoolein is a major component of lipid cubic phase systems. For several SANS studies of lipid cubic phases, deuterated monoolein has been essential for contrast matching the lipid cubic phase to high D2O. Previous works have used contrast matched detergents, but detergent micelles are believed to not sufficiently mimic the natural lipid bilayer environment (Conn et al., 2021).

4.8 Questions of deuteration on the physical characteristics of the molecules

A concern about deuteration is whether the isotopic labeling changes important properties. It has been commonly noted that lipid deuteration is associated with differences in melting transition temperatures, for example, deuterated versus unlabeled molecules: DOPC (Bryant et al., 2019);

phytanoyl monoethanolamide (van 't Hag et al., 2016), monoolein (van 't Hag et al., 2019). However, for phytanoyl monoethanolamide and monoolein, despite a change in melting temperature, deuteration did not alter SANS-observable features, or apparent functionality of the system. Similarly, (Cleveland IV et al., 2021), concluded that deuterated materials did not change the system functionality.

Deuteration of proteins is considered to have negligible chemical or physical effects on proteins. Neutron diffraction, inelastic neutron scattering and Raman studies of the effects of deuteration have gone down to individual amino acids, to reveal no measurable effects at physiological temperatures (de Souza, Freire, Bordallo, & Argyriou, 2007).

Apart from lipid melting temperature, all effects of deuteration appear to be associated with hydrogen bonding. (Waldie et al., 2020) observed that deuteration of the backbone of the lipid headgroup had an effect on lipid exchange into the membrane. They ascribed this to vibration and hydrogen bonding effects, and did not observe an affect correlating with lipid tail deuteration, and note that consideration should be given to these effects in comparisons for deuterated and unlabeled systems. Changing the solvent from H_2O to D_2O is known to affect solubility. Lysozyme is less soluble in D_2O (Broutin, Riès-Kautt, & Ducruix, 1995) and an effect is observed with the dissolution of crystalline hen egg lysozyme (Sigma-Aldrich) in H_2O vs D_2O. In H_2O, lysozyme dissolves readily. In D_2O, when the lysozyme is added to the surface, it becomes a sticky gel. The sometimes-encountered decreased solubility in D_2O can be alleviated by an increase in dissolved salts. Further study of the validity of results in D_2O vs H_2O are desirable.

5. Examples of uses of deuteration for neutron scattering

The table below contains a selection of neutron scattering publications, where deuterated materials were obtained from the NDF or another deuteration facility. These examples may provide useful insight into the choice of molecule for deuteration, level of deuteration, and experimental design of the neutron scattering experiment. Most examples are for the use of SANS. Some uses are for the highly related use in neutron reflectometry (NR), either in combination or alone. In the structural studies of membrane systems, the methods of SANS and NR are frequently combined. One case, Row 11, involved Raman scattering and is included as its use of deuteration is analogous.

6. Deuteration facilities

Deuteration facilities are increasing in number and are developing. Although deuteration for neutron scattering is many decades on, the first record of a proposal for funding of a deuteration user facility was "A laboratory to support the deuteration of biological macromolecules" by Forsyth, Myles, Timmins, and Haertlein in 2001 (Haertlein et al., 2016). In 2017, (Fisher et al., 2017) listed five and in 2022, we are aware of 14 deuteration facilities and laboratories worldwide, listed in the Table 5.

Table 5 Deuteration facilities, 2022.

Facility/laboratory	Where
National Deuteration Facility (NDF), Australian Nuclear Science and Technology Organisation (ANSTO)	Lucas Heights, Australia
Deuteration and Macromolecular Crystallisation (DEMAX) Platform, European Spallation Source (ESS)	Lund, Sweden
Lund Protein Production Platform, LP3 (Lund University)	Lund, Sweden
ISIS Deuteration Facility, ISIS Neutron and Muon Facility, STFC—UKRI	Didcot, United Kingdom
Deuteration Laboratory (D-Lab), Institut Laue-Langevin (ILL)	Grenoble, France
Lipid Deuteration Platform (L-Lab), Institut Laue-Langevin (ILL)	Grenoble, France
Jülich Centre for Neutron Science (JCNS) Deuteration Service	Jülich, Germany
The Biomolecular Labeling Laboratory (BL^2), joint facility of the National Institute of Standards and Technology (NIST) and the University of Maryland (UMD)	Rockville, Maryland, USA
Bio-Deuteration Laboratory (BDL) Center for Structural Molecular Biology (CSMB), Oak Ridge National Laboratory (ORNL)	Oak Ridge, Tennessee, USA

Table 5 Deuteration facilities, 2022.—cont'd

Facility/laboratory	Where
Deuteration Laboratory, Materials and Life Science Facility (MLF), Japan Proton Accelerator Research Complex (J-PARC)	Tokai, Ibaraki, Japan
Deut-Switch, MEXT-JSPS, Kyoto University	Kyoto, Japan
KumatorI Deuteration Station (KIDS), Kyoto University	Kumatori, Osaka, Japan
Gifu Pharmaceutical University	Gifu, Japan
Songshan Lake Deuteration Facility (SLDF), Songshan Lake Materials Laboratory (SLAB)	Guangdong, China

7. Deuteration facility use: Costs, feasibility, and quality of the experiment

Value in the use of a deuteration facility, over doing your own deuteration, lies in efficiencies with specialist equipment, existing stocks of precursors, experience with deuteration and subsequent characterisation, and associated experience with the neutron scattering applications. In most instances molecular deuteration, especially protein deuteration, can be done one-self for the first time in your own laboratory, but in practice, both the costs and control of deuteration level is difficult for the inexperienced (Haertlein et al., 2016; Okuda et al., 2021). Some equipment, such as Parr reactors, and bioreactors, are expensive, non-trivial to setup, and require training and experience for their use. Prior experience in making a wide range of materials for varied applications means that deuteration staff are valuable collaborators in experimental design.

The custom synthesis of deuterated molecules can be laborious and costly to achieve optimum chemical as well as isotopic purity. Users of deuterated molecules may have different requirements depending on the type of experiments they are performing. For example, MS and NMR based research requires certain isotopic enrichment that may not be needed for neutron scattering experiments.

Deuteration laboratories can provide cost-effective, bespoke services to the neutron user community, due to their experience and familiarity with the particular requirements of their associated neutron facility. For example, isotopic enrichment of deuterated fatty acids is performed using hydrothermal reactions in D_2O, which requires 3 exchange cycles to reach >99% deuterated isotopic purity. Reducing the exchange cycles to 2 can reduce the cost of this material by 50% while reducing the isotopic enrichment by only 2–3%, to give 96–97% deuterated fatty acids. This 2–3% difference in deuteration level corresponds to a similarly small difference in nSLD, which is not near a critical value in most neutron experiments.

Custom synthesis of deuterated molecules tailored to the neutron users' need can produce more useful deuterated products. A better choice in deuteration decisions, made in consultation between the user-scientist, the deuteration scientist, and the neutron scattering scientist, can enable cost-effective experiments that are otherwise hindered by the elevated costs of the deuterated molecules and their limited supply and availability.

See Fig. 4 for an illustration of cost versus deuteration level for a typical precursor, palmitic acid.

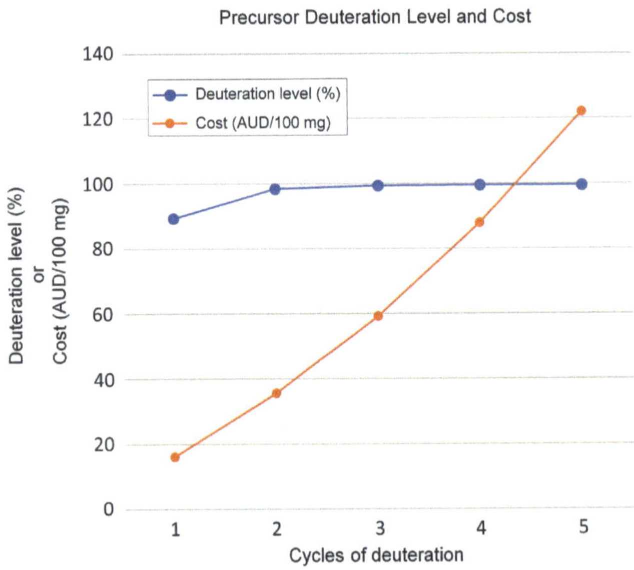

Fig. 4 Costing of the production of a typical fatty acid. The material is palmitic acid, but it is typical of any saturated lipid. Cost per gram, considering staff time costs, equipment costs, and product yields per cycle, results in an increasing cost per gram per cycle, for a rapidly diminishing increase in deuteration level.

8. The naming of Isotopologues: A problem for structural biologists, industry, and databases

Structural biologists and synthetic chemists often use different terminology to describe and trade in molecules. Indeed, chemists themselves are seldom aware of the complexities involved in describing and unambiguously identifying compounds that have been specifically or generally isotopically-substituted (herein referred to as isotopologues).

The principal system by which compounds are unambiguously identified has changed over time. The unregulated naming conventions of early chemistry were recognised as problematic in the mid nineteenth century, resulting in the eventual formation of IUPAC in 1919 and the subsequent standardisation of chemical naming conventions (Fennell, 1994; Verkade & Alderweireldt, 1985). As molecular complexity increased and the scope of scientific endeavour expanded, numerical systems to uniquely identify molecules began to proliferate. The Chemical Abstracts Service Registry (CAS) numbers, first promulgated in the 1960s (Engel & Zass, 2007), have become the standard handle by which a compound contained within a publication or product list is identified, notwithstanding the existence of many unique identifiers such as SMILES, Beilstein Registry Numbers or InChI systems.

The evolution of naming and numbering conventions reflects not only theoretical demands but also practical needs of scientists and those in allied professions. An organic chemist might seek to obtain a compound commercially by searching for it by CAS number, although they name it according to IUPAC conventions in published experimental records. Furthermore, in using one of the numerous structural searches available to them, such as CAS's SciFinder platform, they may be unaware of which unique identifiers the software associated with their drawing to generate a set of results pertinent to the search.

A patent attorney deals with Markush structures (where functionally equivalent substructures are considered together) as a result not of an IUPAC ruling, but rather because of legal precedent (Simmons, 2003). Queries of databases using repeated manual searches and the resultant laborious winnowing of large numbers of results remain a feature of the professional activities of a patent office, despite the proliferation of unique identifiers (Barnard & Wright, 2009).

The established need to unambiguously identify chemical compounds extends to isotopologues – compounds which differ only in their isotopic composition. Nevertheless, as for the patent attorney, the community of scientists who work principally with deuterated compounds do not necessarily employ naming or numbering conventions rigorously, nor are they necessarily well-served by them either in searching the literature, seeking to purchase or sell deuterated compounds or in recording their activities for publication.

8.1 The identification of isotopologues

For the most part, isotopologues are of peripheral interest in chemistry. They are crucial aspects of analysis (e.g., mass spectrometry and NMR), but the isotopic composition of reagents is seldom considered in terms of chemical reactivity, biological activity or the "unique identity" of a compound, except for the well-known and exceptional kinetic isotope effect displayed by hydrogen.

Isotopologues are a feature of molecular deuteration for structural biologists, for their use in Small-Angle Neutron Scattering (SANS) and Neutron Reflectometry (NR) (Mahieu et al., 2020). These experiments require a range of deuterated solvents and biomolecules (Hoogerheide, Forsyth, & Brown, 2020). When a structural biologist wishes to obtain a biomolecule at a particular level of deuteration for contrast matching, they will encounter the limitations of the systems by which chemists describe isotopologues and mixtures thereof.

Taking a common element which has two stable isotopes occurring on earth in an approximately 3:1 ratio, chlorine, may be used to illustrate the issue of isotopologues. A researcher wishing to synthesise a compound of ^{35}Cl, rather than the naturally-occurring mixture of ^{35}Cl and ^{37}Cl, will first determine what ^{35}Cl compounds are commercially-available—assuming they are unwilling or unable to separate or synthesise ^{35}Cl. In this case only one is: $Na^{35}Cl$. This, in itself, is a non-trivial determination to make. It is aided, however, by the fact that $Na^{35}Cl$ has been assigned a different CAS number than native NaCl and thus searches for this compound are relatively straightforward. Suppose, however, that the compound required was $C^{35}Cl_4$, but that the level of isotopic purity required was 70–90%. Thus, $C^{35}Cl_3^{37}Cl$ in the sample would be permissible, with presumably smaller amounts of $C^{35}Cl_n^{37}Cl_{4-n}$ ($n = 0,1,2$). Fig. 5 shows the five isotopologues of CCl_4 where only chlorine isomers are considered. CAS numbers can

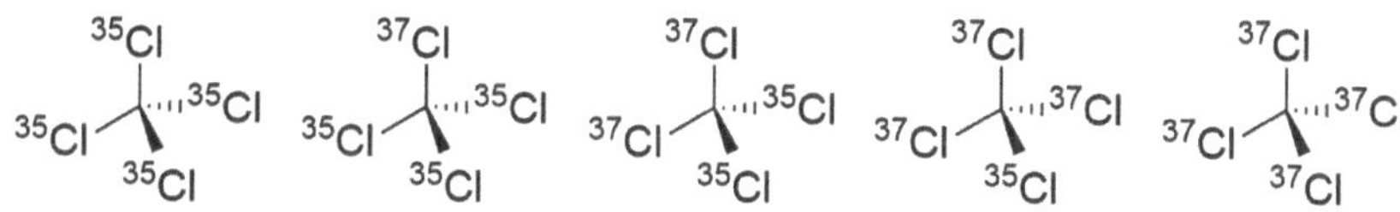

26456-42-8 26456-45-1 37654-12-9 26456-43-9 2742-18-9

Fig. 5 Isotopologues of carbon tetrachloride and their CAS numbers. All of these isotopologues form part of the set of compounds classified under the CAS number 56-23-5 which represents carbon tetrachloride with the native distribution of naturally-occurring isotopes.

be assigned to variably isotopically labeled compounds without specifying the level of labeling. Thus, a search could be performed using the hypothetical CAS number for variably ^{35}Cl-labeled CCl_4. Unfortunately, such a CAS number has not been assigned. Since the degree of labeling is not captured by a CAS number (nor by an InChI or SMILES) the result will be a tendency to characterise compounds subjectively. A researcher using a sample enriched in ^{35}Cl might choose to characterise it using the CAS number for $C^{35}Cl_4$ but noting only 70% isotopic purity (even if, as is likely, the most common isotopologue in the mixture would be $C^{35}Cl_3^{37}Cl_4$), or they might choose to characterise it as mixture of isotopologues, or even an impure sample of $C^{35}Cl_3^{37}Cl_4$. It might be important to this researcher that no $C^{37}Cl_4$ is present and therefore not all variably-labeled mixtures of isotopologues that a hypothetical variably-labeled CAS number for CCl_4 represented would be suitable for their purposes.

This would frustrate anybody seeking information about a specific mix of isotopologues of CCl_4, although this is unlikely to be a large pool of researchers. Reference may have been made to a specific mix of isotopologues of CCl_4, but in terms of CAS numbers each isotopologue has been registered separately and in practice the task of seeking references for a mixture of isotopologues of CCl_4 would be characterised by laborious manual searching of a result pool from a database.

While the issues evident for isotopologues of chlorine are difficult, they represent a simplification of those that occur around deuterated compounds, in which multiple parameters of variability are possible and indeed desirable for contemporary applications in drug discovery, neutron experiments and the study of functional materials.

Fundamentally, these issues arise from the fact that existing unique identifiers do not differentiate the degree of isotopic substitution in variably labeled compounds.

8.2 A biosynthetically-derived material: Deuterated cholesterol

The simplified example of CCl_4 given above did not account for variable levels of site-specific isotopic substitution. The introduction of this factor gives rise to an arbitrarily large number of isotopologues for complex molecules. Unlike the example with the chlorine atoms of CCl_4, it would be impractical to name and number every isotopologue of variably deuterated cholesterol (just cholesterol-d_2 has more than 2000 positional isomers – see Fig. 6, leftmost structure, for one of these isomers).

Demand exists for tuneable deuteration levels to match different neutron scattering length densities. Cholesterol is such a molecule, with relatively uniform, non-site specific labeling at particular levels of deuteration required for a range of neutron experiments (Moulin et al., 2018; Waldie et al., 2019).

Fortunately, IUPAC's naming system captures variable deuteration (Biochemistry, 1979; Nomenclature of Organic Chemistry. Section H: Isotopically Modified Compounds, 1978). The uniform, 'U' and general, 'G' labels may be applied to indicate, respectively, a uniform degree of isotopic substitution and a general isotopic substitution (where labeling at each position varies) (Nomenclature of Organic Chemistry. Section H: Isotopically Modified Compounds, 1978). This system is currently used primarily in the identification of labeled sugars by vendors, and only rarely for other compounds.

Thus, microbial biosynthesis of deuterated cholesterol in heavy water yields [G-$^2H_{46}$]cholesterol. This systematically-named compound corresponds exactly with the CAS Registry Number 350671–55-5 with 100% deuteration (see Fig. 6, middle structure), but since this level of deuteration is neither easily achieved nor is it desired in practice, a CAS number for variably-deuterated cholesterol would be a more useful and accurate

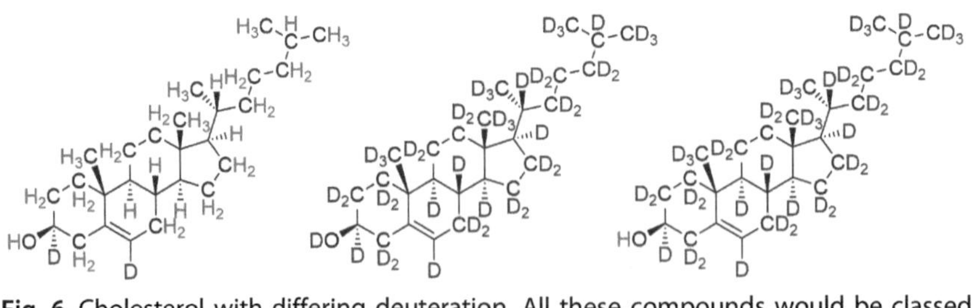

Fig. 6 Cholesterol with differing deuteration. All these compounds would be classed under a CAS number for variably deuterated cholesterol, but only the middle structure has a CAS number at the time of writing.

numerical unique identifier for highly deuterated cholesterol produced biosynthetically than 350671-55-5. The name of this variably deuterated cholesterol in the CAS registry would be "cholesterol, labeled with deuterium."

It may seem counterintuitive for a unique identifier such as a CAS number to represent a mixture of compounds but it is a known expediency—for instance the enzyme family "Alcohol dehydrogenase" share a single CAS number (9031-72-5) (ACS, 2022).

Furthermore, this perdeuterated cholesterol bears ^2H not only on carbon centres but also in the exchangeable position, the alcohol. Arguably, this perdeuterated structure is less useful to vendors and researchers. Indeed, in our own facility the isolation and purification of biosynthetically derived deuterated cholesterol produces the 3-OH rather than the 3-OD structure; no researcher has yet expressed interest in the 3-OD variant. This presents for us the naming issue that we produce a molecule with a site-specific proton, which is not strictly [G-$^2H_{46}$]cholesterol but [G-1,1,2,2,3,4,4,6,7,7,8,9,11,11,12,12,14,15,15,16,16,17,18,18,18,19,19,19,20,21,21,21,22,22,23,23,24,24,25,26,26,26,27,27,27-$^2H_{45}$]cholesterol, a compound with a distinct identity which nevertheless does not currently have a CAS number assigned to it (see Fig. 6, right structure).

A vendor might choose to identify this compound as perdeuterated (including the exchangeable hydrogen) cholesterol at a particular percentage isotopic purity and use the CAS Registry Number 350671-55-5—since most labeled compounds are offered at a specific isotopic purity (including, of course, native compounds). A researcher using this compound might use the same naming convention in a published work. The publication will be indexed as having included compound 350671-55-5. If, later, another researcher treats the 3-OH and 3-OD structures as separate (by means of an experiment or simply in their description) a new CAS number could be filed for the 3-OH but C-perdeuterated variant (ideally with the long 'G' name given above) and all existing records of the use of this product would, in theory, need reindexing, with attention given to whether previous use of 350,671–55-5 was really the 3-OD variant or was non-specific with respect to the isotope distribution at oxygen and therefore was not appropriate to describe as 350671-55-5.

One advantage of the CAS registry as it stands is that it permits users of SciFinder to see only isotopologues that have been indexed according to their inclusion in a published work. Thus, the most useful and interesting isotopologues often have CAS numbers and a reader may be quickly directed toward a publication or vendor of interest to them. There is nevertheless scope for improved accuracy and convergence between IUPAC naming conventions and numerical unique identifiers. Recommendation: avoid the unique identifier corresponding to an isotopologue with deuterium at an exchangeable position unless this structure is actually the subject of study.

8.3 The naming of variously deuterated oleic acid

Deuterated oleic acid is a material frequently sought by structural biologists, as are other deuterated fatty acids. These molecules are highly deuterated overall but are protiated in site-specific positions, both exchangeable positions and one vinylic position in the case of synthetic oleic acid. These site-specific isotopic substitutions in the context of a material that is deuterated overall are an issue in terms of the identity of the compound.

An example which illustrates the problem in associating a deuterated structure with a particular compound reported in the literature or available commercially among the fatty acids is shown in Fig. 7. A CAS number has been registered for octadecanedioic-d_{32} acid-d_2 (76020-14-9—Fig. 7, top left structure) but not octadecanedioic-d_{32} acid (where only the chain is deuterated—Fig. 7, top right structure), whereas the reverse is true for heptadecanedioic-d_{30} acid (130348-91-3—Fig. 7, bottom right structure)—the chain- deuterated acid; heptadecanedioic-d_{30} acid-d_2 (Fig. 7, bottom left structure) is unregistered as of the date of writing.

These saturated fatty acids are relatively simple to name using IUPAC conventions. For instance octadecanoic-d_{32} acid-d_2 would be [U—$^2H_{34}$] octadecanoic acid at 100% deuteration but [G-$^2H_{34}$]octadecanoic acid if there was detectable non-uniformity in the deuteration level. Octadecanoic-d_{32} acid (chain-deuterated octadecanoic acid) would be [G-2,2,3,3,4,4,5,5,6,6,7,7,8,8,9,9,10,10,11,11,12,12,13,13,14,14,15,15,16,16,17,17,18,18,18-$^2H_{32}$]. Both of these compounds could hypothetically be recorded under a CAS number for variably-deuterated octadecanoic acid, but only the latter would correspond to the existing CAS number 17660–51-4.

Since all biosynthetically-derived deuterated compounds are (1) generally but non-uniformly deuterated, with the exception of solvent-dependent exchangeable hydrogen atoms and (2) usually considered in terms of overall deuteration level in neutron experiments, the 'G' notation is particularly useful.

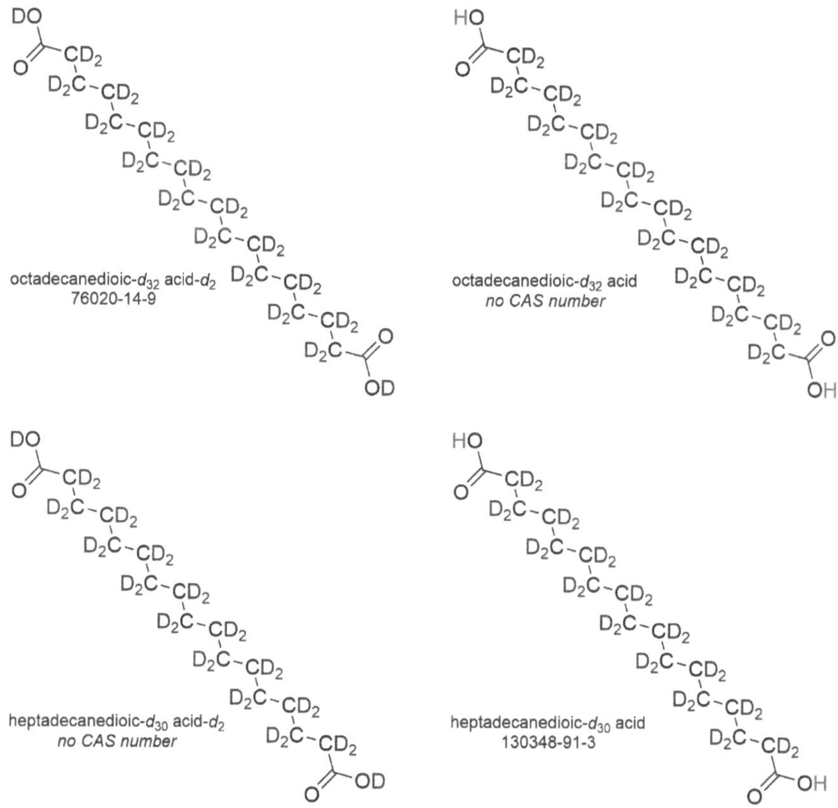

Fig. 7 Variably deuterated alkanoic acids. The CAS registry often omits isotopologues where exchangeable hydrogen differs.

We have encountered these issues with a fatty acid synthesised at the National Deuteration Facility: oleic-d_{32} acid. This fatty acid is synthesised with a single, residual vinylic proton at C-9. This residual proton assists users of this compound in NMR analysis of reaction products and increases the atom-efficiency of the synthesis while reducing the raw material cost. This oleic acid remains suitable for neutron applications. This substance has been assigned a CAS number (1466552-38-4—Fig. 8, left structure) and has the IUPAC name (under the 'G' convention for variably-labeled materials) (*Z*)-9-[G-2,2,3,3,4,4,5,5,6,6,7,7,8,8,10,11,11,12,12,13,13,14,14,15,15,16,16,17,17,18,18,18-$^2H_{32}$]octadecenoic acid.

Neutron users neglect to describe this material with reference to the single, site-specific proton and instead consider this material as perdeuterated oleic acid (Fig. 8, right structure). This is because a subjective evaluation

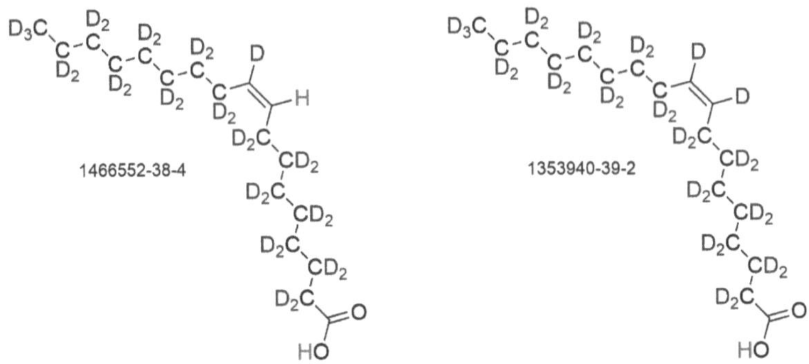

Fig. 8 Oleic acid, proton at C9. The same compound produced by the National Deuteration Facility has been described as both 1466552-38-4 and 1353940-39-2 in the literature.

of this molecule as highly-deuterated oleic acid is used for calculations pertinent to neutron experiments.

The two structures in Fig. 8 would together form part of the set of isotopologues that make up a hypothetical designation for variably-deuterated oleic acid, while the left structure would be implied as part of a set of structures represented by the CAS number 1353940-39-2 (Fig. 8, right structure) when this CAS number was taken to represent the set of isotopologues present in highly-deuterated oleic acid (>95%-*d*).

Recommendation: vendors should register the CAS number for variably-deuterated molecules where it does not yet exist and use it. The name "*Oleic acid, labeled with deuterium*" allows for:

1. Customised changes in deuteration levels by the manufacturer for contrast matching purposes.
2. Avoiding very long names. For instance, the left structure in Fig. **8** that was named over two lines by (CAS: 1466552–38-4) becomes [G-^{2}H] oleic acid when described as "*Oleic acid, labeled with deuterium.*"

In referring to "highly-deuterated" biomolecules it is necessary also to note that the tendency to characterise commercial samples as, e.g., ≥95%-*d* or > 95%-*d* is a practice typically derived from considering the variability between production batches and gives no information about the distribution of isotopologues or the true distribution of masses within a sample (and therefore levels of various isotopologues). Recommendation: the use of names in the format of "*Oleic acid, labeled with deuterium*" should be accompanied by vendors supplying data indicating the measured deuteration level.

Use of the "G" convention for variably-labeled materials is advantageous in this case as even with site-specific protiation the overall deuteration level is high enough that there is no difference in the behaviour of the two compounds depicted in Fig. 8 in neutron experiments. Thus, the "-*d*" notation could be counterproductive where used to advertise compounds suitable for neutron applications since the left structure is oleic-d_{32} acid and the right oleic-d_{33} acid.

This material has indeed been identified as both of the structures in Fig. 8 in publications and used as a reagent to synthesise other deuterated compounds, e.g., van 't Hag et al. (2019). If a fully chain-deuterated oleic acid (Fig. 8, right structure) was used to recreate the syntheses a new set of characteristic data would be obtained. Thus, two conflicting data sets (outside of the usual variability that is accepted for measurement of physical and spectroscopic data) would exist for two molecules that are considered separate compounds according to IUPAC rules but have been described as being identical. An example would be a molecule like PIP_2 or monoolein, which would have substantially different NMR spectra depending on whether 1466552-38-4 or 1353940-39-2 was the reactant used in their synthesis, even when both reactants had the same overall level of deuteration.

8.4 Important lipids that are understood to be mixtures

Phosphatidylinositol (PIP_2) is an example of an important membrane lipid that is often referred to in the literature without reference to numerical unique identifiers, due to its conventionally being a mixture. PIP_2 occurs naturally with a mixture of acyl substitutions at the *sn*-1 and *sn*-2 positions, but predominantly (in mammals) 1-stearoyl-2-arachidonoyl (Tanaka et al., 2003). Thus, "PIP_2" may conventionally refer to either a naturally-occurring mixture or a single compound which dominates that mixture in mammals. This latter substance, 1-stearoyl-2-arachidonoyl-*sn*-glycero-3-phosphoinositol-4,5-bisphosphate, has been assigned the CAS number 245126-95-8.

It would be inaccurate, however, to use this number to index naturally-derived PIP_2 with its mixture of fatty acid chains, since it is a CAS number specific to a single substance. Instead, like alcohol dehydrogenases, a single CAS number assigned to PIP_2 with the understanding of variability in fatty acid content would be a helpful addition to the CAS registry, to complement the array of CAS numbers associated with specific acyl substitutions. Assignment of CAS numbers to specific mixtures of different, specifically acyl substituted PIP_2 molecules each with their own CAS numbers would enhance the capability of the database to direct users to useful information.

PIP_2, being a large and innately variable material, is an example of the type of substance that exists at the edge of the ability to capture the unique identity of a substance through CAS numbers and related unique identifiers.

A strategy used to reduce the ambiguity when dealing with naming these compounds is by using conventions for shorthand notation of fatty acids to identify the acyl substituents. For instance, PIP_2(16:0/18:1(9Z)) has a 1-steroyl-2-oleyl pattern. The compound which dominates mammalian PIP_2, 1-stearoyl-2-arachidonoyl-*sn*-glycero-3-phosphoinositol-4,5-bisphosphate, is analogously PIP_2(16:0/20:4(5Z,8Z,11Z,14Z)). It is shown in Fig. 9 as a tail-perdeuterated isotopologue.

Where PIP_2 of variable composition (mixed acyl substituents or deuteration levels) may have been recorded using a CAS number assigned to a more specific structure there is scope for future confusion. As soon as a researcher or vendor had cause to distinguish materials previously indexed together it would be necessary to reclassify the existing molecules that were previously entered into the CAS Registry. The nature of scientific records, being extensively referenced, suggests considerable work could be required to correct defects in the identification of isotopologues or mixtures of variably acyl-substituted PIP_2.

Looking at Fig. 9 it is possible to envisage, also, a set of variably-labeled registry entries that would allow researchers to seek information about, or to obtain, deuterated PIP_2. Allowing variable deuteration at one, two or three of the labeled subunits (inositol, glycerol and the fatty acid chains) gives a total of seven 'variably-labeled' structures (the fatty acid chains themselves

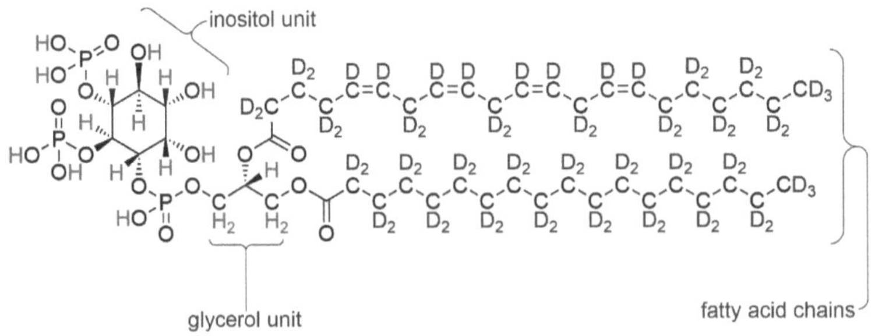

Fig. 9 PIP_2. The structure of tail-deuterated 1-stearoyl-2-arachidonoyl-sn-glycero-3-phosphoinositol-4,5-bisphosphate.

must also be allowed to vary). Researchers could therefore search for labeled PIP_2 where the glycerol and fatty acid but not the inositol substructure were deuterated without manually sifting through a large number of database entries. For this to occur, the structural biology community and the chemists who support them would need to agree on the relevant, variably-deuterated structures to register with unique identifiers and use them according to this convention.

Without the use of general identifiers the issue of uniquely identifying a sample of PIP_2 will generate enormous complexity. Each sample of PIP_2 will likely be composed of various compounds with variable acyl substitutions and variable proportions in the mixture. If some parts of some of these compounds are deuterated the number of database entries required to capture the known and measurable level of complexity becomes impractically large. Attempts to use CAS numbers for each possible isotopologue in the mixture would inhibit rather than facilitate the work of researchers by forcing them to delineate excessive numbers of components.

It is notable also that the degree of ionisation of exchangeable protons of the phosphate moieties in PIP_2, i.e., the headgroup charge, will change the identity of the compound if indexing occurs as the phosphate salts rather than protonated phosphoric acid groups. 1-stearoyl-2-arachidonoyl-*sn*-glycero-3-phosphoinositol-4,5-bisphosphate is already offered for sale both as the ammonium salt and the fully-protonated acid.

A further variable to consider is stereochemistry. The biosynthetically-derived product will generally have variable lipids, but likely near-pure stereochemistry. A synthetic molecule is probably lipid-chain-pure but may not be diastereopure or enantiopure. As this variable is usually notable in crystallographic studies but not otherwise measured it may be prudent to register a CAS number for PIP_2 with variable acyl substitutions and undefined stereochemistry in case future researchers have cause to distinguish PIP_2 molecules on a stereochemical basis.

The scope for variability in natural and synthetic PIP_2 is very high. The scope for confusion in the indexing of the mixtures or pure substances that make up PIP_2 is equally large. Where possible, efforts to use useful naming conventions and variable unique identifiers to reduce this complexity should be standardised. Recommendation: structural biologists and the chemists that support them should cooperate to establish a convention for naming PIP_2 and assigning unique identifiers to molecules being traded or described in publication.

9. Proposals for the registration of deuterated compounds

IUPAC nomenclature allows significant latitude in the way organic compounds are described. This has been characterised by De Gryter: "*a long history of competing demands for flexibility and rigor in organic nomenclature, a history whose product is IUPAC nomenclature itself.*" The implication is that nomenclature is adaptable to the needs of researchers and vendors and is not simply a formalism with whose rules familiarity is increasingly irrelevant as the task of naming compounds is automated.

Similarly, numerical unique identifiers permit exact descriptions of specific isotopologues and mixtures thereof, while simultaneously permitting the assignment of identifiers to mixtures with variance in levels of isotopic substitution. Thus, CAS numbers and similar numerical unique identifiers can work effectively with IUPAC names to characterise deuterated single substances and mixtures when both are used correctly. Vendors and researchers have roles to play in ensuring:

1. Familiarity with the rules governing the entry of substances into, e.g., the CAS Registry
2. Systematic use of naming conventions by chemists, including the "G" and "U" notation for labeled compounds which correspond to these database entries
3. Registering variably-deuterated materials as such where appropriate, rather than choosing an existing registration which implies incorrect information—such as the identity of exchangeable protons or site-specific isotopic substitution versus general substitution.
4. The establishment of conventions among chemists and structural biologists to identify the needs of the latter in order that the former can provide the substances they require, with specifications pertinent to their work.

We hope that increased awareness of these nomenclature considerations will lead to consistent use among structural biologists and deuteration chemists, and will safeguard the future reliability of the CAS database.

10. Concluding remarks

Experience in a deuteration laboratory has come with a few interesting surprises. Many of them involve simple chemistry. It may be at first surprising how straightforward it is to use high temperature and pressure to exchange covalently bonded hydrogen atoms, and then for following steps the surprise swings the other way, with the pursuit of high deuteration levels and particular chemistries requiring unexpected stringency and complexity. For example, unexpected hydrophilicity, such as the first time introducing detergents to catalysts, can result in unexpected aqueous volumes that foam.

In principle, any chemistry can be performed, but the time, equipment and materials costs can explode, and sometimes due to innocuous details of the design that may not even be important for the intended experiment. The cost of deuterated synthesis may be disproportionate to the cost of the unlabeled molecule.

New users' requested deuteration levels often will benefit from discussions between all of the user scientist, the deuteration scientist, and the neutron scattering scientist. The simple default request or "100% deuterated," or "maximum deuteration," may be prohibitive if not impossible, and may not even be better than a lower level of deuteration. In biological deuteration, an ideal level of deuteration, e.g., 100% D_2O match-out, is often achieved using D_2O with no other deuterated material. It has also been learned that the "required deuteration level" is rarely critical; for many experiments, as long as the deuteration level of the supplied material is accurately specified, it will work.

A pleasant surprise is that through direct discussion between chemists and neutron scattering scientists, a very acceptable compromise on deuteration level, and quantity, could make for a high increase in feasibly (cost, and time of delivery), and that these discussions could even lead to a better experimental design for the neutron scattering experiment. For this reason, we strongly advocate early discussion on deuteration needs.

References

ACS. (2022). Alcohol dehydrogenase. *CAS Common Chemistry*.

Akutsu, K., Cagnes, M., Niizeki, T., Hasegawa, Y., & Darwish, T. A. (2018). Penetration behavior of an ionic liquid in thin-layer silica coating: Ionic liquid deuteration and neutron reflectivity analysis. *Physica B: Condensed Matter*, *551*, 262–265. https://doi.org/10.1016/j.physb.2018.01.063.

Akutsu-Suyama, K., Cagnes, M., Tamura, K., Kanaya, T., & Darwish, T. A. (2019). Controlled deuterium labelling of imidazolium ionic liquids to probe the fine structure of the electrical double layer using neutron reflectometry. *Physical Chemistry Chemical Physics*, *21*(32), 17512–17516. https://doi.org/10.1039/C9CP02479D.

Alfonso García, A., Pfisterer, S., Riezman, H., Ikonen, E., & Potma, E. (2015). D38-cholesterol as a Raman active probe for imaging intracellular cholesterol storage. *Journal of Biomedical, Optics(6)*, 061003.

Ashkar, R., Bilheux, H. Z., Bordallo, H., Briber, R. M., Callaway, D. J. E., Cheng, X., et al. (2018). Neutron scattering in the biological sciences: progress and prospects. *Acta Crystallographica Section D: Structural Biology*, *74*(Pt 12), 1129–1168.

Bagatolli, L. A., Ipsen, J. H., Simonsen, A. C., & Mouritsen, O. G. (2010). An outlook on organization of lipids in membranes: Searching for a realistic connection with the organization of biological membranes. *Progress in Lipid Research*, *49*(4), 378–389. https://doi.org/10.1016/j.plipres.2010.05.001.

Barnard, J. M., & Wright, P. M. (2009). Towards in-house searching of Markush structures from patents. *World Patent Information*, *31*(2), 97–103.

Biochemistry, I. U. O. (1979). Enzyme Nomenclature. *European Journal of Biochemistry*, *102*(2), 315–316. https://doi.org/10.1111/j.1432-1033.1979.tb04245.x.

Bremer, A., Kent, B., Hauß, T., Thalhammer, A., Yepuri, N. R., Darwish, T. A., et al. (2017). Intrinsically disordered stress protein COR15A resides at the membrane surface during dehydration. *Biophysical Journal*, *113*(3), 572–579. https://doi.org/10.1016/j.bpj.2017.06.027.

Broutin, I., Riès-Kautt, M., & Ducruix, A. (1995). Lysozyme solubility in H2O and D2O solutions as a function of sodium chloride concentration. *Journal of Applied Crystallography*, *28*(5), 614–617. https://doi.org/10.1107/S0021889895005425.

Bryant, G., Taylor, M. B., Darwish, T. A., Krause-Heuer, A. M., Kent, B., & Garvey, C. J. (2019). Effect of deuteration on the phase behaviour and structure of lamellar phases of phosphatidylcholines – Deuterated lipids as proxies for the physical properties of native bilayers. *Colloids and Surfaces B: Biointerfaces*, *177*, 196–203. https://doi.org/10.1016/j.colsurfb.2019.01.040.

Cleveland, T., IV, Blick, E., Krueger, S., Leung, A., Darwish, T., & Butler, P. (2021). Direct localization of detergents and bacteriorhodopsin in the lipidic cubic phase by small-angle neutron scattering. *IUCrJ*, *8*(1). https://doi.org/10.1107/S2052252520013974.

Conn, C. E., de Campo, L., Whitten, A. E., Garvey, C. J., Krause-Heuer, A. M., & van 't Hag, L. (2021). Membrane protein structures in lipid bilayers; small-angle neutron scattering with contrast-matched Bicontinuous cubic phases. *Frontiers in Chemistry*, *8*-(1220). https://doi.org/10.3389/fchem.2020.619470.

Darwish, T. A., Luks, E., Moraes, G., Yepuri, N. R., Holden, P. J., & James, M. (2013). Synthesis of deuterated [D32]oleic acid and its phospholipid derivative [D64]dioleoyl-sn-glycero-3-phosphocholine. *Journal of Labelled Compounds and Radiopharmaceuticals*, *56*(9–10), 520–529. https://doi.org/10.1002/jlcr.3088.

de Ghellinck, A., Schaller, H., Laux, V., Haertlein, M., Sferrazza, M., Maréchal, E., et al. (2014). Production and analysis of Perdeuterated lipids from Pichia pastoris cells. *PLoS One*, *9*(4), e92999. https://doi.org/10.1371/journal.pone.0092999.

de Souza, J. M., Freire, P. T., Bordallo, H. N., & Argyriou, D. N. (2007). Structural isotopic effects in the smallest chiral amino acid: Observation of a structural phase transition in fully deuterated alanine. *The Journal of Physical Chemistry. B*, *111*(19), 5034–5039. https://doi.org/10.1021/jp070366z.

Delhom, R., Nelson, A., Laux, V., Haertlein, M., Knecht, W., Fragneto, G., et al. (2020). The antifungal mechanism of amphotericin B elucidated in Ergosterol and cholesterol-containing membranes using neutron reflectometry. *Nanomaterials*, *10*(12), 2439. https://doi.org/10.3390/nano10122439.

Duff, A. P., Wilde, K. L., Rekas, A., Lake, V., & Holden, P. J. (2015). Robust high-yield methodologies for (2)H and (2)H/(15)N/(13)C labeling of proteins for structural investigations using neutron scattering and NMR. *Methods in Enzymology*, *565*, 3–25. https://doi.org/10.1016/bs.mie.2015.06.014.

Dunne, O., Weidenhaupt, M., Callow, P., Martel, A., Moulin, M., Perkins, S. J., et al. (2017). Matchout deuterium labelling of proteins for small-angle neutron scattering studies using prokaryotic and eukaryotic expression systems and high cell-density cultures. *European Biophysics Journal*, *46*(5), 425–432. https://doi.org/10.1007/s00249-016-1186-2.

Engel, T., & Zass, E. (2007). 3.13 - Chemical information systems and databases. In J. B. Taylor, & D. J. Triggle (Eds.), *Comprehensive medicinal chemistry II* (pp. 265–291). Oxford: Elsevier.

Engelman, D. M., Moore, P. B., & Schoenborn, B. P. (1975). Neutron scattering measurements of separation and shape of proteins in 30S ribosomal subunit of Escherichia coli: S2-S5, S5-S8, S3-S7. *Proceedings of the National Academy of Sciences of the United States of America*, *72*(10), 3888–3892. https://doi.org/10.1073/pnas.72.10.3888.

Fennell, R. (1994). *History of IUPAC, 1919-1987*. Oxford, Boston: Blackwell Science Ltd.

Fisher, Z., Jackson, A., Kovalevsky, A., Oksanen, E., & Wacklin, H. (2017). Chapter 1 - biological structures. In F. Fernandez-Alonso, & D. L. Price (Eds.), *Vol. 49. Experimental methods in the physical sciences* (pp. 1–75). Academic Press.

Furlong, E. J., Choudhury, H. G., Kurth, F., Duff, A. P., Whitten, A. E., & Martin, J. L. (2018). Disulfide isomerase activity of the dynamic, trimeric Proteus mirabilis ScsC protein is primed by the tandem immunoglobulin-fold domain of ScsB. *The Journal of Biological Chemistry*, *293*(16), 5793–5805. https://doi.org/10.1074/jbc.RA118.001860.

Golub, M., Gätcke, J., Subramanian, S., Kölsch, A., Darwish, T., Howard, J. K., et al. (2022). "Invisible" detergents enable a reliable determination of solution structures of native photosystems by small-angle neutron scattering. *The Journal of Physical Chemistry B*. https://doi.org/10.1021/acs.jpcb.2c01591.

Grosas, A. B., Rekas, A., Mata, J. P., Thorn, D. C., & Carver, J. A. (2020). The aggregation of αB-Crystallin under crowding conditions is prevented by αA-Crystallin: Implications for α-Crystallin stability and Lens transparency. *Journal of Molecular Biology*, *432*(20), 5593–5613. https://doi.org/10.1016/j.jmb.2020.08.011.

Haertlein, M., Moulin, M., Devos, J. M., Laux, V., Dunne, O. M., & Forsyth, V. T. J. M. (2016). Biomolecular Deuteration for Neutron Structural Biology and Dynamics. *Methods in Enzymology*, *566*, 113–157.

Hiruma-Shimizu, K., Shimizu, H., Thompson, G. S., Kalverda, A. P., & Patching, S. G. (2015). Deuterated detergents for structural and functional studies of membrane proteins: Properties, chemical synthesis and applications. *Molecular Membrane Biology*, *32*(5–8), 139–155. https://doi.org/10.3109/09687688.2015.1125536.

Hoogerheide, D. P., Forsyth, V. T., & Brown, K. A. (2020). Neutron scattering for structural biology. *Physics Today*, *73*(6), 36–42. https://doi.org/10.1063/pt.3.4498.

Ibrahim, Z., Martel, A., Moulin, M., Kim, H. S., Härtlein, M., Franzetti, B., et al. (2017). Time-resolved neutron scattering provides new insight into protein substrate processing by a AAA+ unfoldase. *Scientific Reports*, 7(1), 40948. https://doi.org/10.1038/srep40948.

Johansen, N. T., Bonaccorsi, M., Bengtsen, T., Larsen, A. H., Tidemand, F. G., Pedersen, M. C., et al. (2022). Mg(2+)-dependent conformational equilibria in CorA and an integrated view on transport regulation. *eLife*, *11*, e71887. https://doi.org/10.7554/eLife.71887.

Jünemann, R., Wadzack, J., Triana-Alonso, F., Bittner, J. U., Caillet, J., Meinnel, T., et al. (1996). In vivo Deuteration of transfer RNAs: Overexpression and large-scale purification of deuterated specific tRNAs. *Nucleic Acids Research*, *24*, 907–913. https://doi.org/10.1093/nar/24.5.907.

Kawaguchi, T., Hamanaka, T., Kito, Y., & Machida, H. (1991). Structural studies of a homologous series of alkyl sucrose ester micelle by x-ray scattering. *The Journal of Physical Chemistry*, *95*(9), 3837–3846. https://doi.org/10.1021/j100162a073.

Kulkarni, C. V., Wachter, W., Iglesias-Salto, G., Engelskirchen, S., & Ahualli, S. (2011). Monoolein: a magic lipid? *Physical Chemistry Chemical Physics*, *13*(8), 3004–3021. https://doi.org/10.1039/C0CP01539C.

Le Brun, A. P., Clifton, L. A., Holt, S. A., Holden, P. J., & Lakey, J. H. (2016). Deuterium labeling strategies for creating contrast in structure-function studies of model bacterial outer membranes using neutron reflectometry. *Methods in Enzymology*, *566*, 231–252. https://doi.org/10.1016/bs.mie.2015.05.020.

Luchini, A., Micciulla, S., Corucci, G., Batchu, K. C., Santamaria, A., Laux, V., et al. (2021). Lipid bilayer degradation induced by SARS-CoV-2 spike protein as revealed by neutron reflectometry. *Scientific Reports*, *11*(1), 14867. https://doi.org/10.1038/s41598-021-93996-x.

Mahieu, E., Ibrahim, Z., Moulin, M., Härtlein, M., Franzetti, B., Martel, A., et al. (2020). The power of SANS, combined with deuteration and contrast variation, for structural studies of functional and dynamic biomacromolecular systems in solution. *EPJ Web of Conferences*, *236*, 03002.

Maric, S., Skar-Gislinge, N., Midtgaard, S., Thygesen, M. B., Schiller, J., Frielinghaus, H., et al. (2014). Stealth carriers for low-resolution structure determination of membrane proteins in solution. *Acta Crystallographica. Section D, Biological Crystallography*, *70*(Pt 2), 317–328. https://doi.org/10.1107/s1399004713027466.

Maric, S., Thygesen, M. B., Schiller, J., Marek, M., Moulin, M., Haertlein, M., et al. (2015). Biosynthetic preparation of selectively deuterated phosphatidylcholine in genetically modified Escherichia coli. *Applied Microbiology and Biotechnology*, *99*(1), 241–254. https://doi.org/10.1007/s00253-014-6082-z.

McCoy, T. M., King, J. P., Moore, J. E., Kelleppan, V. T., Sokolova, A. V., de Campo, L., et al. (2019). The effects of small molecule organic additives on the self-assembly and rheology of betaine wormlike micellar fluids. *Journal of Colloid and Interface Science*, *534*, 518–532. https://doi.org/10.1016/j.jcis.2018.09.046.

Midtgaard, S. R., Darwish, T. A., Pedersen, M. C., Huda, P., Larsen, A. H., Jensen, G. V., et al. (2018). Invisible detergents for structure determination of membrane proteins by small-angle neutron scattering. *The FEBS Journal*, *285*(2), 357–371. https://doi.org/10.1111/febs.14345.

Moulin, M., Strohmeier, G. A., Hirz, M., Thompson, K. C., Rennie, A. R., Campbell, R. A., et al. (2018). Perdeuteration of cholesterol for neutron scattering applications using recombinant Pichia pastoris. *Chemistry and Physics of Lipids*, *212*, 80–87. https://doi.org/10.1016/j.chemphyslip.2018.01.006.

Mouritsen, O. G., & Zuckermann, M. J. (2004). What's so special about cholesterol? *Lipids*, *39*(11), 1101–1113. https://doi.org/10.1007/s11745-004-1336-x.

Nickels, J. D., Chatterjee, S., Stanley, C. B., Qian, S., Cheng, X., Myles, D. A. A., et al. (2017). The in vivo structure of biological membranes and evidence for lipid domains. *PLoS Biology*, *15*(5), e2002214. https://doi.org/10.1371/journal.pbio.2002214.

Nomenclature of Organic Chemistry. Section H: Isotopically Modified Compounds. (1978). 86(1), 9–25. doi:https://doi.org/10.1111/j.1432-1033.1978.tb12279.x.

Okuda, A., Inoue, R., Morishima, K., Saio, T., Yunoki, Y., Yagi-Utsumi, M., et al. (2021). Deuteration aiming for neutron scattering. *Biophysics and Physicobiology*, *18*, 16–27. https://doi.org/10.2142/biophysico.bppb-v18.003.

Osborne, H. B., & Chabre, M. (1982). Use of hydrogenated and deuterated detergents in the study of rhodopsin by neutron small-angle scattering. *Methods in Enzymology*, *81*, 742–752. https://doi.org/10.1016/s0076-6879(82)81100-2.

Patel, S., Ashwanikumar, N., Robinson, E., Xia, Y., Mihai, C., Griffith, J. P., et al. (2020). Naturally-occurring cholesterol analogues in lipid nanoparticles induce polymorphic shape and enhance intracellular delivery of mRNA. *Nature Communications*, *11*(1), 983. https://doi.org/10.1038/s41467-020-14527-2.

Santamaria, A., Batchu, K. C., Matsarskaia, O., Prévost, S. F., Russo, D., Natali, F., et al. (2022). Strikingly different roles of SARS-CoV-2 fusion peptides uncovered by neutron scattering. *Journal of the American Chemical Society*, *144*(7), 2968–2979. https://doi.org/10.1021/jacs.1c09856.

Sebastiani, F., Yanez Arteta, M., Lerche, M., Porcar, L., Lang, C., Bragg, R. A., et al. (2021). Apolipoprotein E binding drives structural and compositional rearrangement of mRNA-containing lipid nanoparticles. *ACS Nano*, *15*(4), 6709–6722. https://doi.org/10.1021/acsnano.0c10064.

Shotton, M. W., Pope, L. H., Forsyth, T., Langan, P., Denny, R. C., Giesen, U., et al. (1997). A high-angle neutron fibre diffraction study of the hydration of deuterated A-DNA. *Biophysical Chemistry*, *69*(1), 85–96. https://doi.org/10.1016/s0301-4622(97)00090-2.
Simmons, E. S. (2003). Markush structure searching over the years. *World Patent Information*, *25*(3), 195–202. https://doi.org/10.1016/S0172-2190(03)00073-5.
Sonntag, M., Jagtap, P. K. A., Simon, B., Appavou, M.-S., Geerlof, A., Stehle, R., et al. (2017). Segmental, domain-selective Perdeuteration and small-angle neutron scattering for structural analysis of multi-domain. *Proteins*, *56*(32), 9322–9325. https://doi.org/10.1002/anie.201702904.
Souza, C. M., Schwabe, T. M. E., Pichler, H., Ploier, B., Leitner, E., Guan, X. L., et al. (2011). A stable yeast strain efficiently producing cholesterol instead of ergosterol is functional for tryptophan uptake, but not weak organic acid resistance. *Metabolic Engineering*, *13*(5), 555–569. https://doi.org/10.1016/j.ymben.2011.06.006.
Tanaka, T., Iwawaki, D., Sakamoto, M., Takai, Y., Morishige, J.-I., Murakami, K., et al. (2003). Mechanisms of accumulation of arachidonate in phosphatidylinositol in yellowtail. *European Journal of Biochemistry*, *270*(7), 1466–1473. https://doi.org/10.1046/j.1432-1033.2003.03512.x.
van 't Hag, L., de Campo, L., Garvey, C. J., Feast, G. C., Leung, A. E., Yepuri, N. R., et al. (2016). Using SANS with contrast-matched lipid Bicontinuous cubic phases to determine the location of encapsulated peptides, proteins, and other biomolecules. *The Journal of Physical Chemistry Letters*, 7(14), 2862–2866. https://doi.org/10.1021/acs.jpclett.6b01173.
van 't Hag, L., de Campo, L., Tran, N., Sokolova, A., Trenker, R., Call, M. E., et al. (2019). Protein-eye view of the in Meso crystallization mechanism. *Langmuir*, *35*(25), 8344–8356. https://doi.org/10.1021/acs.langmuir.9b00647.
Verkade, P. E., & Alderweireldt, F. C. (1985). *A history of the nomenclature of organic chemistry*. Springer.
Waldie, S., Lind, T. K., Browning, K., Moulin, M., Haertlein, M., Forsyth, V. T., et al. (2018). Localization of cholesterol within supported lipid bilayers made of a natural extract of tailor-deuterated phosphatidylcholine. *Langmuir*, *34*(1), 472–479. https://doi.org/10.1021/acs.langmuir.7b02716.
Waldie, S., Moulin, M., Porcar, L., Pichler, H., Strohmeier, G. A., Skoda, M., et al. (2019). The production of Matchout-deuterated cholesterol and the study of bilayer-cholesterol interactions. *Scientific Reports*, *9*(1), 5118. https://doi.org/10.1038/s41598-019-41439-z.
Waldie, S., Sebastiani, F., Browning, K., Maric, S., Lind, T. K., Yepuri, N., et al. (2020). Lipoprotein ability to exchange and remove lipids from model membranes as a function of fatty acid saturation and presence of cholesterol. *Biochimica et Biophysica Acta (BBA) - molecular and cell biology of Lipids*, *1865*(10), 158769. https://doi.org/10.1016/j.bbalip.2020.158769.
Waldie, S., Sebastiani, F., Moulin, M., Del Giudice, R., Paracini, N., Roosen-Runge, F., et al. (2021). ApoE and ApoE nascent-like HDL particles at model cellular membranes: Effect of protein isoform and membrane composition. *Frontiers in Chemistry*, *9*. https://doi.org/10.3389/fchem.2021.630152.
Whitten, A., Cai, S., & Trewhella, J. (2008). MULCh: Modules for the analysis of small-angle neutron contrast variation data from biomolecular assemblies. *Journal of Applied Crystallography*, *41*. https://doi.org/10.1107/S0021889807055136.
Xu, A. Y., Melton, L. D., Ryan, T. M., Mata, J. P., Rekas, A., Williams, M. A. K., et al. (2018). Effects of polysaccharide charge pattern on the microstructures of β-lactoglobulin-pectin complex coacervates, studied by SAXS and SANS. *Food Hydrocolloids*, 77, 952–963. https://doi.org/10.1016/j.foodhyd.2017.11.045.
Yepuri, N. R., Clulow, A. J., Prentice, R. N., Gilbert, E. P., Hawley, A., Rizwan, S. B., et al. (2019). Deuterated phytantriol – A versatile compound for probing material distribution in liquid crystalline lipid phases using neutron scattering. *Journal of Colloid and Interface Science*, *534*, 399–407. https://doi.org/10.1016/j.jcis.2018.09.022.

Yepuri, N. R., Darwish, T. A., Krause-Heuer, A. M., Leung, A. E., Delhom, R., Wacklin, H. P., et al. (2016). Synthesis of Perdeuterated 1-Palmitoyl-2-oleoyl-sn-glycero-3-phosphocholine ([D82]POPC) and characterisation of its lipid bilayer membrane structure by neutron reflectometry. *ChemPlusChem, 81*(3), 315–321. https://doi.org/10.1002/cplu.201500452.

Yepuri, N. R., Jamieson, S. A., Darwish, T. A., Rawal, A., Hook, J. M., Thordarson, P., et al. (2013). Synthesis of per-deuterated alkyl amines for the preparation of deuterated organic pyromellitamide gelators. *Tetrahedron Letters, 54*(20), 2538–2541. https://doi.org/10.1016/j.tetlet.2013.03.031.

Zhu, X., Ng, S. Y., Gupta, A. N., Feng, Y. P., Ho, B., Lapp, A., et al. (2010). Effect of crowding on the conformation of interwound DNA strands from neutron scattering measurements and Monte Carlo simulations. *Physical Review E, 81*(6), 061905. https://doi.org/10.1103/PhysRevE.81.061905.

CHAPTER FOUR

Planning, executing and assessing the validity of SANS contrast variation experiments

Susan Krueger*
NIST Center for Neutron Research, NIST, Gaithersburg, MD, United States
*Corresponding author: e-mail address: susan.krueger@nist.gov

Contents

Abstract

Small angle neutron scattering (SANS) along with contrast variation (CV) can provide key information that is used to determine the shape and structure of biological complexes in solution. The successful SANS CV experiment is usually a result of judicious planning, careful execution and meticulous scrutiny of the resultant SANS data. A workflow for planning, executing and, importantly, assessing the validity of SANS CV data is presented here, along with tips to follow in order to perform a successful SANS CV experiment. Some knowledge of the basics of small angle scattering is assumed, including data reduction and standard analysis to obtain model independent parameters such as the radius of gyration and forward scattering intensity, and SANS CV theory is not covered in detail. Rather, the focus is on the SANS CV workflow from

Methods in Enzymology, Volume 677
ISSN 0076-6879
https://doi.org/10.1016/bs.mie.2022.08.023

in silico experiment planning and execution to obtaining the SANS CV data, assessing its validity, and determining a model structure or ensemble of structures using the SANS CV data as constraints.

1. Introduction

Small angle neutron scattering (SANS) is a valuable tool that allows the determination of the structure of biological macromolecules in solution, which can lead to insights into biological function, improvements in the formulation of biopharmaceuticals and better ways to preserve foods, to name a few examples. When combined with the contrast variation (CV) technique, SANS can provide unique information that is especially helpful in finding the structure of biological complexes. The SANS CV technique has been described in a number of recent review papers (Jacques & Trewhella, 2010; Jeffries et al., 2016; Krueger, 2017; Mahieu & Gabel, 2018; Zaccai et al., 2016). Additionally, a review of recent studies using SANS CV (Krueger, 2022) shows the wide range of applications of the technique and breadth of knowledge that can be obtained. A brief description of the main concepts of SANS CV is included here.

The coherent SANS intensity from *N* monodisperse, randomly oriented biological macromolecules in a dilute solution can be written as

$$I(q) = n\,\Delta\rho^2\,V^2\left\langle\left|F\left(\vec{q}\right)\right|^2\right\rangle = n\,\Delta\rho^2\,V^2P(q), \tag{1}$$

where $\left|F\left(\vec{q}\right)\right|^2$ is the form factor of the molecule that is dependent on its structure, $\vec{q}$ is the momentum transfer vector with a magnitude defined as

$$q = \frac{4\pi\,\sin\left(\theta\right)}{\lambda} \tag{2}$$

with neutron wavelength, λ, at a scattering angle of 2θ, n is the number density of molecules in the solution, V is the volume of the molecule, and $\Delta\rho = \rho - \rho_s$ is the neutron scattering contrast, with ρ defined as the neutron scattering length density (SLD) of the molecule and ρ_s defined as the neutron SLD of the solvent. The brackets represent an averaging over all orientations of the molecule. The rotationally averaged form factor is sometimes also called $P(q)$. Note that there is also an incoherent scattering component (Krueger, 2017) that mainly arises from hydrogen atoms in the buffer solution. This scattering is isotropic and is often referred to as the incoherent background or simply, the background. Here, it is assumed that scattering from the buffer has been subtracted from that of the macromolecule and that the incoherent scattering from hydrogen in the molecule itself is negligible in the dilute limit.

An important parameter is the forward scattering intensity at $q=0$, which can be written as

$$I(0) = n\,\Delta\rho^2 V^2. \tag{3}$$

The number density, in turn, can be written as

$$n = \frac{cN_A}{M_w}, \tag{4}$$

and the molecular volume can be written as

$$V = \frac{\overline{v}M_w}{N_A}, \tag{5}$$

where c is the concentration of the molecule in the solution, M_w is its molecular mass (molecular weight), $\overline{v}$ is its partial specific volume, and N_A is Avogadro's number. An important assumption here is that the sample is monodisperse and dilute. Thus, the samples must be well-characterized ahead of time to verify that there is no aggregation and no interactions between the molecules.

For a biological complex consisting of more than one component, Eq. (3) can be extended as

$$I(0) = n\left(\sum_i \Delta\rho_i V_i\right)^2, \tag{6}$$

where the neutron scattering contrast and volume now apply to the individual components in the molecule. Note that the number density still applies to the entire complex. From here, $I(0)$ for a two-component complex can be written as

$$I(0) = n(\Delta\rho_1 V_1 + \Delta\rho_2 V_2)^2 = \frac{cM_w}{N_A}(f_1\Delta\rho_1\overline{v}_1 + f_2\Delta\rho_2\overline{v}_2)^2, \tag{7}$$

where $\Delta\rho_1=\rho_1-\rho_s$ and $\Delta\rho_2=\rho_2-\rho_s$ are the neutron scattering contrasts for each component, V_1 and V_2 are the volumes of each component and

$$f_1 = \frac{M_1}{M_1+M_2},\ f_2 = \frac{M_2}{M_1+M_2} \tag{8}$$

are the mass fractions of each component. Here, ρ_1, ρ_2 and ρ_s are the scattering length densities of the components and the solvent. The number density of scatterers, n, was written as in Eq. (4). However, now $c = c_1 + c_2$ is the total concentration of the complex and $M_w = M_1 + M_2$ is the total molecular weight of the complex. Furthermore, the volumes have been written as

$$V_1 = \frac{\overline{v}_1 M_1}{N_A}, \quad V_2 = \frac{\overline{v}_2 M_2}{N_A}, \tag{9}$$

where $\overline{v}_1$ and $\overline{v}_2$ are the partial specific volumes of each component. Thus, $I(0)$ can be written in terms of the individual molecular weights, contrasts and partial specific volumes of each component. Eq. (7) is important when calculating expected $I(0)$ values for model structures.

Typically, a SANS CV experiment is performed by varying the SLD of the molecule or the solvent by substitution of deuterium (D) for hydrogen (H) in order to obtain data at several different contrast points. As the contrast changes, different parts of the molecule contribute to the scattering more than others. The data are then analyzed simultaneously to obtain parameters such as the size and shape of the complex and its components as well as the spatial arrangement of the components with respect to each other. Fig. 1 shows neutron SLD as a function of the percentage of D_2O (%D_2O) in the solvent for some molecules of biological interest. The lines have a non-zero slope for proteins and nucleic acids due to the exchange of labile H atoms in the molecule with D atoms in the solvent (H—D exchange). The perdeuterated proteins and nucleic acids have all the aliphatic (non-exchangeable) H atoms replaced with D during synthesis. The contrast, $\Delta\rho$, between the protein and water is shown for the case where there is 10% D_2O and 90% H_2O. Similarly, the contrast can be defined for all the molecules shown for any desired %D_2O in the solvent. Note that $\Delta\rho = 0$ at the point on the graph where the lines for the molecule and solvent intersect. This is called the contrast match point and the scattering intensity also is zero under this condition (Eqs. 1 and 3). Thus, a multicomponent complex has a match point for each component where the scattering from that component is effectively zero (Eqs. 6 and 7) and it is said to be matched out. There can also be a match point for which the entire complex is matched, although there may be measurable residual scattering since the SLDs of the components are not uniform as is assumed in the above equations.

It is important to mention here that the complementary technique, small-angle X-ray scattering (SAXS), can also be used to obtain the

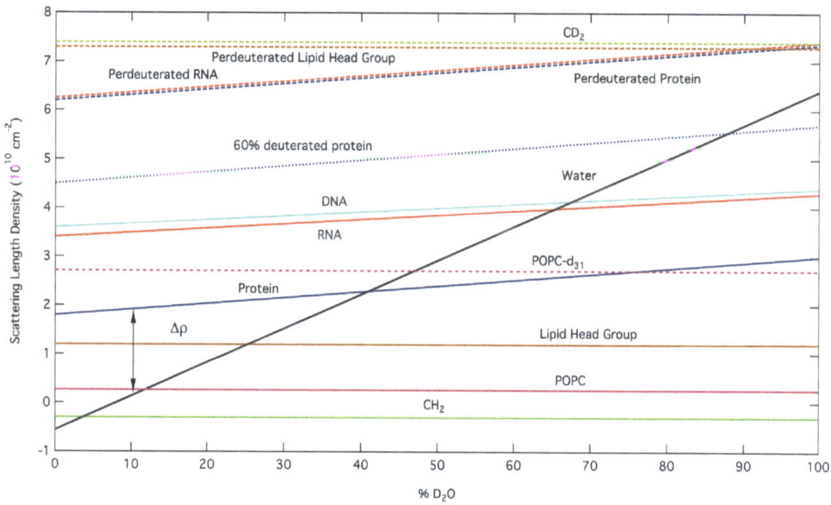

Fig. 1 Plot of neutron SLDs of several biological molecules as a function of $\%D_2O$ in the solvent, assumed to be water in this case, along with the SLD of water for comparison. The double-sided arrow illustrates the contrast, $\Delta\rho$, between a protein and water in 10% D_2O solvent. *Taken from Krueger, S. (2022). Small-angle neutron scattering contrast variation studies of biological complexes: Challenges and triumphs. Current Opinion in Structural Biology, 74, 102375.*

scattering intensity from biological complexes in solution. However, the X-ray SLDs of H and D are identical, meaning that there is essentially only one contrast that can be measured with X-rays unless more radical changes are made to the molecule such as heavy-atom substitution, or to the solvent such as the addition of sugars. Both substitutions can alter the structure and/or interactions of the molecules. By using neutrons, the simpler and less intrusive substitution of D for H can be used to advantage to access all the contrast options shown in Fig. 1. Together, SAXS would be used to obtain information on the overall structure of the complex and SANS to obtain unique structural information about each component as it exists in the complex.

An earlier tutorial (Krueger, 2017) described SANS CV experiment planning, execution and data analysis of a multicomponent biological complex. This chapter will concentrate on the workflow necessary to perform a successful SANS CV experiment, which can include multiple rounds of experiment planning, data collection, data analysis to assess feasibility, and model structure refinement. Fig. 2 shows this potential workflow, with emphasis on performing the experiment *in silico* as part of the experiment

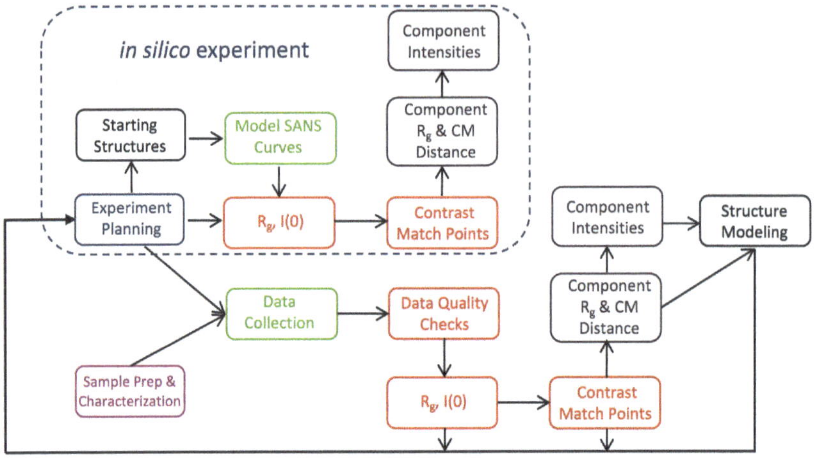

Fig. 2 SANS Contrast Variation Experiment Workflow. *Adapted from Krueger, S. (2017). Designing and performing biological solution small-angle neutron scattering contrast variation experiments on multi-component assemblies. In B. Chaudhuri, I. G. Muñoz, S. Qian, & V. S. Urban (Eds.), Biological small angle scattering: Techniques, strategies and tips (pp. 65–85). Singapore: Springer Singapore.*

planning process. Experiment planning will be discussed first, including constructing starting model structures and model SANS CV data, calculating expected neutron scattering contrasts and $I(0)$ values, and choosing the optimal contrast conditions (%D_2O in the solvent) for which to perform the measurements. The next steps are to collect the SANS data as planned and to perform a few critical analyses to assess the validity of the data and to compare the results to those from the model data in order to determine whether more data need to be collected and/or the model structure(s) needs to be refined. Finally, more advanced SANS contrast variation data will be discussed in context of refining the model structure(s).

2. Experiment planning

Of major importance for a successful SANS CV experiment is the planning and execution of the experiment *in silico* in order to have as much information as possible prior to the experiment, including model SANS data when possible. The portion of Fig. 2 surrounded by the dashed blue line illustrates the steps involved in the *in silico* experiment. This includes constructing starting model structures for the molecule. Here, the focus will be on all-atom model structures, although other types of models such as coarse-grained, geometrical and hybrid, many of which are referenced in a recent review of SANS and SAXS used with constraints from complementary biophysical techniques (Grishaev, 2017), can also be built in order to calculate model SANS curves for comparison with the data. The availability of a model structure allows for the calculation of expected radius of gyration, R_g, and maximum distance, D_{max}, values, which can be very helpful in choosing instrument configuration parameters. Another important part of experiment planning is the calculation of neutron SLDs, contrasts and expected $I(0)$ values. This information is beneficial in determining optimal sample concentrations and contrast conditions for the CV series of measurements. It is also helpful to keep in mind the recent publication guidelines for biomolecular SAS experiments (Jacques, Guss, Svergun, & Trewhella, 2012; Jacques, Guss, & Trewhella, 2012; Trewhella et al., 2017) during the experiment planning process, as they provide useful strategies and resources for the entire experimental process from experiment planning to publication of results, some of which are beyond the scope of this work.

2.1 Contrast and $I(0)$ calculations

The calculation of neutron SLDS, contrasts and expected $I(0)$ values is model independent and does not require knowledge of the molecular structure. Rather, it depends on the chemical composition of the complex. Thus, sequence information for proteins and nucleic acids as well as the chemical composition of additional components must be available, along with an estimate of the H—D exchange for each component.

Two web-based programs that are available to aid in the calculation of neutron SLDs and contrasts as a function of the amount of D_2O in the solvent are MULCh (MULCh, n.d.; Whitten, Cai, & Trewhella, 2008) and the Contrast Calculator module (Sarachan, Curtis, & Krueger, 2013) in SASSIE-web (Curtis, Raghunandan, Nanda, & Krueger, 2012; SASSIE-web, n.d.). Both take H—D exchange into account when performing the calculations and both provide the contrast match point of the complex as well as their components. X-ray SLDs and contrasts are also provided in both cases, as SAXS data are often obtained not only to check for deuterium effects but also as another contrast point for structure modeling. MULCh limits the number of components in the complex to two, while Contrast Calculator allows for an unlimited number of components and it predicts $I(0)$ values for a given complex concentration in addition to SLDs and contrasts.

Once neutron contrasts and match points have been calculated, how does one decide which contrasts to measure? This depends on the concentration of the complex as well as the contrasts and volume fractions of the components. It has been suggested that, for a two-component system, there should be a minimum of two contrast points on either side of the match point of the entire complex (Whitten et al., 2008) to encompass a wide distribution of contrasts sufficient to obtain accurate size and shape information for both components. Thus, if the two components are proteins, it has been recommended that the deuteration level of the deuterated component should be adjusted such that the match point of the complex is between 50% and 70% D_2O. The measured contrasts could then be 0%, 20%, 40%, 80% and 100% D_2O, provided that the scattering intensity is measurable at 40% D_2O, which is the match point of the protiated component (Whitten et al., 2008). If one of the components is present in significantly lower volume fraction than the other, size and shape information for that component will only be obtained for high signal-to-noise data, which is often not the case over the entire range of contrasts for a variety of reasons including aggregation and interparticle interference effects as the concentration is increased.

While the analysis described in Whitten et al. (2008) can be expanded to complexes with more than two components, the number of contrasts required to solve the set of simultaneous equations quickly makes the experiment increasingly difficult in practice. Thus, for multicomponent complexes, the contrasts chosen are often those that are at the match points of the individual components, while avoiding those that are too close to the match point of the entire complex, in addition to 0% and 100% D_2O, in order to have a wide range of contrasts for which to compare model structures.

A related question is how does one decide what concentrations are needed at each contrast? This can be estimated from the calculated $I(0)$ values, taking into account the minimum measurable scattering intensity under each contrast condition, as the incoherent scattering will be higher for contrasts closer to 0% D_2O. This value is somewhat facility- and sample-dependent (Krueger, 2017), but a good rule of thumb is that a $I(0)$ value of $\approx 0.01\ \mathrm{cm}^{-1}$ is likely measurable in 100% D_2O in a reasonable amount of time (up to a few hours for good statistics), but a higher $I(0)$ value would be needed in 0% D_2O. This can be translated to a sample concentration at each contrast and an assessment can be made as to whether that concentration is attainable in practice. Higher concentrations of the complex will lead to better quality data, provided that there are no aggregation or interparticle interference effects. For this reason, a concentration series of measurements should ideally be performed prior to the CV experiment in order to determine the highest concentration for which there are no such effects.

2.2 Starting model structures

In the case of complexes consisting of compact structures such as globular proteins or condensed nucleic acids, a model-independent lower limit on R_g can be calculated based on partial specific volume and molecular weight. The molecular volume of each component can be obtained from Eq. (5) and then set equal to the volume of an equivalent sphere to obtain the radius, R, and then subsequently, R_g, using the relation $R_g = \sqrt{\frac{3}{5}}R$. This approximation isn't as useful for extended structures. However, similar relationships can be used for ellipsoid and cylinder shapes. A model structure for which a SANS curve can be calculated will provide a better estimate of R_g and how it varies as a function of %D_2O in the solvent, provided that the different parts of the complex can be modeled by an all-atom, geometrical, coarse-grained, etc. structure or a hybrid combination thereof where scattering lengths, SLDs or contrasts can be assigned to each atom or component. Since the focus here is on all-atom starting model structures, other types of starting structures are not discussed further.

All-atom starting model structures are specified in files containing the atomic coordinates of each atom in the molecule, usually in Protein Data Bank (PDB) or the newer PDBx/mmCIF format (PDB101, n.d.), where the amino acid residues are also specified. All-atom starting structures are typically obtained using high-resolution X-ray crystallography, NMR and, in some cases, cryo-EM. A full model structure may need to be assembled from multiple coordinate files. With the advent of the AlphaFold database of protein structures (Jumper et al., 2021; Varadi et al., 2022) that provides theoretically-generated 3-D structures from protein amino acid sequences, all-atom starting model structures are now more available than ever. Note that in order to obtain model SANS curves that accurately represent the planned SANS CV experiment, the coordinate files should contain the same amino acid residues, nucleic acid bases, lipids, glycans, etc. that are in the samples. Missing residues such as flexible linkers and disordered loops should be included. Resources to add missing regions and mutations include, for example, Phyre2 (Kelley, Mezulis, Yates, Wass, & Sternberg, 2015; Kelley & Sternberg, 2009), PSIPRED (Buchan, Minneci, Nugent, Bryson, & Jones, 2013; Jones, 1999) and MODELLER (Webb & Sali, 2016). In addition to adding missing regions, PSFGEN (Phillips et al., 2005) and CHARMM-GUI (Jo et al., 2017) prepare PDB files for MD simulation, the latter facilitating the creation of input files (Lee et al., 2016) for a number of programs including CHARMM (Best et al., 2012; Brooks et al., 2009; MacKerell et al., 1998), NAMD (Phillips et al., 2005), GROMACS (Abraham et al., 2015; Pronk et al., 2013) and AMBER (Cornell et al., 1995; Lee et al., 2020). Starting structures of complexes may have to be built from structures of the individual components using a combination of docking and MD simulations based on geometry or known binding regions. Docking software such as RosettaDock (Lyskov & Gray, 2008; Lyskov et al., 2013), which performs a local docking search, can be used for this purpose. Components need to be placed near each other either manually using a visualization tool such as VMD (Humphrey, 1996) or PyMol (Schrödinger, n.d.) or using docking software that performs a global docking search, e.g., ClusPro (Kozakov et al., 2017), ZDOCK (Pierce et al., 2014), GRAMM-X (Tovchigrechko & Vakser, 2006).

If the starting model structure contains flexible linkers that allow for a range of conformations, it may be useful to generate an ensemble of structures in order to obtain an expected range of R_g values. One way to generate ensembles of structures that cover a wide range of conformation space is using Monte Carlo (MC) simulation techniques. The Monomer Monte Carlo, Complex Monte Carlo and Torsion Angle Monte Carlo (TAMC) modules in SASSIE-web (SASSIE-web, n.d.), for example, allow for the generation of such ensembles. In addition, the Energy Minimization and Torsion Angle MD (TAMD) modules in SASSIE-web allow for MD simulations using NAMD as well as the TAMD (Chen, Im, & Brooks, 2005) module of CHARMM, respectively. If MC or MD simulations are to be performed, the coordinate files need to be in the proper format for the method being used. If using CHARMM, for example, atoms in the coordinate files need to be named according to the CHARMM naming convention. Once a structural ensemble is generated, it might be useful to cluster like conformations in order to reduce the size of the ensemble and test the ensemble for convergence of the structures in conformation space (Zhang et al., 2017) in order to obtain a range of R_g values that represents the entire conformation space.

2.3 Optimal sample concentrations

Following the portion of the flow chart surrounded by dashed lines in Fig. 2, SLD, contrast and *I*(0) values can be calculated without the need for a model structure, although the Contrast Calculator module will accept both sequence information and coordinate files. As an example, the neutron SLDs, contrasts and *I*(0) values were calculated as a function of %D_2O in the solvent for a complex consisting of a protiated 50 kDa protein and a 36 kDa protein with 50% deuteration. The calculated match points were 44% D_2O for the 50 kDa protein and 80% D_2O for the deuterated 36 kDA protein, while that for the complex was 60% D_2O. The match points of the two components are well separated and sufficiently different from the match point of the entire complex. Thus, it should be possible to follow the contrast series recommended in Section 2.1 (0%, 20%, 40%, 80% and 100% D_2O) and to make measurements at or near the match points of the individual components provided that the *I*(0) values are measurable at those contrasts. For a total complex concentration of 1 mg/mL, the calculated *I*(0) value is 0.009 cm^{-1} in 40% D_2O and 0.01 cm^{-1} in 80% D_2O. While these values satisfy the rule of thumb value above, better data would be obtained if the concentration could be increased to 2 mg/mL, especially in 40% D_2O where the incoherent scattering will be higher. The calculated *I*(0) values for the other contrasts in the series, i.e., 0%, 20% and 100% D_2O for a concentration of 1 mg/mL are 0.086 cm^{-1}, 0.038 cm^{-1} and 0.04 cm^{-1}, respectively. Again, while a measurable signal could be obtained at 1 mg/mL, a concentration of 2 mg/mL would result in better quality data. The take home message is that this experiment is doable. The higher the concentration, the higher the scattering intensity and, thus, the better the chance of obtaining information on the individual components in the complex. A separate issue is whether aggregation effects will be observed as a function of concentration and/or %D_2O in the solvent. Severe aggregation could become evident during the sample preparation process. However, more subtle aggregation effects might only become evident through SAXS or SANS measurements of the complex at several concentrations and in 0% and 100% D_2O.

2.4 Model SANS intensity curves

Once SLD, contrast and $I(0)$ calculations have been made and a starting model structure or structural ensemble is at hand, model SANS data can be calculated at each planned $\%D_2O$ and *in silico* data analysis can be performed. For all-atom model structures, software such as CRYSON (Svergun, Barberato, & Koch, 1995), which is part of the ATSAS software suite (Manalastas-Cantos et al., 2021), Pepsi-SANS (Grudinin, Garkavenko, & Kazennov, 2017; Pepsi-SANS, n.d.) or the SasCalc module (Watson & Curtis, 2013) in SASSIE-web (Curtis et al., 2012; SASSIE-web, n.d.) can be used to calculate model SANS curves for multicomponent complexes from atomic coordinates. CRYSON, pepsi-SANS and SasCalc take neutron contrast into account, including H—D exchange, in order to generate model SANS curves that vary properly as a function of contrast. The methods also allow the specification of deuteration fraction for individual components in a complex. CRYSON uses implicit hydrogen atoms in the calculation of SLDs, while pepsi-SANS and SasCalc require the hydrogen atoms to be present in the coordinate file. Because the SASSIE-web software suite contains modules for both MC and MD simulations, SasCalc requires that the atoms in the coordinate files must also be specified using the CHARMM (Brooks et al., 2009) naming convention.

Model SANS curves should be calculated at each planned experimental contrast, using the estimated $I(0)$ values corresponding to each, in order to obtain a complete model SANS CV data set that can be plotted on a log-lin or log-log scale for comparison to the SANS CV data. Note that the calculated model SANS curves represent only the coherent scattering component from a dilute solution of biological macromolecules (Krueger, 2017) and should be compared to the buffer-subtracted SANS data.

2.5 R_g and D_{max} for the complex and components

There are two ways to calculate R_g and D_{max} for all-atom structures. The first is to perform Guinier (Guinier & Fournet, 1955) and/or distance distribution, $P(r)$, analysis (by means of an indirect Fourier transform method (Glatter, 1977; Moore, 1982; Semenyuk & Svergun, 1991; Svergun, 1992)) to obtain them from the calculated model SANS curves. The second is to calculate them directly from the atomic coordinates. The first method will produce contrast-dependent values corresponding to the contrast conditions at which the model SANS curves were calculated. The second method will not produce contrast-dependent values unless the contrast is explicitly taken into account during the calculation. For a single-component molecule, the contrast-independent and contrast-dependent values are usually similar. However, the two values can vary greatly for a complex, especially when calculated at a contrast that is near the match point of one of the components.

Most software packages for SANS and SAXS data reduction and analysis include the ability to perform both Guinier and *P*(*r*) analyses on the model SANS curves. SasCalc provides contrast-dependent and contrast-independent R_g values for the calculated SANS curves from the atomic coordinates. For two-component systems, it is also useful to calculate from the atomic coordinates the contrast-independent R_g values of the individual components, the center of mass (CM) coordinate of each component and the distance between the centers of mass of the two components (D_{CM}). These values are directly comparable to those obtained from the SANS CV data analysis and will aid in the refinement of the model structures.

2.6 Optimal instrument configuration(s)

The R_g and D_{max} values calculated from the model SANS CV set of curves will be helpful for the determination of the q range needed for the entire CV data set, and thus the instrument configuration(s) needed for the SANS experiment. If a structural ensemble has been generated, the range of possible R_g and/or D_{max} values will be equally helpful. How does one estimate the q range needed for their SANS CV experiment? A visual inspection of the model SANS curves plotted on a log-log scale will provide a clue, as the model curves can be calculated to q_{min} values that are close to $q = 0$. Thus, it should be easy to determine the Guinier region, as it will be the flat portion of the model curves in the low-q region. This region can be compared to the q_{min} values obtainable at various instrument configurations, which will be instrument and facility-dependent, and a determination can be made as to which configuration would provide a desirable q_{min} value while still providing sufficient intensity to perform the experiment. One can also perform a similar analysis using estimated contrast-dependent R_g values derived either from Guinier analysis of the model SANS data or directly from the model structures. For example, if the contrast-dependent R_g values for a protein-deuterated protein complex range from 35 to 60 Å, one can use the criterion for validity of the Guinier approximation, i.e., $qR_g \sim 1$ to obtain an estimate for q_{min}. If one desires $q_{min} = 1/R_g$ to be approximately in the middle of the Guinier range, the largest R_g value can be used such that $1/60\,\text{Å} = 0.016\,\text{Å}^{-1}$. Thus, $q_{min} \leq 0.01\,\text{Å}^{-1}$ is a reasonable starting point from which to determine an appropriate instrument configuration.

On the other hand, the model SANS curves are not as reliable for determining the best value for q_{max}, since they do not include the incoherent scattering contribution. A value of $q_{max} \geq 0.4\,\text{Å}^{-1}$ is generally sufficient to obtain a good match with the buffer scattering in the high q region in order to obtain a good buffer subtraction. Again, these rules of thumb are somewhat instrument and facility dependent. However, judicious calculation of model SANS curves and size parameters will greatly aid the instrument scientists in determining the most optimal instrument configurations for a particular biomolecular complex.

2.7 R_g and decomposition analyses

For two component systems, the *in silico* experiment can be completed by performing the same analysis on the model SANS CV data as will be performed on the measured data. This will help determine whether the planned measured contrast points are well-chosen and will produce the expected results when compared to those from the model structures. This is especially useful if one or more planned contrast conditions had to be eliminated due to concentration or D_2O effects (discussed below). This exercise provides an idea of the results that can be expected from ideal model SANS data with no noise. In the case of flexible systems where a structural ensemble has been generated, this exercise can be performed with just one structure, since it is a consistency check of the model SANS CV data with respect to its corresponding model structure.

Both a Stuhrmann (Ibel & Stuhrmann, 1975) and parallel axis theorem (Engelman & Moore, 1975) analysis can be performed on the model SANS data in order to obtain contrast-independent R_g values for the two components, R_1 and R_2, as well as D_{CM}. Both analyses can be performed using the MULCh software package. The inputs are the fitted R_g values and calculated $\Delta\rho$ values for each component. The resultant R_1 and R_2 values for each component and D_{CM} should agree with the contrast-independent values obtained from the atomic coordinates if everything is consistent and the contrast points are well-chosen.

For the Stuhrmann analysis, R_g^2 can be written as a function of $\Delta\rho$ of the complex as

$$R_g^2 = R_m^2 + \frac{\alpha}{\Delta\rho} - \frac{\beta}{\Delta\rho^2}, \tag{10}$$

where R_m is the radius of gyration of the complex with a homogeneous SLD, α is the second moment of the SLD fluctuations and β is the square of the first moment of the SLD fluctuations. R_m, α and β can be written in terms of R_1, R_2, and D_{CM} (Moore, 1982; Olah, Rokop, Wang, Blechner, & Trewhella, 1994), giving three linear equations that can be solved for R_1, R_2, and D_{CM}. α is related to the distribution of SLDs relative to the CM and of the complex and β is related to the separation of the CM of the two components. Stuhrmann plots of R_g^2 vs $\frac{1}{\Delta\rho}$ for two different β conditions are shown in Fig. 3.

For the parallel axis theorem analysis, R_g^2 at each contrast can be written as,

$$R_g^2 = \frac{\Delta\rho_1 V_1}{\Delta\rho V} R_1^2 + \frac{\Delta\rho_2 V_2}{\Delta\rho V} R_2^2 + \frac{\Delta\rho_1 \Delta\rho_2 V_1 V_2}{(\Delta\rho V)^2} D_{CM}^2, \tag{11}$$

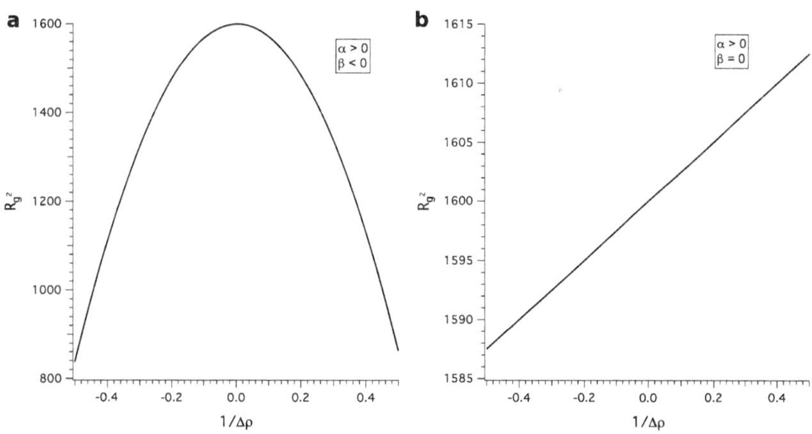

Fig. 3 Stuhrmann plots for (A) $\beta > 0$, indicating that the centers of mass of the two components are not concentric, and (B) $\beta = 0$, indicating that the center of mass of the two components are concentric. In this case, $\alpha > 0$ indicates that the component with higher SLD is on the exterior of the complex. *Taken from Krueger, S. (2017). Designing and performing biological solution small-angle neutron scattering contrast variation experiments on multi-component assemblies. In B. Chaudhuri, I. G. Muñoz, S. Qian, & V. S. Urban (Eds.), Biological small angle scattering: Techniques, strategies and tips (pp. 65–85). Singapore: Springer Singapore.*

where $\Delta\rho_1$, $\Delta\rho_2$, V_1 and V_2 refer to the individual components and

$$\Delta\rho V = \Delta\rho_1 V_1 + \Delta\rho_2 V_2. \tag{12}$$

Thus, R_1, R_2, and D_{CM} are calculated directly from the parallel axis theorem while they are obtained from the equations for R_m, α and β for the Stuhrmann analysis.

The SANS intensity from a two-component system at each contrast can be approximated by a relation involving the intensities from the individual components, $I_{11}(q)$ and $I_{22}(q)$ as well as a cross term, $I_{12}(q)$ (Olah et al., 1994; Whitten et al., 2008) as

$$I(q) = \Delta\rho_1^2 I_{11}(q) + \Delta\rho_1 \Delta\rho_2 I_{12}(q) + \Delta\rho_2^2 I_{22}(q). \tag{13}$$

Thus, the scattering intensities from the two components and the cross term can, in theory, be separated out from the total scattering measured at three or more contrasts. This information can be useful in the structure modeling of the individual components, as it can provide information beyond that from the Stuhrmann and parallel axis theorem analyses. However, in practice, it is not always possible to obtain all three intensities with good statistics from the measured SANS CV data. Success depends on several factors including the signal-to-noise in the data and the chosen contrast points. For a more in-depth discussion and examples for data with varying noise levels and differing numbers of contrasts, see (Whitten et al., 2008). However, it may still be useful to perform the decomposition analysis on the model data to determine $I_{11}(q)$, $I_{22}(q)$ and $I_{12}(q)$ under ideal conditions at the planned experimental contrasts and sample concentrations. If all three intensities are not adequately determined under these conditions, then the planned contrasts and/or sample concentrations may need to be revisited.

3. Data collection and assessment of validity

3.1 Data collection

Following the flow chart in Fig. 2, the data collection is performed following careful sample preparation, (non-SAS) characterization (beyond the scope of this work) and experiment planning (Jacques & Trewhella, 2010; Krueger, 2017). Prior to performing the SANS CV experiment, it is desirable to have already conducted an initial set of measurements of the complex at multiple concentrations in both 0% and 100% D_2O buffers to test for aggregation or interparticle interference effects. If a complex contains a deuterated component, it would also be helpful to measure the fully protiated complex for comparison. SAXS can be used for these experiments, if possible, as SAXS experiments typically don't require as much sample volume and the measurements can be made at higher throughput. SAXS is very useful for detecting structural changes due to H—D substitution since the X-ray SLDs for H and D are the same. While SANS data obtained from a complex in 0% and 100% D_2O buffers could be quite different, SAXS data obtained on the same two samples should be identical. SAXS data also serve as an additional contrast point for structure modeling. It is assumed that SAXS and/or SANS CV data are obtained on an absolute scale, usually in units of cm^{-1}.

If initial experiments reveal something unexpected, such as concentration-dependent or D_2O-induced aggregation, be prepared to adjust expectations and re-evaluate the previously-planned contrast points. It may be necessary to revisit the estimated $I(0)$ results and adjust the planned measurements if some are no longer feasible at the planned concentrations and contrasts. The complete *in silico* experiment should be performed again to determine if any critical information will be lost by having to eliminate one or more of the originally planned contrast points. The R_g and decomposition analysis may reveal that it won't be possible to obtain detailed information on one of the components. But it might still be possible to obtain R_g at match point of other component, for instance—at least under ideal conditions. Once any re-evaluations and adjustments have been made, the SANS CV experiment can be performed if it is determined that useful results might still be obtained.

During the SANS experiment, be prepared for the unexpected. If something was missed in the contrast calculations, which is more common for multicomponent systems in which the stoichiometry may not be exactly known, the match point may not be where it is expected, resulting in $I(0)$ values that are different from the calculated values. In this case, the total concentration of the complex also may not be the same as that assumed when performing the calculations. However, if there are no obvious signs of aggregation or interparticle interference, an approximation for the stoichiometry of the complex can possibly be obtained from the complex match point. Thus, the SANS CV data may still provide some useful information for the next iteration of the experiment.

If the model structure is not a good representation of the measured sample, the R_g values and model SANS curves may be different than initially expected, especially if the sample has significant flexibility. This shouldn't be a cause for alarm if the $I(0)$ values are still roughly as expected (discussed further below). In this case, the SANS CV experiment can be completed and any discrepancies can hopefully be resolved during the data analysis and model refinement stages.

3.2 Assessment of validity

Once the SANS CV data have been collected and no obvious aggregation or interparticle interference effects are evident, the flow chart in Fig. 2 is followed to analyze the data and perform data quality checks that will allow assessment of the validity of the data. Note that this analysis can—and should—be performed on any preliminary SANS or SAXS data obtained prior to the SANS CV experiment as well.

The main assessment for validity of the SANS CV data is to determine whether the experimental $I(0)$ values agree with those calculated from the chemical composition of the sample. This assumes the correct stoichiometry was used in calculating $I(0)$. Guinier and $P(r)$ analyses should be performed in order to obtain R_g and $I(0)$ at each contrast, as well as D_{max} in the case of the latter analysis. Note that the data must have the buffer subtracted for the proper calculation of R_g and $I(0)$, unless a background term is included in the data fitting. Since the buffer-subtracted data are also needed for comparison to model SANS data, it is useful to just use buffer-subtracted data at this stage in any case. For folded globular molecules, improper background subtraction can be detected by not constraining $P(r)=0$ at $r=0$ when performing the $P(r)$ analysis. If $P(0)\neq 0$, adjust the background subtraction and try the analysis again. Examination of the Kratky plot (Glatter & Kratky, 1982) should be performed prior to this analysis to be sure that the molecule is well folded.

The first test is whether R_g and $I(0)$ are consistent from both analyses. Since the Guinier analysis only uses the data at low q values where $qR_g \sim 1$, the R_g and $I(0)$ values could be different than those obtained from the $P(r)$ analysis, which uses all of the data. If the values obtained from the two methods are sufficiently different at one or more contrasts, this could be an indication that subtle aggregation or interparticle interference effects may be present.

Next, the measured $I(0)$ (cm^{-1}) values should be compared to the calculated values at each contrast to determine whether they are reasonable. One might rightfully ask what is reasonable. Eq. (7) shows the relationship between $I(0)$ and the complex M_w and c, as well as the stoichiometry of the components and their partial specific volumes. The M_w is calculated from the chemical composition of the complex as is the stoichiometry. However, the partial specific volumes may be less certain, especially for RNA (Sarachan et al., 2013). The concentration is a measured value that also has an uncertainty that is not always easy to quantify and depends on the method used. However, a rule of thumb has been to assume it is around 5–10% of the measured value. Thus, an error can be put on the value of $I(0)/c$ by propagating the errors on $I(0)$ and c so that the experimental and calculated values can be properly compared. Note that the standard error on $I(0)$ from the Guinier or $P(r)$ fit may not represent the total error, as the fitted values also depend on chosen q-range and D_{max} values. Although this is an important consideration, it is not discussed further here. While it is difficult to quantify what it means for the experimental and calculated values to be the same, it can be deemed acceptable if they agree to within about 10–20%. A factor of two is clearly not acceptable and even a factor of 1.5 could indicate a significant mixture of monomers and dimers, for instance.

Some software packages calculate the M_w from $I(0)$, which is another way to determine whether the experimental values of $I(0)$ are the same. This is acceptable if contrast is taken into account when calculating the M_w, which is not always the case, especially for software packages that were developed mainly for SAXS data analysis of proteins. Contrast is not an issue for SAXS data of protein–protein complexes regardless of deuteration, but it needs to be considered for protein–DNA, protein–detergent or protein–lipid complexes since the X-ray contrast between protein and DNA, detergents and lipids are different. Another method often used with SAXS data is to calculate the M_w from the Porod volume, V_p, obtained from the Porod invariant (Porod, 1951) or the volume of correlation, V_c (Rambo & Tainer, 2013). These volumes are obtained from the integrated area of the data. Thus, the data do not need to be on an absolute scale. However, they do need to be of high quality over the entire q-range with highly accurate background subtraction, which is why this method isn't typically used with SANS data.

A related assessment of validity of the SANS CV data is the match point analysis. The contrast match point for the complex was presumably calculated from its chemical composition prior to the contrast variation experiment. It should also be determined from the SANS data for comparison. From Eqs. (4) and (6), it is apparent that $I(0)$ is proportional to c and $\Delta\rho^2$, which in turn is proportional to $\%D_2O$ in the solvent. If a linear fit to $\sqrt{\frac{I(0)}{c}}$ vs $\%D_2O$ is performed using the fitted $I(0)$ values and measured c values, the x-intercept is the match point of the complex. This analysis can be performed with any linear fitting program or using the MULCh software package. Fig. 4 shows an example of the match point analysis for a 1 mg/mL complex consisting of a protiated 50 kDa protein and a 36 kDa protein with 50% deuteration. Note that, in order to obtain a linear plot, the signs of the $\sqrt{\frac{I(0)}{c}}$ values for which $\%D_2O > 60$ (the complex match point) are switched so that they are negative. In practice, the match point isn't exactly known in advance, but an initial guess can be made based on the calculated match point or by fitting $I(0)/c$ vs $\%D_2O$ to a second

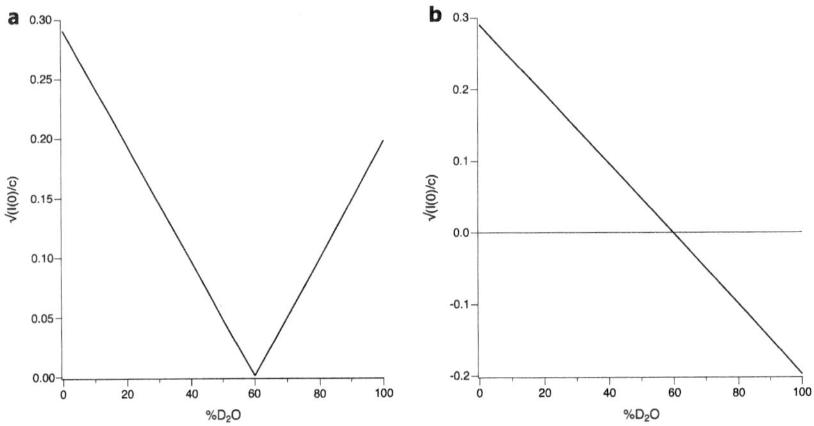

Fig. 4 Match point analysis plots of $\sqrt{\frac{I(0)}{c}}$ vs $\%D_2O$ for a 1 mg/mL complex consisting of a protiated 50 kDa protein and a 36 kDa protein with 50% deuteration. The complex match point is 60% D_2O. (A) All $\sqrt{\frac{I(0)}{c}}$ values are positive. (B) After making $\sqrt{\frac{I(0)}{c}}$ values negative for $\%D_2O$ values beyond the initially-guessed contrast match point in (A) to facilitate linear fitting.

degree polynomial to obtain the minimum of the curve. After the appropriate signs are changed, the linear fit provides a more exact value for the match point.

The $I(0)/c$ consistency checks are very important, especially for complexes with flexible components, where a wide range of R_g values may be possible. Whether the complex contains flexible components or not, the experimental R_g value may not agree with that from the starting model structure. However, if the $I(0)/c$ values and match point analysis show agreement, this provides more confidence that the data can be trusted and it is then a matter of finding a structure or ensemble of structures that fits the SANS CV data. On the other hand, if one or both of the $I(0)/c$ assessments doesn't provide the expected results based on the chemical composition, it is time to seriously consider that the sample is not as originally thought. Perhaps additional (non-SAS) characterization is needed prior to moving forward? Perhaps the stoichiometry is not as assumed? The match point analysis can provide clues as to the stoichiometry of the sample. Once further characterization and/or assessments have been made, jump back into the *in silico* experiment workflow to calculate the expected contrasts and $I(0)$ values and new model SANS curves based on the new information.

4. Further data analysis and model refinement

Once validity is established, the SANS CV data analysis can be continued using the Stuhrmann and parallel axis theorem analyses to obtain R_1, R_2, D_{CM} and decomposition analysis to obtain $I_{11}(q)$, $I_{22}(q)$ and $I_{12}(q)$, if possible, for comparison to the *in silico* results. This information will serve as a guide during the refinement of the starting model structure or in the search for a structural ensemble that best agrees with the complete set of SANS CV data.

The specifics of structure model refinement and structural ensemble analysis are beyond the scope of this work. For examples of structure refinement using MC or MD methods that utilize SANS and/or SAXS data as constraints, see (Castellanos, Howell, Gallagher, & Curtis, 2018; Chen et al., 2019; Clark et al., 2013; Datta et al., 2007; Gupta et al., 2012; Hermann & Hub, 2019; Potrzebowski, Trewhella, & Andre, 2018;

Puster et al., 2019; Wright, Elliston, Hui, & Perkins, 2019; Zaccai et al., 2016). With respect to the workflow, the next step is to identify model structures that are consistent with the SANS data.

For a single starting model structure, the calculated model SANS curves are compared to the corresponding SANS data at each contrast. It is important to note that this is not the same as fitting the data to a structure model that contains adjustable parameters. Rather, the model SANS data are often compared to the experimental data using a goodness of fit criterion such as the reduced χ_2, defined here as

$$\chi^2 = \frac{1}{(N_p - 1)} \sum_q \frac{\left(I_{\exp}(q) - I_{\text{calc}}(q)\right)^2}{\sigma_{\exp}(q)^2}, \tag{14}$$

where $I_{\exp}(q)$ is the experimentally determined SANS curve, $I_{\text{calc}}(q)$ is the calculated model SANS curve and $\sigma_{exp}(q)$ is the q-dependent error of the $I_{\exp}(q)$ values. The sum is taken over N_p independent data points. The Chi-Square Filter module in SASSIE-web provides several goodness of fit criteria such as reduced χ_2, χ_2, Pearson χ_2 and R factor, that can be used to compare model and experimental SANS curves. In Hura et al. (2013), a volatility ratio between model and experimental SANS curves serves as a goodness of fit test. Yet another is defined by χ^2_{free} (Rambo & Tainer, 2013). While often the regions of the data that do not match the calculated SANS curve well can be seen "by eye," application of the correlation map goodness of fit technique (Franke, Jeffries, & Svergun, 2015) provides quantitative information on where in q-space the data and model SANS curves diverge. This will not necessarily provide direct information on what parts of the model structure are not correct. However, it can provide some insight into the length scales where the model structures are not in good agreement with the data—and at which contrasts.

For an ensemble of structures, those structures producing the model SANS curves that best fit the SANS data need to be found. This process can be partially automated by plotting χ^2 vs R_g at each contrast to identify the structures that match the experimental R_g values as well as best agree with the SANS CV data. This can serve as an initial screening to reduce the number of single structures that best agree with the SANS CV data. The model structure that agrees best with the data at one contrast may not agree as well at other contrasts. Therefore, the search for structures with the best agreement to the data must be performed globally at all contrasts.

One means to accomplish this is to find ensembles of structures that have the lowest χ^2 with respect to the SANS data at each contrast and then identify the subset of structures that is present in all of the ensembles. This method was used in Puster et al. (2019) to find structures for the intrinsically disordered domain of vitronectin in a complex consisting of deuterated plasminogen activator inhibitor-1 and protiated vitronectin. The subset of structures was then further evaluated to assess agreement with the R_1, R_2 and D_{CM} values determined from the Stuhrmann and parallel axis theorem data analyses. Another method that is perhaps easier to automate is to determine an average χ^2 value for each structure over all contrasts measured and then identify an ensemble of structures with the lowest average χ^2. This strategy was employed in the study of a complex of the chaperone, Skp, and deuterated unfolded outer membrane A (OmpA) (Zaccai et al., 2016). Other types of ensemble analyses include the ensemble optimization method (EOM) (Bernadó, Mylonas, Petoukhov, Blackledge, & Svergun, 2007), minimal ensemble search (MES) (Pelikan, Hura, & Hammel, 2009), Bayesian ensemble estimation from SAS (BEES) (Bowerman, Curtis, Clayton, Brookes, & Wereszczynski, 2019) and ensemble refinement of SAXS (EROS) (Różycki, Kim, & Hummer, 2011).

Finally, it is important to remember that structure modeling is an iterative process as discussed in Krueger (2017). In the case of the Skp-OmpA complex, the model structure for Skp was held fixed in the conformation that produced good agreement to SANS data obtained on another complex of Skp with deuterated unfolded outer membrane protein W (OmpW). However, even the structure of Skp-OmpA with the lowest average χ^2 did not agree well with the 85% D_2O data, which is near the match point of the deuterated OmpA. Therefore, the structure for Skp was refined in order to obtain better agreement with the 85% D_2O data, as illustrated in Fig. 5. A larger ensemble of Skp structures would have enabled a more thorough search of conformation space for Skp, perhaps resulting in even better-suited structures. The next step to finding structures in better agreement with the entire SANS CV data set would be to calculate a new structure ensemble of the complex using the refined Skp conformation. The take home message is that the model structure(s) should be rethought if suitable agreement to the SANS CV data isn't found during the first round of structure refinement. It is then time to cycle back into the workflow to attempt to find structures that better represent the SANS CV data.

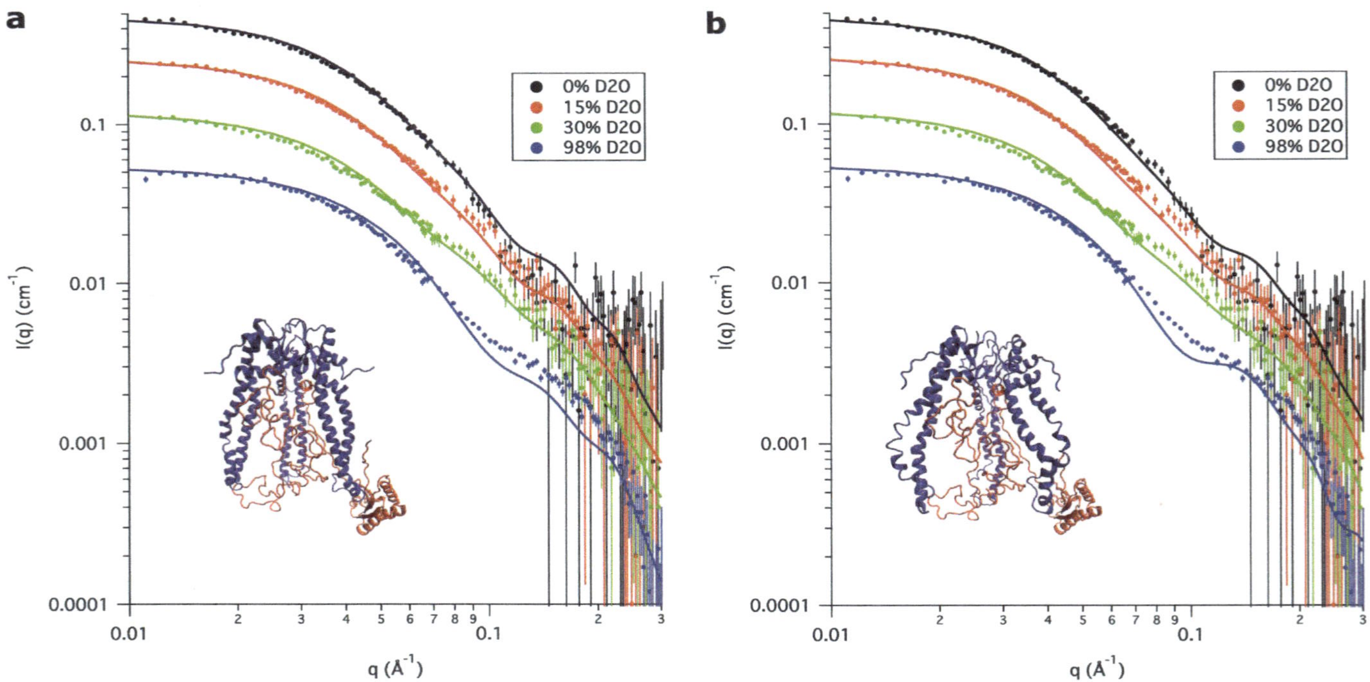

Fig. 5 SANS contrast variation data for a Skp-OmpA complex. Solid lines represent the calculated SANS curves from the model structures in the inset assuming 60% deuteration for OmpA. Skp is shown in blue and OmpA is shown in red. (A) The model curves were calculated from the Skp-OmpA structure described in (Zaccai et al., 2016). (B) Best-matched model curve calculated using the same OmpA structure as in panel (A) paired with a more symmetric Skp structure. Error bars represent the standard error of the mean with respect to the number of pixels used in the data averaging. *Adapted from Krueger, S. (2017). Designing and performing biological solution small-angle neutron scattering contrast variation experiments on multi-component assemblies. In B. Chaudhuri, I. G. Muñoz, S. Qian, & V. S. Urban (Eds.), Biological small angle scattering: Techniques, strategies and tips (pp. 65–85). Singapore: Springer Singapore.*

5. Concluding remarks

A workflow for performing successful SANS CV experiments is offered here with emphasis on experiment planning and execution *in silico* and validation of the SANS CV data prior to more advanced data analysis and structure modeling. It is well worth the effort to plan the experiment as much as possible in order to choose the contrasts that will give the most useful information for a given biological complex. The importance of assessing the validity of the data before testing structure models can't be overstated. This is even more important for flexible systems where the size of the starting structure is likely to be different than the average size determined from the SANS CV data. Once the validity of the data has been established, then and only then, it is appropriate to proceed with advanced data analysis, if applicable, and model structure refinement using the SANS CV data as constraints.

Acknowledgments

The author thanks Joseph E. Curtis and Susana C. M. Teixeira for their critical reading of this chapter and their helpful comments.

Disclaimer

Certain commercial equipment, materials, software or suppliers are identified in this paper to foster understanding. Such identification does not imply recommendation or endorsement by the National Institute of Standards and Technology, nor does it imply that the materials or equipment identified are necessarily the best available for the purpose.

References

Abraham, M. J., Murtola, T., Schulz, R., Páll, S., Smith, J. C., Hess, B., et al. (2015). GROMACS: High performance molecular simulations through multi-level parallelism from laptops to supercomputers. *SoftwareX*, *1–2*, 19–25.

Bernadó, P., Mylonas, E., Petoukhov, M. V., Blackledge, M., & Svergun, D. I. (2007). Structural characterization of flexible proteins using small-angle X-ray scattering. *Journal of the American Chemical Society*, *129*, 5656–5664.

Best, R. B., Zhu, X., Shim, J., Lopes, P. E. M., Mittal, J., Feig, M., et al. (2012). Optimization of the additive CHARMM all-atom protein force field targeting improved sampling of the backbone ϕ, ψ and side-chain χ_1 and χ_2 dihedral angles. *Journal of Chemical Theory and Computation*, *8*, 3257–3273.

Bowerman, S., Curtis, J. E., Clayton, J., Brookes, E. H., & Wereszczynski, J. (2019). BEES: Bayesian ensemble Estimation from SAS. *Biophysical Journal*, *117*, 399–407.

Brooks, B. R., Brooks, C. L., Mackerell, A. D., Nilsson, L., Petrella, R. J., Roux, B., et al. (2009). CHARMM: The biomolecular simulation program. *Journal of Computational Chemistry*, *30*, 1545–1614.

Buchan, D. W. A., Minneci, F., Nugent, T. C. O., Bryson, K., & Jones, D. T. (2013). Scalable web services for the PSIPRED protein analysis workbench. *Nucleic Acids Research*, *41*, W349–W357.

Castellanos, M. M., Howell, S. C., Gallagher, D. T., & Curtis, J. E. (2018). Characterization of the NISTmAb reference material using small-angle scattering and molecular simulation: Part I: Dilute protein solutions. *Analytical and Bioanalytical Chemistry*, *410*, 2141–2159.

Chen, J., Im, W., & Brooks, C. L. (2005). Application of torsion angle molecular dynamics for efficient sampling of protein conformations. *Journal of Computational Chemistry*, *26*, 1565–1578.

Chen, P., Shevchuk, R., Strnad, F. M., Lorenz, C., Karge, L., Gilles, R., et al. (2019). Combined small-angle X-ray and neutron scattering restraints in molecular dynamics simulations. *Journal of Chemical Theory and Computation*, *15*, 4687–4698.

Clark, N. J., Zhang, H., Krueger, S., Lee, H. J., Ketchem, R. R., Kerwin, B., et al. (2013). Small-angle neutron scattering study of a monoclonal antibody using free-energy constraints. *The Journal of Physical Chemistry B*, *117*, 14029–14038.

Cornell, W. D., Cieplak, P., Bayly, C. I., Gould, I. R., Merz, K. M., Ferguson, D. M., et al. (1995). A second generation force field for the simulation of proteins, nucleic acids, and organic molecules. *Journal of the American Chemical Society*, *117*, 5179–5197.

Curtis, J. E., Raghunandan, S., Nanda, H., & Krueger, S. (2012). SASSIE: A program to study intrinsically disordered biological molecules and macromolecular ensembles using experimental scattering restraints. *Computer Physics Communications*, *183*, 382–389.

Datta, S. A. K., Curtis, J. E., Ratcliff, W., Clark, P. K., Crist, R. M., Lebowitz, J., et al. (2007). Conformation of the HIV-1 Gag protein in solution. *Journal of Molecular Biology*, *365*, 812–824.

Engelman, D., & Moore, P. B. (1975). Determination of quaternary structure by small-angle neutron scattering. *Quarterly Reviews of Biophysics*, *4*, 219–241.

Franke, D., Jeffries, C. M., & Svergun, D. I. (2015). Correlation map, a goodness-of-fit test for one-dimensional X-ray scattering spectra. *Nature Methods*, *12*, 419–422.

Glatter, O. (1977). A new method for the evaluation of small-angle scattering data. *Journal of Applied Crystallography*, *10*, 415–421.

Glatter, O., & Kratky, O. (1982). *Small-angle X-ray scattering*. New York: Academic Press.

Grishaev, A. V. (2017). Hybrid applications of solution scattering to aid structural biology. In B. Chaudhuri, I. G. Muñoz, S. Qian, & V. S. Urban (Eds.), *Biological small angle scattering: Techniques, strategies and tips* (pp. 215–227). Singapore: Springer Singapore.

Grudinin, S., Garkavenko, M., & Kazennov, A. (2017). *Pepsi-SAXS*: An adaptive method for rapid and accurate computation of small-angle X-ray scattering profiles. *Acta Crystallographica Section D Structural Biology*, *73*, 449–464.

Guinier, A., & Fournet, G. (1955). *Small-angle scattering of X-rays*. New York: John Wiley & Sons, Inc.

Gupta, K., Curtis, J. E., Krueger, S., Hwang, Y., Cherepanov, P., Bushman, F. D., et al. (2012). Solution conformations of prototype foamy virus integrase and its stable synaptic complex with U5 viral DNA. *Structure*, *20*, 1918–1928.

Hermann, M. R., & Hub, J. S. (2019). SAXS-restrained ensemble simulations of intrinsically disordered proteins with commitment to the principle of maximum entropy. *Journal of Chemical Theory and Computation*, *15*, 5103–5115.

Humphrey, W. (1996). VMD: Visual molecular dynamics. *Journal of Molecular Graphics*, *14*, 33–38.

Hura, G. L., Budworth, H., Dyer, K. N., Rambo, R. P., Hammel, M., McMurray, C. T., et al. (2013). Comprehensive macromolecular conformations mapped by quantitative SAXS analyses. *Nature Methods*, *10*, 453–454.

Ibel, K., & Stuhrmann, H. B. (1975). Comparison of neutron and X-ray scattering of dilute myoglobin solutions. *Journal of Molecular Biology, 93*, 255–265.

Jacques, D. A., Guss, J. M., Svergun, D. I., & Trewhella, J. (2012). Publication guidelines for structural modelling of small-angle scattering data from biomolecules in solution. *Acta Crystallographica, Section D: Biological Crystallography, 68*, 620–626.

Jacques, D. A., Guss, J. M., & Trewhella, J. (2012). Reliable structural interpretation of small-angle scattering data from bio-molecules in solution—The importance of quality control and a standard reporting framework. *BMC Structural Biology, 12*, 9.

Jacques, D. A., & Trewhella, J. (2010). Small-angle scattering for structural biology—Expanding the frontier while avoiding the pitfalls. *Protein Science: A Publication of the Protein Society, 19*, 642–657.

Jeffries, C. M., Graewert, M. A., Blanchet, C. E., Langley, D. B., Whitten, A. E., & Svergun, D. I. (2016). Preparing monodisperse macromolecular samples for successful biological small-angle X-ray and neutron-scattering experiments. *Nature Protocols, 11*, 2122–2153.

Jo, S., Cheng, X., Lee, J., Kim, S., Park, S.-J., Patel, D. S., et al. (2017). CHARMM-GUI 10 years for biomolecular modeling and simulation. *Journal of Computational Chemistry, 38*, 1114–1124.

Jones, D. T. (1999). Protein secondary structure prediction based on position-specific scoring matrices. *Journal of Molecular Biology, 292*, 195–202.

Jumper, J., Evans, R., Pritzel, A., Green, T., Figurnov, M., Ronneberger, O., et al. (2021). Highly accurate protein structure prediction with AlphaFold. *Nature, 596*, 583–589.

Kelley, L. A., Mezulis, S., Yates, C. M., Wass, M. N., & Sternberg, M. J. E. (2015). The Phyre2 web portal for protein modeling, prediction and analysis. *Nature Protocols, 10*, 845–858.

Kelley, L. A., & Sternberg, M. J. E. (2009). Protein structure prediction on the web: A case study using the Phyre server. *Nature Protocols, 4*, 363–371.

Kozakov, D., Hall, D. R., Xia, B., Porter, K. A., Padhorny, D., Yueh, C., et al. (2017). The ClusPro web server for protein–protein docking. *Nature Protocols, 12*, 255–278.

Krueger, S. (2017). Designing and performing biological solution small-angle neutron scattering contrast variation experiments on multi-component assemblies. In B. Chaudhuri, I. G. Muñoz, S. Qian, & V. S. Urban (Eds.), *Biological small angle scattering: Techniques, strategies and tips* (pp. 65–85). Singapore: Springer Singapore.

Krueger, S. (2022). Small-angle neutron scattering contrast variation studies of biological complexes: Challenges and triumphs. *Current Opinion in Structural Biology, 74*, 102375.

Lee, J., Cheng, X., Swails, J. M., Yeom, M. S., Eastman, P. K., Lemkul, J. A., et al. (2016). CHARMM-GUI input generator for NAMD, GROMACS, AMBER, OpenMM, and CHARMM/OpenMM simulations using the CHARMM36 additive force Field. *Journal of Chemical Theory and Computation, 12*, 405–413.

Lee, J., Hitzenberger, M., Rieger, M., Kern, N. R., Zacharias, M., & Im, W. (2020). CHARMM-GUI supports the Amber force fields. *The Journal of Chemical Physics, 153*, 035103.

Lyskov, S., Chou, F.-C., Conchúir, S.Ó., Der, B. S., Drew, K., Kuroda, D., et al. (2013). Serverification of molecular modeling applications: The Rosetta online server that includes everyone (ROSIE). *PLoS One, 8*, e63906.

Lyskov, S., & Gray, J. J. (2008). The RosettaDock server for local protein-protein docking. *Nucleic Acids Research, 36*, W233–W238.

MacKerell, A. D., Bashford, D., Bellott, M., Dunbrack, R. L., Evanseck, J. D., Field, M. J., et al. (1998). All-atom empirical potential for molecular modeling and dynamics studies of proteins. *The Journal of Physical Chemistry B, 102*, 3586–3616.

Mahieu, E., & Gabel, F. (2018). Biological small-angle neutron scattering: Recent results and development. *Acta Crystallographica Section D Structural Biology, 74*, 715–726.

Manalastas-Cantos, K., Konarev, P. V., Hajizadeh, N. R., Kikhney, A. G., Petoukhov, M. V., Molodenskiy, D. S., et al. (2021). *ATSAS 3.0*: Expanded functionality and new tools for small-angle scattering data analysis. *Journal of Applied Crystallography*, *54*, 343–355.

Moore, P. B. (1982). Small-angle scattering techniques for the study of biological macromolecules and macromolecular aggregates. In G. Ehrenstein, & H. Lecar (Eds.), *Vol. 20. Methods of experimental physics* (pp. 337–390). New York: Academic Press.

MULCh. (n.d.). Retrieved March 18, 2022, from MULCH: ModULes For The Analysis of Contrast Variation Data website: https://smb-research.smb.usyd.edu.au/NCVWeb/index.jsp.

Olah, G. A., Rokop, S. E., Wang, C.-L. A., Blechner, S. L., & Trewhella, J. (1994). Troponin I encompasses an extended troponin C in the Ca2+-bound complex: A small-angle X-ray and neutron scattering study. *Biochemistry*, *33*, 8233–8239.

PDB101 (n.d.): Learn: Guide to Understanding PDB Data: Beginner's Guide to PDB Structures and the PDBx/mmCIF Format. Retrieved April 12, 2022, from RCSB: PDB-101 website: https://pdb101.rcsb.org/learn/guide-to-understanding-pdb-data/beginner%E2%80%99s-guide-to-pdb-structures-and-the-pdbx-mmcif-format.

Pelikan, M., Hura, G., & Hammel, M. (2009). Structure and flexibility within proteins as identified through small angle X-ray scattering. *General Physiology and Biophysics*, *28*, 174–189.

Pepsi-SANS. (n.d.). Retrieved April 12, 2022, from https://team.inria.fr/nano-d/software/pepsi-sans/.

Phillips, J. C., Braun, R., Wang, W., Gumbart, J., Tajkhorshid, E., Villa, E., et al. (2005). Scalable molecular dynamics with NAMD. *Journal of Computational Chemistry*, *26*, 1781–1802.

Pierce, B. G., Wiehe, K., Hwang, H., Kim, B.-H., Vreven, T., & Weng, Z. (2014). ZDOCK server: Interactive docking prediction of protein-protein complexes and symmetric multimers. *Bioinformatics*, *30*, 1771–1773.

Porod, G. (1951). Die Röntgenkleinwinkelstreuung von dichtgepackten kolloiden Systemen. *Kolloid Zeitschrift*, *124*, 83–114.

Potrzebowski, W., Trewhella, J., & Andre, I. (2018). Bayesian inference of protein conformational ensembles from limited structural data. *PLoS Computational Biology*, *14*, e1006641.

Pronk, S., Páll, S., Schulz, R., Larsson, P., Bjelkmar, P., Apostolov, R., et al. (2013). GROMACS 4.5: A high-throughput and highly parallel open source molecular simulation toolkit. *Bioinformatics*, *29*, 845–854.

Puster, L. O., Stanley, C. B., Uversky, V. N., Curtis, J. E., Krueger, S., Chu, Y., et al. (2019). Characterization of an extensive interface on vitronectin for binding to plasminogen activator Inhibitor-1: Adoption of structure in an intrinsically disordered region. *Biochemistry*, *58*, 5117–5134.

Rambo, R. P., & Tainer, J. A. (2013). Accurate assessment of mass, models and resolution by small-angle scattering. *Nature*, *496*, 477–481.

Różycki, B., Kim, Y. C., & Hummer, G. (2011). SAXS ensemble refinement of ESCRT-III CHMP3 conformational transitions. *Structure*, *19*, 109–116.

Sarachan, K. L., Curtis, J. E., & Krueger, S. (2013). Small-angle scattering contrast calculator for protein and nucleic acid complexes in solution. *Journal of Applied Crystallography*, *46*, 1889–1893.

SASSIE-web. (n.d.). Retrieved June 19, 2015, from SASSIE-web website: https://sassie-web.chem.utk.edu/sassie2/.

Schrödinger *The PyMOL Molecular Graphics System, Version 2.1, Schrödinger, LLC.* (n.d.).

Semenyuk, A. V., & Svergun, D. I. (1991). GNOM—A program package for small-angle scattering data processing. *Journal of Applied Crystallography*, *24*, 537–540.

Svergun, D. I. (1992). Determination of the regularization parameter in indirect-transform methods using perceptual criteria. *Journal of Applied Crystallography*, *25*, 495–503.

Svergun, D., Barberato, C., & Koch, M. H. J. (1995). CRYSOL—A program to evaluate X-ray solution scattering of biological macromolecules from atomic coordinates. *Journal of Applied Crystallography*, *28*, 768–773.

Tovchigrechko, A., & Vakser, I. A. (2006). GRAMM-X public web server for protein-protein docking. *Nucleic Acids Research*, *34*, W310–W314.

Trewhella, J., Duff, A. P., Durand, D., Gabel, F., Guss, J. M., Hendrickson, W. A., et al. (2017). 2017 publication guidelines for structural modelling of small-angle scattering data from biomolecules in solution: An update. *Acta Crystallographica Section D Structural Biology*, *73*, 710–728.

Varadi, M., Anyango, S., Deshpande, M., Nair, S., Natassia, C., Yordanova, G., et al. (2022). AlphaFold protein structure database: Massively expanding the structural coverage of protein-sequence space with high-accuracy models. *Nucleic Acids Research*, *50*, D439–D444.

Watson, M. C., & Curtis, J. E. (2013). Rapid and accurate calculation of small-angle scattering profiles using the golden ratio. *Journal of Applied Crystallography*, *46*, 1171–1177.

Webb, B., & Sali, A. (2016). Comparative protein structure modeling using MODELLER. *Current Protocols in Bioinformatics*, *54*. https://doi.org/10.1002/cpbi.3.

Whitten, A. E., Cai, S., & Trewhella, J. (2008). *MULCh*: Modules for the analysis of small-angle neutron contrast variation data from biomolecular assemblies. *Journal of Applied Crystallography*, *41*, 222–226.

Wright, D. W., Elliston, E. L. K., Hui, G. K., & Perkins, S. J. (2019). Atomistic modeling of scattering curves for human IgG1/4 reveals new structure-function insights. *Biophysical Journal*, *117*, 2101–2119.

Zaccai, N. R., Sandlin, C. W., Hoopes, J. T., Curtis, J. E., Fleming, P. J., Fleming, K. G., et al. (2016). Deuterium labeling together with contrast variation small-angle neutron scattering suggests how Skp captures and releases unfolded outer membrane proteins. In Z. Kelman (Ed.), *Vol. 566. Methods in enzymology* (pp. 159–210). Elsevier.

Zhang, W., Howell, S. C., Wright, D. W., Heindel, A., Qiu, X., Chen, J., et al. (2017). Combined Monte Carlo/torsion-angle molecular dynamics for ensemble modeling of proteins, nucleic acids and carbohydrates. *Journal of Molecular Graphics and Modelling*, *73*, 179–190.

CHAPTER FIVE

Technical considerations for small-angle neutron scattering from biological macromolecules in solution: Cross sections, contrasts, instrument setup and measurement

Zuzanna Pietras[a], Kathleen Wood[b], Andrew E. Whitten[b,*], and Cy M. Jeffries[c,*]

[a]Department of Physics, Chemistry, and Biology, Linköping University, Linköping, Sweden
[b]Australian Nuclear Science and Technology Organisation, New Illawarra Rd, Lucas Heights, NSW, Australia
[c]European Molecular Biology Laboratory (EMBL), Hamburg, Germany
*Corresponding authors: e-mail address: andrew.whitten@ansto.gov.au; cy.jeffries@embl-hamburg.de

Contents

Methods in Enzymology, Volume 677
ISSN 0076-6879
https://doi.org/10.1016/bs.mie.2022.08.008

Abstract

Small angle scattering affords an approach to evaluate the structure of dilute populations of macromolecules in solution where the measured scattering intensities relate to the distribution of scattering-pair distances within each macromolecule. When small angle neutron scattering (SANS) with contrast variation is employed, additional structural information can be obtained regarding the internal organization of bio-macromolecule complexes and assemblies. The technique allows for the components of assemblies to be selectively 'matched in' and 'matched out' of the scattering profiles due to the different ways the isotopes of hydrogen—protium ^{1}H, and deuterium ^{2}H (or D)—scatter neutrons. The isotopic substitution of ^{1}H for D in the sample enables the controlled variation of the scattering contrasts. A contrast variation experiment requires trade-offs between neutron beam intensity, q-range, wavelength and q-resolution, isotopic labelling levels, sample concentration and path-length, and measurement times. Navigating these competing aspects to find an optimal combination is a daunting task. Here we provide an overview of how to calculate the neutron scattering contrasts of dilute biological macromolecule samples prior to an experiment and how this then informs the approach to configuring SANS instruments and the measurement of a contrast variation series dataset.

1. Introduction

Neutron scattering has diverse applications across the hard and soft matter sciences to probe the atomic and magnetic structure, internal dynamics, and bulk organization of materials (Frielinghaus, 2012; Pynn, 2009). In the specific case of using small-angle neutron scattering (SANS) for the structural analysis of biological macromolecules (Fitter, Gutberlet, & Katsaras, 2006; Lombardo, Calandra, & Kiselev, 2020; Mahieu & Gabel, 2018; Svergun, Koch, Timmins, & May, 2013b), it is the measurement of elastic coherent-scattering events that yields structural information. The coherently scattered neutron emanating from the nucleus of each isotope in a sample may be viewed as a set of convoluted waves where the constructive and destructive interference pattern of the amplitudes relates to the spatial distribution of scattering centers, i.e., the internal structure. The pattern of the squared scattering amplitudes generates scattering intensities as a function of angle. It is from these measured intensity profiles that structural parameters are extracted, and models developed, to describe the structures of biomacromolecules, their complexes and assemblies in solution (see chapter "Data analysis and modelling of small-angle neutron scattering data with contrast variation from bio-macromolecular complexes" by Jeffries and Whitten in Methods in Enzymology, vol. 678, Part B).

As with all successful experiments it is necessary to understand some key aspects relating to the samples, the technique, and instrument prior to an experiment (Hammouda, 2012; Schober, 2009). In relation to the samples used for 'bio-SANS,' they are typically very dilute, monodisperse systems where the mass fraction of the macromolecules is extremely low in the surrounding aqueous solvent. The dilute aspect of the sample combined with the *very* low flux of a SANS instrument generates very weak coherent scattering profiles. Therefore, prior to an experiment is crucial to: (i) Obtain information about the propensity of biomacromolecules and the supporting aqueous environment of the sample to scatter neutrons; (ii) How the composition of the sample can be manipulated, using a process called contrast variation, to increase the information content of a SANS experiment; (iii) How a SANS instrument is configured to detect the very weak coherent biomolecule scattering intensities above the dominant scattering from the aqueous solvent, and; (iv) How to approach data collection and processing to yield interpretable results.

2. Nuclear cross sections

The propensity of a sample to scatter a neutron is dependent on the 'strength' of any one specific isotope nucleus-neutron interaction. The magnitude of neutron-isotope scattering is described by the differential cross section that quantifies the scattering rate (dR), or the number of neutron scattering events per unit time per unit area of sample, through any given (differential) solid scattering angle ($d\mathbf{\Omega}_{(2\theta,\varphi)}$), normalized to the incident flux (ψ_i) of a collimated neutron beam (i.e., a beam with a directed incident neutron wave-vector):

$$\frac{d\sigma}{d\mathbf{\Omega}_{(2\theta,\varphi)}} = \frac{dR}{\mathbf{\Omega}_{(2\theta,\varphi)}} / \psi_i \tag{1}$$

Each pixel on a detector, with finite size, and corresponding solid angle (Ω), can be defined by the scattering angle, 2θ, and its corresponding azimuth angle, φ, relative to the incident beam (Fig. 1A). By measuring the number

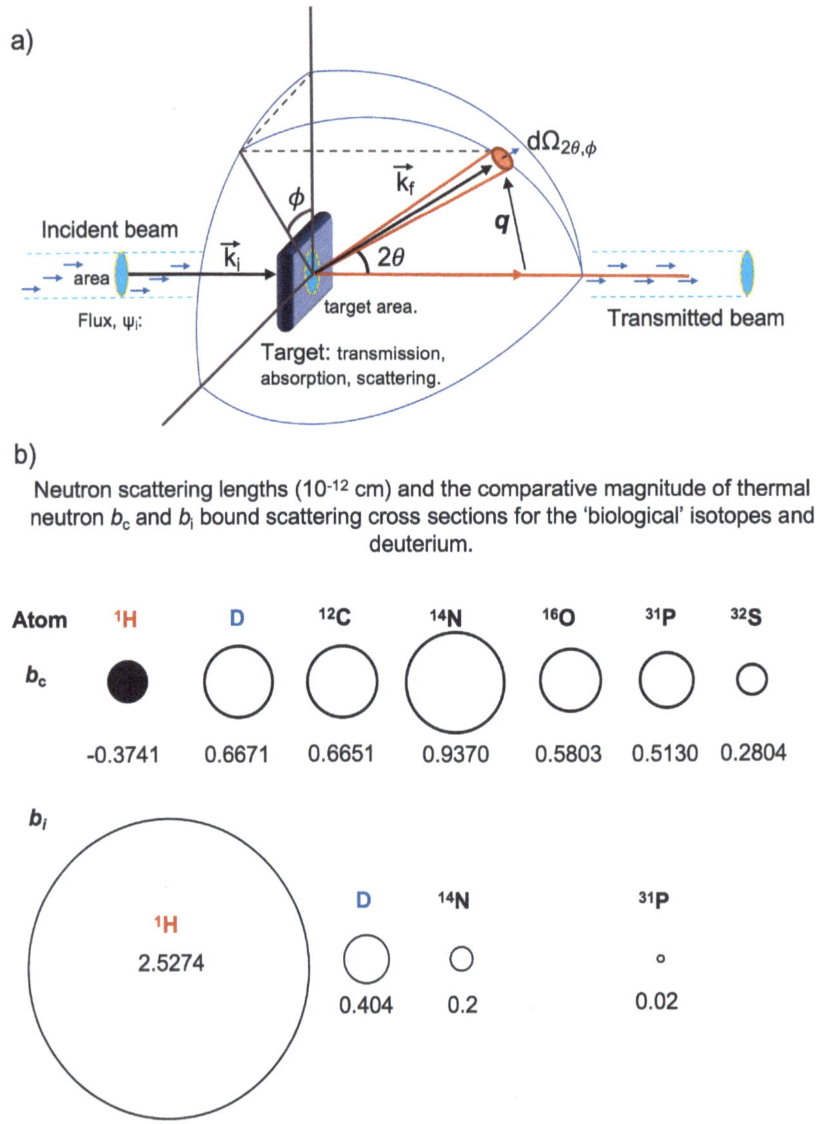

Fig. 1 (A) A representation of the differential scattering cross section that is a measure of the capacity of a nucleus to scatter an incident neutron beam of a particular energy and direction through a given solid angle. The momentum transfer **q** relates to the scattering angle 2θ. (B) The bound thermal neutron coherent and incoherent scattering lengths for the common isotopes found in biological samples, and the relative magnitude of the subsequent cross sections represented as conceptual areas. The ^{12}C, ^{16}O and ^{32}S isotopes do not have an incoherent scattering cross section due to the nuclear spin-state.

of scattered neutrons across all pixels (each with a given 2θ, φ,and Ω) it is possible to obtain total cross section, σ_c, of each nuclear isotope at a given energy that quantifies the probability of a nucleus-neutron interaction:

$$\sigma_c = 4\pi b^2 \tag{?}$$

where b is the scattering length (cm). The total scattering length maybe sub-divided into a number of components including the coherent scattering length, b_c, and the incoherent scattering length, b_i, which are relevant to bio-macromolecule solution investigations (Rauch & Waschkowski, 2003; Sears, 1992).

2.1 The coherent scattering length-b_c

To obtain structural information from biomolecules using SANS, b_c plays an important role as it quantifies the probability of coherent scattering events. Neutrons maybe viewed in terms of wave-particle duality as described by the de Broglie relation that links a neutrons wavelength to its momentum, **p**:

$$\lambda = \frac{h}{|\mathbf{p}|} \tag{3}$$

where h is Planck's constant. Coherent scattering events, where the velocity/momentum/energy/wavelength of the neutron remain unchanged, are referred to as elastic, that is, for elastic-coherent neutron scattering only the direction of the wave vector changes relative to the wave vector of the incident neutron beam. In 'particle terms' there is a momentum transfer with no loss or gain in kinetic energy to the outgoing neutron. Consequently, if the distance between any two scattering centers within a macromolecule is preserved over time, the neutrons that interact elastically with each scattering center will result in constructive interference of coherently scattered wave amplitudes. For a macromolecule, the average magnitude of the coherent scattering wave amplitudes will depend on the isotope-specific coherent scattering length of each nucleus, b_c, per unit volume, V, i.e., the average scattering length density of the macromolecule, $\overline{\rho}_m$:

$$\overline{\rho}_m = \frac{\sum b_c}{V} \tag{4}$$

In a SANS measurement, the square of the coherent scattering amplitudes manifest as the scattering intensities, I. However, this only occurs if there is a difference between the average scattering length of the solvent, $\overline{\rho}_s$, and the average scattering length density of the macromolecules, in the sample:

$$(\overline{\rho}_m - \overline{\rho}_s) \leq 0 \geq (\overline{\rho}_m - \overline{\rho}_s). \tag{5}$$

This difference in scattering length density—known as the neutron scattering contrast, $\Delta\rho$,—maybe positive or negative. If $\Delta\rho = 0$, the scattering length density will be near homogeneous throughout the sample ($\overline{\rho}_m \cong \overline{\rho}_s$), except for scattering length density fluctuations, $\rho_i(\boldsymbol{r})$, internal to the macromolecules:

$$\Delta\rho_m(\boldsymbol{r}) = \Delta\rho_m\rho_{sh}(\boldsymbol{r}) + \rho_i(\boldsymbol{r}). \tag{6}$$

where $\rho_{sh}(\boldsymbol{r})$ represents the shape of the molecule. Thus, Eq. (6) show that at high contrast, the scattering is related to the shape of the molecule, while at low contrast, the scattering is related to the density fluctuations (H. B. Stuhrmann, 2012). Subsequently, at $\Delta\rho = 0$ the coherent (shape) scattering amplitudes get 'matched out' with the scattering contributions from the solvent (Glinka, 2011). Therefore, to obtain structural information there must be an *excess* contrast within or between the components of the sample where the absolute magnitude of the neutron scattering contrast is greater than zero, $|\Delta\rho| > 0$. To extract the coherent scattering contributions of the macromolecules from the sample scattering requires subtracting the scattering contributions of the solvent. SANS with contrast variation on biomacromolecules in solution is therefore a subtractive technique, requiring the preparation of samples and solvent blanks with the same $\overline{\rho}_s$.

The predominant 'biological isotopes' include ^{12}C, ^{16}O, ^{14}N, ^{31}P, ^{32}S and ^{1}H (Fig. 1B). The majority of these isotopes have a positive b_c, a terminology that is ascribed to a 180° phase shift in the out-going scattered neutron amplitude. However, and importantly, the most common biological isotope, ^{1}H, has a negative b_c that preserves the phase of the incident neutron as it is scattered. Consequently, the average number of ^{1}H per unit volume has a dramatic influence on the $\Delta\rho$. For example, protein and polynucleotides have naturally different chemical, and thus isotopic, compositions, and very different average ^{1}H per unit volume resulting in different average scattering length densities and subsequent scattering contrasts in identical aqueous conditions (Fig. 2A). Importantly, the substitution of ^{1}H for deuterium (^{2}H or D)—that has a positive b_c similar to the other 'biological isotopes'—allows for the controlled manipulation of the $\Delta\rho$ of the sample, both in the macromolecules themselves via non-exchangeable D-labelling or in the aqueous medium by swapping out $^{1}H_2O$ with D_2O at different volume fraction ratios.

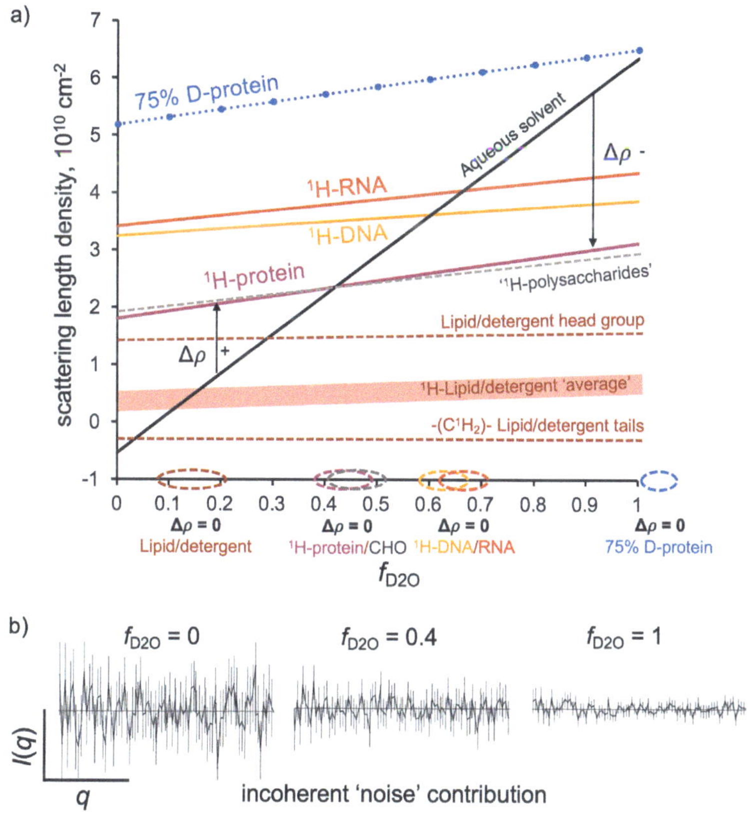

Fig. 2 (A) Scattering length densities of different classes of ^{1}H biopolymer classes in an aqueous solution as a function of the volume fraction of D_2O in the solvent, including a deuterated protein example ($f_D = 0.75$). Lipids and detergents have numerous varying scattering length densities due to distinct differences in the composition of the lipid/detergent heads compared to the aliphatic (-C^1H_2-) tails. Th Dr. refers to the contrast of relative for a ^{1}H protein at below ($\Delta\rho$ +) and above the contrast match point ($\Delta\rho$ −). The contrast match point occurs at that f_{D_2O} where $\Delta\rho = 0$, i.e., when the scattering length density of the respective biomolecule crosses the average scattering length density of the aqueous solvent line. (B) Nominal incoherent scattering 'noise' that is primarily derived from ^{1}H, adds to the SANS $I(q)$ vs. q profiles which decreases significantly in high volume fractions of D_2O.

2.2 The incoherent scattering length—b_i

The second major neutron scattering length component that affects a SANS experiment, and particularly aqueous ^1H_2O solvents, is caused by incoherent scattering (Glinka, 2011; Hammouda, n.d.-b, Chapter 9; Harroun, Wignall, & Katsaras, 2006). All neutrons are scattered by nuclei with a given spin. For example, ^{1}H has two very distinct scattering lengths for each of its up and down spin states, b_+ 1.085 × 10^{-12} cm, and $b- = -4.751 \times 10^{-12}$ cm. The average coherent scattering length is a weighted average of $b+$ and $b-$, and the difference between these scattering lengths determines the magnitude of incoherent scattering length, b_i. When compared to the other isotopes, ^1H has a large difference, and the resulting incoherent b_i and subsequent σ_i cross section is comparatively enormous compared to the other 'biological isotopes' and deuterium. In most cases, SANS measurements are performed using unpolarized neutrons and in sample environments where the nuclear spins of the sample are not aligned. Therefore, when viewed from the perspective of a 'chaotic spin system,' the scattering amplitudes will be uncorrelated, i.e., incoherent. The effect of incoherent scattering—and especially the incoherent scattering from ^1H—is to produce severe fluctuations across all angles of the coherent scattering amplitudes, thereby increasing the variance of the measured scattering intensities. Incoherent spin-scattering yields significant noise and impacts data quality, and, in turn, influences the experimental approach to sample preparation and measurement especially in samples with high volume fractions of ^1H_2O (Fig. 2B).

2.3 Neutron scattering contrast

Small angle neutron scattering detects and quantifies the angular-dependent intensity of scattered neutrons expressed as $I(\mathbf{q})$ vs $\mathbf{q}$. For systems such as dilute tumbling macromolecules in solution which generate isotropic scattering profiles, the scattering intensity depends only the magnitude of $\mathbf{q}$, or the momentum transfer:

$$q = \frac{4\pi}{\lambda} \sin\theta \tag{7}$$

where 2θ is the scattering angle, and λ the neutron wavelength. For SANS, the $I(q)$ are typically placed on an absolute scale, cm^{-1}, so they can be transformed to quantify the size, volume and distribution of scattering pair distances internal to the macromolecules from which structural models may be developed (see chapter "Data analysis and modelling of small-angle neutron scattering data with contrast variation from bio-macromolecular complexes" by Jeffries and Whitten in Methods in Enzymology, vol. 678, Part B).

SANS measures the summed contribution of the scattering from each individual molecule i, of a population. After all background contributions have been subtracted, the scattering intensities from a population of macromolecules may be described by:

$$I(q) = S(q) \sum_{i}^{n} \left[(\Delta \rho_i V_i)^2 P_i(q) \right] \tag{8}$$

where V is the volume of each particle, $\Delta \rho$ the respective contrasts and $P_i(q)$ the form factor that represents the angular-dependent coherent scattering intensities arising from the distribution of distances between spatially correlated scattering centres within each particle of the population. If the macromolecules of a population are structurally identical and non-interacting, i.e., under very dilute conditions where $S(q)$ approaches 1, then Eq. (8) simplifies to

$$I(q) = N\left[(\Delta \rho V)^2 P(q) \right] \tag{9}$$

that describes the scattering from identical, non-interacting homogeneous tumbling particles with a number-density (concentration) of N. Such ideal monodisperse samples can be analyzed in terms of single-molecule particle scattering. The squared product of $\Delta \rho V$ dictates the magnitude of the scattering intensities. For example, small particles with large contrast may generate similar scattering intensities compared to larger particles with low contrast. Through extension, if the $\Delta \rho$ is systematically manipulated via ^{1}H—D substitution to alter the scattering length density of the solvent or macromolecules of the sample, it becomes possible to control the magnitude of $\Delta \rho$ and the resulting SANS intensities. Both contrast matching, where $\Delta \rho = 0$, and contrast variation, where multiple datasets are measured under various $\Delta \rho$ conditions, are especially pertinent to SANS investigations as such experiments allow for the structural interrogation of biological samples with *different* regions of average scattering length density (Stuhrmann, 2012; Stuhrmann, 2007). This includes, but is not limited to, protein-RNA and protein-DNA complexes, lipid-protein, and detergent-protein assemblies, conjugated bioorganic metal nanoparticles, lipid membranes, vesicles, filamentous assemblies, and even intact cellular machines, such as ribosomes. For protein-protein complexes, it is necessary to isotopically label one of the components with D to affect a scattering length density separation between the components because proteins have, on average, very similar ^{1}H per unit volume and thus similar $\Delta \rho$ relative to an aqueous solvent.

The capacity to collect SANS with contrast matching or variation data yields the structural disposition of the intact assemblies, and importantly, the structures, arrangement, and organization between the individual components of such systems. For a monodisperse biomacromolecular complex consisting of two distinct average scattering length density regions (components 1 and 2), and assuming that internal scattering length density fluctuations are negligible, the $I(q)$ maybe approximated as:

$$I(q) \approx N\left[(\Delta\rho_1 V_1)^2 P_1(q) + (\Delta\rho_2 V_2)^2 P_2(q) + 2\Delta\rho_1 V_1 \Delta\rho_2 V_2 P_{1,2}(q)\right] \tag{10}$$

at any one contrast condition. The SANS profile will be the summed contribution of the scattering intensities from either component of the complex, plus the cross-term ($2\Delta\rho_1 V_1 \Delta\rho_2 V_2 P_{1,\,2}(q)$) which describes the contrast-weighted scattering contributions from pair distances between the different components of the complex. If the contrast is systematically changed to a point where the scattering contribution of component 1 approaches 0, i.e., at the component 1 match point, $\Delta\rho_1 = 0$, the scattering profile will be dominated by contributions from the *second* component:

$$I(q)_{\Delta\rho_1=0} \approx N\left[(\Delta\rho_2 V_2)^2 P_2(q)\right] \tag{11}$$

from which $P_2(q)$, its structural parameters and shape, may be determined. Conversely, at the contrast match point of the second component, $\Delta\rho_2 = 0$, the coherent scattering will primarily be derived from component 1.

2.4 Estimating neutron scattering contrast

Dilute macromolecular samples for SANS are typically prepared between 5 and 10 mg/mL. Therefore 99–99.5% of the sample by mass will be aqueous solvent. To obtain the scattering contributions from the macromolecules of interest 'above' the dominant solvent scattering, relies on estimating what the contrast of a sample is, or how it may be modified to improve the coherent scattering signal. Neutron exposure times, sample concentration, the quantity material, and the number of contrast variation points, may be tailored and amended accordingly based on contrast estimates prior to an experiment.

The average neutron scattering contrast of a macromolecule is determined from the respective molecular volumes, concentrations, and isotopic scattering length composition of the solvent components (buffers, salts, small molecule additives, H_2O and D_2O) and the isotopic scattering length composition and molecular volumes of the sample macromolecules. Contrast calculations account for the possible isotopic labelling a given fraction of the non-exchangeable D in the sample macromolecule (f_D), in addition to the number of ^{1}H—D exchangeable positions (n_{xch1H}) and extent of ^{1}H—D exchange ($f_{access1H}$) between the macromolecule and solvent (Whitten, Cai, & Trewhella, 2008):

$$\Delta\bar{\rho}_m = \frac{\sum b_c + (n_{1H} - n_{xch1H}) f_D (b_D - b_{1H}) + n_{xch1H} f_{access1H} f_{D_2O} (b_D - b_{1H})}{V_m} - \frac{2b_{1H} + b_O + 2f_{D_2O}(b_D - b_{1H})}{V_{H_2O}} \quad (12)$$

Taking a two-component macromolecular complex as an example, the contrast calculations provide (Fig. 3):

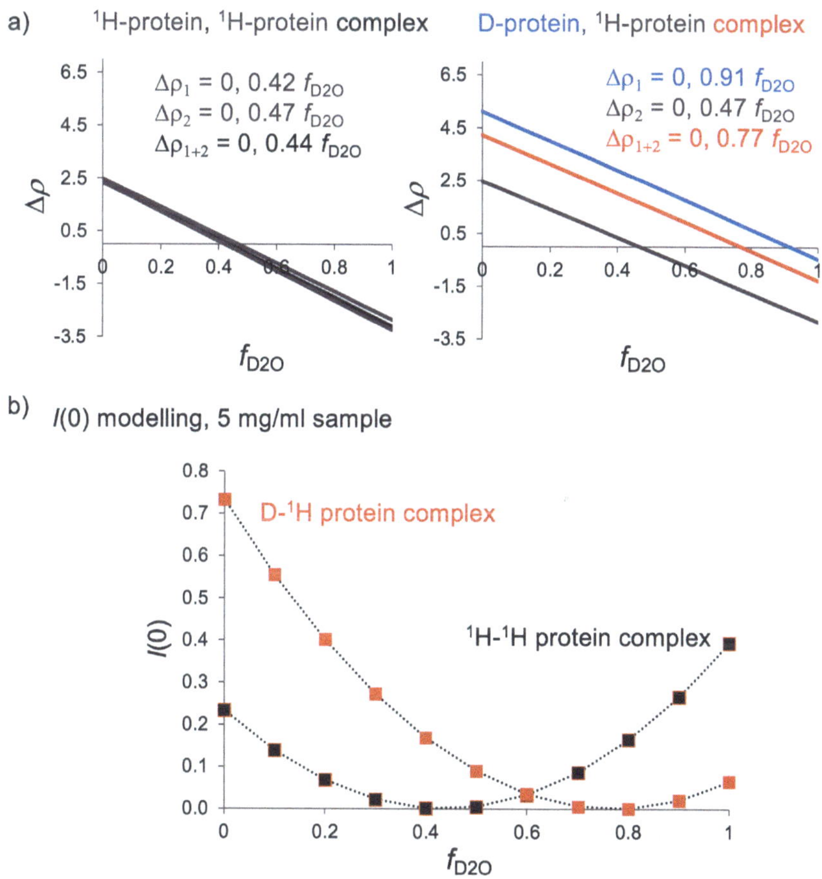

Fig. 3 (A) *Left*: A plot of the individual component and whole complex neutron scattering contrasts and match points for a ^{1}H—^{1}H protein-protein complex obtained from *Contrast* calculation outputs. There is little to no separation between the contrasts and match points at any f_{D_2O} in the solvent. *Right*: The effect on the neutron scattering contrasts and match points when one of the protein components of the complex is isotopically labeled with non-exchangeable deuterium to an average level of f_D=0.6 (60%). (B) The calculated $I(0)$ estimate of the unlabeled and labeled samples at 5 mg/mL as if measured through a number of contrast variation points at 0.1 f_{D_2O} increments.

(1) Information on the relative SANS intensity gains or losses as $\Delta\rho$ is varied as a consequence of altering the volume fraction of D_2O, f_{D_2O}, in the solvent

(2) Estimates of the individual component match points, i.e., the f_{D_2O} where $\Delta\rho_1$ or $\Delta\rho_2 = 0$

(3) Estimates of the whole complex match point, $\Delta\rho_{complex} = 0$

(4) Estimates of the separation of the three match points in terms of f_{D_2O}. Poor match point separation of the component match points relative to that of the whole complex indicates small differences between the average scattering length densities of either component within the complex. If poor match point separation occurs, e.g., within a +/− 10% f_{D_2O} range, and especially at lower f_{D_2O} where ^1H incoherent scattering dominates, the low internal contrast between components makes it exceedingly difficult to measure quality coherent SANS data at, or near, contrast matching conditions. It is preferable to measure contrast variation data under sample conditions with high contrast, for both the individual components and for the complex

(5) Assess of whether non-exchangeable D-labelling is required to increase the internal contrast between the individual components to affect f_{D_2O} match point separation. Intuitively, ^1H components will match out in higher volume fractions of ^1H_2O while deuterated components with higher f_D match out in higher volume fractions of D_2O. Therefore, where practical, consider deuterium-labelling the larger component of the complex. The rationale for this is the smaller unlabeled component will have a high contrast in high f_{D_2O} (> 90%) where the incoherent scattering from the solvent is low. Conversely, the larger deuterium-labeled component will have a high contrast in lower f_{D_2O} which will help to 'overcome' incoherent scattering from the ^1H_2O in the solvent

2.5 Procedure: Calculation of neutron scattering contrast

The following procedure describes the calculation of coherent neutron scattering length densities and contrasts from a dilute aqueous two-component protein-protein complex using the *Contrast* module of MULCh https://aws-smb-research.smb.usyd.edu.au/NCVWeb/ (Whitten et al., 2008). There are alternative calculators available including the SASSIE Contrast Calculator (SCC), that may be used to assess single, two-component and multi-component contrast systems https://sassie-web.chem.utk.edu/sassie2/ (Sarachan, Curtis, & Krueger, 2013) and the Biomolecular Scattering Length Density Calculator http://psldc.isis.rl.ac.uk/Psldc/. Element/isotope specific neutron scattering lengths and cross sections and neutron activation data are provided by the NIST Center for Neutron Research (https://www.nist.gov/ncnr/planning-your-experiment) and the National Nuclear Data Center at Brookhaven National Laboratory (https://www.nndc.bnl.gov). More specific details of the implementation of MULCh are made available in chapter "Data analysis and modelling of small-angle neutron scattering data with contrast variation from bio-macromolecular complexes" by Jeffries and Whitten in Methods in Enzymology, vol. 678, Part B.

1. Have the following information available for the *Contrast* calculation:
   - **a.** A nominal project title
   - **b.** The molecular formulae and respective concentrations in mol/l of all dissolved small molecule (M) components in the sample buffer (e.g., NaCl 0.15; C4H10O2S2 0.01; C8H18N2O4S 0.02, which is equivalent to 150 mM NaCl, 10 mM DTT, 20 mM HEPES).
   - **c.** The one letter code of component 1 (subunit 1) of the complex and the one letter code of component 2 (subunit 2) of the complex. The one letter codes refer to the sequences of proteins (P), DNA (D) or RNA (R) in simple text format
   - **d.** The expected stoichiometry ($N_{\mathrm{molecules}}$) of each macromolecule component in the complex, for example component 1 is expected to be a tetramer ($N_{\mathrm{molecules}} = 4$) and component 2 a dimer ($N_{\mathrm{molecules}} = 2$).
   - **e.** For macromolecules with bound metals, cofactors, etc., the atomic formulae of these bound species that are inputted as an additional M component(s) of the corresponding subunit
2. The *Contrast* module will automatically write an input text file summarizing the sample information, that maybe downloaded for future (re)use or modification:

```
Project title MIE
0
3
0.15 M  NaCl 0.0
0.01 M  C4H10O2S2  0.0
0.02 M  C8H18N2O4S  0.0
0.95
0
1
2  P    SMVNPDLMPALMAVFQHVRTRIQSELDC......... 0.0
0.95
0.0
2
4  P    GSSGSSGHLYYCSQCHYSSITKNCLKRHVIQKHSNIL  0.0
8  M    Zn  0.0
```

The consecutive lines `0.95` and `0` refer to the fraction of exchangeable hydrogen positions accessible by the solvent, f_{access1H} (typically set between 0.9 and 1.0) and the respective fraction of non-exchangeable D, i.e., the average level of deuteration f_{D}, on the first P component and second P component as listed in the subsequent lines. In this example, both proteins are not labeled with deuterium and are fully protonated. The first protein component `1` is listed as a dimer (`2 P`) followed by its amino acid sequence, while protein component `1` is listed as a tetramer (`4 P`) that also contains eight bound zinc ions (`8M`).

3. The *Contrast* module generates a *Tabulated scattering length densities and contrasts* output table. The last three columns of the table list the NEUTRON $\Delta\rho$ for (component) 1, 2 and the 'Total' complex in increments of 0.1 $f_{\text{D}_2\text{O}}$ in the solvent. *Contrast* also outputs the individual component and whole-complex $f_{\text{D}_2\text{O}}$ match points (note i).
4. Alter the f_{D} parameter of one of the components to assess the effect on $\Delta\rho$ and the match point separation. Plot $f_{\text{D}_2\text{O}}$ vs $\Delta\rho$ is to visualize the changes in component and whole complex contrasts and the subsequent 'match point splitting' as a component f_{D} is increased
5. As a rule of thumb, aim to increase the match point separation to as high as is achievable by altering the f_{D} parameter so that the deuterated component matches out between 90 and 100% $f_{\text{D}_2\text{O}}$ (note ii).
6. Obtain and estimate of the zero-angle scattering intensity, $I(0)$, at each contrast point (Fig. 3B) at a typical concentration, c, used for dilute biomacromolecule SANS, e.g., 5–10 mg/mL (note iii). The $I(0)$ in cm^{-1} of the complex at each $f_{\text{D}_2\text{O}}$ can be modeled using the $\Delta\rho$ 'Total' numbers from the *Contrast* module (in $10^{10}\,\text{cm}^{-2}$) via the following relationship:

$$I(0)_{f_{D_2O}} \approx \frac{\text{MW}c\left(\Delta\rho_{\text{Total}_{f_{D_2O}}}\nu\right)^2}{N_A} \qquad (13)$$

where N_A is Avogadro's number, c is in units of g/cm^3, and MW is the molecular weight in Da (g/mol) of the complex calculated from the amino acid sequence, taking into account the subunit stoichiometry. The partial specific volume, υ in cm^3/g, can be obtained from the *Contrast* output, SEDNTERP (Cole, Lary, Thomas, & Laue, 2008; Laue, Shah, Ridgeway, & Pelletier, 1992), SEDFIT (Schuck, 2000), or NucProt (Voss & Gerstein, 2005).

Notes

i. For this example (Fig. 3A), there is no discernible difference in the contrast of the components or the whole complex and consequently there is no significant match point separation

ii. It is often the case that deuterating the larger component of a complex to match out at high f_{D_2O} will enable the collection of data from the smaller component in conditions with low incoherent scattering. However, this assumes that the larger component is amenable to isotopic labelling and that the complex, irrespective of deuteration level, remains monodispersed in high percentage D_2O buffers

iii. The $I(0)$ of the whole complex will be the highest background-corrected scattering intensity for each contrast point. If the $I(0)$ of the whole complex at a component match point is predicted to be small, for example to 10% of the $I(0)$ estimated under high contrast conditions ($f_{D_2O}=0$), it may be difficult to obtain quality scattering data especially in lower f_{D_2O} (0.3–0.6). An alternative D-labelling approach may have to be investigated, e.g., perdeuteration (100% isotopic-labeling) of component 1 to increase the magnitude of the whole complex $I(0)$ at the component 2 match point, or vice versa

3. Small angle neutron scattering instruments

The samples used for biomolecule SANS are mostly prepared in the dilute regime where the target macromolecules occupy a very low-mass fraction of the sample. Therefore, it is necessary to deliver as many neutrons as possible to the samples. However, even at the highest flux facilities, the number of neutrons produced is many orders of magnitude less compared to X-ray sources. Dilute samples combined with inherently low flux requires the appropriate configuration of SANS instruments to optimize neutron delivery and detection. Here we will focus on continuous flux nuclear reactor-based SANS instruments, as opposed to a pulsed neutron sources found at spallation facilities. The Quokka SANS instrument located at the Australian Nuclear Science and Technology Organization (ANSTO) will be used as an example (Wood et al., 2018).

3.1 Neutron wavelengths and guides

High-energy neutrons produced in the core of a nuclear reactor are slowed down by a process called moderation, after which produces thermal neutrons with wavelengths of around 0.1 nm. To investigate length scales on the tens of nanometers probed using SANS requires access to very-low *q*-vectors and therefore longer wavelength neutrons are required on the order of 0.4–2 nm. The thermal neutrons originating from the core are subsequently slowed even further via moderation through a cold source, for example cryogenic liquid deuterium, to produce longer wavelength neutrons that are then guided to the SANS instrument (Fig. 4A,i).

The 'cold neutrons' emerge from the source at multiple wavelengths. To select a specific wavelength, a neutron velocity selector is placed in the flightpath that rotates at tens of thousands of rotations per minute (Fig. 4A,ii). Only those neutrons of the correct speed are transmitted through the selector; those traveling at other speeds, either too slow or too fast, are absorbed by the selector drum. If the velocity selector is set to stringently, it may be the case that not many neutrons at one specific wavelength are all allowed to transmit, severely compromising the flux. Therefore, the velocity selector can be tilted relative to the beam to allow neutrons to pass through as a limited band-width distribution. This distribution is generally triangular, with a peak maximum at the selected wavelength and a typical full width at half maximum height of $\Delta\lambda/\lambda$. An instrument

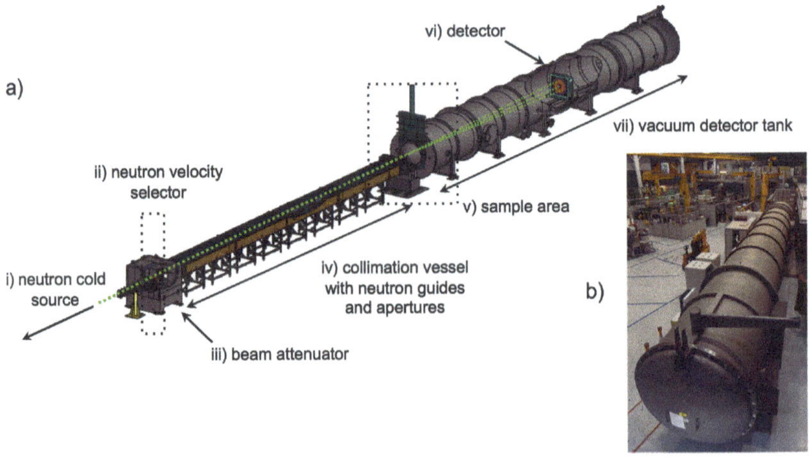

Fig. 4 (A) Diagram of the Quokka small angle neutron scattering instrument with different instrument components labeled. (B) The photograph of Quokka, taken from the back of the detector tank, looking toward the reactor face.

setting with a large $\Delta\lambda/\lambda$ will transmit a large neutron flux to the sample, but a large spread of neutron wavelengths will result in poor q-resolution (in Eq. (7), q is inversely proportional to wavelength, thus, any uncertainty in the wavelength creates uncertainty in the calculation of q), which has the effect of smearing the measured scattering data. Conversely, an instrument setting with a small $\Delta\lambda/\lambda$ will transmit a lower neutron flux to the sample but will result in a superior q-resolution. Consequently, there may be a requirement to compromise between the total number of neutrons allowed through the velocity selector and $\Delta\lambda/\lambda$. For many dilute biological macromolecule solution samples, the isotropic scattering profiles monotonically decrease, and in many circumstances poor wavelength resolution doesn't significantly impact the result. Maximizing the flux is often a more important consideration for these types of samples compared to q-resolution to obtain improved signal-to-noise ratios in the data. Therefore, and typically for 'bioSANS,' a $\Delta\lambda/\lambda$ of ~10–15% is selected, i.e., if a wavelength of 0.6 nm is reported, the bulk of the wavelength distribution would lie between 0.54 and 0.66 nm. This $\Delta\lambda/\lambda$ represents a reasonable compromise between flux and q-resolution for most dilute sample investigations.

After leaving the velocity selector, the neutron beam is defined in the so-called 'collimation vessel' (Fig. 4A,iv). The collimation vessel contains a series of guides (supermirrors that reflect divergent neutrons) on translation stages that can either be moved into, or out of, the beam path along the 20 m of the collimation vessel. When a beam with low divergence is required, neutron guides are removed to allow a series of apertures in the collimation tank to define the beam, leading to a lower neutron flux at the sample position, but a more parallel beam. If a more divergent beam is acceptable, neutron guides are inserted to transport all the neutrons up to the sample position. At the end of the last guide position, an aperture is placed in the beam, called the source aperture for the experiment. The source aperture diameter maybe opened to transmit the bulk of the neutron flux from the guides. If no guides are inserted, the source aperture is located at the start of the collimation tank. The distance between the source to sample apertures is often noted as SSD (for source-to-sample distance) or $L1$. The sample aperture, the last beam defining element neutrons pass through before interacting with the sample.

As the neutron beam passes through the sample, a percentage of neutrons are scattered by the sample, either coherently or incoherently, and the rest pass through as the transmitted beam. A position sensitive 2-dimensional neutron detector housed within a detector vessel is placed after the sample at a known sample to detector distance, $L2$, to record the neutron scattering events. A beam stop is placed in the beam path to protect the detector from the intense transmitted beam. The neutron detector may be translated relative to the sample position to access different $L2$ values, and hence different angular ranges for scattering measurements, where longer $L2$ values record data to lower q. Instruments may also have multiple detectors to enable scattering data to be collected over a wider angular range simultaneously (J. Barker et al., 2022; Dewhurst et al., 2016; Sokolova et al., 2019). To minimize scattering from the air, as much of the neutron path as practical is maintained under vacuum including the collimation tanks and detector vessels, with typically only the sample area exposed to air.

3.2 Choosing instrument configurations, q_{min}

At the start of the SANS experiment, the instrument is set up in a certain configuration(s) to measure data over a suitable q-range. The smallest q required, $q_{\min}$, relates to the maximum dimension of the particles of interest (D_{max}) that requires the measurement to at least $q_{\min} < \pi/D_{\max}$ (Jacques & Trewhella, 2010), or preferably $q_{\min} < 1/D_{max}$ (Svergun, Koch, Timmins, & May, 2013a). The $q_{\min}$ in a certain instrument configuration is determined by several factors:

(1) The beam diameter, that is defined by the dimensions of the source and sample apertures, and the distance between the source and sample apertures $L1$. The sample aperture is chosen as the largest diameter that will fully illuminate the sample while not illuminating the edges of the sample holder

(2) The sample-to-detector distance, SSD, often called $L2$, where smaller $q_{\min}$ are accessed at longer $L2$

(3) The neutron λ, where smaller $q_{\min}$ become accessible using longer wavelength neutrons

Instrument associated macros named SASCalc enables optimization of the instrument configurations based on similar instruments at the National Institute for Standards and Technology (Kline, 2006). Configurations suitable to measure proteins in solution are detailed in (Wood et al., 2018). Once the configuration of the $L2$ sample to detector distance is selected to obtain the desired $q_{\min}$, the $L1$ source-to-sample distance is optimized. The $L1$ defines the number of guides to be inserted in the beam path in the collimation tank. The shorter the $L1$, the more guides are inserted into the collimation tank. As guides are used to transport neutrons up to the source aperture prior to the sample, the more guides inserted in the collimation tank will produce more flux through the source aperture with a divergent beam. For dilute biological macromolecule scattering experiments, $L1$ values are kept as short as possible to maximize the flux at the sample position, where and in general the optimal configuration has $L1 = L2$ (Mildner & Carpenter, 1984). The beam dimensions and beam divergence affect the primary beam-spot size on the detector, and thus contribute to accessing a working q_{min} on the detector. It is necessary to have a beam divergence that produces a spot-size less than the dimensions of the beam stop.

The choice of $L1$ must also be balanced against the number of neutrons transmitted through the velocity selector at a selected wavelength. The characteristic neutron spectrum of a cold source defines the neutron flux passing through the velocity selector at each wavelength setting. For selected wavelengths between 0.5 and 1 nm, the neutron flux is highest at 0.5 nm, and relative to this, the flux at 1 nm is almost an order of magnitude lower (Fig. 5A). Naively, it might seem that data should always be collected at 0.5 nm because this gives the maximum neutron flux through the velocity selector, but in finding the optimal configuration for a SANS experiment there is significant interplay between $L1$, $L2$, and the neutron wavelength. Firstly, the incident beam intensity is proportional to $1/(L1)^2$, which means halving the collimation length increases intensity by a factor of 4. Secondly, the solid angle of a detector pixel is proportional to $1/(L2)^2$, so while the beam intensity doesn't change, this means the halving the sample-to-detector distance will increase the counts measured per pixel by a factor of 4. Thus, if a given q_{min} value was reached using a configuration of $\lambda = 0.5$ nm, and $L1 = L2 = 10$ m, a similar q-range could be covered by $\lambda = 0.6$ nm, and $L1 = L2 = 8$ m, with a factor of $1.7\times$ improvement in the counts measured per pixel in this configuration (Fig. 5B). In fact, even greater improvements can be found at higher wavelengths. However, there are trade-offs, as the configuration used to measure higher angle data usually has the detector as close to the sample as is reasonable (thus, $L2$ cannot be shortened), and typically the maximum number of guides are already in place (thus, $L1$ cannot be shortened). Hence, choosing a longer wavelength will mean the number of counts per pixel will drop relative to $\lambda = 0.5$ nm. Further, the accessible q_{min} falls as wavelength increases, meaning that potentially important q-ranges cannot be measured. In general, $\lambda = 0.6$ nm provides a good balance between these competing factors.

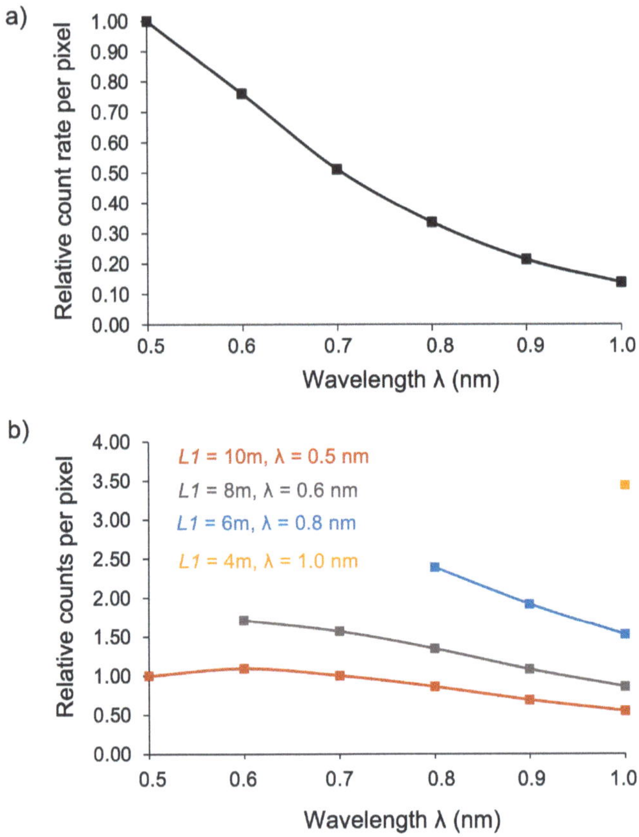

Fig. 5 (A) The relative flux for range of different cold-source neutron wavelengths from the Quokka instrument. (B) Relative signal for a fixed q_{min} in different $L1 = L2$ configurations for a number of neutron wavelengths.

3.3 Procedure: Optimization of configurations using SASCalc

1. Insert the appropriate sample and source apertures in the SASCalc window. The simulated scattering intensity is displayed
2. Chose a sample to detector distance (*L2*) and wavelength that enables the desired q_{min} to be measured as shown in the 'Trial Configuration' panel. SASCalc will calculate a figure of merit for the configuration, the higher the figure of merit the more intense the signal. The figure of merit can be optimized by changing *L2* and the wavelength
3. Change the source to sample (*L1* to match *L2*). If the beam will become too large for the available beam stops, a warning will appear
4. Once the desired low-*q* configuration is determined, it is possible to freeze the configuration and begin to choose the second (high-*q*) configuration

5. Change the detector position to have a smaller *L2* and enable measurements to higher q values to measure the desired q_{max}. *L*1 can also be changed to a smaller value to increase flux, but there is no requirement for *L*1 to match *L*2. The simulated scattering intensity for the higher q configuration will appear overlayed on the figure. Ensure that the high-q configuration overlaps with the low-q configuration with at least 10 points in the overlap region

3.4 Choosing instrument configurations, q_{max}

For the majority of SANS experiments on dilute macromolecules the wavelength is maintained, and the variable factor is to measure at different sample-to-detector distances to cover a wide q-range. When using an instrument with only one detector, a second, higher-q configuration using a shorter *L2* is often selected to ensure data are measured out to sufficiently high angles, typically to $q_{max} = 5\text{–}6\,\text{nm}^{-1}$. The q_{max} at a given *L2* is dependent on the size of the detector and whether the beam is centred or offset toward the detector edge. The data collected at the long- and short *L2* detector positions are then merged to encompass the complete form factor scattering of the sample macromolecules. Depending on the size of the macromolecules in the sample, more than two *L2* configurations maybe required to capture the full form factor scattering q-range.

3.5 q-Resolution

The q-values calculated in a SANS experiment have uncertainties associated with them, called the q-resolution (Barker & Pedersen, 1995; Grillo, 2001; Jaksch, Pipich, & Frielinghaus, 2021; Mildner & Carpenter, 1984; Pedersen & S., 1993; Pedersen, Posselt, & Mortensen, 1990), and the data calculated from a model can be 'smeared' to take into account the instrument resolution function. The major contributions to the smearing of SANS data include:

1. Instrument geometry: As the instrument is made up of apertures of a certain diameter, when a neutron is recorded in a certain pixel, there is an inherent uncertainty on the calculation of the scattering angle, related to the diameters of the source and sample apertures, and the distances *L*1 and *L2*, sample thickness and pixel size
2. Wavelength distribution: Ideally, when measured, the sample is exposed to neutrons with a defined wavelength within a narrow distribution. For most SANS experiments, it is beneficial to have an increased wavelength

distribution, on the order of $\Delta\lambda/\lambda \approx 10\%$, to increase the flux of neutrons on the sample, while still providing adequate q-resolution

3. A third effect is gravity that becomes more of an issue at long neutron wavelengths and longer detector positions. For a typical bio-SANS experiment ($\lambda = 0.5$–0.6 nm; $L2_{max} = 5$–6 m) the effects of gravity on the resolution function are generally minimal. The full calculation of the resulting uncertainty on the q-values is given in Chapter 15 of 'The SANS toolbox' (Hammouda, n.d.-c, Chapter 15).

The Quokka instrument for example is a pinhole collimated SANS instrument, using a detector with finite pixel size and narrow $\Delta\lambda/\lambda$, where all these factors produce pinhole smeared data. At a single detector position, the geometric effects dominate at low-q, while at high-q it is the wavelength spread that is the most important.

3.6 *q*-Calibration

To calculate the q-values from a pixel where neutrons are recorded on the detector, the neutron wavelength and the geometric parameters of the instrument must be known. Geometric parameters such as the distance between the sample and the detector, the position of the primary beam on the detector are simply measured, while detector pixel size is given by the manufacturer. The neutron wavelength is well defined by the rotation speed of the velocity selector, and once the velocity selector is calibrated the wavelength may be reliably chosen by simply selecting the correction rotation speed. On Quokka to calibrate the velocity selector, a chopper is placed at the sample position using a certain velocity selector rotation speed and the neutron wavelength calculated using the time-of-flight method (Wood et al., 2018). The q-calibration may be verified with a standard sample such as silver behenate.

3.7 Procedure: *q*-Calibration

1. Place a chopper at the sample position
2. Set chopper frequency to 80 Hz
3. Set detector to record data as a time histogram, where neutron counts are on y-axis and x-axis represents time channel numbers
4. Pulses of neutrons created by the opening of the chopper travel to the detector and are recorded as peaks as a function of time
5. Collect data at a minimum of three different detector positions

6. Fit the Gaussian function to estimate the values of the peak positions and their standard deviations
7. The peak maxima provide information about the average flight times—wavelengths, and standard deviations corresponding to the wavelength spread
8. Plot the peak maxima positions as a function of the detector distance
9. A linear fit of the peak positions versus detector distance gives the neutron velocity in meters per second [v] and the wavelength in nm is then given by the relation: $\lambda = 395.6/v$ (Hammouda, n.d.-a, Chapter 2).

4. The measurement approach

Two different types of measurements are performed during the experiment: transmission measurements and scattering measurements, depicted in Fig. 6.

4.1 Transmission measurements

Sample transmissions are an important component of biological macromolecule SANS experiments as they are used for data normalization, correction. The transmission of neutrons through the predominantly aqueous samples and corresponding matched solvent blanks are exceptionally sensitive to the strong incoherent scattering from ^1H, that attenuates the neutron beam. Thus, transmission measurements may be used to confirm that the f_{D_2O} in the sample and a corresponding matched solvent are identical. The transmission, T, is related to the number of molecules, N, of ^1H_2O and D_2O per cm^3 via (Svergun et al., 2013a):

$$T = e^{\left(-N_{H_2O}\sigma_{H_2O} - N_{D_2O}\sigma_{D_2O}\right)P_L}$$

The transmission is dependent on the cross sections of both ^1H_2O and D_2O and their number density and the sample pathlength, P_L, or sample thickness, that is typically restricted to 1 mm for solution samples (2 mm may be used for high $f_{D_2O} > 50\%$). The transmission will depend on the temperature, which affects the mass density of the solvent, thus altering the number of solvent molecules per unit volume, and the neutron wavelength, which affects the magnitude of inelastic scattering (e.g., for ^1H_2O transmission decreases with increasing λ (Lindner, 2000). Transmissions must be experimentally determined for each sample and each matched solvent spanning a contrast variation experiment, where the log of the transmission normalized

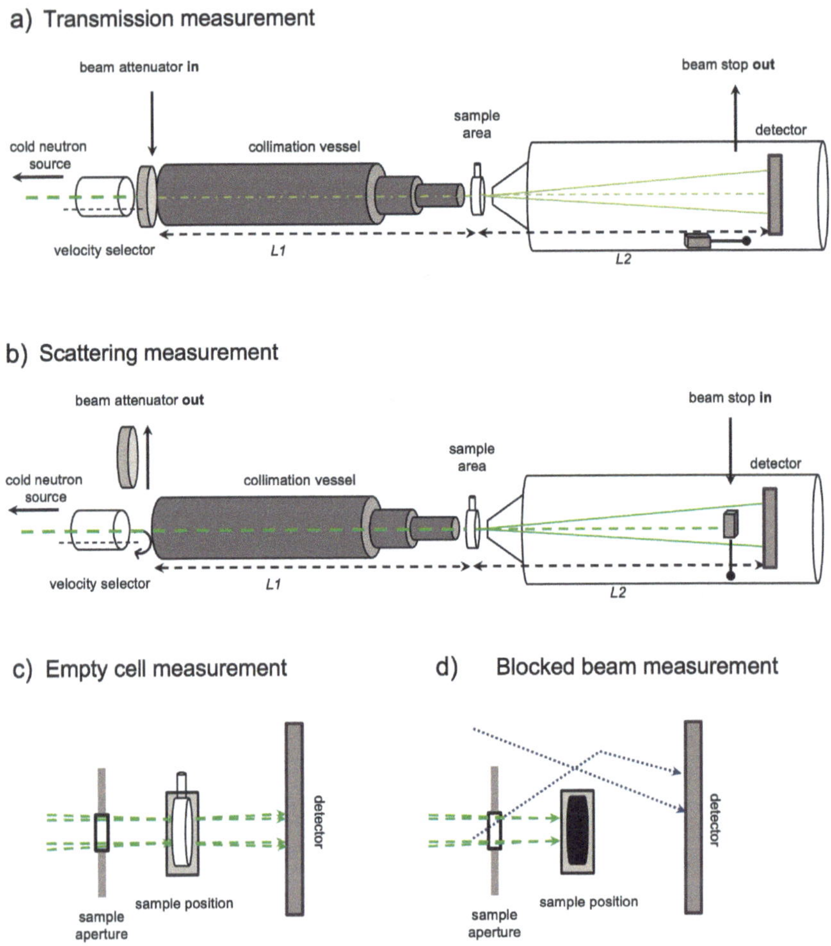

Fig. 6 Schematic representation of the types of measurements performed during a small angle neutron scattering experiment. (A) Scattering measurement, where there is no beam attenuator in position and the beam stop prevents the unscattered part of the neutron beam from hitting the detector. (B) Transmission measurement, where an attenuator is inserted into the neutron beam to reduce the number of neutrons entering the instrument, and the beam stop is removed to allow the unscattered part of the beam to be measured on the detector. (C) An empty cell measurement, where neutrons that have passed through the sample position, represented as dashed green arrows, and measured by placing an empty sample holder at the sample position. (D) Blocked beam measurement, where neutrons that have not passed through the sample position, represented as dotted blue arrows, and measured by placing a blocked beam at the sample position. The second contribution is very small, but both are measured as the data reduction takes into consideration that only the first is attenuated by the sample.

to the pathlength, $-\frac{1}{P_L}\ln(\mathrm{T})$, versus the mol-fraction of D_2O produces a negative linear plot (Svergun et al., 2013a).

The transmission measurements are short, typically of the order of a minute, which will yield transmission values with an error of about 1% and only need to be measured at one instrument configuration. Transmission measurements record high count rates of neutrons over a small number of detector pixels and are for many detectors are best done in the configuration with the detector the furthest away from the sample position (longest *L2* configuration). With the detector further away from the sample, the beam is spread out across more detector pixels, the count rates on individual pixels are lower, and the measurements are less likely to be affected by non-linear detector response. To perform a transmission measurement, the beam stop is removed, and an attenuator is inserted to reduce the neutron flux to prevent detector damage by the direct beam (Fig. 6A).

In addition to being used to calculate the transmission, the empty beam transmission measurement is used often used to calculate the absolute scale of the scattering data. For neutrons, the data are typically placed on an absolute scale by dividing the scattered intensity by the beam intensity, sample thickness, and solid angle. Thus, intensities on an absolute scale represent the probability of an incident neutron being scattered into a pixel with position $\mathbf{q}$ and solid angle $\boldsymbol{\Omega}$ per unit thickness of the sample. When placed on an absolute scale, the mass and volume data may be obtained from $I(0)$ (Eqs. 9 and 13) and the data recorded from the same sample on one instrument or set of instrument configurations maybe directly compared to each other.

4.2 Procedure: Transmission measurement

1. Set up the instrument in transmission mode, by inserting a beam attenuator and removing the beam stop from the direct beam path
2. Measure the empty beam, empty cell, samples and solvents in transmission mode

Notes

i. The data in transmission geometry are often measured for one or 2 min

4.3 Scattering measurements

To determine the scattering from a sample requires the measurement of background neutron scattering events that do not originate from the sample and correcting the scattering measurement data accordingly. There are two main components to the background: scattered neutrons from the quartz cuvette and cell holder of the sample stage, or other up-stream instrument components that pass through the sample (a.k.a 'empty cell measurements') and; neutrons that arrive at the detector that do not pass through the sample. Scattering measurements are performed without the attenuator and with the beam-stop in place (Fig. 6B).

4.4 Procedure: Scattering measurements

1. Move instrument to scattering geometry, by inserting the beam stop to cover the direct beam and removing all attenuation
2. Measure scattering from the sample and solvent at two sample-to-detector positions
3. Measure background contributions to the scattering:
   a. an empty cell measurement to measure the background through the sample position with the quartz cuvette in place (Fig. 6C). The scattering arises not only from the empty cell, but also the instrument itself
   b. a blocked beam measurement is to evaluate those neutrons that that do not pass through the sample position. In this case we record the background given by stray neutrons and electronic noise. The beam is blocked using a neutron absorber, for example a strip of cadmium or boron-carbide (Fig. 6D) and the measurements is generally more important to perform at higher-q configurations when the detector is close to the sample position
4. Detector sensitivity run is performed by measuring a Plexiglass or water in a sample holder. It gives information about the detector response and efficiency

Notes

i. The blocked beam and empty cell measurements are typically shorter than the scattering runs from the samples and buffers themselves, which are measured on the timescale of hours

ii. Detector sensitivity measurement is generally done periodically by the instrument scientist and is not performed as part of every experiment

4.5 Procedure: Measurement approach

For a SANS study of a protein complex at five different contrasts, at two detector positions using a 1 mm quartz cuvettes:

1. Place the samples S1, S2, S3, S4, S5 and the matched solvent blanks B1, B2, B3, B4, B5 into the sample changer making sure the samples and

matched solvents are degassed and temperature equilibrated to avoid dissolved gas bubble formation over time and prevent condensation from the atmosphere on the outside of the cuvettes

2. Have one slot in the sample changer completely empty, another with an empty quartz cuvette for the empty cell measurement, and another with blocked beam material
3. Long detector position: Perform sequential transmission and scattering measurements on all samples, buffers, the empty cell, and blocked beam using the empty slot for the reference for the transmission measurements. Measuring the empty cell and blocked beam first will enable the data reduction of the samples and buffers as they are collected
4. Short detector position: Perform sequential scattering measurements on all samples, buffers, quartz cuvette and blocked beam

Notes

i. Use the $I(0)$ estimation at each contrast point calculated from the *Contrast* module of MULCh to guide exposure times (refer to **2.5 PROCEDURE: Calculating of neutron scattering contrasts**). Samples with higher $I(0)$ estimates, e.g., where both components of a two-component system contribute to the scattering (large V^2) maybe exposed for less time compared to measurements performed at the respective component match points (low V^2). Samples in higher f_{D_2O} may also be measured for less time due to the decrease in incoherent ^1H scattering

ii. The detector position $L2$ and the detector efficiency also affects the exposure times. Halving $L2$ generates in a four-fold increase in the number of neutron counts per unit area on the detector (the solid angle of a detector pixel scales to $1/(L2)^2$). To improve the counting statistics, longer exposure times are typically required at longer L2, lower q_{min} configurations

iii. The selected wavelength, $\Delta\lambda/\lambda$ and the size of the neutron beam affects exposure times. Reactor-based bio-SANS instruments employ cold neutrons at the wavelength range from 0.4–0.6 nm, where the highest neutron flux is encountered from the cold source. The beam size diameter is in the order of 1–2 cm at the sample position

iv. If, on average, if it takes 2 h per measurement for long detector position measurements, and 0.5 h at the short detector position the total approximate time for a five-point contrast variation experiment is around 30 h: $2 \times 11 + 0.5 \times 11 = 27.5$ h for the scattering measurements, plus a total transmission time of around 30 min for all samples, solvent blanks and the empty cell, plus detector reconfiguration times

4.6 Procedure: Data reduction (brief overview)

Once scattering data are measured, they are corrected for instrument specific effects and azimuthally averaged to produce a one-dimensional scattering pattern on an absolute scale in a process commonly referred to as data reduction (Brûlet, Lairez, Lapp, & Cotton, 2007). A schematic of how data is reduced is given in Fig. 7 and described in (Kline, 2006).

1) The information required for data reduction includes:

a. Detector mask: Regions of the detector that are artifacts and need to be ignored during azimuthal averaging (e.g., dead pixels, beam stop region).

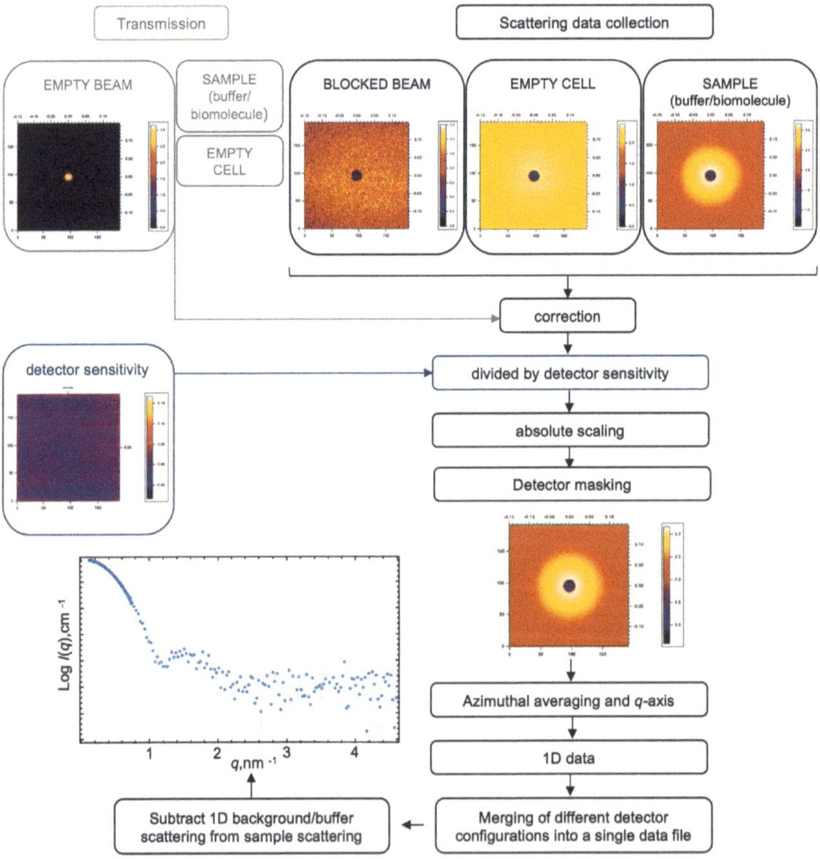

Fig. 7 Representation of a basic data reduction protocol, showing how the different measurements are used to produce a data file representing the scattering of the sample macromolecules, corrected for all instrument and background scattering effects.

   b. Detector sensitivity: Provides a measurement of the pixel efficiency
   c. Detector distortion/manufacturers corrections, detector dead times
   d. Sample to detector distances, with corrections for any detector offsets to obtain the exact sample-to-detector distance
   e. Sample thickness
   f. Measurement time, for normalization
   g. A calibrated q-axis for each detector position
   h. Absolute scaling information
   i. Sample and matched-solvent transmissions
   j. Blocked beam and empty cell measurements
   k. Sample and matched-solvent scattering measurements
2) The sample and matched-solvent transmissions are calculated as the ratio of the sum of neutrons in the direct beam with and without sample. For example, to calculate the transmission for sample S1, two transmission measurements are used, that of S1 and that of the empty beam. A sum of neutrons in the direct beam is performed for the two data files, and the transmission of S1 equal to one divided by the other
3) For each different detector configuration:
   a. The scattering data are corrected for the instrument background, using the following: Corrected Sample Scattering = Sample – blocked beam – [T_{sam}/T_{emp}](Empty Cell – Blocked Beam), where T_{sam}, T_{emp} are the transmission values of the sample and empty cell respectively
   b. The background corrected file is then divided by the detector sensitivity, to consider the fact that the detector response is not completely uniform across the whole detector
   c. The data are then put on an absolute scale, often using the number of neutrons in the direct beam
   d. Any pixels in the detector that need to be masked are masked
   e. The data are azimuthally averaged to extract a 1-dimensional file
4) The different configurations are then merged into a single data file for the sample, often given in the format of columns, q, $I(q)$, $\sigma I(q)$, q-resolution, where $\sigma I(q)$ are the errors on the intensities. A detailed description of q-resolution resolution is given in (Hammouda, n.d.-c, Chapter 15).

5. Summary

The calculation of neutron scattering contrasts of biomacromolecules prior to a SANS with contrast variation experiment ultimately dictates nearly every aspect of the sample preparation, instrument configuration and measurement protocols. The calculations can be used assess what the relative magnitudes of the scattering intensities will be under different sample conditions, to make decisions on whether isotopic labelling is required to effect match point separation, how many contrast points need to be measured and at what concentrations of D_2O. The component contrasts combined with the dilute nature of the samples, incoherent scattering from ^1H, and the inherently low flux of SANS instruments, all go to inform what quantity of unlabeled or potentially labeled components needs to be prepared, that maybe upwards of 10 mg of pure, well characterized material. Although there are additional challenges, for example assessing the effects of component solubility and component affinities in D_2O over time (due to the necessary long measurement times), SANS with contrast variation does allow for more creative aspects of structural biology to be investigated. Aside from monodisperse macromolecular complexes and assemblies in solution, other types of samples and sample environments are amenable to SANS which other structural techniques struggle with, for example the analysis of macromolecule-conjugated nanoparticles, nanoplastics, lipid membranes, gels, macromolecular phases, emulsions, and systems under pressure, shear, or time-evolving systems (Banc et al., 2016; Ibrahim et al., 2017; Kihara et al., 2019; Mehan, Chinchalikar, Kumar, Aswal, & Schweins, 2013; Oliver, Rolband, Hutchinson-Lundy, Afonin, & Krueger, 2019; Semeraro, Marx, Frewein, & Pabst, 2021; Urban, Heller, Katsaras, & Bras, 2021; Yadav, Kumar, Aswal, & Kohlbrecher, 2014). As such, SANS will remain as a component of the structural biology tool kit. However, irrespective of what is being investigated, from simple to more complicated systems, what remains a consistent mode of practice is an appreciation of the neutron scattering contrasts, how the contrast can be manipulated, and the conceptual link between contrast and the configuration of a SANS instrument to optimally record quality data from biological samples.

Acknowledgments

The authors acknowledge the support of the Australian Centre for Neutron Scattering, Australian Nuclear Science and Technology Organization, the Australian Government through the National Collaborative Research Infrastructure Strategy, the European Union's Horizon 2020 research and innovation programme 'iNEXT Discovery' grant agreement No 871037, and the Swedish Foundation for Strategic Research (SSF) through the national Graduate School (SwedNess).

References

Banc, A., Charbonneau, C., Dahesh, M., Appavou, M. S., Fu, Z., Morel, M. H., et al. (2016). Small angle neutron scattering contrast variation reveals heterogeneities of interactions in protein gels. *Soft Matter, 12*(24), 5340–5352. https://doi.org/10.1039/C6SM00710D.

Barker, J., Moyer, J., Kline, S., Jensen, G., Cook, J., Gagnon, C., et al. (2022). The very small angle neutron scattering instrument at the National Institute of Standards and Technology. *Journal of Applied Crystallography, 55*(2), 271–283. https://doi.org/10.1107/S1600576722000826.

Barker, J. G., & Pedersen, J. S. (1995). Instrumental smearing effects in radially symmetric small-angle neutron scattering by numerical and analytical methods. *Journal of Applied Crystallography, 28*(2), 105–114. https://doi.org/10.1107/S0021889894010095.

Brûlet, A., Lairez, D., Lapp, A., & Cotton, J.-P. (2007). Improvement of data treatment in small-angle neutron scattering. *Journal of Applied Crystallography, 40*(1), 165–177. https://doi.org/10.1107/S0021889806051442.

Cole, J. L., Lary, J. W., Thomas, P. M., & Laue, T. M. (2008). Analytical ultracentrifugation: sedimentation velocity and sedimentation equilibrium. *Methods in Cell Biology, 84*, 143–179. Academic Press.

Dewhurst, C. D., Grillo, I., Honecker, D., Bonnaud, M., Jacques, M., Amrouni, C., et al. (2016). The small-angle neutron scattering instrument D33 at the Institute Laue-Langevin. *Journal of Applied Crystallography, 49*(1), 1–14. https://doi.org/10.1107/S1600576715021792.

Fitter, J., Gutberlet, T., & Katsaras, J. (2006). *Neutron scattering in biology, techniques and applications*. Berlin, Heidelberg: Springer.

Frielinghaus, H. (2012). Small angle scattering and large scale structures. In T. B. M. Angst, D. Richter, & R. Zorn (Eds.), *Lecture notes of the 43rd IFF spring school 'scattering methods for condensed matter research: Towards novel applications at future sources'*. Forschungszentrum Jülich: Forschungszentrum Jülich GmbH. D1.2–D1.27.

Glinka, C. (2011). Incoherent neutron scattering from multi-element materials. *Journal of Applied Crystallography, 44*(3), 618–624. https://doi.org/10.1107/S0021889811008223.

Grillo, I. (2001). *Instrumental resolution and polydispersity*. Grenoble, France: Retrieved from Institut Laue-Langevin (ILL), ILL technical report ILL01GR08T. https://www.ill.eu/fileadmin/user_upload/ILL/3_Users/Scientific_groups/Large_Scale_Structures/People/Isabelle_Grillo/resolution.pdf.

Hammouda, B. (2012). *A tutorial on small-angle neutron scattering from polymers*. Retrieved from https://www.ncnr.nist.gov/programs/sans/pdf/polymer_tut.pdf.

Hammouda, B. n.d.-a (Chapter 2). Part A. Neutron Sources and Neutron Flux: The Neutron Probe (pp. 10–11). Gaithersburg, MD: National Institute of Standards and Technology, Center for Neutron Research.

Hammouda, B. n.d.-b (Chapter 9). The SANS Toolbox, Part B. Neutron Scattering: Coherent and Incoherent Neutron Scattering (pp. 62-83). Gaithersburg, MD: National Institute of Standards and Technology, Center for Neutron Research.

Hammouda, B. n.d.-c (Chapter 15). *The SANS Toolbox, Part D. SANS Resolution and Smearing: The SANS Instrumental Resolution* (pp. 148–165). Gaithersburg, MD: National Institute of Standards and Technology, Center for Neutron Research Retrieved from http://www.ncnr.nist.gov/staff/hammouda/the_SANS_toolbox.pdf.

Harroun, T. A., Wignall, G. D., & Katsaras, J. (2006). Neutron scattering for biology. In J. Fitter, T. Gutberlet, & J. Katsaras (Eds.), *Neutron scattering in biology: Techniques and applications* (pp. 1–18). Berlin, Heidelberg: Springer Berlin Heidelberg.

Ibrahim, Z., Martel, A., Moulin, M., Kim, H. S., Härtlein, M., Franzetti, B., et al. (2017). Time-resolved neutron scattering provides new insight into protein substrate processing by a AAA+ unfoldase. *Scientific Reports*, 7(1), 40948. https://doi.org/10.1038/srep40948.

Jacques, D. A., & Trewhella, J. (2010). Small-angle scattering for structural biology--expanding the frontier while avoiding the pitfalls. *Protein science : a publication of the Protein Society, 19*(4), 642–657. https://doi.org/10.1002/pro.351.

Jaksch, S., Pipich, V., & Frielinghaus, H. (2021). Multiple scattering and resolution effects in small-angle neutron scattering experiments calculated and corrected by the software package MuScatt. *Journal of Applied Crystallography, 54*(6), 1580–1593. https://doi.org/10.1107/S1600576721009067.

Kihara, S., van der Heijden, N. J., Seal, C. K., Mata, J. P., Whitten, A. E., Köper, I., et al. (2019). Soft and hard interactions between polystyrene Nanoplastics and human serum albumin protein Corona. *Bioconjugate Chemistry, 30*(4), 1067–1076. https://doi.org/10.1021/acs.bioconjchem.9b00015.

Kline, S. R. (2006). Reduction and analysis of SANS and USANS data using IGOR Pro. *Journal of Applied Crystallography, 39*, 895–900. https://doi.org/10.1107/S0021889806035059.

Laue, T. M., Shah, B. D., Ridgeway, T. M., & Pelletier, S. L. (1992). Computer-aided interpretation of analytical sedimentation data for proteins. In S. Harding, A. Rowe, & J. Horton (Eds.), *Analytical ultracentrifugation in biochemistry and polymer science* (pp. 90–125). Cambridge: Royal Society of Chemistry.

Lindner, P. (2000). Water calibration at D11 verified with polymer samples. *Journal of Applied Crystallography, 33*(3 Part 1), 807–811. https://doi.org/10.1107/S0021889800099763.

Lombardo, D., Calandra, P., & Kiselev, M. A. (2020). Structural characterization of biomaterials by means of small angle X-rays and neutron scattering (SAXS and SANS), and light scattering experiments. *Molecules, 25*(23), 5624.

Mahieu, E., & Gabel, F. (2018). Biological small-angle neutron scattering: Recent results and development. *Acta Crystallogr D Struct Biol, 74*(Pt 8), 715–726. https://doi.org/10.1107/S2059798318005016.

Mehan, S., Chinchalikar, A. J., Kumar, S., Aswal, V. K., & Schweins, R. (2013). Small-angle neutron scattering study of structure and interaction of nanoparticle, protein, and surfactant complexes. *Langmuir, 29*(36), 11290–11299. https://doi.org/10.1021/la402977b.

Mildner, D. F. R., & Carpenter, J. M. (1984). Optimization of the experimental resolution for small-angle scattering. *Journal of Applied Crystallography, 17*(4), 249–256. https://doi.org/10.1107/S0021889884011468.

Oliver, R. C., Rolband, L. A., Hutchinson-Lundy, A. M., Afonin, K. A., & Krueger, J. K. (2019). Small-angle scattering as a structural probe for nucleic acid nanoparticles (NANPs) in a dynamic solution environment. *Nanomaterials, 9*(5), 681.

Pedersen, J. S. (1993). Resolution effects and analysis of small-angle neutron scattering data. *Journal de Physique IV France, 03*(C8). C8-491-C498-498.

Pedersen, J. S., Posselt, D., & Mortensen, K. (1990). Analytical treatment of the resolution function for small-angle scattering. *Journal of Applied Crystallography, 23*(4), 321–333. https://doi.org/10.1107/S0021889890003946.

Pynn, Roger. (2009). Neutron scattering—A non-destructive microscope for seeing inside matter. In Liyuan Liang, Romano Rinaldi, & Helmut Schober (Eds.), Neutron applications in earth, energy and environmental sciences (pp. 15–36). Boston, MA: Springer US.

Rauch, H., & Waschkowski, W. (2003). Neutron Scattering Lengths. In A.-J. Dianoux, & G. Lander (Eds.), *ILL Neutron Data Booklet* (2nd edition). Grenoble, France: Institut Laue-Langevin (pp. 1.1–1 to 1.1–17).

Sarachan, K. L., Curtis, J. E., & Krueger, S. (2013). Small-angle scattering contrast calculator for protein and nucleic acid complexes in solution. *Journal of Applied Crystallography, 46*(6), 1889–1893. https://doi.org/10.1107/S0021889813025727.

Schober, H. (2009). Neutron scattering instrumentation. In L. Liang, R. Rinaldi, & H. Schober (Eds.), *Neutron applications in earth, energy and environmental sciences* (pp. 37–104). Boston, MA: Springer US.

Schuck, P. (2000). Size-distribution analysis of macromolecules by sedimentation velocity ultracentrifugation and Lamm equation modeling. *Biophysical Journal*, *78*(3), 1606–1619. https://doi.org/10.1016/S0006-3495(00)76713-0.

Sears, V. F. (1992). Neutron scattering lengths and cross sections. *Neutron News*, *3*(3), 26–37.

Semeraro, E. F., Marx, L., Frewein, M. P. K., & Pabst, G. (2021). Increasing complexity in small-angle X-ray and neutron scattering experiments: From biological membrane mimics to live cells. *Soft Matter*, *17*(2), 222–232. https://doi.org/10.1039/C9SM02352F.

Sokolova, A., Whitten, A. E., De Campo, L., Christoforidis, J., Eltobaji, A., Barnes, J., et al. (2019). Performance and characteristics of the bilby timeof- flight small-angle neutron scattering instrument. *Journal of Applied Crystallography*, *52*, 1–12. https://doi.org/10.1107/S1600576718018009.

Stuhrmann, H. (2007). Contrast variation in X-ray and neutron scattering. *Journal of Applied Crystallography*, *40*(s1), s23–s27. https://doi.org/10.1107/S0021889807003718.

Stuhrmann, H. B. (2012). Contrast variation application in small-angle neutron scattering experiments. *Journal of Physics: Conference Series*, *351*, 012002. https://doi.org/10.1088/1742-6596/351/1/012002.

Svergun, D., Koch, M., Timmins, P., & May, R. (2013a). Experimental practice and data processing. In *In small angle X-ray and Neutron scattering from solutions of biological macromolecules* (pp. 65–92). Oxford: Oxford University Press.

Svergun, D., Koch, M., Timmins, P., & May, R. (2013b). *Small angle X-ray and Neutron scattering from solutions of biological macromolecules*. Oxford: Oxford University Press.

Urban, V. S., Heller, W. T., Katsaras, J., & Bras, W. (2021). Soft matter sample environments for time-resolved small angle neutron scattering experiments: A review. *Applied Sciences*, *11*(12), 5566.

Voss, N. R., & Gerstein, M. (2005). Calculation of standard atomic volumes for RNA and comparison with proteins: RNA is packed more tightly. Journal of Molecular Biology, 346(2), 477–492. doi:https://doi.org/https://doi.org/10.1016/j.jmb.2004.11.072.

Whitten, A. E., Cai, S., & Trewhella, J. (2008). MULCh: Modules for the analysis of small-angle neutron contrast variation data from biomolecular assemblies. *Journal of Applied Crystallography*, *41*(1), 222–226. https://doi.org/10.1107/S0021889807055136.

Wood, K., Mata, J. P., Garvey, C. J., Wu, C. M., Hamilton, W. A., Abbeywick, P., et al. (2018). QUOKKA, the pinhole small-angle neutron scattering instrument at the OPAL research reactor, Australia: Design, performance, operation and scientific highlights. *Journal of Applied Crystallography*, *51*(2), 294–314. https://doi.org/10.1107/S1600576718002534.

Yadav, I., Kumar, S., Aswal, V. K., & Kohlbrecher, J. (2014). Small-angle neutron scattering study of differences in phase behavior of silica nanoparticles in the presence of lysozyme and bovine serum albumin proteins. *Physical Review E*, *89*(3), 032304. https://doi.org/10.1103/PhysRevE.89.032304.

CHAPTER SIX

Size exclusion chromatography coupled small angle X-ray scattering with tandem multiangle light scattering at the SIBYLS beamline

Daniel J. Rosenberg[a,b], Greg L. Hura[a,c], and Michal Hammel[a,*]
[a]Molecular Biophysics and Integrated Bioimaging, Lawrence Berkeley National Laboratory, Berkeley, CA, United States
[b]Graduate Group in Biophysics, University of California Berkeley, Berkeley, CA, United States
[c]Chemistry and Biochemistry Department, University of California Santa Cruz, Santa Cruz, CA, United States
*Corresponding author: e-mail address: mhammel@lbl.gov

Contents

Methods in Enzymology, Volume 677
ISSN 0076-6879
https://doi.org/10.1016/bs.mie.2022.08.031

Abstract

Small-angle X-ray Scattering (SAXS) has been a versatile technique for studying biomolecules in solution for several decades now. Developments in SAXS techniques that integrate *in situ* purification with a high-throughput, multimodal design philosophy have revolutionized the reach and tempo of BioSAXS experiments. The current zenith of the field comes in the form of size exclusion chromatography coupled SAXS with in-line multiangle light scattering (SEC-SAXS-MALS). This technique has been a considerable focus at the **S**tructurally **I**ntegrated **B**iolog**Y** for **L**ife **S**ciences (SIBYLS) beamline at the Advanced Light Source (ALS) in Berkeley, California, over the last 5 years and continues to be a point of active development. In this chapter, we describe the design of the SEC-SAXS-MALS system and general guidelines for collecting, processing, and analyzing SEC-SAXS-MALS data at the SIBYLS beamline.

1. Introduction

Small-angle X-ray Scattering (SAXS) is a convenient and versatile technique for studying the structure, complexation, and activity of biomolecules in solution (Putnam, Hammel, Hura, & Tainer, 2007). The method utilizes X-ray radiation to observe changes in the electron density of macromolecules as they exist in solution at near biological conditions. While commercial, "table top" SAXS instruments are becoming more widely available, the most popular method still involves the use of synchrotron facilities like the Advanced Light Source (ALS) in Berkeley, California, due to the significantly increased flux of the X-ray source. This increased brightness consequently reduces the exposure times needed from minutes or hours down to seconds. Leveraging the critical reduction in data collection time, SAXS beamlines at synchrotrons worldwide have been able to expand upon traditional SAXS collection methods to develop new techniques with a high-throughput, multimodal design philosophy.

In 2001, the **S**tructurally **I**ntegrated **B**iolog**Y** for **L**ife **S**ciences (SIBYLS) beamline (bl12.3.1) (Classen et al., 2013) was commissioned at the ALS to be the flagship BioSAXS (Putnam et al., 2007) beamline at Lawrence Berkeley National Laboratory (LBNL). This beamline specializes in technologies for the characterization of biological molecules and has been a staging ground for the development of integrated biophysical techniques including size exclusion chromatography coupled SAXS with in-line multiangle light scattering (SEC-SAXS-MALS) (Curran et al., 2020; Foster et al., 2019; Horst et al., 2019; Hura et al., 2019; Knott et al., 2019; Topolska-Woś et al., 2020) and high-throughput SAXS (HT-SAXS) (Dyer et al., 2014; Hura et al., 2009). Both systems are capable of automated data collection, with SEC-SAXS-MALS having the added benefits of providing *in situ* purification along with in-line complementary spectroscopic and light scattering techniques. These techniques provide immediate validation for the analogous values they provide. The molecular weight calculated from the MALS (MW_{MALS}) is analogous to the molecular weight calculated from the SAXS intensity at the zero angles, I(0), (MW_{SAXS}). One angle from the MALS is isolated to generate quasi-elastic light scattering (QELS, a.k.a. dynamic light scattering) data providing values for the radius of hydration (Rh) analogous to the radius of gyration (Rg) calculated from the SAXS. The continued development of SEC-SAXS-MALS builds upon an expanding pool previous work (Berthaud, Manzi, Pérez, & Mangenot, 2012; Brosey & Tainer, 2019; Bucciarelli et al., 2018; Graewert et al., 2015, Graewert et al., 2020; Inoue et al., 2019; Mathew, Mirza, & Menhart, 2004; Pérez & Vachette, 2017; Watanabe & Inoko, 2009). Importantly, this technology allows the definition of different functional assembly states in hours or days compared to their being determined over years by X-ray crystallography and other biophysics methods, such as was seen for the cell cycle regulatory adaptor CKS (Bourne et al., 1996; Parge, Arvai, Murtari, Reed, & Tainer, 1993), the proliferation and repair adaptor GRB2 (Ahmed et al., 2015; Ye et al., 2021), the ATPase DNA break repair adaptor Rad50 (Deshpande et al., 2014; Williams et al., 2011), the base to nucleotide excision repair adaptor Alt1 (Latypov et al., 2012; Tubbs et al., 2009), the WRN exonuclease (Perry et al., 2006; Perry et al., 2010), or protein mimicry of damaged DNA in repair complexes (Mol et al., 1995; Putnam et al., 1999).

In this chapter, we first describe the design of the current SEC-SAXS-MALS system at the SIBYLS beamline in detail. Then we discuss the collection and processing of SEC-SAXS-MALS data specific to the SIBYLS beamline and outline general protocols for the analysis of SEC-SAXS-MALS data.

2. Endstation design

The design of each beamline and endstation at an accelerator facility is uniquely designed and often meant to serve a very specific function, striving to enable experiments that often cannot be accomplished in any other way (Cowieson et al., 2020; Da Vela & Svergun, 2020; Trewhella, 2022). The SIBYLS beamline is no exception to this general design philosophy striving principally to achieve high collection speeds and robust automatization to study structural biology *via* SAXS. For the SIBYLS beamline to achieve automated high-throughput data collection, the conventional SAXS endstation had to be modified to include several significant design changes (Fig. 1). Firstly, a modern shutterless detector (Dectris PilatusX3 2M, Baden, Switzerland) was required to increase framerates and expand the detector area. Additionally, to incorporate the changes to the larger detector area and increase the sample-to-detector distance for higher resolution at lower scattering angles, a new vacuum flight tube was designed. This flight tube includes a 2 × 2 cm X-ray transparent 20 μm potassium aluminosilicate (Muscovite Mica) window (Goodfellow Corp., Coraopolis, PA) pressure sealed with rubber o-rings to cover the 1 cm diameter entrance to the flight tube and a sheet of 127 μm polyimide (Kapton) film (McMaster-Carr, Santa Fe Springs, CA) pressure sealed with 318 mm thick oil-resistant nitrile (Buna-N) rubber (McMaster-Carr, Santa Fe Springs, CA) to cover the 29.5 × 16.6 cm exit. In order to slow the inevitable deflection of the Kapton it would experience being subjected to ~4900 N and to reduce the air gap between the exit of the flight tube and the detector, horizontal bars are built into the flight tube which are identical in size and align to the inter-module spaces on the detector face (Fig. 1, bottom inset).

The X-ray energy is typically set at 10 keV (λ = 1.24 Å) but can be changed between ~7 and 16 keV with 110 (E/ΔE) resolution using a monochromator with dual Mo/B_4C multilayers as previously described (Classen et al., 2013). The standard working energy of 10 keV and an increased sample-to-detector distance of 2075 mm, the resulting scattering vector (q) ranges from 0.008 to 0.46 $Å^{-1}$, which correspond to the 800–14 Å resolution in real space. The scattering vector is defined as $q = 4\pi sin\theta/\lambda$, where 2θ is the scattering angle. Second, the integration of a modified liquid handling robot with custom-designed liquid handling needles was leveraged to establish the current HT-SAXS system (Dyer et al., 2014; Hura et al., 2009). Finally, a system designed from carefully selected commercial instruments (see Section 2.1 and Video 1 in the online version at https://doi.org/10.1016/bs.mie.2022.08.031) was adapted in conjunction with a purpose-built sample flow cell (see Section 2.2 and Video 2 in the online version at https://doi.org/10.1016/bs.mie.2022.08.031) to establish the current SEC-SAXS-MALS system (Fig. 1).

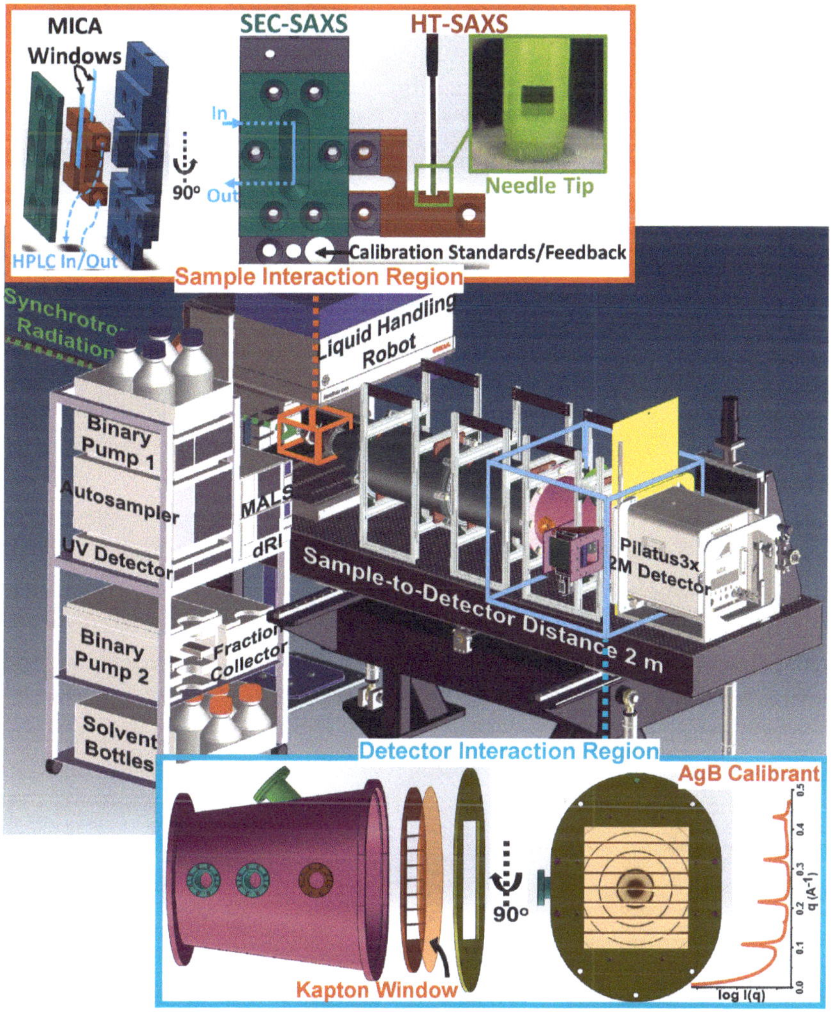

Fig. 1 Beamline layout showing a schematic representation of the SIBYLS beamline with SEC-SAXS-MALS and HT-SAXS systems. Top inset (orange) shows the sample interaction region with SEC-SAXS flow cell and HT-SAXS needle designs. Bottom inset (blue) shows the detector interaction region with flight tube design and representative Silver Behenate calibration showing scattering vector.

2.1 SEC-SAXS-MALS layout

The SEC-SAXS-MALS system has both a primary flow path for the active experiment and a secondary flow path to simultaneously equilibrate the system for the next running conditions (Fig. 2). Both flow paths are modular but under standard running configuration, both flow paths consist of four available buffer slots. Buffer selection and flow rate of each flow path is controlled by binary pumps from an Infinity series high-performance liquid chromatography (HPLC) system (Agilent, Santa Clara, CA). In the primary flow path, a sample is injected from a standard 96-well plate using a -dual-needle, 1290 Infinity series thermostat-controlled autosampler (Agilent, Santa Clara, CA) set to 7 °C. The elution is separated by a selected size-exclusion column and the absorbance is monitored by an in-line 1290 Infinity II Diode Array Detector (DAD) (Agilent, Santa Clara, CA). Subsequently, in-line MALS and QELS experiments are performed using an 18-angle DAWN HELEOS II (Wyatt Technology, Goleta, CA) light scattering detector with integrated QELS detector. The sample then travels to the custom flow cell (see Section 2.2, Fig. 1, and Video 2 in the online version at https://doi.org/10.1016/bs.mie.2022.08.031) for SAXS measurements then back to an Optilab T-rEX (Wyatt Technology, Goleta, CA) refractometer to measure the differential refractive index (dRI) (Wyatt Technology, Goleta, CA) needed for proper MALS analysis. Finally, the sample is either sent to waste or recollected in a 1260 series thermostat-controlled fraction collector (Agilent, Santa Clara, CA) set to 4 °C. As the primary flow path experiment proceeds, the secondary flow path has been equilibrating another column for the subsequent experimental conditions and upon completion of the active experiment, switches over to being the primary flow path through the control of a custom dual-valve system (Agilent, Santa Clara, CA) outlined in Fig. 2.

2.2 SEC-SAXS flow cell design

When collecting SEC-SAXS-MALS data under standard operating conditions, a custom, modular flow cell is used (Fig. 1 top inset and Video 1 in the online version at https://doi.org/10.1016/bs.mie.2022.08.031). This flow cell utilizes X-ray transparent 20 μm potassium aluminosilicate (Muscovite Mica) windows (Goodfellow Corp., Coraopolis, PA) held on either side of a 1.2 × 1.2 × 20 mm channel by pressure and sealed with 100 μm polytetrafluoroethylene (PTFE, Teflon) gaskets. The flow cell is under continued development and the current version is designed to operate as a standard flow cell with a single inlet and outlet with a calculated volume of 28.8 μL.

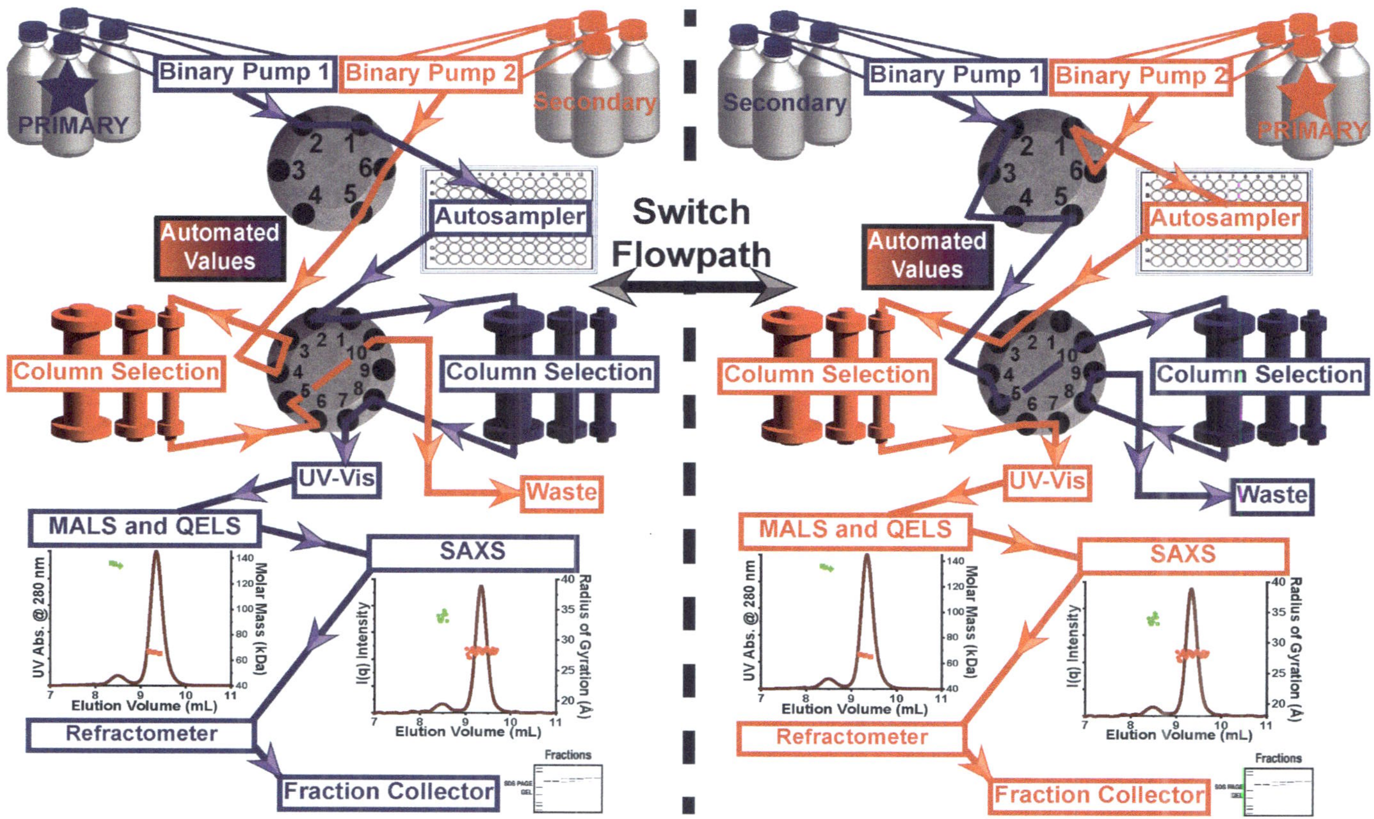

Fig. 2 Detailed SEC-SAXS-MALS layout demonstrating the two switchable flow paths made possible by the automated valve assembly.

3. Sample preparation for SEC-SAXS-MALS

The overall pipeline for the mail-in SEC-SAXS-MALS program, data collection, processing, and deposition is shown in Fig. 3. This process begins with users booking a slot on our website and preparing samples to be received and collected by our beamline staff. A major advantage of the SEC-SAXS-MALS over other traditional SAXS techniques is the *in situ* purification, accomplished through SEC, however this comes at the cost of sample dilution. Thus, the addition of a pre-purification step, specific to the macromolecule of interest, helps preserve the integrity of the system and allows for a more accurate initial concentration measurement (see Note 1). While generating and purifying macromolecules is case-specific, some general recommendations on sample purity and concentration in addition to selecting the elution buffer and column conditions should be followed when preparing for a SEC-SAXS-MALS experiment (Graewert & Jeffries, 2017; Hammel et al., 2020; Hodge et al., 2021; Horst et al., 2019; Knott et al., 2019; Topolska-Woś et al., 2020; Tsutakawa et al., 2020; Wilamowski et al., 2021).

3.1 Equipment

- Pipette
- Pipette tips
- 96-well plate (see Note 2)
- 96-well plate adhesive pierceable polyethylene films
- Eppendorf tubes
- Parafilm®
- 50 mL falcon tubes
- Temperature controlled centrifuge compatible with Eppendorf tubes (see Note 3)
- 500 mL Agilent HPLC compatible solvent bottles
- Solvent bottle filtration device
- Solvent bottle 0.22 μm membrane filters

3.2 Reagents

- A sample with a minimum volume of 50 μL (see Note 4) at proper concentration (see Note 1).
- A minimum of 50 mL of 10 × SEC buffer (see Note 5)
- Deionized water (diH_2O)

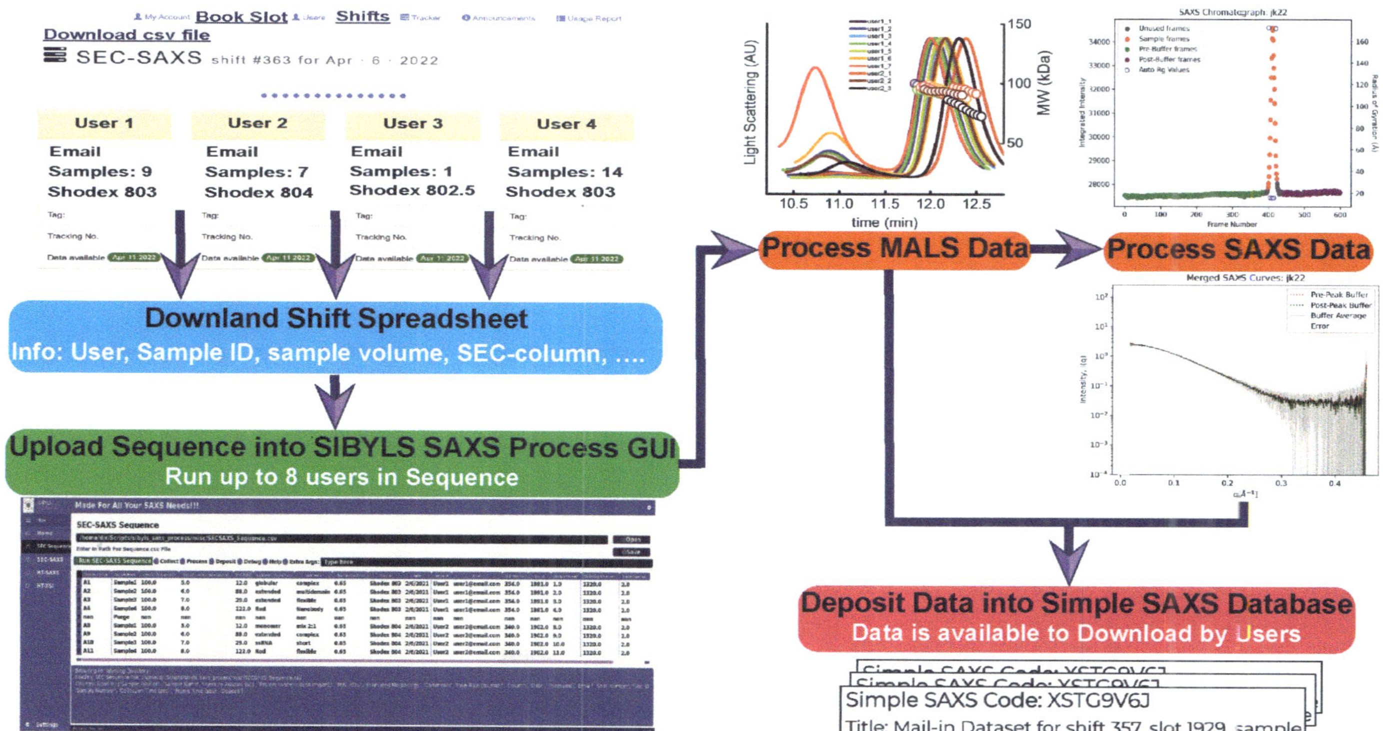

Fig. 3 Representation of the automated multiuser sequence data collection, processing, and deposition pipeline.

3.3 Procedure

1. To collect SEC-SAXS-MALS data at the SIBYLS beamline, a user will first book a time slot on the SIBYLS mail-in website (see Note 6) using the "Book Slot" link on the top of the page (Fig. 3). The number of samples and buffer conditions are subject to limitations and availability (see Note 7).
2. Macromolecules are produced and purified by sample-specific methods
3. The appropriate sample concentrations and volumes are pipetted into individual Eppendorf tubes and sealed with Parafilm® and placed in 50 mL falcon tubes for shipment
4. A minimum of 50 ml of 10 × of desired SEC buffer (see Note 5) is prepared in 50 mL falcon tubes and sealed with Parafilm® for shipment
5. Sample, buffer, and column information (see Note 8) is added to booked slot
6. Samples and buffer are packed for shipping to the ALS (see Note 9).
7. Once samples are received, the 50 mL of 10 × of desired SEC buffer is added to 450 mL diH_2O to prepare the 1 × buffer used as the SEC mobile phase
8. 1 × SEC buffer is filtered with 0.22 µm membrane filters into a 500 mL Agilent HPLC compatible solvent bottle
9. Received samples are thawed rapidly at room temperature if frozen and then immediately placed in a temperature-controlled centrifuged at 4 °C (unless samples require a different storage temperature).
10. Samples are spun at 8000 rpm for 5 min at 4 °C to remove macromolecule precipitation that might have occurred during transport
11. Centrifuged samples are added sequentially to the 96-well plate in the order of the user sample information, sealed with 96-well polyethylene films, and placed in the Agilent autosampler at 7 °C (unless samples require a different storage temperature) before data collection

3.4 Notes

1. The sample concentration should be 3–10 mg/mL with smaller molecules requiring higher concentrations. The general rule of thumb for concentration uses the following general formula for proteins and about half the concentration for nucleic acids:

$$concentration\ (mg/mL) = \frac{300}{molecularweight\ (kDa)} \quad (1)$$

For example, a 100 kDa protein = 3 mg/mL minimum recommended concentration.

2. For a 96-well plate, we recommend Axygen® 96-well Polypropylene PCR Microplate, but others should be fine if they are listed as compatible with a 1290 series Agilent Autosampler
3. If smaller Eppendorf tubes or PCR tubes are used, centrifuge adaptors may be required
4. To avoid bubbles in the system and account for inconsistencies in pipetting the sample into the 96-well plates, 5 μL less is injected than is put into each well of the 96-well plate. The maximum injection volume is 100 μL limited by the autosampler loop size
5. Special consideration should be taken when preparing the buffer for a SAXS experiment (Graewert & Jeffries, 2017). Ideally, the sample should be prepared in a buffer that is identical or very similar to the SEC buffer but this is not required as the buffer will exchange on the column as the elution proceeds. For SEC buffers, salt concentrations $\leq$500 mM and pH $\leq$ 8 should be maintained when using the standard Shodex columns provided by the SIBYLS beamline as per the manufacture's recommendations. The concentration of glycerol should not exceed 2% and the concentrations of other stabilizers or additives should be kept to a minimum to reduce the contribution to the background scattering and avoid high column pressure
6. SIBYLS mail-in website and most current instructions can be found here: https://bl1231.als.lbl.gov/htsaxs/
7. Each beamtime slot is currently limited to 15 samples using 1 column/buffer condition, but these numbers are subject to change and can be modified upon request
8. Please specify the columns to use in your experiments. If standard columns do not meet the needs of the experiment, other columns may be available or may need to be sent with samples. A list of available columns with a summary of their properties is shown in Table 1
9. We recommend that samples are shipped in standard Eppendorf PCR tubes together with 10 × buffer at 4 °C with cold packs or frozen with dry ice in a Styrofoam shipping container. DO NOT ship wet ice. Clearly label the outside of your shipping box with the temperature of your samples. This is extremely important if a shipping delay occurs. Lastly, please provide us with a shipping tracking number. See SIBYLS mail-in website (see Note 6) for address

Table 1 Available size-exclusion columns.

Product name	Exclusion limit (Protein kDa)	Particle size (μm)	Maximum pore size (Å)	Column size (mm) ID × L
PROTEIN KW-802.5	150	5	400	8 × 300
PROTEIN KW-803	700	5	400	8 × 300
PROTEIN KW-804	1000	7	1500	8 × 300

4. Data acquisition

Once samples are prepared, plated, and placed in the Agilent autosampler, SEC-SAXS-MALS collection can proceed as an individual sample, a single user sequence, or a multiple user sequence. Collection, processing, and deposition for SEC-SAXS-MALS at the SIBYLS beamline can all be managed through beamline specific controls software called SIBYLS SAXS Process following the pipeline shown in Fig. 3.

4.1 Equipment

- Pilatus 3 × 2 M Detector (Dectris, Baden, Switzerland)
- 1290 Infinity Multisampler with Sample Loop-Flex 100 μL and InfinityLab Sample Thermostat (Agilent, Santa Clara, CA)
- 1290 Infinity II High Speed Pump (Agilent, Santa Clara, CA)
- 1290 Infinity II Diode Array Detector (DAD) (Agilent, Santa Clara, CA)
- 1260 Infinity Fraction Collector with Sample Thermostat (Agilent, Santa Clara, CA)
- 1260 Infinity Binary Pump with four solvent head (Agilent, Santa Clara, CA)
- In-line degasser (Agilent, Santa Clara, CA)
- 2pos/6port valve head (Agilent, Santa Clara, CA)
- 2pos/10port Quick Change valve head (Agilent, Santa Clara, CA)
- 18-angle DAWN HELEOS II multi-angle light scattering (MALS) detector with WyattQELS unit installed (Wyatt Technology, Goleta, CA)
- Optilab T-rEX differential Refractive Index (dRI) detector (Wyatt Technology, Goleta, CA)

- Trident LC Column in-line filter (Restek, Bellefonte, PA)
- Three standard Shodex KW series columns (Showa Denko, Tokyo, Japan)
- Agilent OpenLab (version Rev. C.01.10 or later recommended) (Agilent, Santa Clara, CA)
- Wyatt ASTRA software (version 7.1 or later recommended) (Wyatt Technology, Goleta, CA)

4.2 Reagents

- A minimum of 500 mL of 1 × SEC buffer
- Deionized water (diH_2O)

4.3 SEC-SAXS-MALS sequence setup

1. Once user samples are received, Agilent and Wyatt systems are turned on and the modules with lasers are allowed to "warm up" (see Note 1).
2. SEC is then performed using one of three standard Shodex KW series columns (Showa Denko, Tokyo, Japan) depending on the expected size range of the macromolecules in question (see Section 3 Note 8). This column is connected from a selection of columns found in the chilled column chamber (Fig. 2) using standard practices (see Note 2).
3. The column is equilibrated ahead of time with a mobile phase (1 × buffer) at a standard flow rate of 0.65 mL/min for about 1 h until the differential refractive index (dRI) signal stabilizes
4. On mail-in SAXS website (see Section 3 Note 6), navigate to the current shift using the "Shift" link on the top of the page (Fig. 3) and download the spreadsheet with all user and sample information as a comma-separated value file (.csv) using the "Download csv file" link in the upper right corner of the shift page (Fig. 3).
5. Once the system is at equilibrium, use the downloaded shift spreadsheet to set up complimentary sample sequences in the Agilent OpenLab software, the Wyatt ASTRA software (see Note 3), and the SIBYLS SAXS Process GUI (see Section 4.4).
6. The sample is separated by SEC and the elution is monitored with up to eight discreet wavelengths (see Note 4) by an in-line DAD to monitor sample concentration as it elutes off the column
7. The sample then travels through the in-line MALS to obtain MW_{MALS} and Rh measurements and onto the SAXS sample interaction region

8. After the sample passes through the X-ray beam, it travels to the in-line refractometer to measure reflectivity index (dRI) required to calculate the MW_{MALS} from the MALS data (see Note 5).
9. Finally, the sample will either be sent to waste or recollected in the fraction collector as the experiment demands (see Note 6).
10. A typical sample will require approximately 25 min to complete data acquisition and, if needed, an additional 35 min system purge to establish new equilibrium conditions for each unique user or buffer condition

4.4 Setting up the SIBYLS SAXS process GUI

1. Once the system is at equilibrium and the complimentary sample sequences are set up in the Agilent OpenLab software, the Wyatt ASTRA software (see Note 3), the SIBYLS SAXS Process GUI (see Note 7) is used to set up the automated multisample sequence data collection, processing, and deposition
2. After loading the SIBYLS SAXS Process GUI, the main side menu is expanded (Fig. 4(1)) and the "SEC Sequence" tab is selected to open the multi-sample sequence widget
3. The "Open" button (Fig. 4(2)) is used to load a pre-made sample sequence spreadsheet (see Note 8) into the editable sequence table widget (Fig. 4(3)). For details on a typical sequence setup, see Note 9
4. Once loaded, the sequence can be edited in the table widget (Fig. 4(3)) and saved using the save button (Fig. 4(4)). If changes are made to the sequence, they must be saved before continuing (see Note 10).
5. If multiple users with different mobile phases or column conditions are in a single sequence, "Purge" is placed in the appropriate row in the "Sample Name" column, so the program knows to purge and equilibrate the column between samples (see Note 11).

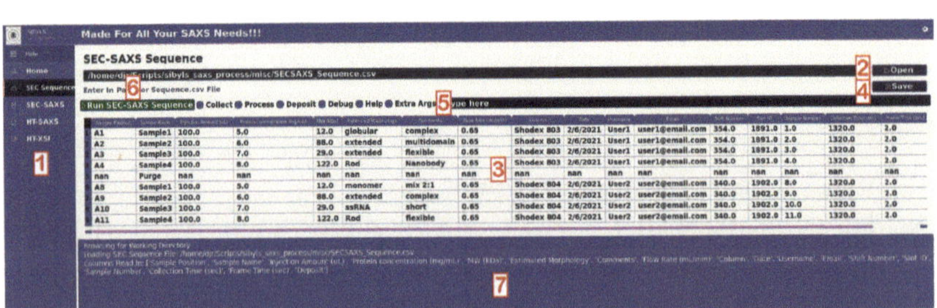

Fig. 4 SIBYLS SAXS Process GUI showing the SEC-SAXS automated multisample sequence data collection, processing, and deposition panel.

6. The option buttons (Fig. 4(5)) select for specific actions or arguments during the sequence run (see Note 12). For example, the user can select to "Collect," "Process," or "Deposit" data using the respective option buttons
7. To accurately automate SEC-SAXS data processing for multiple users, a mask file and an experimental parameters file must first be created (see Note 13).
8. Once the sequence is saved and the options are selected, the sequence can be initiated by hitting "Run SEC-SAXS Sequence" (Fig. 4(6)). Do this only when the complementary sequences are prepared and started on the in-line Agilent and Wyatt systems (see Note 14). The run can be followed in the output terminal (Fig. 4(7)).
9. Once the sequence has completed and all data has been collection and processed, the results are organized in the following file hierarchy for each Shift Date → Username → Sample:

```
Date_Folder (ie /data/secsaxs/041622):
    └─ Results
        └─ 2022_04_16_SEC-SAXS_ Username1
            └─ Sample1
                ├─ Sample1_Peak_Finding (Auto Peak Finder Results)
                |   ├─ Sample1_LC_Trace.png (SEC-SAXS Chromatograph)
                |       ├─ Sample1_Merged_Curves.png (Merged SAXS Curves)
                |       ├─ Subtracted (Subtracted SAXS Curves for Each Frame)
                |       |   ├─ Full_Buffer
                |       |   ├─ Post_Buffer
                |       |   └─ Pre_Buffer
                |       └─ Unsubtracted (Unsubtracted SAXS Curves)
                |           ├─ Buffer_Frames
                |           └─ Sample_Frames
                └─ Sample1_Results (All Unsubtracted SAXS Curves)
```

4.5 Notes

1. The Agilent Diode Array Detector (DAD) and the Wyatt MALS detector should take approximately 30 min and 60 min, respectively, to "warm up" as recommended by the manufacturer's instructions. The MALS is ready when the 90-degree angle fluctuates $< \pm 0.002$ V
2. Initialize Agilent pumps prior to connecting desired column in order to bleed air out of the lines and establish a wet connection

3. See user manuals for details on setting up sequences on Agilent OpenLab (version Rev. C.01.10 or later recommended) controlling the Agilent HPLC modules and Wyatt ASTRA software (version 7.1 or later recommended) controlling the Wyatt instruments
4. In a typical structural biology experiment, proteins and RNA are monitored at 280 nm, and DNA is monitored a 260 nm, however some macromolecules may require additional or alternative wavelengths to be monitored
5. The dRI is necessary to determine fluctuations in the mobile phase, which may lead to differences in the motion of the particles measured by MALS and is thus needed to make corrections in the MW_{MALS} calculations
6. Fraction collection is not a standard feature of a typical SEC-SAXS-MALS experiment and is usually only enabled upon request
7. The SIBYLS SAXS Process package is available on our GitLab repository (https://git.bl1231.als.lbl.gov/saxs/sibyls_saxs_process).
8. Pre-made sample sequence spreadsheets can be loaded or saved either a comma-separated value file (.csv) or an excel workbook file (.xlsx).
9. A standard template should be followed for a typical sequence, which comes with the SIBYLS SAXS Process package found on our git repository (see Note 7) following the path: /sibyls_saxs_process/misc./SECSAXS_Sequence.csv
10. If changes are made to the sequence, the program will not allow you to start the sequence until all changes have been saved. The terminal will prompt you if there is a conflict (Fig. 4(7)).
11. A minimum system purge time of 35 min is typically required for the system to reach equilibrium when a change in mobile phase or column is required. These "purge" breaks need to be set up independently in the Agilent OpenLab software, the Wyatt ASTRA software (see Note 3), and the SIBYLS SEC-SAXS Process GUI
12. To add additional, non-standard arguments the "Extra Args" options button can be selected, and extra inputs typed into the text box next to it. Use the help option for more details on other arguments that can be used
13. To automate data processing, the "Process" option button must be selected (Fig. 4(5)). Additionally, a mask for the unused regions of the detector (mask.ASC) and an experimental parameters file (ExptParam) containing information on the X-ray wavelength (Å) and the detector metrology must be created. These files are generated using the FIT2D software(Hammersley, 1997). These two files must be

placed in the date directory for the shift where all current user data is being collected (i.e., /data/{*account collecting data*}/{*shift date*}). If no files are found at that location the program with default to pulling the necessary files from the last ones generated but this might lead to processing error and the data may need to be re-processed

14. The Wyatt ASTRA, the Agilent OpenLab (see Note 3), and the SIBYLS SAXS process sequences should be started in this order respectively

5. Data processing

In a typical SEC-SAXS-MALS experiment, the depth and detail of the analysis may vary but the most basic processing and analysis of the raw data typically remains the same. Indirectly the complementary spectroscopy and light scattering measurements are used to monitor the concentration and homogeneity of the samples in real time to ensure the quality of the sample immediately before SEC-SAXS measurements are taken. More directly, however, some rudimentary tools are included in the SIBYLS SAXS Process GUI to automate SEC-SAXS data processing and display the most critical information to provide rapid feedback on data quality. Certain standard practices should be taken to ensure the quality and accuracy of SEC-SAXS-MALS data and both sources of feedback end up being crucial towards meeting this goal.

5.1 Processing MALS, QELS, and UV data

1. System normalization and calibration are performed for each new experimental condition with bovine serum albumin (BSA) (see Note 1) or some other protein of known MW
2. The UV–vis signal from the DAD is used to calculate the concentration of each eluted sample by setting up a standard curve in either Agilent OpenLab or Wyatt ASTRA software (see Section 4.5 Note 3).
3. The MALS and dRI signals are used to determine MW_{MALS} and the QELS measurements are used to determine the Rh across the principal peaks in the SEC analysis all processed and analyzed with the Wyatt ASTRA software (see Section 4.5 Note 3).
4. All spectroscopy and light scattering measurements can be used to monitor the concentration and homogeneity of the sample across the elution peaks in real-time and serve as an early prediction of the quality of the SEC-SAXS data being collected

5.2 Processing SEC-SAXS data

1. Beamline-specific software built into the SIBYLS SAXS Process GUI is used to process raw SAXS-detector images for individual samples by expanding the main side menu (Fig. 5(1)) and selecting the "SEC-SAXS" tab to open the SEC-SAXS processing widget. Alternatively, SEC-SAXS data processing can be run in batch or automated to proceed each sample as data is collected using the "SEC-SAXS sequence" tab and selecting the "Process" option as described (see Sections 4.4 and 4.5 Note 13).
2. To process data individually within the "SEC-SAXS" tab, the "Open" button (Fig. 5(2)) is used to load the directory containing the raw SAXS detector images (.cbf file format).
3. To accurately process SAXS data, a mask file and an experimental parameters file must first be created (see Section 4.5 Note 13).
4. After selecting the working directory (Fig. 5(2)), clicking the "Process SEC-SAXS" button (Fig. 5(3)) with the desired optional argument buttons (Fig. 5(4)) (see Section 4.5 Note 12) will produce SAXS curves for every exposure frame by radially integrating and normalizing each raw detector image, as previously described (Dyer et al., 2014; Hura et al., 2009). The output of this process is a results folder with the name *samplename*_Results, within the loaded working directory (Fig. 5(2)) (see Note 2).

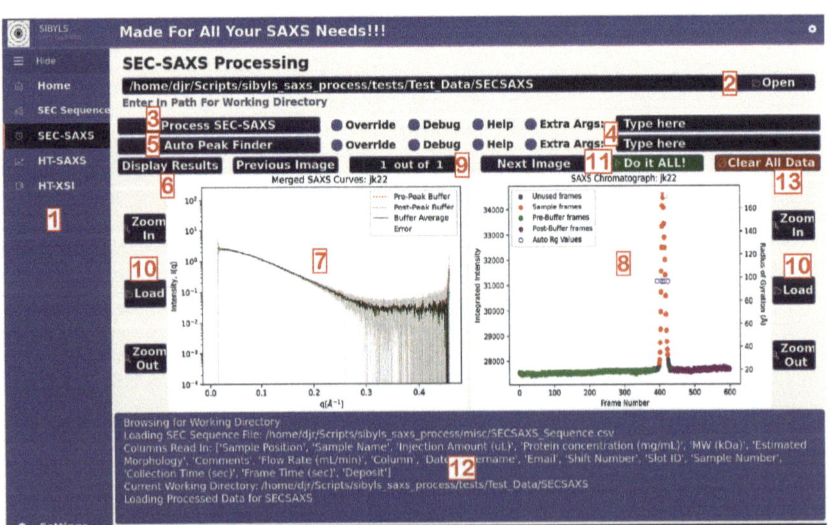

Fig. 5 SIBYLS SAXS Process GUI showing the SEC-SAXS single sample data processing panel with automated peak finding.

5. The data files produced by the "Process SEC-SAXS" step each represent a single time point or frame, usually 1–3 s long, during the SEC elution. The "Auto Peak Finder" button (Fig. 5(5)) sequentially integrates the SAXS curve for each frame across the elution and display the results as a SEC-SAXS chromatogram (Fig. 5(8)). The automatic peak finder function by default then takes each frame selected as the primary sample elution peak, subtracts an averaged selection of both pre- and post-peak buffers from each frame, and merges the results leading to three final curve (see Note 3). Alternatively, to the default automated peak finding, users can customize the peak finding parameters by modifying the number of peaks found, sample frames merged, and buffer frames averaged by using the associated optional arguments (Fig. 5(4)) (see Section 4.5 Note 12).
6. The "Display Newest Results" button (Fig. 5(6)) shows the latest results from the "Auto Peak Finder" function as a set of two corresponding plots showing the three final subtracted SAXS curves (see Note 3) (Fig. 5(7)) and the SEC-SAXS chromatogram (Fig. 5(8)) that produced them. All sets in the same directory can be browsed with the "Previous" and "Next" image buttons (Fig. 5(9)) and older sets can be searched and displayed with the "Load" buttons (Fig. 5(10)).
7. To automate the whole SEC-SAXS data processing pipeline for a single sample, use the "Do it ALL!" button (Fig. 5(11)). The output of each step is displayed in the terminal below (Fig. 5(12)).
8. Any processed data in the working directory (Fig. 5(2)) can also be removed by using the "Clear All Data" button (Fig. 5(13)) and answering "yes" to confirm when prompted in the pop-up window (see Note 4).

5.3 Notes

1. BSA is typically collected by injecting 50 μL of 7 mg/mL sample in the same mobile phase and on the same column as the experimental conditions demand. The use of a protein of known MW as a standard is particularly critical for normalizing the MALS system for the new mobile phase and column but is also useful to ensure the quality of the acquired SEC-SAXS data before starting to run important samples
2. The radially integrated and normalized data files (.dat) are in the form of three-column, space separated files. Column one is the q values (Å^{-1}), column two the Intensity values I(q), and column three the error in the I(q) (Putnam et al., 2007).

3. The three final curves produced by the default automatic peak finder are: (1) pre-peak buffer subtracted, (2) post-peak buffer subtracted, and (3) an average of the two previous curves
4. The data removed with the "Clear All Data" button (Fig. 5(13)) is not permanently removed and are instead moved to a temporary "Trash Can" folder within the working directory (Fig. 5(2)) in case the user wished to reverse the removal. This is only done once however and the subsequent removal of data by the same name will override it

6. Data validation and analysis

The steps shown previously in Section 5 are meant for preliminary data processing and quickly displaying results to monitor data quality while running the SEC-SAXS-MALS system. While the SIBYLS SAXS Process interface is currently designed for running the system and feedback during the experiments, continued development is underway to expand the automated data processing pipeline. There are several computational tools available now to further analyze SEC-SAXS data including SCÅTTER (Tully, Tarbouriech, Rambo, & Hutin, 2021), RAW (Hopkins, Gillilan, & Skou, 2017), and CHROMIXS (Panjkovich & Svergun, 2018). These software options are not discussed here, and the focus is instead on the theory of how to extract, validate, and analyze the information obtained from a typical SEC-SAXS-MALS experiment.

Whether you are using SIBYLS SAXS Process or any other software for analyzing SEC-SAXS-MALS data, certain factors must be kept in mind when assessing the quality and validity of your measurements. For SEC-SAXS-MALS data, this validation generally starts with the assessment of the SEC-SAXS chromatogram, and then by analyzing the final buffer subtracted, merged SAXS curves. Subsequently, the SEC-SAXS-MALS data is further validated by comparing analogous values from the complementary spectroscopic and light scattering techniques collected in-tandem with the SAXS. In combination, the in-line UV, MALS, and QELS measurements, provide an accurate assessment of sample aggregation, heterogeneity, and the oligomerization state of macromolecules across the SEC elution and can often aid in SEC-SAXS data interpretation.

6.1 Self-validation and analysis of SEC-SAXS data

For SEC-SAXS data, validation generally starts by selecting baseline frames corresponding to the buffer to use as subtraction for the sample peaks.

The baseline of a SEC-SAXS chromatogram can vary significantly depending on many factors including fluctuations in the X-ray beam intensity or position, sample concentration, and the sensitivity of the sample to radiation. For a well-behaved sample, run on a fully equilibrated system (Fig. 6A), the choice of buffer subtraction is relatively straightforward a small, stable buffer region (See Note 1), typically immediately prior to the sample peak is selected and merged to obtain an adequate baseline SAXS profile. The user must be more cautious when selecting buffer frames for less optimized experiments to ensure accurate baseline subtraction.

One common example of an improper experimental setup is a linear rise in the baseline of the SEC-SAXS chromatogram (Fig. 6B). This is typically caused by improperly equilibrating the system before sample injection (See Note 2), or unstable conditions such as temperature causing X-ray optics to move and the beam to shift (See Note 3). Often a reasonable SAXS profile may still be extracted by taking an equal number of frames from before and after the peak for subtraction. If the fluctuations in the SAXS baseline are linear or uniform, it may be possible to correct for it mathematically. Recently, programs like REGALS (Meisburger, Xu, & Ando, 2021) are being utilized for this purpose.

A typical example of a poorly behaved sample (Fig. 6C), demonstrated as a stepwise rise in the overall background intensity after the main sample peak. This baseline shift commonly occurs from fouling or deposition of sample on the windows of the sample flow cell resulting from an overly concentrated or radiation-sensitive sample. In the case of such an increase in the baseline, which occurs continuously across the elution peak as sample deposits on the flow cell windows, it is often not sufficient to use frames for the baseline subtraction form before or after the sample peak. Instead, an averaging of an equal number of frames from before and after the peak can produce a more suitable baseline for subtraction and can partially eliminate artifacts of the capillary fouling (Fig. 6C and D).

Many factors go into determining the quality of the final SAXS profiles but a linear fitting, of low scattering angles, using the Guinier approximation (Guinier, 1939), $I(q) = I(0)\ \exp\left(\frac{-q^2 Rg^2}{3}\right)$ with the limit $qRg < 1.3$, provides a quick assessment of the aggregation state of the sample, as established previously (Putnam et al., 2007). In the case of SEC-SAXS data, the non-linearity of the Guinier region can also indicate the contribution of capillary fouling. A linear Guinier region validates proper buffer subtraction in the final merged data. As example, only when an equal number of buffer

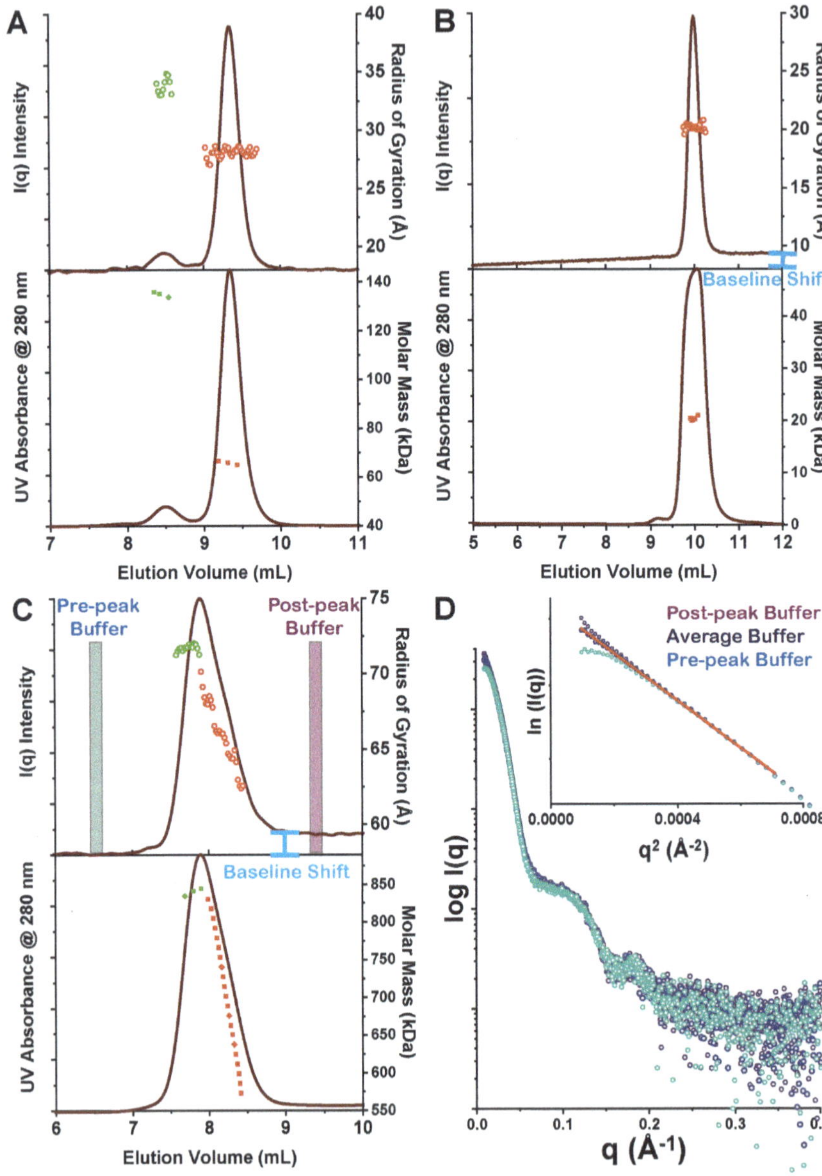

Fig. 6 Examples of SAXS (top) chromatograms with calculated Rg values and MALS (bottom) chromatograms with calculated MW_{MALS} values for (A) a well-behaved sample of bovine serum albumin (BSA) showing dimer and monomer peaks, (B) an incorrectly equilibrated sample demonstrating a linearly increasing baseline, and (C) a highly concentrated sample showing a stepwise increase in the baseline due to fouling of the flow cell windows. Stability in the protein sample is represented by (A) stable Rg and MW_{MALS} values across the peaks, and (C) linear change in Rg and MW_{MALS} representing a transient protein complex or confirmation. (D) Examples of different buffer regions selected for subtraction to produce SAXS profiles as selected from the stable Rg region in (C). Insert shows the linear fit of the Guinier region of each curve plotted as $\ln(I(q))$ against q^2.

frames before and after the sample peak are used in subtraction can artifacts from capillary fouling be eliminated as shown by the linear Guinier region in the inset of Fig. 6D.

An estimate of the Rg value, based on the linearity of Guinier region, for each frame across the sample peak can automatically be calculated and plotted by most SEC-SAXS data processing software, including SIBYLS SAXS Process (Fig. 5(8)). These estimated Rg vs frame number plot can guide the selection of the best frames across the SEC elution to merge for a final SAXS profile. For well-behaved samples, these estimated Rg values across the sample peak will remain stable (Fig. 6A and B) merged to produce the highest quality SAXS profile possible for that sample (See Note 4). It is also common to obverse fluctuation in Rg values across a single peak, especially when analyzing proteins and protein complexes. These changes in Rg values usually signify a transient behavior of the protein complex. In this case, multiple regions across the peak may be selected and merged independently to produce multiple unique SAXS profiles. This sort of analysis usually constitutes a balancing act between the number of SAXS profiles to produce out of the peak and the number of frames to merge for each one. Several tools are available now to assist in this sort of analysis, like the Evolving Factor Analysis tool (Meisburger et al., 2016) included in the RAW (Hopkins et al., 2017) software package or Gaussian decomposition as in EFAMIX (Konarev et al., 2022) and US-SOMO (Brookes, Vachette, Rocco, & Pérez, 2016). Suppose the frames across the peak cannot be divided due to a low concentration of scattering intensity. In that case, the whole peak can instead be merged and consequently analyzed by multistate modeling (Pelikan, Hura, & Hammel, 2009; Schneidman-Duhovny & Hammel, 2018; Schneidman-Duhovny, Hammel, Tainer, & Sali, 2016).

6.2 Sample validation through SEC-MALS

One factor that makes SEC-SAXS-MALS such a powerful technique is that the SAXS data can be validated by the complementary spectroscopy techniques collected in tandem. The UV–vis signal is used to calculate the sample concentration. Still, the wavelengths recorded can be changed to provide additional information like the protein/nucleic acid ratio (Reyes, Schwartz, Tainer, & Rambo, 2014; Wilamowski et al., 2021) or the oxidation state of a protein-heme moiety (Horst et al., 2019). The UV–vis signal is also typically the most sensitive measurement and the best real-time monitor for the elution of samples during an experiment. The MALS and QELS measurements can provide a wealth of information on their own, most notably MW_{MALS} and Rh, respectively. The MW_{MALS} compliments the MW_{SAXS} derived from SAXS (Rambo & Tainer, 2013) and tends to be more accurate, while the Rh from the QELS is analogous to the Rg from the SAXS with the notable difference that the Rh includes the solvation layer surrounding the macromolecule in the calculation (Kok & Rudin, 1981). These complementary in-tandem spectroscopic and light scattering techniques all serve to assess the overall quality of the SAXS measures.

6.3 Notes

1. A small, stable buffer region should be defined as a region including about 1–3% of the total frames where the baseline is relatively flat, and no sample is eluting. Typically for well-behaved samples, this region is taken immediately before the sample peak, but immediately after or a combination of both can be taken as circumstances necessitate
2. For optimal equilibrating of the SEC-SAXS-MALS system, typically be equilibration times at 0.65 mL/min are about 90 min when the column is not pre-equilibrated and about 35 min if it is prior to sample injection. The status of the equilibrating is best observed with the MALS and dRI detector following the advised quality checks in the user manuals for those systems (see Section 4.5 Note 1).
3. Great care should be taken to minimize vibration and temperature fluctuations which can cause optics to shift the beam to drift. Even movements on the order of 10 μm can be enough to cause baseline fluctuation and aberrant data. This baseline shift is not always apparent in the UV, MALS, or dRI chromatograms (Fig. 6B).
4. Several factors play into the signal-to-noise ratio in a fully merged SAXS profile, but generally more frames merged across a stable Rg region led to higher signal-to-noise and hence higher data quality

7. Summary and conclusion

SEC-SAXS-MALS measurements have become widely preferred over traditional SAXS measurements for studying structural biology primarily due to its versatility and *in situ* purification. In this chapter, we first describe the intricacies of the current SEC-SAXS-MALS system design at the SIBYLS beamline which incorporates a general design philosophy of high collection rates and robust automatization. We then discuss the mail-in user program starting with general sample preparation and ending with the use of the SEC-SAXS-MALS data collection, processing, and deposition pipeline simplified by our new SIBYLS SAXS Process beamline control software. Finally, we suggest some general protocols and guidelines for the analysis of SEC-SAXS-MALS data collected at the SIBYLS beamline or elsewhere.

The SEC-SAXS-MALS technique as well as other high-throughput SAXS measurement platforms has been a primary focus of the SIBYLS beamline over the last 5 years and continues to be a point of active development. As we continue to refine our techniques and streamline our data analysis pipeline, we look to the future of brighter light sources, improved sample delivery options, and advanced computational tools which may be the next revolution in the use of SAXS.

Acknowledgments

Efforts to apply SAXS to characterize biomolecules are supported in part by National Cancer Institute grants for Structural Biology of DNA Repair (SBDR) CA092584 and CA220430. This work was conducted at the Advanced Light Source (ALS), a national user facility operated by Lawrence Berkeley National Laboratory on behalf of the Department of Energy, Office of Basic Energy Sciences, through the Integrated Diffraction Analysis Technologies (IDAT) program, supported by DOE Office of Biological and Environmental Research. Additional support comes from the National Institute of Health project ALS-ENABLE (P30 GM124169) and a High-End Instrumentation Grant S10OD018483.

References

Ahmed, Z., Timsah, Z., Suen, K. M., Cook, N. P., Lee, G. R., IV, Lin, C. C., et al. (2015). Grb2 monomer-dimer equilibrium determines normal versus oncogenic function. *Nature Communications*, *6*, 7354. https://doi.org/10.1038/ncomms8354.

Berthaud, A., Manzi, J., Pérez, J., & Mangenot, S. (2012). Modeling detergent organization around Aquaporin-0 using small-angle X-ray scattering. *Journal of the American Chemical Society*, *134*(24), 10080–10088. https://doi.org/10.1021/ja301667n.

Bourne, Y., Watson, M. H., Hickey, M. J., Holmes, W., Rocque, W., Reed, S. I., et al. (1996). Crystal structure and mutational analysis of the human CDK2 kinase complex with cell cycle-regulatory protein CksHs1. *Cell*, *84*(6), 863–874. https://doi.org/10.1016/s0092-8674(00)81065-x.

Brookes, E., Vachette, P., Rocco, M., & Pérez, J. (2016). *US-SOMO* HPLC-SAXS module: Dealing with capillary fouling and extraction of pure component patterns from poorly resolved SEC-SAXS data. *Journal of Applied Crystallography*, *49*(5), 1827–1841. https://doi.org/10.1107/S1600576716011201.

Brosey, C. A., & Tainer, J. A. (2019). Evolving SAXS versatility: Solution X-ray scattering for macromolecular architecture, functional landscapes, and integrative structural biology. *Current Opinion in Structural Biology*, *58*, 197–213. https://doi.org/10.1016/j.sbi.2019.04.004.

Bucciarelli, S., Midtgaard, S. R., Nors Pedersen, M., Skou, S., Arleth, L., & Vestergaard, B. (2018). Size-exclusion chromatography small-angle X-ray scattering of water soluble proteins on a laboratory instrument. *Journal of Applied Crystallography*, *51*(6), 1623–1632. https://doi.org/10.1107/S1600576718014462.

Classen, S., Hura, G. L., Holton, J. M., Rambo, R. P., Rodic, I., McGuire, P. J., et al. (2013). Implementation and performance of SIBYLS: A dual endstation small-angle X-ray scattering and macromolecular crystallography beamline at the advanced light source. *Journal of Applied Crystallography, 46*(Pt. 1), 1–13. https://doi.org/10.1107/S0021889812048698.

Cowieson, N. P., Edwards-Gayle, C. J. C., Inoue, K., Khunti, N. S., Doutch, J., Williams, E., et al. (2020). Beamline B21: High-throughput small-angle X-ray scattering at diamond light source. *Journal of Synchrotron Radiation, 27*(5), 1438–1446. https://doi.org/10.1107/S1600577520009960.

Curran, S. C., Pereira, J. H., Baluyot, M.-J., Lake, J., Puetz, H., Rosenburg, D. J., et al. (2020). Structure and function of BorB, the type II Thioesterase from the Borrelidin biosynthetic gene cluster. *Biochemistry, 59*(16), 1630–1639. https://doi.org/10.1021/acs.biochem.0c00126.

Da Vela, S., & Svergun, D. I. (2020). Methods, development and applications of small-angle X-ray scattering to characterize biological macromolecules in solution. *Current Research in Structural Biology, 2*, 164–170. https://doi.org/10.1016/j.crstbi.2020.08.004.

Deshpande, R. A., Williams, G. J., Limbo, O., Williams, R. S., Kuhnlein, J., Lee, J. H., et al. (2014). ATP-driven Rad50 conformations regulate DNA tethering, end resection, and ATM checkpoint signaling. *The EMBO Journal, 33*(5), 482–500. https://doi.org/embj.201386100 [pii] 10.1002/embj.201386100.

Dyer, K. N., Hammel, M., Rambo, R. P., Tsutakawa, S. E., Rodic, I., Classen, S., et al. (2014). High-throughput SAXS for the characterization of biomolecules in solution: A practical approach. In Y. W. Chen (Ed.), *Structural genomics: General applications* (pp. 245–258). Humana Press. https://doi.org/10.1007/978-1-62703-691-7_18.

Foster, B. M., Rosenberg, D., Salvo, H., Stephens, K. L., Bintz, B. J., Hammel, M., et al. (2019). Combined solution and crystal methods reveal the electrostatic tethers that provide a flexible platform for replication activities in the bacteriophage T7 replisome. *Biochemistry, 58*(45), 4466–4479. https://doi.org/10.1021/acs.biochem.9b00525.

Graewert, M. A., Da Vela, S., Gräwert, T. W., Molodenskiy, D. S., Blanchet, C. E., Svergun, D. I., et al. (2020). Adding size exclusion chromatography (SEC) and light scattering (LS) devices to obtain high-quality small angle X-ray scattering (SAXS) data. *Crystals, 10*(11), 975. https://doi.org/10.3390/cryst10110975.

Graewert, M. A., Franke, D., Jeffries, C. M., Blanchet, C. E., Ruskule, D., Kuhle, K., et al. (2015). Automated pipeline for purification, biophysical and X-ray analysis of biomacromolecular solutions. *Scientific Reports, 5*(1), 10734. https://doi.org/10.1038/srep10734.

Graewert, M. A., & Jeffries, C. M. (2017). Sample and buffer preparation for SAXS. In B. Chaudhuri, I. G. Muñoz, S. Qian, & V. S. Urban (Eds.), *1009. Biological small angle scattering: Techniques, strategies and tips* (pp. 11–30). Singapore, Singapore: Springer Singapore. https://doi.org/10.1007/978-981-10-6038-0_2.

Guinier, A. (1939). La diffraction des rayons X aux très petits angles: Application à l'étude de phénomènes ultramicroscopiques. *Annales de Physique, 11*(12), 161–237. https://doi.org/10.1051/anphys/193911120161.

Hammel, M., Rosenberg, D. J., Bierma, J., Hura, G. L., Thapar, R., Lees-Miller, S. P., et al. (2020). Visualizing functional dynamicity in the DNA-dependent protein kinase holoenzyme DNA-PK complex by integrating SAXS with cryo-EM. *Progress in Biophysics and Molecular Biology*. S0079610720300912 https://doi.org/10.1016/j.pbiomolbio.2020.09.003.

Hammersley, A. P. (1997). FIT2D: An Introduction and Overview'. *ESRF Internal Report, ESRF97HA02T*.

Hodge, C. D., Rosenberg, D. J., Wilamowski, M., Joachimiak, A., Hura, G. L., & Hammel, M. (2021). Rigid monoclonal antibodies improve detection of SARS-CoV-2 nucleocapsid protein. *Biophysics*. https://doi.org/10.1101/2021.01.13.426597.

Hopkins, J. B., Gillilan, R. E., & Skou, S. (2017). *BioXTAS RAW*: Improvements to a free open-source program for small-angle X-ray scattering data reduction and analysis. *Journal of Applied Crystallography*, *50*(5), 1545–1553. https://doi.org/10.1107/S1600576717011438.

Horst, B. G., Yokom, A. L., Rosenberg, D. J., Morris, K. L., Hammel, M., Hurley, J. H., et al. (2019). Allosteric activation of the nitric oxide receptor soluble guanylate cyclase mapped by cryo-electron microscopy. *eLife*, *8*, e50634. https://doi.org/10.7554/eLife.50634.

Hura, G. L., Hodge, C. D., Rosenberg, D., Guzenko, D., Duarte, J. M., Monastyrskyy, B., et al. (2019). Small angle X-ray scattering-assisted protein structure prediction in CASP13 and emergence of solution structure differences. *Proteins: Structure, Function, and Bioinformatics*, *87*(12), 1298–1314. https://doi.org/10.1002/prot.25827.

Hura, G. L., Menon, A. L., Hammel, M., Rambo, R. P., Poole, F. L., II, Tsutakawa, S. E., et al. (2009). Robust, high-throughput solution structural analyses by small angle X-ray scattering (SAXS). *Nature Methods*, *6*(8), 606–612. https://doi.org/10.1038/nmeth.1353.

Inoue, R., Nakagawa, T., Morishima, K., Sato, N., Okuda, A., Urade, R., et al. (2019). Newly developed laboratory-based size exclusion chromatography small-angle x-ray scattering system (La-SSS). *Scientific Reports*, *9*(1), 12610. https://doi.org/10.1038/s41598-019-48911-w.

Knott, G. J., Cress, B. F., Liu, J.-J., Thornton, B. W., Lew, R. J., Al-Shayeb, B., et al. (2019). Structural basis for AcrVA4 inhibition of specific CRISPR-Cas12a. *eLife*, *8*, e49110. https://doi.org/10.7554/eLife.49110.

Kok, C. M., & Rudin, A. (1981). Relationship between the hydrodynamic radius and the radius of gyration of a polymer in solution. *Die Makromolekulare Chemie. Rapid Communications*, *2*(11), 655–659. https://doi.org/10.1002/marc.1981.030021102.

Konarev, P. V., Graewert, M. A., Jeffries, C. M., Fukuda, M., Cheremnykh, T. A., Volkov, V. V., et al. (2022). EFAMIX, a tool to decompose inline chromatography SAXS data from partially overlapping components. *Protein Science*, *31*(1), 269–282. https://doi.org/10.1002/pro.4237.

Latypov, V. F., Tubbs, J. L., Watson, A. J., Marriott, A. S., McGown, G., Thorncroft, M., et al. (2012). Atl1 regulates choice between global genome and transcription-coupled repair of O(6)-alkylguanines. *Molecular Cell*, *47*(1), 50–60. https://doi.org/10.1016/j.molcel.2012.04.028.

Mathew, E., Mirza, A., & Menhart, N. (2004). Liquid-chromatography-coupled SAXS for accurate sizing of aggregating proteins. *Journal of Synchrotron Radiation*, *11*(4), 314–318. https://doi.org/10.1107/S0909049504014086.

Meisburger, S. P., Taylor, A. B., Khan, C. A., Zhang, S., Fitzpatrick, P. F., & Ando, N. (2016). Domain movements upon activation of phenylalanine hydroxylase characterized by crystallography and chromatography-coupled small-angle X-ray scattering. *Journal of the American Chemical Society*, *138*(20), 6506–6516. https://doi.org/10.1021/jacs.6b01563.

Meisburger, S. P., Xu, D., & Ando, N. (2021). *REGALS*: A general method to deconvolve X-ray scattering data from evolving mixtures. *IUCrJ*, *8*(2), 225–237. https://doi.org/10.1107/S2052252521000555.

Mol, C. D., Arvai, A. S., Sanderson, R. J., Slupphaug, G., Kavli, B., Krokan, H. E., et al. (1995). Crystal structure of human uracil-DNA glycosylase in complex with a protein inhibitor: Protein mimicry of DNA. *Cell*, *82*(5), 701–708. https://doi.org/10.1016/0092-8674(95)90467-0.

Panjkovich, A., & Svergun, D. I. (2018). CHROMIXS: Automatic and interactive analysis of chromatography-coupled small-angle X-ray scattering data. *Bioinformatics*, *34*(11), 1944–1946. https://doi.org/10.1093/bioinformatics/btx846.

Parge, H. E., Arvai, A. S., Murtari, D. J., Reed, S. I., & Tainer, J. A. (1993). Human CksHs2 atomic structure: A role for its hexameric assembly in cell cycle control. *Science*, *262*(5132), 387–395. https://doi.org/10.1126/science.8211159.

Pelikan, M., Hura, G. L., & Hammel, M. (2009). Structure and flexibility within proteins as identified through small angle X-ray scattering. *General Physiology and Biophysics*, *28*(2), 174–189.

Pérez, J., & Vachette, P. (2017). A successful combination: Coupling SE-HPLC with SAXS. In B. Chaudhuri, I. G. Muñoz, S. Qian, & V. S. Urban (Eds.), *Vol. 1009. Biological small angle scattering: Techniques, strategies and tips* (pp. 183–199). Springer Singapore. https://doi.org/10.1007/978-981-10-6038-0_11.

Perry, J. J., Asaithamby, A., Barnebey, A., Kiamanesch, F., Chen, D. J., Han, S., et al. (2010). Identification of a coiled-coil in WRN that facilitates multimerization and promotes exonuclease processivity. *The Journal of Biological Chemistry*. https://doi.org/M110.124941 [pii] 10.1074/jbc.M110.124941.

Perry, J. J., Yannone, S. M., Holden, L. G., Hitomi, C., Asaithamby, A., Han, S., et al. (2006). WRN exonuclease structure and molecular mechanism imply an editing role in DNA end processing. *Nature Structural & Molecular Biology*, *13*(5), 414–422. https://doi.org/10.1038/nsmb1088.

Putnam, C. D., Hammel, M., Hura, G. L., & Tainer, J. A. (2007). X-ray solution scattering (SAXS) combined with crystallography and computation: Defining accurate macromolecular structures, conformations and assemblies in solution. *Quarterly Reviews of Biophysics*, *40*(03). https://doi.org/10.1017/S0033583507004635.

Putnam, C. D., Shroyer, M. J., Lundquist, A. J., Mol, C. D., Arvai, A. S., Mosbaugh, D. W., et al. (1999). Protein mimicry of DNA from crystal structures of the uracil-DNA glycosylase inhibitor protein and its complex with Escherichia coli uracil-DNA glycosylase. *Journal of Molecular Biology*, *287*(2), 331–346. https://doi.org/10.1006/jmbi.1999.2605.

Rambo, R. P., & Tainer, J. A. (2013). Accurate assessment of mass, models and resolution by small-angle scattering. *Nature*, *496*(7446), 477–481. https://doi.org/10.1038/nature12070.

Reyes, F. E., Schwartz, C. R., Tainer, J. A., & Rambo, R. P. (2014). Methods for using new conceptual tools and parameters to assess RNA structure by small-angle X-ray scattering. In *549. Methods in Enzymology* (pp. 235–263). Elsevier. https://doi.org/10.1016/B978-0-12-801122-5.00011-8.

Schneidman-Duhovny, D., & Hammel, M. (2018). Modeling structure and dynamics of protein complexes with SAXS profiles. *Methods in Molecular Biology (Clifton, N.J.)*, *1764*, 449–473. https://doi.org/10.1007/978-1-4939-7759-8_29.

Schneidman-Duhovny, D., Hammel, M., Tainer, J. A., & Sali, A. (2016). FoXS, FoXSDock and MultiFoXS: Single-state and multi-state structural modeling of proteins and their complexes based on SAXS profiles. *Nucleic Acids Research*, *44*(Web Server issue), W424–W429. https://doi.org/10.1093/nar/gkw389.

Topolska-Woś, A. M., Sugitani, N., Cordoba, J. J., Le Meur, K. V., Le Meur, R. A., Kim, H. S., et al. (2020). A key interaction with RPA orients XPA in NER complexes. *Nucleic Acids Research*, *48*(4), 2173–2188. https://doi.org/10.1093/nar/gkz1231.

Trewhella, J. (2022). Recent advances in small-angle scattering and its expanding impact in structural biology. *Structure*, *30*(1), 15–23. https://doi.org/10.1016/j.str.2021.09.008.

Tsutakawa, S. E., Sarker, A. H., Ng, C., Arvai, A. S., Shin, D. S., Shih, B., et al. (2020). Human XPG nuclease structure, assembly, and activities with insights for neurodegeneration and cancer from pathogenic mutations. *Proceedings of the National Academy of Sciences*, *117*(25), 14127–14138. https://doi.org/10.1073/pnas.1921311117.

Tubbs, J. L., Latypov, V., Kanugula, S., Butt, A., Melikishvili, M., Kraehenbuehl, R., et al. (2009). Flipping of alkylated DNA damage bridges base and nucleotide excision repair. *Nature*, *459*(7248), 808–813. https://doi.org/10.1038/nature08076.

Tully, M. D., Tarbouriech, N., Rambo, R. P., & Hutin, S. (2021). Analysis of SEC-SAXS data via EFA deconvolution and scatter. *JoVE*, *167*, e61578. https://doi.org/10.3791/61578.

Watanabe, Y., & Inoko, Y. (2009). Size-exclusion chromatography combined with small-angle X-ray scattering optics. *Journal of Chromatography A*, *1216*(44), 7461–7465. https://doi.org/10.1016/j.chroma.2009.02.053.

Wilamowski, M., Hammel, M., Leite, W., Zhang, Q., Kim, Y., Weiss, K. L., et al. (2021). Transient and stabilized complexes of Nsp7, Nsp8, and Nsp12 in SARS-CoV-2 replication. *Biophysical Journal*, *120*(15), 3152–3165. https://doi.org/10.1016/j.bpj.2021.06.006.

Williams, G. J., Williams, R. S., Williams, J. S., Moncalian, G., Arvai, A. S., Limbo, O., et al. (2011). ABC ATPase signature helices in Rad50 link nucleotide state to Mre11 interface for DNA repair. *Nature Structural & Molecular Biology*, *18*(4), 423–431. https://doi.org/nsmb.2038, 10.1038/nsmb.2038. [pii].

Ye, Z., Xu, S., Shi, Y., Bacolla, A., Syed, A., Moiani, D., et al. (2021). GRB2 enforces homology-directed repair initiation by MRE11. *Science Advances*, 7(32). https://doi.org/10.1126/sciadv.abe9254.

CHAPTER SEVEN

SEC-SAXS: Experimental set-up and software developments build up a powerful tool

Javier Pérez[a], Aurélien Thureau[a], and Patrice Vachette[b,*]
[a]Orme des Merisiers, Saint-Aubin, France
[b]Université Paris-Saclay, CEA, CNRS Institute for Integrative Biology of the Cell, Gif-sur-Yvette, France
*Corresponding author: e-mail address: patrice.vachette@i2bc.paris-saclay.fr

Contents

Abstract

A monodispersed and ideal solution is a key requirement of BioSAXS to allow one to extract structural information from the recorded pattern. On-line size exclusion chromatography (SEC) marked a major breakthrough, separating particles present in solution according to their size. Scattering curves with identical shape under an elution peak can be averaged and further processed free from contamination. However, this is not always straightforward, separation is often incomplete. Software have been developed to deconvolve the contributions from the different species (molecules or oligomeric forms) within the sample. In this chapter, we present the general workflow of a SEC-SAXS experiment. We present recent instrumental and data analysis developments that have improved the quality of recorded data, extended the potential of SEC-SAXS and turned it into a mainstream approach. We report a comparative analysis

Methods in Enzymology, Volume 677
ISSN 0076-6879
https://doi.org/10.1016/bs.mie.2022.08.024

of two macromolecular systems using various deconvolution approaches that have been developed over the last years. Parallel analysis appears to be the best cross-validation method to assess the reliability of the reconstructed isolated species patterns that can safely be used as a support for meaningful molecular modeling.

Abbreviations

EFA evolving factor analysis
REGALS regularized alternating least-squares
SANS small angle neutron scattering
SAXS small angle X-ray scattering
SEC size-exclusion chromatography
SVD singular value decomposition
US-SOMO UltraScan solution modeler

1. Introduction

This review draws heavily upon a previous chapter written by two of us (Perez & Vachette, 2017) by permission of the publisher but has been thoroughly updated with references to more recent work and the addition of a comparative analysis of SEC-SAXS datasets using various methods developed over the last years.

For meaningful subsequent 3D modeling of the studied particle, BioSAXS requires a monodispersed and ideal solution. In the case of a mixture, the composition may be fully defined by a given set of coefficients that are determined by non-negative least-square approaches in the case where the form factors of each species are known. When proteins are prone to aggregation, this can bar any valid BioSAXS study, since the intensity scattered at very small angles[a] by a particle is roughly proportional to the square of its molecular mass, so that even a very low fraction of oligomers or aggregates present in the solution seriously distorts the scattered intensity profile in the small angle region, thereby preventing any interpretation in structural terms. Elution through an upstream size exclusion chromatography column allows the separation of oligomers from the monomeric particle under investigation, enabling data collection of physically meaningful Small Angle X-ray Scattering (SAXS) curves. The present chapter is an

[a] The position of a scattered photon is actually defined not directly by the scattering angle but by the momentum transfer $q=(4\pi\sin\theta)/\lambda$ where 2θ is the scattering angle and λ the wavelength of the X-rays; q is expressed in $Å^{-1}$ or nm^{-1}.

attempt to present the implementation of the SEC-SAXS approach around an actual instrument and to describe data acquisition and processing.

To reach monodispersity in the biochemistry laboratory, Size Exclusion (aka Gel Filtration) Chromatography is a routine and very efficient method. Its coupling online with the flow-through measuring cell on synchrotron radiation SAXS beamlines marked a clear breakthrough in Biosaxs. It was first introduced by "external users" at APS beamline BIOCAT in 2004 (Mathew, Mirza, & Menhart, 2004) and at the Photon Factory BL10C in 2008 (Watanabe & Inoko, 2009) but it was only in 2008 that it became available to users as a routine mode of data acquisition on the SWING beamline at the SOLEIL synchrotron (David & Pérez, 2009). Indeed, the instrument had been designed from its inception with an SE-HPLC setup as an integral part of the SAXS instrument. Since then, this arrangement has spread to a number of other instruments worldwide (Table 1).

2. The instrument

The SEC-SAXS combination basically resembles the popular SEC-MALS combination now available in most biochemistry labs, but with the scattering of visible light "replaced" by the scattering of an X-ray synchrotron beam. A schematic simplified layout is represented in Fig. 1.

For BioSAXS measurements, a thermalized flowing cell made of a cylindrical thin walled (ca. 10 μm) quartz capillary is the usual compromise between rigidity, low X-ray scattering background and cleaning ease. The choice of the capillary diameter results from a compromise between the wish to optimize the ratio between scattering and absorption that depends on the X-ray energy and that of keeping the sample volume required as low as possible. On SWING beamline, this led us to use 1.5 ± 0.05 mm diameter capillaries and 12 keV X-rays. The capillary should best be inserted in a vacuum chamber to avoid shadowing effects from air and window scattering. The chamber should itself be positioned on a horizontal motorized stage to select the zone of the capillary with the lowest scattering background. The tubing between the UV-Vis spectrophotometer of the SEC set-up and the downstream SAXS cell should be kept as short and as narrow as possible to minimize band broadening. Communication between the

Table 1 List of SAXS and SANS instruments at synchrotron radiation facilities and neutron reactors offering SEC-SAXS (SEC-SANS) capabilities.

Beamline and SR facility	Uniform resource locator	SEC-SAS availability	Available SEC columns
P12-EMBL, Petra-3, Germany	http://www.embl-hamburg.de/biosaxs/p12/	On request Agilent 1260 Infinity Bio-Inert HPLC	GE Superdex 200 Increase 10/300 GE Superdex 200 Increase 5/150 GE Superdex 75 Increase 10/300 GE Superdex 75 Increase 5/150 GE Superose 6 Increase 10/300 GE Superose 6 Increase 5/150 GE Superose 6 10/300 Wyatt WTC-015S5 5μm 150+, 7.8 × 300 mm Wyatt WTC-030S5 5μm 300+, 7.8 × 300 mm
B21, DLS, UK	http://www.diamond.ac.uk/Beamlines/Soft-Condensed-Matter/small-angle/B21.html	Routinely Agilent 1200 HPLC	Cytiva Superdex 200, 2.4 mL Cytiva Superose 6, 2.4 mL SHODEX KW-402.5, 4.6 mL SHODEX KW-403, 4.6 mL SHODEX KW-404, 4.6 mL SHODEX KW-404, 4.6 mL
SWING, Synchrotron SOLEIL, France	http://www.synchrotron-soleil.fr/Recherche/LignesLumiere/SWING	Routinely Agilent 1200 HPLC	Agilent AdvBio-Sec-2.7-130, 4.6 mL Waters BioResolve 2.5-200 Agilent AdvBio-Sec-2.7-300, 4.6 mL Agilent Bio-Sec-3-300, 4.6 mL Agilent Bio-Sec-5-500, 4.6 mL
BM29, ESRF, France	http://www.esrf.eu/home/UsersAndScience/Experiments/MX/About_our_beamlines/bm29/beamline-setup/hplc.html	Routinely Shimadzu HPLC	User provided only

SAXS/WAXS, Australian Synchrotron, Australia	https://www.ansto.gov.au/user-access/instruments/australian-synchrotron-beamlines/saxs-waxs	Routinely HPLC System	User provided only
BioSAXS TPS-13A, NSRRC, Taiwan	http://tpsbl.nsrrc.org.tw/bd_page.aspx?lang=en&port=13A&pid=1069	Routinely Agilent 1200 HPLC	Agilent Bio SEC-3 series SuperdexTM 200 Increase 5/150 GL TSKgel G6000PWXL
BL19U2, SSRF, China	http://e-ssrf.sari.ac.cn/beamlines/bl19u2/	On request Agilent 1260 Infinity	Superdex 75 Increase (10/300GL) Superdex 200 Increase (10/300GL)
BL-10C and BL-15A2, Photon Factory, Japan	http://pfwww.kek.jp/saxs/manual/Devices20220217.pdf	Shimazu	Available on both beamlines: Superdex 200 Increase 10/300 GL or 3.2/300 (cytiva) Superdex 75 Increase 10/300 GL or 3.2/300 (cytiva) Superose 6 Increase 10/300 GL or 3.2/300 (cytiva) WTC-030S5 (Wyatt)
BioSAXS, MacCHESS, USA	https://www.chess.cornell.edu/users/biosaxs-hp-bio-beamline	Routinely AKTA Purifier	CytivaSuperdex 200 3.2/300 Increase 2.4 mL Cytiva Superdex 200 5/150 GL, 3.2 mL Cytiva Superdex 200 10/300 GL, 24 mL Cytiva Superdex 75 3.2/300 (RNA work only), 2.4 mL
LiX,NSLS-II, USA	https://sites.google.com/view/lixbeamline/	Shimazu	GE Superdex 200 Increase 5/150 GL and 10/300 GL

Continued

Table 1 List of SAXS and SANS instruments at synchrotron radiation facilities and neutron reactors offering SEC-SAXS (SEC-SANS) capabilities.—cont'd

Beamline and SR facility	Uniform resource locator	SEC-SAS availability	Available SEC columns
BL4-2, SSRL, USA	http://www-ssrl.slac.stanford.edu/~saxs/sample_environ/fplc_saxs.htm	On request Akta Ettan FPLC	Cytiva Superdex 200 5/150 3.2 mL Cytiva Superose 6 5/150 3.2 mL
SIBYLS, ALS, USA	https://bl1231.als.lbl.gov/	See chapter "SEC-SAXS: Experimental set-up and software developments build up a powerful tool" by Pérez et al. in this book	See chapter "Size exclusion coupled small angle X-ray scattering with tandem multiangle light scattering at the SIBYLS beamline" by Rosenberg et al. in this book
BioCAT, APS, USA	https://www.bio.aps.anl.gov/pages/about-saxs.html	Routinely	Cytiva Superdex 75 Increase 10/300 24 mL Cytiva Superdex 75 5/150 5/150 3.2 mL Cytiva Superdex 200 Increase 10/300 24 mL Superdex 200 Increase 5/150 5/150 3.2 mL Cytiva Superose 6 Increase 10/300 24 mL Cytiva Superose 6 5/150 3.2 mL
D22, ILL, France	https://www.ill.eu/users/instruments/instruments-list/d22/sans-sample-environment/sec-sans/1-column-set-up	On request Home made instrument	Superdex 200 and Superdex 75 (10/300) (dilution 7x) Superdex 200 and Superdex 75 (5/150) (dilution 2x) Users can also bring their own column

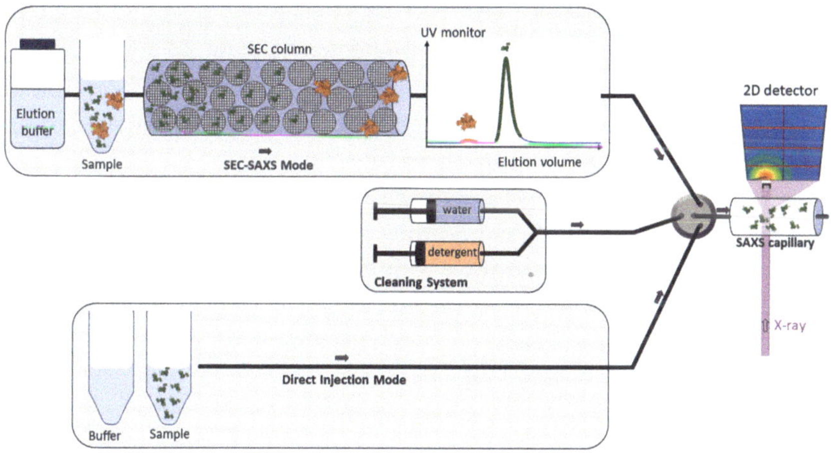

Fig. 1 Schematic representation of the SEC-SAXS set-up at SWING. In the three frames from top to bottom SEC-SAXS elution circuit, cleaning device and direct injection mode. The selection of the circuit connected to the SAXS measuring cell is ensured by a rotating valve. The cleaning system is meant to remove any deposit on the walls of the SAXS measure cell. Cleaning, injection, and data acquisition are usually synchronized by a master control system to simplify the user task. Complementary features such as a second switchable column, or a split way for other biophysical characterizations, can be found at specific beamlines.

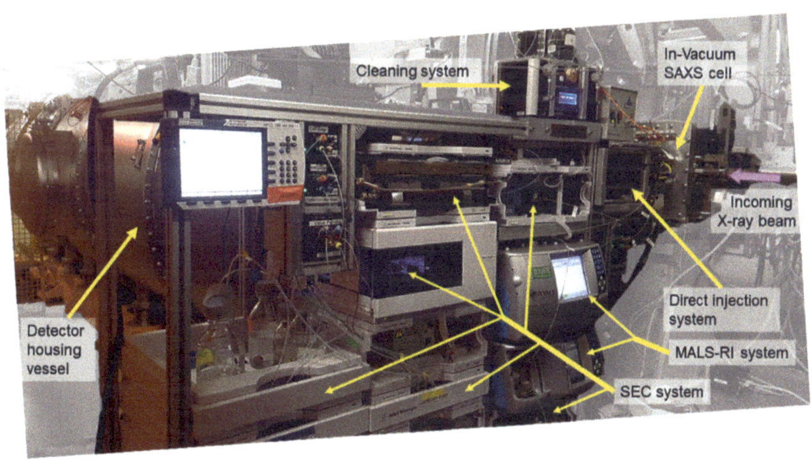

Fig. 2 View of the SEC-SAXS setup installed on the SWING beamline at the Soleil synchrotron.

chromatography instrument and the beamline control should ensure proper synchronization of protein elution and SAXS frames acquisition through an exchange of electronic triggering signals between the two set-ups, with no user intervention (Fig. 2) (Thureau, Roblin, & Perez, 2021).

A dual column system is useful to optimize duty cycle in case the SAXS acquisition finishes before elution completion. Switching columns allows injection and data collection on column B while column A completes the return to equilibrium with the elution buffer. An automated cleaning procedure of the SAXS measuring cell should be implemented and performed after each elution, often using a detergent emulsion followed by water rinsing (Fig. 1). A full compatibility with an automated direct injection set-up is a strong asset for fast switching between both modes, the latter being mostly useful for better statistics at larger q-values where the scattered intensity is not significantly affected by a small amount of oligomers. Finally, the chromatography set-up can be positioned on a lockable rolling table that allows easy removal and repositioning, which is especially convenient on those SAXS beamlines that are not uniquely dedicated to biological solutions.

SEC-SAXS is often complemented by other parameter monitoring for a more complete sample characterization beyond the UV spectrometer at the column outlet. For instance, a refractometer has been installed downstream of the SAXS measuring cell on SWING in the case of membrane protein studies (Baranowski & Perez, 2020; Berthaud, Manzi, Pérez, & Mangenot, 2012). However, due to protein diffusion resulting in band broadening between the various measuring cells, the protein concentration and the quality of particle separation vary, which makes it difficult to combine the observations. In particular, the refractometer must usually be at the end of the circuit whereas the SAXS cell, with its large section, produces a marked band broadening. For these reasons, a second, parallel circuit can conveniently be used for other biophysical characterization. Hence, a significant step forward has been made at the EMBL Outstation in Hamburg by introducing a split valve at the outlet of the SE column that creates two parallel and identical paths for the eluting sample, one flowing to the SAXS measuring cell while the other one flows successively through the UV cell, the refractometer and the viscometer (Graewert et al., 2015). A similar set-up has been installed at the LiX beamline at NSLS-II (Yang et al., 2020).

2.1 Columns

The first attempts at SEC-SAXS were the work of biochemist users who used lab columns with a 7.5 mm diameter, involving larger loaded protein amounts and longer elution times (Mathew et al., 2004). Nowadays, 4.6 mm diameter analytical columns are used practically everywhere as can be seen in Table 1. On SWING beamline, we typically favor matrices of hydrophilic polymer-coated silica beads, from Agilent™ [(Advance) BioSec series], due to their very high resolving power. However, for buffers with pH values higher than 8 or 8.5, organic polymer or carbohydrate-based matrices have to be used, e.g. Superdex series from Cytiva™. Column volumes and column internal diameters should be minimized to avoid excessive dilution and band broadening, e.g., around 5 mL and ca 5 mm, respectively. However, this should not be detrimental to resolution. Depending on the case (close peaks or not) resolution might be preferred and longer columns, or combination of two small columns in series, could be used. Column integrity should be regularly checked by eluting a mixture of reference samples. Filtering of buffers during elution is not as stringent a requirement as for MALS, but it is still recommended to insert a 0.2 μm prefilter upstream of the pump. Even when columns are available on the beamline, users are encouraged to bring their own previously verified column. This limits problems of cross-contamination between various samples while avoiding the discovery on the beamline of unexpected interactions of the sample with the column matrix, a waste of resources and efforts.

2.2 Experimental protocol

Choose the right column for the sample under study. The criteria for choice are fairly obvious: the range of molecular mass in which effective separation is achieved by the column compared with that of the sample; the sample Rh, the pH to be used (silica based usually cannot stand pH higher than 8 or 8.5). If the column resolution is not enough to clearly separate two components, it is advisable to use a column with finer granulometry of the solid phase (the matrix). It can also be useful to combine two columns in series to improve the resolution, even if the resulting concentration at the peak top is slightly decreased.

Choose (if possible) the UV-Vis wavelength for elution monitoring. For instance, 260 and 280 nm for a nucleoprotein complex are compulsory. A remote wavelength (400 nm) can be used to check the lamp stability. More specific wavelengths can be used in case chromophores are present to help distinguish particles containing them from the others and to better estimate their concentration.

Equilibration should be reached, usually checked by several stable UV-Vis profiles. The rule of thumb on SWING beamline for a 4.6 mL column is an absorbance variation at 280 nm lower than 5 mAU over 10 min at 300 μL/min.

At third generation synchrotrons, injection volumes and concentrations should be such that a concentration of the order of 1 mg/mL is reached at the elution peak head. Typically, the injection volume corresponds at most to 1% of the column volume, beyond which resolution can be affected. For instance, 50 μL at 4 mg/mL can be loaded on top of a Biosec-3 300 Agilent 4.6 mL column, resulting in a fivefold dilution at the peak head.

In general, the resulting concentration is low enough to make any potential coulombic repulsion effects negligible. However, when very high concentration solutions are loaded, it could be advisable to proceed to the "usual" dilution series (meaning the concentration at the top of the peak), and in that case, it is much simpler and effective to reduce the injected volume accordingly than to dilute the initial solution. By reducing the injected volume, the band broadening is reduced and so the resolution of the peak is increased. Furthermore, too large an amount of sample may saturate the solid phase resulting in band broadening. Ideally, a maximum of 200 μg of injected sample is recommended for 5 mL columns by column manufacturers; actually, 400 to 500 μg is commonly injected into 5 mL columns to increase the signal to noise ratio.

Buffer frames have to be collected before the void volume and cover a significant fraction of the time between injection and void volume to assess beam and elution stability and provide buffer scattering data with good statistics.

Data acquisition does not need to cover the entire elution process, typically lasting between 15 and 30 min. In particular, shining X-rays on aggregates only increases the risk of capillary fouling for no gain in useful information. It may be wiser to restrict the X-ray exposure to the time period of particle elution. In contrast it is good practice to collect data well after the last peak has eluted, in order to gain information on capillary fouling effects that could then be accounted for (see below).

The acquisition frame rate should be compatible with the typical time widths of the elution peaks. About one frame per second allows more than adequate sampling of a one-minute broad peak.

It is also possible to collect data continuously with ca 10 ms frames when using photon counters as detectors. This "collect all" approach is simple and effective at the cost of (automatically) processing thousands of 2D images to yield 1D frames, most of which might contain no useful information.

3. Data processing

Data reduction, the first stage of data processing, is done using an automated pipeline converting 2D frames into 1D scattering profiles. Buffer frames are examined, their similarity assessed by the absence of monotonic evolution (Thureau et al., 2021) or using some more general statistical tools such as CorMap (Franke, Jeffries, & Svergun, 2015) before averaging identical frames. The average buffer profile is subtracted from all frames of the elution dataset properly speaking (Hopkins, Gillilan, & Skou, 2017; Panjkovich & Svergun, 2018). Then I(0) and R_g are evaluated by applying the Guinier analysis to all buffer-subtracted curves. Two major cases can then be distinguished, a simple one and a more complex situation. However, before considering these situations in detail, we must examine the potentially important consequences of radiation damage.

3.1 The capillary fouling issue

Radiation damage can be a real issue and it has actually been the object of various studies, notably a very interesting work (Hopkins & Thorne, 2016) to which we refer the reader for an in-depth study of the phenomenon. Let us just mention a couple of simple advices: Tris-HCl acts as a radical scavenger and its wide use can only be recommended. In contrast, phosphate buffer seems to increase radical production and must be avoided as far as possible when planning a SAXS study (see red circles in Fig. 3). It has also become common practice to add from 2% to 5% (v/v) glycerol to sample solution as this limits the consequences of irradiation (Kuwamoto, Akiyama, & Fujisawa, 2004) but one should keep in mind that this could affect the conformation of the protein, primarily those that are partly unstructured as glycerol is known to stabilize compact, folded protein conformations. The consequence of radiation damage on the SEC-SAXS data is the appearance of a small but increasing contribution to scattering at very low angles that can be directly attributed to progressive protein adsorption onto the irradiated part of the capillary walls. This parasitic scattering causes an increase of all apparent radii of gyration, and most spectacularly on frames after the elution peak, where the protein concentration vanishes as shown in Fig. 3 (red circles).

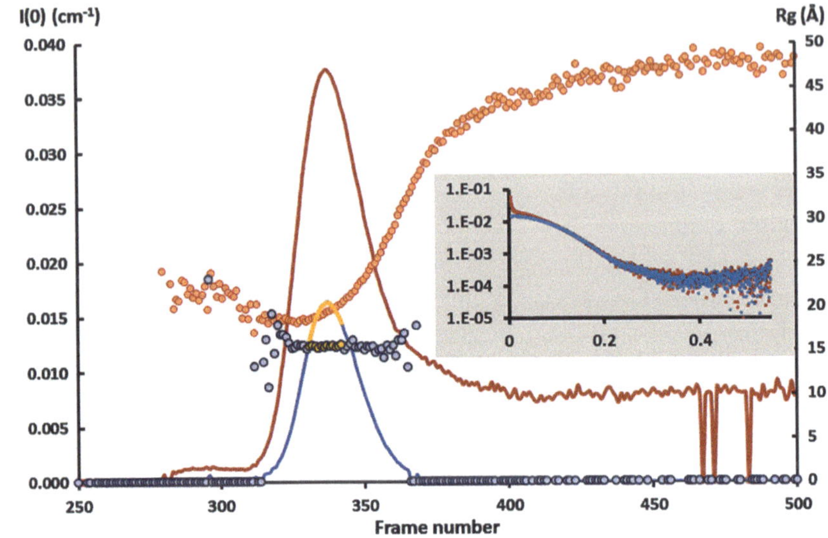

Fig. 3 SEC-SAXS elution of a radiation sensitive sample with or without the CoFlow system. SEC-SAXS elution profile of a sample of RNaseA in PBS buffer. 50 μL of a 6.7 mg/mL solution were injected on an Agilent™ BioSec3-300 column with and without the CoFlow system (blue and red, respectively). For each time frame (990 ms counting time/10 ms dead time), Guinier analysis yielded I(0) and Rg values represented as line and circles, respectively. Without the CoFlow system, high Rg values (~18.3 Å) are obtained right from the start of the main peak and continue to rise sharply even over the falling edge of the I(0) peak to reach a plateau at the end of elution (~46.6 Å) while the I(0) profile does not return to baseline and stays at a high level (0.008). In contrast, use of the CoFlow system yields a lower R_g value that remains constant over the entire peak while the I(0) peak nicely returns to baseline. The cost of the system is a 55% reduction in scattering intensity due to a shorter optical path.

This specific issue was addressed in the HPLC module of the US-SOMO package (Brookes, Vachette, Rocco, & Perez, 2016). In particular, the conversion of the experimental curves into chromatograms[b] for each q-value clearly demonstrates that the parasitic intensity remains constant as soon as the protein elution is completed. This strongly supports that protein progressive deposition and not instantaneous aggregation is the origin of the unwanted scattering. The algorithm in the US-SOMO package estimates the time evolution of this effect at each q-value and corrects for it using an iterative procedure, finally delivering a set of corrected frames

[b] We call chromatogram the function measuring the evolution with time of the scattered intensity at a given q-value.

as illustrated in figure S5 in the Supporting information for Brookes et al. (2016). The experimental requisite for this algorithm to work is that data must be recorded well after the last peak elution. This algorithm has also been implemented in the Bioxtas RAW package for BioSAXS data analysis (Hopkins et al., 2017).

Instrumental responses to the fouling issue are also possible. Kirby and coll. at the Small Angle Scattering beamline of the Australian Synchrotron came up with a totally new design of a flow cell where the main stream of protein solution is separated from the capillary walls by a thin laminar flow of buffer, which almost entirely eliminates any apparent radiation effect on the scattering pattern (Kirby et al., 2016). A schematic view of this so-called Co-Flow" cell as compared to a standard flow cell is shown in Fig. 4 (panels A and B). This was a technical "tour de force" but the obvious advantages of the approach are such that it has since been also implemented at the BioCAT beamline at APS (see https://www.bio.aps.anl.gov/pages/about-saxs.html) and at the SWING beamline (SOLEIL, France) (see Fig. 4, panel C), where the first satisfactory results have recently been obtained, using for instance a solution of RNaseA in PBS: we have seen that, in a standard flow-cell, radiation damage caused a major aggregation phenomenon. Using the Co-Flow system, the phenomenon disappears altogether, identical frames are recorded across the elution peak, the top of the peak are averaged and yield a curve in blue giving a reliable R_g value 15.3 Å vs 18.3 Å at the start of the elution peak without the Co-Flow that rises sharply during peak elution (Fig. 3).

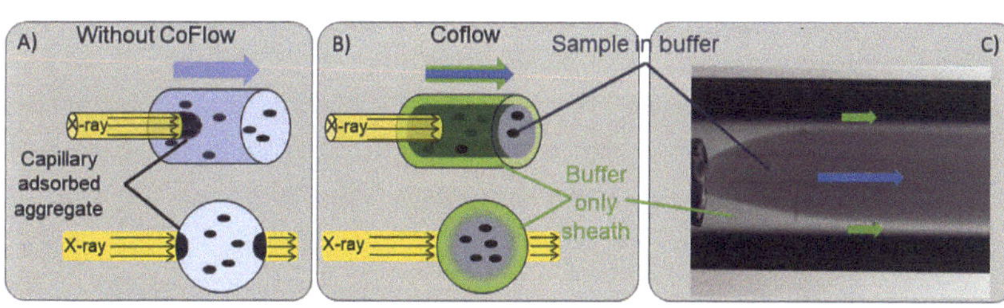

Fig. 4 Schematic representation of CoFlow principle and photograph of the cell. Macromolecules are represented as black ellipses. (A) Under X-ray irradiation, without CoFlow and under certain conditions macromolecules may adsorb on to capillary walls. (B) With CoFlow, the macromolecule solution (blue) is surrounded by the corresponding buffer solution (green), thereby preventing the direct interaction of macromolecules with capillary walls. (C) A colored solution comes out of a small tubing introduced into the capillary. A colorless solution surrounds the flux of solute. Capillary walls are hardly discernible due to the projected shadow of the capillary holder.

Finally, some simple choices regarding the X-ray beam characteristics can help with the problem of radiation damage. Firstly, using a shorter wavelength reduces absorption by the solution and thus the number of free radicals created in the sample that are the origin of radiation damage. Better use 1 Å X-rays than longer, less energetic X-rays that are more readily absorbed, even if the brilliance is significantly lower. Similarly, slightly defocusing the X-ray beam at the sample position distributes the incident flux over a larger number of molecules in the flowing sample solution without any detectable adverse effect on the spatial distribution of scattered intensities. This is a simple and very effective step in trying to mitigate or eliminate altogether the adverse effects of X-ray irradiation.

3.2 Simple case: Well-resolved peaks

In a favorable case, the species of interest elutes as a well-resolved peak far from the larger oligomers or aggregates that were present in the original solution. After performing the tasks mentioned above, one examines the evolution of R_g with time (frame number) through the elution peak. The absence of monotonic variation across the peak suggests that the shapes of the successive scattering curves (frames) are identical. It is possible to confirm this by a subsequent analysis of the ensemble of frames (e.g. using CorMap in PrimusQt) after scaling all frames to the same integral over a given q-range. Such an example is shown in Fig. 5.

In this example, the frames around the peak maximum show no evolution of their R_g value and are practically identical after scaling. They can then safely be averaged for subsequent analysis and molecular modeling. The user-friendly program DATASW (Shkumatov & Strelkov, 2015) provides an option for automatic averaging using a sliding window to increase statistics if individual frames are too noisy and will yield for each peak the average of all frames under the peak and of frames $\pm 10\%$ around the maximum. This is fine as long as peaks are well resolved. Similar features are present in the CHROMIXS routine (Panjkovich & Svergun, 2018) available within the ATSAS tools suite that provides an easy-to-use graphical interface for data analysis. The initial version included procedures for basic data reduction and automatic buffer and sample region prediction (Franke et al., 2017). The last release within ATSAS 3.0 includes the calculation of MW and R_g estimates for the selected sample-peak elution frames, as well as the ability to load and visualize other time-course data, e.g. UV absorbance (Manalastas-Cantos et al., 2021). Finally, a program with a user-friendly graphical interface has been developed at the Photon Factory

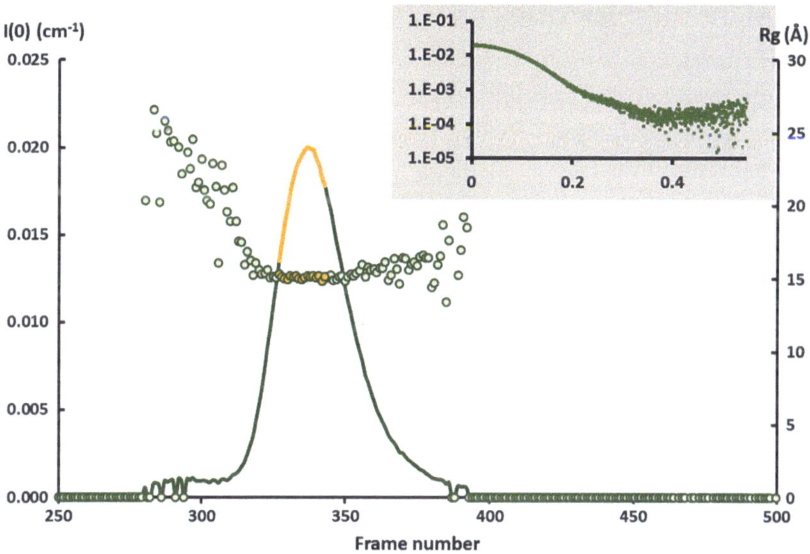

Fig. 5 SEC-SAXS study of a protein well separated from aggregates. SEC-SAXS elution profile of a sample of RNaseA in Hepes buffer (20 mM Hepes, 150 mM NaCl, pH 7.2). 50 μL of a 6.7 mg/mL solution were injected on an Agilent™ BioSec3-300 column. For each time frame (990 ms counting time/10 ms dead time), Guinier analysis yielded I(0) and R_g values represented as line and circles, respectively. Frames exhibiting a constant R_g value (highlighted in yellow) are automatically selected and averaged to yield the final curve (insert).

SAXS beamlines BL-10C and BL-15A2 that automatically processes the data obtained from SEC-SAXS with UV-Vis spectroscopy. It performs baseline correction and extrapolation to infinite dilution (Yonezawa, Takahashi, Yatabe, Nagatani, & Shimizu, 2019).

If one wishes to derive an estimate of the molecular mass from the value of I(0)/c at the peak top, one should take the band broadening effect into account due to the path between the UV detector of the HPLC instrument and the SAXS measuring cell. A simple and effective approximation is to divide the measured OD by the ratio SAXS-over-OD of the peak widths. More sophisticated approaches have been developed in data processing packages for the similar situation of SEC-MALS data, in which the OD peak is convolved with a function mimicking the band-broadening process to match the MALS peak shape. These approaches could easily be implemented for SEC-SAXS. The band broadening issue can also be addressed experimentally by directly measuring the OD within the SAXS measuring cell, but at the cost of slightly worse statistics, due to a smaller optical path length.

3.3 More complex cases

More complex situations are often encountered, in which peaks are not baseline resolved and most elution frames still contain contributions from two or more components. A first situation is that where the main elution peak exhibits an unresolved shoulder at the foot of its rising edge, probably from a fraction of small oligomers. Accordingly, a systematic decrease of R_g is observed along the rising edge of the peak, but still, frames with stable R_g values are obtained on the falling side. These frames can then be used for averaging and further analysis, after thorough check that they are identical once scaled. However, since the data were extracted from the flank of a non-perfectly monodispersed peak, it is crucial to confirm that the curve corresponds to the species of interest. This should be done by estimating its molecular mass, best using approaches not requiring the knowledge of the concentration, e.g. using SaxsMoW (Fischer, de Oliveira Neto, Napolitano, Polikarpov, & Craievich, 2010) or the V_c approach proposed by Rambo and Tainer (Rambo & Tainer, 2013). A specific tool offering a variety of MM determination methods with a consensus Bayesian inference approach providing a MM estimate together with a credibility interval is available within PrimusQt (Hajizadeh, Franke, Jeffries, & Svergun, 2018).

One may also want to retrieve more information from a set of unresolved patterns. To try and overcome the difficulties arising from poorly separated components, specific software was developed. The first implementation of a deconvolution protocol was introduced as a specific module within the US-SOMO package for macromolecular modeling using hydrodynamic and scattering data (Brookes et al., 2013). This module offers classical SVD analysis of the data set that informs the choice of the number m of species necessary to account for the ensemble of frames. These frames form a 2D scattering dataset I(q,t) as clearly illustrated in figure 5 from (Perez & Vachette, 2017). The specificity of the approach developed in US-SOMO lies in its use of transposed vectors from I(q,t) where chromatograms $I_q(t)$ display the time evolution of intensity at a given q-value during elution. Elution peaks are modeled as Gaussians (see Fig. 7 in the "comparative analysis of BSA" section below), and the entire dataset is globally fitted to yield the resulting m Gaussians together with the m coefficients associated with each frame. Finally, the program extracts the m scattering patterns for each proposed species by attributing a fraction (the relative contribution of each Gaussian to intensity) of the total intensity using either experimental intensity or the fitting intensity (sum of all m Gaussians). In the latter mode a unique scattering pattern for each species is obtained, linear combinations of which yield reconstructed frames that are smoothed versions of the original, more noisy experimental data. In contrast, the first, default mode, yields as many reconstructions as frames and the program averages these reconstructions that come from frames within 5% of the top frame intensity. Alternatively, the user can examine a larger frame selection for subsequent averaging of those deemed identical. Finally, the same decomposition can be applied to the absorbance profile at 280nm. Provided that the band broadening effect between UV and SAXS measuring cell be taken into account as previously mentioned, a quantitative estimate of the concentrations is obtained that allows us to determine the molecular mass of each species from the I(0)/c value since intensities have been put on an absolute scale using normalization from water scattering (intensities in cm^{-1}). A few years later, a much-improved version became available that introduced, beyond the capillary fouling associated baseline correction, the use of modified Gaussians to describe asymmetric peak profiles where the asymmetry reflects the interaction of the eluting macromolecule with the solid phase of the column, a frequently encountered case (Brookes et al., 2016).

At about the same time, an article of great interest was published that reported SEC-SAXS analysis of an allosteric protein in an inactive form and in an allosterically activated form (Meisburger et al., 2016). The authors show that, although a single elution peak is visible on the elution profile, the radius of gyration is not constant across the peak, an indication that several species elute without being resolved. The authors also utilize standard SVD analysis to determine the minimum number of species compatible with each dataset before using the so-called Evolving Factor Analysis (EFA), a variant of SVD that determines the peak range associated with each species before retrieving the true SAXS pattern of each species from the singular vectors by imposing a zero concentration outside the peak range. Starting estimates of each species boundaries are automatically provided by the program that must be manually refined to minimize the χ^2 values between each frame and its reconstruction using the proposed SAXS patterns and fractional concentrations. More recently, the same group published a new approach dubbed REGALS for Regularized-Alternating Least-Squares that can use EFA as a starting point to determine the time range of presence of a given species, but this is not mandatory (Meisburger, Xu, & Ando, 2021). The protocol, that also involves a manual refinement step of each species boundaries similar to EFA followed by a REGALS run, yields the scattering patterns together with well-defined monomodal concentration profiles of each isolated species. Of note, the approach can be applied to datasets that do not satisfy the validity condition of EFA, namely limited time-range for the presence of each species, such as IEC-SAXS data (in line Ion Exchange Chromatography coupled to SAXS) and even Time-Resolved SAXS datasets, in which the evolution of a system is monitored following an initial perturbation (mixing, temperature or pH jump, etc). Both approaches have been implemented in the BioXtas RAW package (Hopkins et al., 2017). Finally, a second implementation of the EFA approach has been reported in a very recent article describing the corresponding program EFAMIX (Konarev et al., 2022).

4. Comparative analysis of SEC-SAXS datasets using the various available programs

As mentioned above, several methods have been developed to address the deconvolution of the various species only partially separated during the column elution. Each one uses specific hypotheses and set of constraints or restraints. The calculations yield data of variable quality for each molecular species scattering patterns. It is difficult to assess *a priori* the validity of the resulting curves. The most convincing approach seems to rely on parallel analyses of a dataset of interest using the various methods. The (dis)agreement between the results provides an indirect assessment of their validity. To illustrate the interest of such an approach, we present the analyses of two datasets using the (modified) Gaussian decomposition in the HPLC module of US-SOMO (https://somo.aucsolutions.com/index.php) (Brookes et al., 2016), EFA implementation in RAW (Hopkins et al., 2017; Meisburger et al., 2016) and the REGALS approach also implemented in RAW (https://bioxtas-raw.readthedocs.io/en/latest/install.html) (Meisburger et al., 2021). To date, EFAMIX is only available as a preliminary version with no interactive graphical interface that does not allow a fair comparison with the other more advanced programs. It was not therefore included into the comparison.

4.1 BSA

BSA (Sigma-Aldrich A7906) was dissolved in Hepes 20 mM pH 7.4 NaCl 150 mM. The solution was stored at 5 °C for a few days before measurements, conditions that allow for a well-documented self-association of the protein into dimers, trimers and higher order oligomers. 50 µL of 10.8 mg/mL solution were loaded onto a BioResolve SEC column 200 Å, 2.5 µm, 4.6 x 300 mm from Waters™.

Following background subtraction, the resulting subtracted data were examined. No signal was detected beyond frame 250, demonstrating that no capillary fouling had taken place. Accordingly, data analysis was restricted to the ensemble of the first 250 frames. All 250 subtracted data files together with the results of their analysis using the three methods have been deposited in the Small Angle Scattering Biological Data Bank (SASBDB), a curated repository of SAS experimental data (Kikhney, Borges, Molodenskiy, Jeffries, & Svergun, 2020) with the entry code SASDPH3. They can be freely downloaded at the following URL: https://www.sasbdb.org/data/SASDPH3/.

Chromatograms $I_q(t)$ for small angle q-values all show poorly resolved eluting species before the monomer peak. This is confirmed by the Guinier analysis performed for all subtracted frames that gave both I(0) and R_g values for each frame. I(0) displays a complex evolution with time similar to the $I_q(t)$ chromatograms just mentioned, while R_g shows no region of constant value but a general decrease suggesting that most frames contain more than one scattering species (Fig. 6). SVD analysis performed in US-SOMO or RAW suggests that four species are necessary to account for the complete dataset.

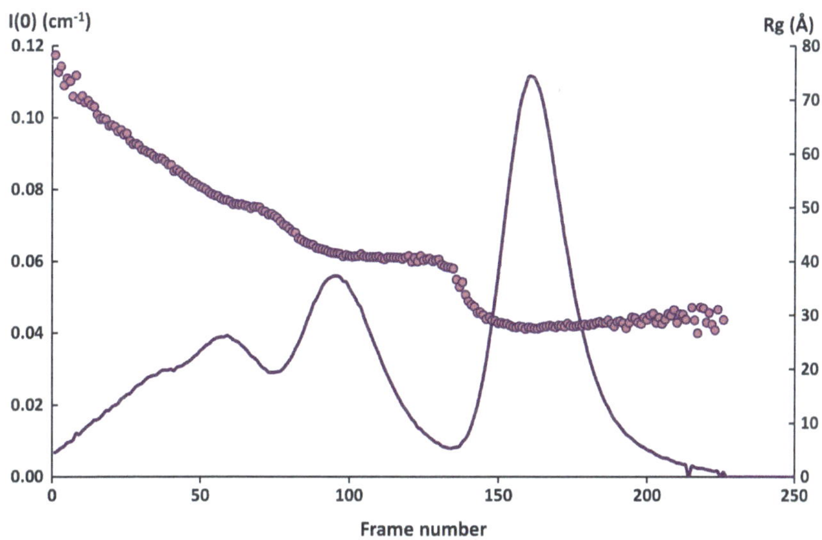

Fig. 6 SEC-SAXS elution profile of a sample of BSA in Hepes buffer (20 mM Hepes, 150 mM NaCl, pH 7.2). **50** µL of a 10.8 mg/mL solution were injected on a Waters™ BioResolve Sec 2.5-200 column. For each time frame (990 ms counting time/10 ms dead time), Guinier analysis yielded I(0) and R_g values represented as line and circles, respectively.

4.1.1 US-SOMO HPLC module

We chose $q = 0.0200\,Å^{-1}$ to try and fit the intensity profile vs time by a combination of four modified Gaussians. The choice is the result of a compromise between high intensity (small angle) and low experimental uncertainty. The $I(q)/\sigma(q)$ ratio is highest around $0.03\,Å^{-1}$–$0.04\,Å^{-1}$ but the contribution of high molecular weight species is larger at smaller angles. The resulting fit can be seen in Fig. 7. The fit is generalized to all $I_q(t)$ profiles and to the absorbance profile at 280 nm after the necessary time shift and a correction accounting for band broadening between UV and SAXS profiles. At that point, the program reconstructs the SAXS profile of each isolated species (peak1 (monomer) to peak4) as well as the concentration profile of each species.

4.1.2 EFA analysis

The analysis was performed within the BioXtas RAW package, v. 2.1.1. We essentially followed the very clear and detailed workflow provided in the associated tutorial. The complete results of the analysis were saved in a .csv file for further use as well as the scattering profiles of each isolated species.

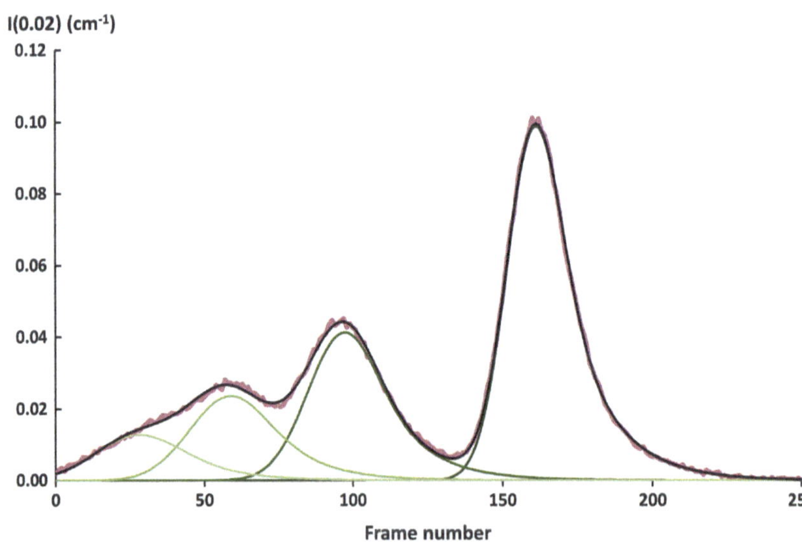

Fig. 7 SEC-SAXS elution profile with the Gaussian decomposition of a sample of BSA. The experimental intensity profile for $q = 0.0200\,Å^{-1}$ is in purple, the four modified Gaussian components are in shades of green while their sum, fitting experimental intensities, is shown in black.

4.1.3 REGALS analysis

The analysis was also performed within the Bioxtas RAW package, v. 2.1.1. We essentially followed the workflow provided in the associated tutorial. The analysis started from a preliminary EFA run since the dataset satisfies its requirements. The complete results of the analysis as well as the scattering profiles of each isolated species were saved. Both EFA and REGALS number the four contributing species from 0 to 3, starting from the first eluting species (largest molecular mass). For the sake of clarity in presenting the three parallel analyses, we renumbered all species from 1 to 4 where 1 designates the BSA monomer, 2 the dimer and so on.

4.1.4 Results

The distributions of χ^2 values for each time frame obtained using each approach are presented in Fig. 8. The EFA derived values appear (blue line) to be globally higher than the US-SOMO results (green line) that are in turn larger than the REGALS values (red line). The reconstructed curves for each of the isolated species are reported in the four panels of Fig. 9 after scaling to the intensity at $q_0 = 0.005\,\text{Å}^{-1}$, a proxy for I(0). Two reconstructions are displayed for US-SOMO: in orange, the curve using the intensity value given by the modified Gaussian fit and in green, the profile obtained using the experimental intensity value. The corresponding

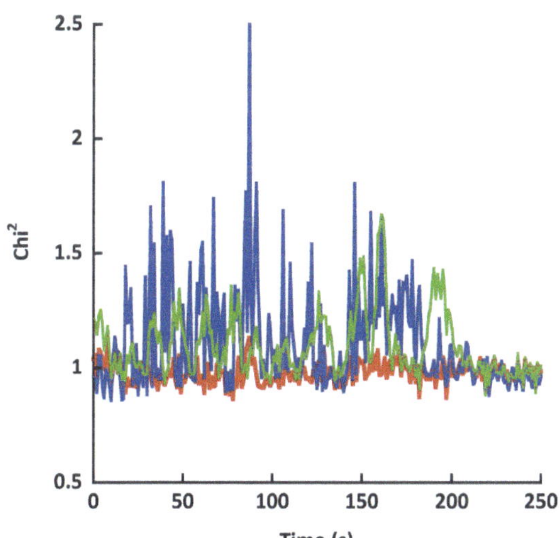

Fig. 8 Distributions of χ^2 values as a function of frame number for each method. Color code: Blue line: EFA; Green line: US-SOMO; Red line: REGALS.

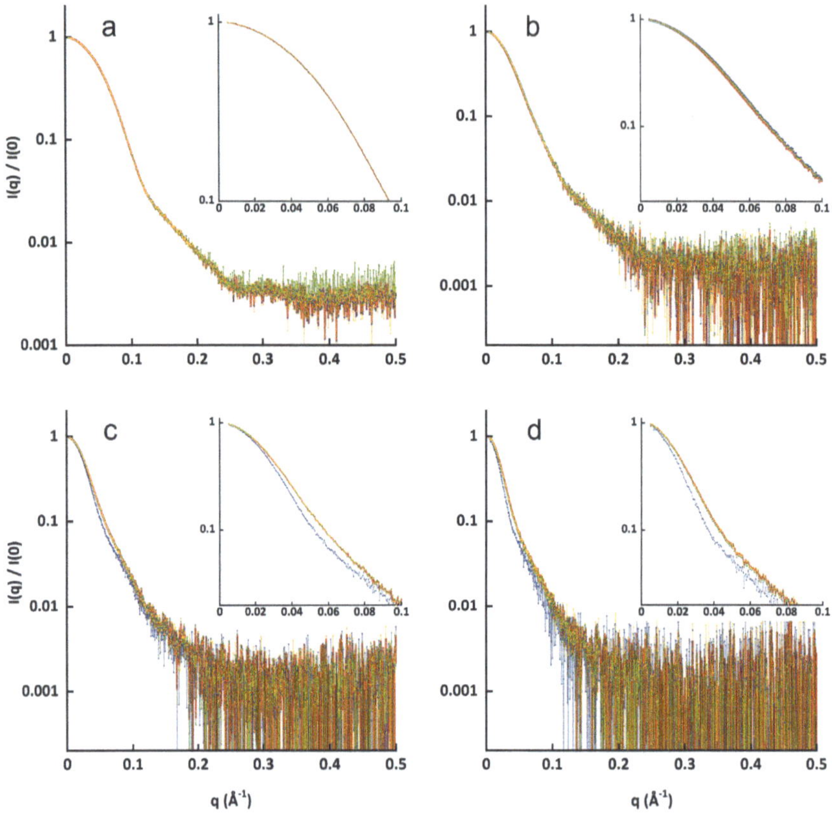

Fig. 9 Comparison of the various reconstructions of the scattering patterns of all four species using each method. In each panel, all four reconstructions have been scaled to the intensity at $q = 0.005\ \text{Å}^{-1}$, a proxy for I(0). Inserts: zoom on the smallest-angle region. Panel A: monomer; Panel B: dimer; panel C: trimer; panel D: oligomer. Color code: blue: EFA. green: US-SOMO, Full Gaussian; orange: US-SOMO, exp. Intensities; red: REGALS.

(i.e. after scaling to I(0)) elution profiles of each species are shown in Fig. 10. Decomposition of the absorbance profile at 280 nm yields a quantitative estimate of the concentration of each species as a function of time, thereby allowing us to determine the corresponding molecular mass from the I(0)/c value. Results are reported in Table 2 together with the V_c estimate (independent from the concentration) as well as the Bayesian estimate provided within Primus. Table 2 also present the R_g values for each species and each method derived from the Guinier analysis and from the p(r) determined by Gnom (Svergun, 1992) ran within Primus as well as the maximum distance D_{max}.

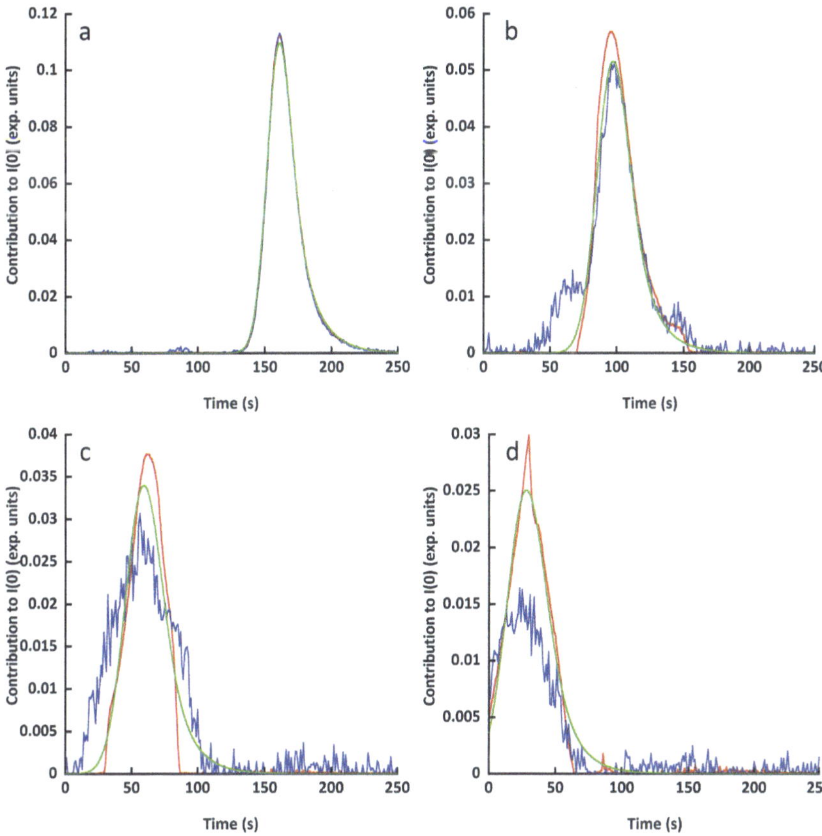

Fig. 10 Elution profiles of each species obtained using the various methods after scaling to I(0). Color code: Blue: EFA; Red: REGALS; Green: US-SOMO.

This comparative analysis of the same dataset using different methods reveals that the results are comparable between the three methods (Gaussian deconvolution, EFA and REGALS) with the best agreement observed between the Gaussian approach and REGALS.

While the three elution profiles of the monomer are essentially identical, EFA profiles are fairly noisy, exhibiting spurious side-peaks (dimer) and broader peaks (trimer, oligomers). In contrast, REGALS profiles are much smoother due to the alternating least-square procedure at the price of somewhat unrealistically sharp boundaries for all three associated species, while US-SOMO profiles are analytical (modified Gaussian) by design.

All reconstructions of the monomer of BSA are practically identical. The situation is similar for the dimer pattern, with hardly detectable differences

Table 2 Structural parameters derived from the reconstructed curves obtained using all approaches for each species resolved in BSA SEC-SAXS dataset.

pK1:monomer	**R_g Guinier (Å)**	**R_g p(r) (Å)**	**D_{max} (Å)**	**MW Bayesian estimate (KDa)**	**MW I(0)/c US-SOMO (KDa)**	**MW Vc (KDa)**
US-SOMO Gaussian	27.9	27.84	80	58.2	63.7	61.3
US-SOMO	27.7	27.79	80	59.5	63.9	61.9
EFA_RAW	27.8	27.78	80	59.5	NA	61.1
REGALS	27.9	27.84	80	59.5	NA	61.3
pK2:dimer	**R_g Guinier (Å)**	**R_g p(r) (Å)**	**D_{max} (Å)**	**MW Bayesian estimate (KDa)**	**MW I(0)/c US-SOMO (KDa)**	**MW Vc (KDa)**
US-SOMO Gaussian	41.5	42.1	139	130.9	127.0	130.0
US-SOMO	41.3	41.8	140	124.4	120.6	129.0
EFA_RAW	40.8	41.2	137	130.9	NA	128.0
REGALS	42.7	43.3	146	130.9	NA	134.9
pK3:trimer	**R_g Guinier (Å)**	**R_g p(r) (Å)**	**D_{max} (Å)**	**MW Bayesian estimate (KDa)**	**MW I(0)/c US-SOMO (KDa)**	**MW Vc (KDa)**
US-SOMO Gaussian	52.1	53.1	177	208	179.0	215.6
US-SOMO	51.8	52.7	176	242.6	185.3	221.4
EFA_RAW	56.2	57.0	187	318.4	NA	267.1
REGALS	52.8	53.9	184	208.0	NA	215.1
pK4:oligomer	**R_g Guinier (Å)**	**R_g p(r) (Å)**	**D_{max} (Å)**	**MW Bayesian estimate (KDa)**	**MW I(0)/c US-SOMO (KDa)**	**MW Vc (KDa)**
US-SOMO Gaussian	74.5	76.1	270	318.4	349	422.5
US-SOMO	72.8	74.6	265	318.4	355.2	453.7
EFA_RAW	82.5	83.7	290	714.8	NA	605.7
REGALS	72.9	74.7	263	318.4	NA	422.7

in the small angle range (see the insert in panel 2). For the last two curves, the trimer of BSA and the higher order oligomer (most likely an unresolved mixture), the US-SOMO and REGALS reconstructions appear essentially identical while the EFA curve exhibits significant differences. This is confirmed by the very similar parameter values derived from US-SOMO and REGALS analyses. Of note, the MM estimates derived from I(0)/c values in US-SOMO are comparable to the V_c estimates (independent from the concentration) and close to, though slightly lower than, the expected values for monomer, dimer and trimer of BSA (66.7, 133.4 and 200.1 KDa, respectively).

We must underline that all three approaches present at least one step that involves choices from the user. Slightly different results from those presented here could be obtained by another user but the main features would be very similar. In the case of US-SOMO, the initial and critical step is the choice of a proper and physically meaningful set of (modified) Gaussians that fits the I(t) curve at a given q-value (chromatogram). In the case of both EFA and REGALS, manual intervention is required for the iterative refinement procedure of the boundaries for the presence of each species at a given time. Although the mathematical tools used are very different, the information used as constraints in the reconstruction process of scattering profiles is strong enough to yield very similar results regardless of the protocol details.

In conclusion, the remarkable agreement between the results of the modified Gaussian decomposition and REGALS, using entirely different mathematical tools and making different assumptions, provides a precious cross validation, suggesting that the scattering patterns of isolated species are reliable reconstructions and can safely be used for further analysis and modeling.

4.2 Aldolase

A similar comparative analysis of SEC-SAXS data of a solution of aldolase has been performed and the results are reported in Supplementary Information in the online version at https://doi.org/10.1016/bs.mie.2022.08.024. Although detailed observations may be different the global picture is similar to that observed in the case of BSA.

5. Conclusion

Whenever possible SEC-SAXS should be used in addition to or even instead of direct injection measurements. Despite the clear drawback of sample dilution during the elution process, the advantages in terms of monodispersity are overwhelming, and avoid drawing wrong conclusions from unreliable data. Several computing and analytical tools are currently being developed to help retrieve the structural information contained in these 2D datasets. These advantages are such that a group at the ILL (Grenoble, France), a high-flux neutron reactor facility has installed a similar set-up on one of the SANS beamlines (Jordan et al., 2016). Progresses in laboratory X-ray sources with the advent of the Metaljet generator have even made possible SEC-SAXS data collection using a laboratory instrument (Bucciarelli et al., 2018). Moreover, the throughput of the method, though apparently much lower than that of "high-throughput," direct measurements at synchrotron beamlines, is in actual fact only a factor of the order of two to five slower due to (i) the need to record several concentrations of a given molecule with the direct injection strategy, (ii) the often associated need of multiple buffer recordings interleaved between the various concentrations and (iii) the time needed to clean and dry the measuring cell after each acquisition, all factors that significantly increase the duty cycle for conventional measurements. Table 1 illustrates the coming of age of SEC-SAXS that has now become a mainstream approach available at many places worldwide. Some specific applications use SEC-SAXS beyond the mere separation of the molecule of interest from aggregates or various oligomeric forms. This is primarily the case of membrane proteins for which an optimized protocol has been developed for data collection yielding data appropriate for structural interpretation (Berthaud et al., 2012; Wojtowicz et al., 2016). A detailed description of how to carry SEC-SAXS experiments from membrane proteins can be found in (Baranowski & Perez, 2020). In a related field, a very recent article reports the global fitting of multiple data frames from SEC–SAXS to investigate the structure of next-generation nanodiscs (Barclay, Tidemand Johansen, Tidemand, Arleth, & Pedersen, 2022). It presents a novel model refinement strategy in which a single model is globally refined against many frames from the same dataset.

Finally, the appearance over the last few years of several approaches addressing the question of mixture deconvolution to restore the scattering patterns of isolated species testifies to the interest of the approach. Our comparative analysis shows that reliable patterns can be extracted not only for the main species (monomer) that is the direct object of the study and for which the separating column has been selected but also for higher molecular mass entities. This can be important in the case of complex formation for instance. The availability of several deconvolution methods using very different premises and mathematical tools is key to validate the restored scattering patterns and legitimate subsequent structural parameter calculations and modeling. We hope this short survey will have contributed to increase the awareness of SEC-SAXS potential among biochemists and molecular biologists.

Acknowledgments

This chapter used parts of the following contribution by two of us: Adv Exp Med Biol, 1009, Biological Small Angle Scattering: Techniques, Strategies and Tips, Editors: B. Chaudhuri, I. G. Muñoz, S. Qian and V.S. Urban, 183-199. doi:10.1007/978-981-10-6038-0_11 by J. Pérez and P. Vachette (2017), by permission from Springer Nature Customer Service Centre GmbH publisher.

We thank our colleagues in charge of SAXS beamlines at synchrotron facilities who kindly provided us with detailed information that constitutes the basis of Table 1. We are very grateful to Mattia Rocco for exchanges about several aspects of US-SOMO deconvolution protocol. Finally, A. Thureau and J. Pérez thank the various groups of users who came to SWING with experimental challenges that prompted several of the reported developments.

References

Baranowski, M., & Perez, J. (2020). Solution X-ray scattering for membrane proteins. *Methods in Molecular Biology*, *2168*, 177–197. https://doi.org/10.1007/978-1-0716-0724-4_8.

Barclay, A., TidemandJohansen, N., Tidemand, F. G., Arleth, L., & Pedersen, M. C. (2022). Global fitting of multiple data frames from SEC-SAXS to investigate the structure of next-generation nanodiscs. *Acta Crystallographica. Section D, Structural Biology*, *78*(Pt 4), 483–493. https://doi.org/10.1107/S2059798322001838.

Berthaud, A., Manzi, J., Pérez, J., & Mangenot, S. (2012). Modeling detergent organization around aquaporin-0 using small-angle X-ray scattering. *Journal of the American Chemical Society*, *134*(24), 10080–10088. https://doi.org/10.1021/ja301667n.

Brookes, E., Pérez, J., Cardinali, B., Profumo, A., Vachette, P., & Rocco, M. (2013). Fibrinogen species as resolved by HPLC-SAXS data processing within the UltraScan solution modeler (US-SOMO) enhanced SAS module. *Journal of Applied Crystallography*, *46*, 1823–1833. https://doi.org/10.1107/S0021889813027751.

Brookes, E., Vachette, P., Rocco, M., & Perez, J. (2016). US-SOMO HPLC-SAXS module: Dealing with capillary fouling and extraction of pure component patterns from poorly resolved SEC-SAXS data. *Journal of Applied Crystallography*, *49*(Pt 5), 1827–1841. https://doi.org/10.1107/S1600576716011201.

Bucciarelli, S., Midtgaard, S. R., Nors Pedersen, M., Skou, S., Arleth, L., & Vestergaard, B. (2018). Size-exclusion chromatography small-angle X-ray scattering of water soluble proteins on a laboratory instrument. *Journal of Applied Crystallography*, *51*(Pt. 6), 1623–1632. https://doi.org/10.1107/S1600576718014462.

David, G., & Pérez, J. (2009). Combined sampler robot and high-performance liquid chromatography: A fully automated system for biological small-angle X-ray scattering experiments at the Synchrotron SOLEIL SWING beamline. *Journal of Applied Crystallography*, *42*, 892–900. https://doi.org/10.1107/S0021889809029288.

Fischer, H., de Oliveira Neto, M., Napolitano, H. B., Polikarpov, I., & Craievich, A. F. (2010). Determination of the molecular weight of proteins in solution from a single small-angle X-ray scattering measurement on a relative scale. *Journal of Applied Crystallography*, *43*, 101–109.

Franke, D., Jeffries, C. M., & Svergun, D. I. (2015). Correlation map, a goodness-of-fit test for one-dimensional X-ray scattering spectra. *Nature Methods, 12*(5), 419–422. https://doi.org/10.1038/nmeth.3358.

Franke, D., Petoukhov, M. V., Konarev, P. V., Panjkovich, A., Tuukkanen, A., Mertens, H. D. T., et al. (2017). ATSAS 2.8: A comprehensive data analysis suite for small-angle scattering from macromolecular solutions. *Journal of Applied Crystallography, 50*(Pt. 4), 1212–1225. https://doi.org/10.1107/S1600576717007786.

Graewert, M. A., Franke, D., Jeffries, C. M., Blanchet, C. E., Ruskule, D., Kuhle, K., et al. (2015). Automated pipeline for purification, biophysical and X-ray analysis of biomacromolecular solutions. *Scientific Reports, 5*, 10734. https://doi.org/10.1038/srep10734.

Hajizadeh, N. R., Franke, D., Jeffries, C. M., & Svergun, D. I. (2018). Consensus Bayesian assessment of protein molecular mass from solution X-ray scattering data. *Scientific Reports, 8*(1), 7204. https://doi.org/10.1038/s41598-018-25355-2.

Hopkins, J. B., Gillilan, R. E., & Skou, S. (2017). BioXTAS RAW: Improvements to a free open-source program for small-angle X-ray scattering data reduction and analysis. *Journal of Applied Crystallography, 50*(Pt. 5), 1545–1553. https://doi.org/10.1107/S1600576717011438.

Hopkins, J. B., & Thorne, R. E. (2016). Quantifying radiation damage in biomolecular small-angle X-ray scattering. *Journal of Applied Crystallography, 49*(Pt. 3), 880–890. https://doi.org/10.1107/S1600576716005136.

Jordan, A., Jacques, M., Merrick, C., Devos, J., Forsyth, V. T., Porcar, L., et al. (2016). SEC-SANS: Size exclusion chromatography combined in situ with small-angle neutron scattering. *Journal of Applied Crystallography, 49*(Pt. 6), 2015–2020. https://doi.org/10.1107/S1600576716016514.

Kikhney, A. G., Borges, C. R., Molodenskiy, D. S., Jeffries, C. M., & Svergun, D. I. (2020). SASBDB: Towards an automatically curated and validated repository for biological scattering data. *Protein Science, 29*(1), 66–75. https://doi.org/10.1002/pro.3731.

Kirby, N., Cowieson, N., Hawley, A. M., Mudie, S. T., McGillivray, D. J., Kusel, M., et al. (2016). Improved radiation dose efficiency in solution SAXS using a sheath flow sample environment. *Acta Crystallography. Section D, Structural Biology, 72*(Pt. 12), 1254–1266. https://doi.org/10.1107/S2059798316017174.

Konarev, P. V., Graewert, M. A., Jeffries, C. M., Fukuda, M., Cheremnykh, T. A., Volkov, V. V., et al. (2022). EFAMIX, a tool to decompose inline chromatography SAXS data from partially overlapping components. *Protein Science, 31*(1), 269–282. https://doi.org/10.1002/pro.4237.

Kuwamoto, S., Akiyama, S., & Fujisawa, T. (2004). Radiation damage to a protein solution, detected by synchrotron X-ray small-angle scattering: Dose-related considerations and suppression by cryoprotectants. *Journal of Synchrotron Radiation, 11*(Pt. 6), 462–468. https://doi.org/10.1107/S0909049504019272.

Manalastas-Cantos, K., Konarev, P. V., Hajizadeh, N. R., Kikhney, A. G., Petoukhov, M. V., Molodenskiy, D. S., et al. (2021). ATSAS 3.0: Expanded functionality and new tools for small-angle scattering data analysis. *Journal of Applied Crystallography, 54*(Pt. 1), 343–355. https://doi.org/10.1107/S1600576720013412.

Mathew, E., Mirza, A., & Menhart, N. (2004). Liquid-chromatography-coupled SAXS for accurate sizing of aggregating proteins. *Journal of Synchrotron Radiation, 11*, 314–318. https://doi.org/10.1107/S0909049504014086.

Meisburger, S. P., Taylor, A. B., Khan, C. A., Zhang, S., Fitzpatrick, P. F., & Ando, N. (2016). Domain movements upon activation of phenylalanine hydroxylase characterized by crystallography and chromatography-coupled small-angle X-ray scattering. *Journal of the American Chemical Society, 138*(20), 6506–6516. https://doi.org/10.1021/jacs.6b01563.

Meisburger, S. P., Xu, D., & Ando, N. (2021). REGALS: A general method to deconvolve X-ray scattering data from evolving mixtures. *IUCrJ*, *8*(Pt. 2), 225–237. https://doi.org/10.1107/S2052252521000555.

Panjkovich, A., & Svergun, D. I. (2018). CHROMIXS: Automatic and interactive analysis of chromatography-coupled small-angle X-ray scattering data. *Bioinformatics*, *34*(11), 1944–1946. https://doi.org/10.1093/bioinformatics/btx846.

Perez, J., & Vachette, P. (2017). A successful combination: Coupling SE-HPLC with SAXS. *Advances in Experimental Medicine and Biology*, *1009*, 183–199. https://doi.org/10.1007/978-981-10-6038-0_11.

Rambo, R. P., & Tainer, J. A. (2013). Accurate assessment of mass, models and resolution by small-angle scattering. *Nature*, *496*(7446), 477–481. https://doi.org/10.1038/nature12070.

Shkumatov, A. V., & Strelkov, S. V. (2015). DATASW, a tool for HPLC-SAXS data analysis. *Acta Crystallographica. Section D, Biological Crystallography*, *71*(Pt. 6), 1347–1350. https://doi.org/10.1107/S1399004715007154.

Svergun, D. (1992). Determination of the regularization parameter in indirect-transform methods using perceptual criteria. *Journal of Applied Crystallography*, *25*(4), 495–503. https://doi.org/10.1107/S0021889892001663.

Thureau, A., Roblin, P., & Perez, J. (2021). BioSAXS on the SWING beamline at Synchrotron SOLEIL. *Journal of Applied Crystallography*, *54*(6), 1698–1710. https://doi.org/10.1107/S1600576721008736.

Watanabe, Y., & Inoko, Y. (2009). Size-exclusion chromatography combined with small-angle X-ray scattering optics. *Journal of Chromatography. A*, *1216*(44), 7461–7465. https://doi.org/10.1016/j.chroma.2009.02.053.

Wojtowicz, H., Prochnicka-Chalufour, A., de Amorim, G. C., Roudenko, O., Simenel, C., Malki, I., et al. (2016). Structural basis of the signalling through a bacterial membrane receptor HasR deciphered by an integrative approach. *The Biochemical Journal*, *473*(14), 2239–2248. https://doi.org/10.1042/BCJ20160131.

Yang, L., Antonelli, S., Chodankar, S., Byrnes, J., Lazo, E., & Qian, K. (2020). Solution scattering at the life science X-ray scattering (LiX) beamline. *Journal of Synchrotron Radiation*, *27*(Pt. 3), 804–812. https://doi.org/10.1107/S1600577520002362.

Yonezawa, K., Takahashi, M., Yatabe, K., Nagatani, Y., & Shimizu, N. (2019). Software for serial data analysis measured by SEC-SAXS/UV-Vis spectroscopy. *AIP Conference Proceedings*, *2054*(1), 060082. https://doi.org/10.1063/1.5084713.

CHAPTER EIGHT

Stopped-flow-time-resolved SAXS for studies of ligand-driven protein dimerization

Henning Tidow[a,b,*] and Inokentijs Josts[a,b,*]
[a]The Hamburg Advanced Research Center for Bioorganic Chemistry (HARBOR), Hamburg, Germany
[b]Department of Chemistry, Institute for Biochemistry and Molecular Biology, University of Hamburg, Hamburg, Germany
*Corresponding authors: e-mail address: tidow@chemie.uni-hamburg.de; josts@chemie.uni-hamburg.de

Contents

Abstract

Protein function is highly dependent on conformational changes and association or dissociation into numerous oligomeric states. Stopped-flow approaches are suitable for probing transient kinetics in proteins, and combining this approach with small-angle X-ray scattering offers an excellent probe into the structural kinetics of protein function. In this chapter we describe in detail the methodological aspects of our recent investigation of ATP-driven dimerization of nucleotide-binding domains from the bacterial transporter MsbA using stopped-flow small-angle X-ray scattering experiments. Despite extensive studies into the structure and function of MsbA, the structural-temporal insights into the conformational rearrangements and transient intermediates along the pathway in this transporter are missing. In our stopped-flow experiments we observe the rapid formation of a transient protein dimer and subsequent dimer decay over hundreds of seconds. Thus, this approach can be used to detect kinetic parameters associated with conformational changes over a wide range of time-scales for soluble and membrane proteins.

Methods in Enzymology, Volume 677
ISSN 0076-6879
https://doi.org/10.1016/bs.mie.2022.08.032

Abbreviations

ABC ATP-binding cassette
ATP adenosine 5′-triphosphate
NBD nucleotide-binding domain
SAXS small-angle X-ray scattering
SEC size-exclusion chromatography
SF stopped-flow
SVD singular value decomposition
TCEP tris(2-carboxyethyl)phosphine
TEV tobacco etch virus

1. Introduction

Proteins are dynamic entities that function at the centre of all biological processes in cells. This functionality stems from their ability to sample a multitude of conformational ensembles and assembly states, often involving transient intermediates. The timescales of these motions and association processes can span several orders of magnitude ranging from femtoseconds to seconds and minutes and underpin all biochemical processes inside the living cells (Levantino, Yorke, Monteiro, Cammarata, & Pearson, 2015). These protein motions are critical for numerous biological phenomena such as protein folding, association/dissociation and enzymatic catalysis. A number of structural techniques offer only static representations of protein structures and protein association interfaces, with little insights into the temporal or kinetic information into these biologically important processes. Therefore, probing and capturing the structural intermediates and encounter complexes and their lifetimes is important for our understanding of numerous fundamental aspects of protein function. Small-angle X-ray scattering (SAXS) is one of the methods to probe the structure of proteins in solution since it provides direct structural information on the biochemical states of particles exposed to X-rays such as the radius of gyration, protein folding and oligomeric state and real-space structural information from the indirect Fourier Transform of the scattering data (Da Vela & Svergun, 2020).

Stopped-flow (SF) technique involves rapid mixing of solutions, which are injected into a cuvette or capillary that is coupled with an optical detection system to follow the reaction kinetics. SF methods allow access to time-scales ranging from micro/milliseconds (depending on the deadtime of the SF instrument) up to seconds/minutes and has provided crucial insights into the kinetics of numerous biological processes, for example probing protein association and dissociation, conformational changes and

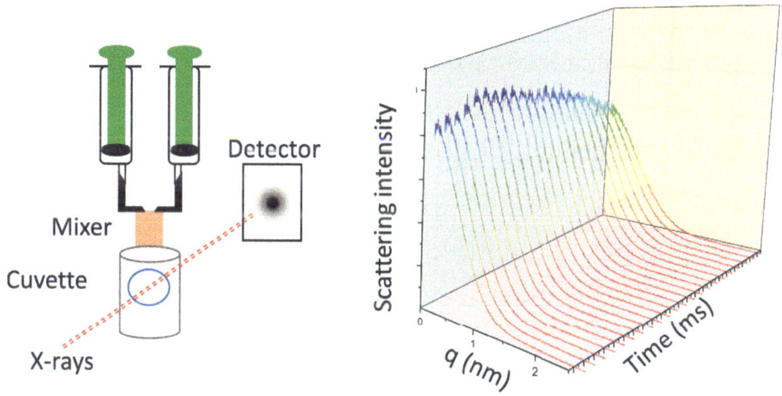

Fig. 1 Schematic overview of a stopped-flow time-resolved SAXS (SF-TR-SAXS) experiment, which provides opportunities for probing numerous ligand-driven kinetic processes in proteins.

protein refolding or unfolding (Jahn & Radford, 2005; Roder, Maki, & Cheng, 2006; Royer, 2006; Schreiber, 2002). Combining the stopped-flow approach with SAXS offers opportunities for quantifying and characterising structural changes and kinetics in solution for many biomolecular reactions (Fig. 1).

In this chapter, we focus on our investigation of ATP-driven dimerization of nucleotide-binding domains from the ATP-binding cassette (ABC) transporter MsbA (NBD_{MsbA}) from *Escherichia coli* (Zhou, White, Polissi, Georgopoulos, & Raetz, 1998). MsbA is involved in the biogenesis of the bacterial outer membrane and is an important pharmacological target for the design of novel antimicrobial compounds. MsbA functions as a flippase that transports lipid A and some glycolipids across the bacterial inner membrane leaflets (Doerrler, Gibbons, Christian, & Raetz, 2004; Thélot et al., 2021). It functions as a homodimer through an alternating-access 'power stroke' mechanism driven by ATP hydrolysis. ATP-binding at the NBDs initiates the transport mechanism by forming a closed NBD dimer, followed by rigid motion in the transmembrane domains that allows the substrate to exit at the opposite side of the bilayer. Full molecular details of the ATP-driven substrate translocation are still not fully understood. Previously, we were able to capture the dimerization step of NBD_{MsbA} using caged ATP and time-resolved solution scattering experiments (Josts et al., 2018). Our latest time-resolved SAXS experiments employing the SF technique on the isolated NBDs from MsbA allowed us to characterise the kinetics of the full ATP-driven catalytic cycle (Josts et al., 2021).

2. Sample preparation

The NBD (residues 336–579) from *Escherichia coli msbA* gene was amplified from genomic DNA and cloned into a pnEK vector with an N-terminal His_6-tag using *Nde*I and BamHI digestion sites following the standard T4 ligase ligation protocol. Protein expression was explored in several different growth media (lysogeny broth (LB), autoinduction ZY media, 2xYT and terrific broth (TB)) with BL21 (Gold) *E. coli* cells. Protein production in TB was found to be most optimal, with protein yields of 200–300 mg per litre of cell expression and purity above 95% as judged by SDS-PAGE. *E.coli* cells overexpressing NBD_{MsbA} were resuspended in Tris-buffered saline (TBS) buffer with 1 mM $MgCl_2$ 10% glycerol, 1 mg/mL lysozyme and 0.1 mg/mL DNase and lysed using the high-pressure homogeniser (EmulsiFlex-C3, Avestin). Cell debris was cleared by high-speed centrifugation (40,000 g) and lysate was loaded onto a 5 mL HisTrap™ column. Following 10 column volume wash with TBS supplemented with 40 mM imidazole, NBD_{MsbA} was eluted with 300 mM imidazole and tobacco etch virus (TEV) protease was added to the eluted samples (1:100 protease: protein ratio). Purified protein with His_6-tagged TEV were dialysed overnight against TBS buffer supplemented with 1 mM dithiothreitol, and re-purified via the 5 mL HisTrap™ column to remove TEV and undigested protein. The collected flow-through was concentrated to 20–40 mg/mL and stored at −80 °C until needed. Functional ATPase activity of the soluble domain was tested using the Baginski method. The details of the ATPase assay are provided in more detail in the chapter "Studying integral membrane protein by SANS using stealth reconstitution systems" in this edition of *Methods in Enzymology*.

2.1 Additional sample characterisation

Protein quality for SF-SAXS was assessed using static SAXS experiments, where we screened a range of protein concentrations (5–20 mg/mL) to ensure that there is no sample aggregation and no interparticle interactions at high protein concentrations, which could affect the data in the low-q range of the scattering curve. NBD dimerisation was also observed by size-exclusion chromatography (SEC) by adding 1 mM Mg^{2+}-ATP into the column equilibration buffer as well as by SEC-SAXS.

2.2 Protein concentration considerations

The intensity of the scattering signal is directly proportional to the concentration and size of the particle in solution, therefore for proteins with smaller molecular weights, higher concentrations are recommended to achieve the desirable signal-to-noise ratio. Alternatively, if high protein concentrations lead to protein instability, aggregation or interparticle interactions, then averaging many frames with corresponding time-delays is also possible. The scattering curve from the samples must provide a confident readout of the kinetic change, for example the change in R_g of 1–2 Å must be confidently derived. If the scattering data in the Guinier region are noisy from a sample that is too dilute then the measurement of the kinetics will be challenging. We recommend using 5–10 mg/mL solutions for proteins ranging 10–30 kDa in size and 2–5 mg/mL for proteins ranging from 60 to 120 kDa. Lower concentrations can be used to some degree if the signal change occurring during the reaction is sufficiently large. In this case, frame averaging can compensate for the loss of signal, however resulting in higher overall sample consumption.

3. Stopped-flow experiment: The measurement

Prior to the experiment, concentrated protein is dialysed against the desired buffer (20 mM Tris-HCl, 200 mM NaCl, 5 mM $MgCl_2$) overnight. Next day, the dialysis buffer is filtered and degassed extensively. This buffer is also used to prepare ATP, which serves as the ligand in these experiments. Concentrated NBD_{MsbA} at 20 mg/mL (corresponding to 1 mM protein) in TBS + 0.95 mM Mg^{2+}-ATP are loaded into the SF syringes. Minimum loading volume for the SFM 400 SF mixer (Bio-logic, Seyssinet-Pariset, France) machine is 3 mL of solution per pneumatic syringe. The sample and ligand are injected into a quartz capillary (0.8 mm inner diameter) with the desired ratio. We used the injection ratio of 1:1 for measuring all the mixing kinetics for time-resolved scattering curves, injecting 110 μL of NBD_{MsbA} protein and 110 μL of Mg^{2+}-ATP. In the case of NBD_{MsbA} the overall experiment required approximately 10 mL of protein solution. To gain some prior insights into the time-scales of the protein association reaction, stopped-flow measurements could be carried out using fluorescence (e.g., tryptophan fluorescence) or absorbance as a detection readout method prior to the

SF-SAXS experiments. This could help in selecting an optimal delay time and number of points to sample the kinetics of association in the SF-SAXS experiment.

At the start of the experiment, scattering data derived from the appropriate buffers used in the data processing steps must be collected. Generally, this involves scattering the buffer without the ligand injection and buffer with ligand mixed in for subtracting from the scattering curves of proteins at a later stage. If the ligand is added at high concentrations, this is necessary to obtain scalable datasets for comparison with the apo protein. The mixing injections must be performed in the same ratio as would be performed for the time-resolved kinetic experiment with the protein, i.e., mixing 1:1 in the case of NBD_{MsbA} and Mg^{2+}-ATP.

Once a sufficient number of buffer frames have been collected, the protein can be mixed with ligand-free buffer for measuring scattering data corresponding to the resting state of the molecule prior to the initiation of kinetics measurements. At this stage, subtraction of the buffer frames from apo protein scattering datasets can performed and buffer-subtracted SAXS data can be compared with the measurements from static SAXS experiments or with CRYSOL-calculated SAXS curves from protein structure models, if a 3D model of the protein of interest is available (Svergun, Barberato, & Koch, 1995). This step helps to identify the presence and onset of radiation damage in the consecutive frames during X-ray exposure. If the sample is sensitive to X-rays then the length of X-ray exposure and transmission can be optimised to minimise the damage to the samples. During our measurements of NBD_{MsbA} dimerization kinetics, the protein suffered from radiation damage after approximately 100 ms of X-ray exposure with subsequent frames containing radiation damaged particles, which would obscure the kinetic measurement and data interpretation (Fig. 2). Our final data collection parameters were 35 millisecond exposure per frame, which corresponded to 3–4 frames or approximately 100 milliseconds of kinetic data per injection. The most convenient way to check for radiation damage onset is to examine the frames visually (Fig. 2A) or to use a statistical test for assessing similarity of the scattering curves for all the frames, for example with CorMap software as part of the ATSAS package (Franke, Jeffries, & Svergun, 2015). If radiation damage is severe even in the initial frames then radical scavengers such as dithiothreitol or tris(2-carboxyethyl)phosphine (TCEP) could be added to all the samples and buffer.

The previously characterised ATPase cycle of isolated NBD_{MsbA} domains, measured using TR-Fourier transform infrared spectroscopy, was longer than 100 s (Syberg, Suveyzdis, Kötting, Gerwert, & Hofmann, 2012).

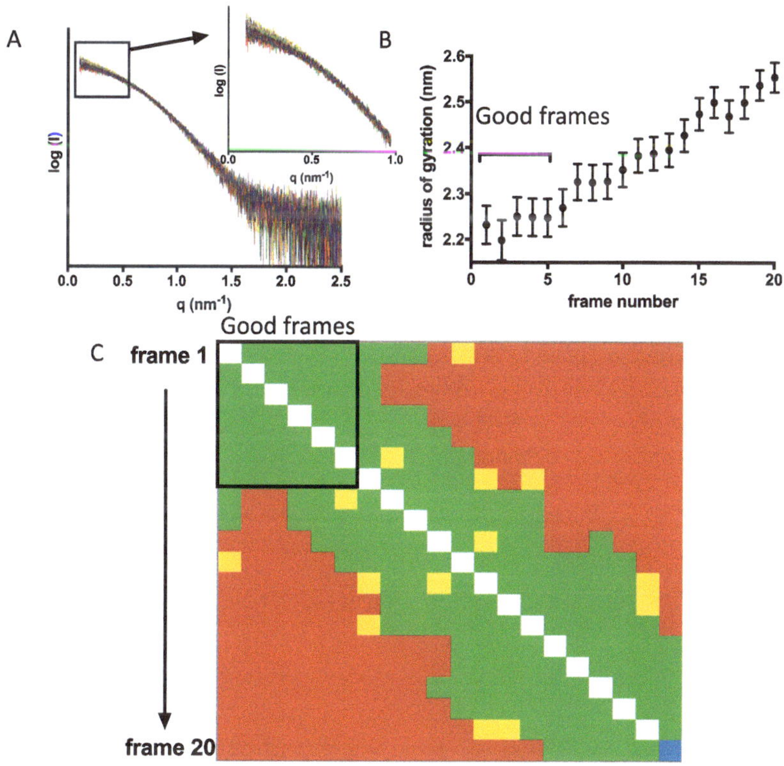

Fig. 2 Various ways to detect radiation damage in SF-SAXS. (A) Overlay of subtracted scattering curves, showing an increase in the very low-q region. (B) Plot of R_g with increasing exposure time (frame number). (C) Correlation map comparing the first 20 frames of the time-series, highlighting the deviations arising from radiation damage.

In our SF-SAXS experiments each NBD_{MsbA} injection only allowed us to collect 70–100 ms of radiation damage-free kinetic data. Therefore, in order to collect a full time-series of the ATP-driven NBD association and subsequent dissociation, we opted for setting delay times between the X-ray shutter and start of the mixing reaction, which allowed us to probe the reaction kinetics for up to 120 s and could potentially be used to collect even longer delays if necessary. This method in principle is reminiscent of the pump-probe setup used to collect laser-light triggered TR scattering data.

4. Data analysis and interpretation

A kinetic reaction could involve changes in the overall protein shape or size and its oligomeric state. In our experiments with NBD_{MsbA} we expect the proteins to form homodimers after binding Mg^{2+}-ATP.

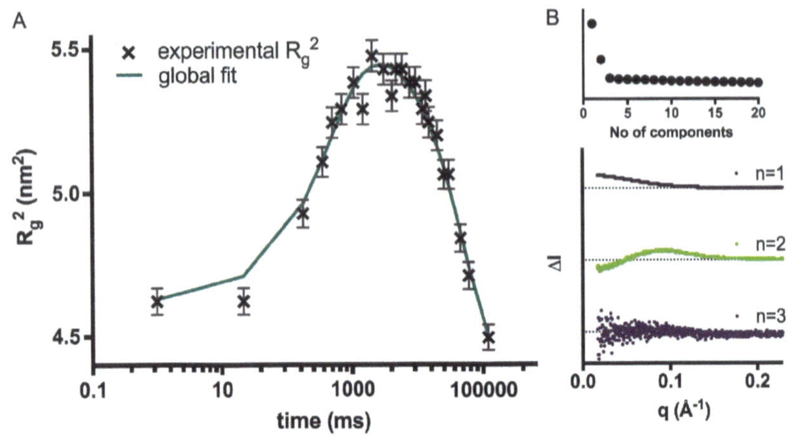

Fig. 3 Kinetics of NBD_{MsbA} dimerization and SVD analysis of the SF-SAXS data. (A) Plot of R_g^2 as a function of time with a fitted stretched exponential function to obtain the time constants of NBD_{MsbA} dimerization and decay. (B) Model-independent SVD analysis of the dimerization time series, showing the number of significant components (S matrix) within the data and first three independent left basis vectors (U matrix). This analysis indicates that there are two components present within our time-resolved experiment.

To analyse the kinetics of dimer formation we plotted the R_g values derived from the Guinier region of the SAXS curves against the corresponding time delays. In our experiments, we observed that dimerization occurred after approximately 700 milliseconds, with R_g values reaching a maximum over 1 to 10 s and eventually returning to the starting value by 120 s reflected by the dimer dissociation (Fig. 3A) (Josts et al., 2018). Other parameters such a I(0) and Porod volume can also be analysed to visualise changes in the molecular weight of the protein during the reaction and thus inform of any transient oligomerisation. Examples of using I(0) and R_g readouts to monitor conformational changes and transient oligomerisation kinetics include ferritin assembly and cytochrome *c* refolding kinetics (Sato et al., 2016; Segel et al., 1999).

5. Singular value decomposition and OLIGOMER analysis of time-resolved SAXS data

SAXS curves from multicomponent mixtures, for example consisting of several oligomeric species, can be considered as volume-weighed linear combinations of scattering intensities of all contributing components. Therefore, complete analysis of such data requires the deconvolution of the number of species and the scattering contribution of their populations to the observed scattering intensity over the course of the kinetic reaction. Singular value decomposition (SVD) is a model-independent matrix factorisation analysis in linear algebra. The kinetic scattering data are represented as a matrix of scattering profiles with q scattering angles over a time-series. SVD analysis returns left and right singular matrices (U and V) and a diagonal matrix D. The largest singular values (from D matrix) and non-random left singular vectors (from U matrix) reflect the number of independent components within the time series. The only prerequisite of SVD analysis is that the number of independent components must be smaller than the number of observed scattering profiles (i.e., the time-series must be well-sampled). SVD of TR-SAXS has been applied to numerous systems and processes undergoing a multistep change such as photoinduced transitions in rhodopsin, myoglobin and haemoglobin, insulin association, fluorescent proteins, phytochrome photosensors and intermediates in protein folding processes (Chen, Hodgson, & Doniach, 1996; Cho et al., 2016; Choa et al., 2010; Levantino et al., 2015; Rimmerman et al., 2017; Takala et al., 2014).

SVD analysis and volumetric fraction calculation of monomer-dimer NBD states can be calculated using the program OLIGOMER as part of the ATSAS package (Konarev, Volkov, Sokolova, Koch, & Svergun, 2003). Deconvolution of our SF data yielded two independent left basis vectors (U matrix), which suggested that there are two major species in our TR experiment (Fig. 3B). We speculated that these two species could be monomeric and dimeric NBDs. Due to the availability of high-resolution models of NBD monomer and NBD dimer (as part of ADP-vanadate trapped MsbA crystal structure) we employed OLIGOMER to calculate the scattering contribution of volumetric fractions of NBD monomer and dimer over the course of our time-series (Ward, Reyes, Yu, Roth, & Chang, 2007). OLIGOMER uses available form-factors (e.g., from PDB models or from experimental intensities of static SAXS measurements) to estimate the volumetric concentrations each component in a mixture. From this analysis OLIGOMER estimates that dimeric NBD_{MsbA} is populated up to 60% of total particles in solution at the peak (1–10s) of our time-series.

If high-resolution models are not available, it is possible to deconvolute the scattering data using SVD in combination with well-sampled molecular dynamics simulations and obtain some structural insights into transient intermediates, however such analyses are outside the scope of this chapter (Björling, Niebling, Marcellini, van der Spoel, & Westenhoff, 2015; Chen & Hub, 2015) .

6. Fitting a kinetic model to extract rate constants from SF-SAXS data

For fitting the evolution of R_g during our kinetic process we considered a four-state rate equation, assuming three transient states: ATP-bound NBD, dimeric NBD and the final post-hydrolysis NBD state. Due to the broad maximum of the R_g distribution in our time-series, the curve was fitted with a stretched exponential relaxation to describe the entire kinetic reaction of ATP-driven dimerization and decay. The derived time constants for ATP-dependent dimerization were 0.5 ± 0.1 s and 57 ± 15 s (Fig. 3A).

7. Summary and conclusions

Stopped-flow mixing combined with SAXS has allowed us to study the transient kinetics of ATP-driven dimerization and decay of NBD_{MsbA} on millisecond to second time-scales. We could extract the time constants for the dimerization event and the subsequent dimer dissociation and combine these with structural modelling of individual states in our kinetic reaction. Assessing the radiation damage onset is crucial for conducting successful stopped-flow SAXS experiments and in this chapter we have discussed several strategies for identifying the frames where radiation damage starts to affect the sample. We show that SF-SAXS experiments can be used for mapping conformational changes and oligomeric states of proteins on millisecond time-scales, making this approach widely accessible for numerous other biological systems.

Acknowledgments

We are grateful to the staff at beamline P12 (EMBL, Hamburg). We thank Tobias W Gräwert, Clement E Blanchet, Martin A Schroer and Dmitri I Svergun for their assistance with the SF experiments and members of the Tidow and Pearson labs for helpful discussions. We acknowledge access to the Sample Preparation and Characterisation Facility of EMBL, Hamburg. This research was funded by the Cluster of Excellence "The Hamburg Centre for Ultrafast Imaging" of the Deutsche Forschungsgemeinschaft (DFG), EXC 1074, project ID 194651731 and by the Cluster of Excellence "Advanced Imaging of Matter" of the Deutsche Forschungsgemeinschaft (DFG), EXC 2056, project ID 390715994.

References

Björling, A., Niebling, S., Marcellini, M., van der Spoel, D., & Westenhoff, S. (2015). Deciphering solution scattering data with experimentally guided molecular dynamics simulations. *Journal of Chemical Theory and Computation*, *11*(2), 780–787. https://doi.org/10.1021/ct5009735.

Chen, L., Hodgson, K. O., & Doniach, S. (1996). A lysozyme folding intermediate revealed by solution X-ray scattering. *Journal of Molecular Biology, 261*(5), 658–671. https://doi.org/10.1006/JMBI.1996.0491.

Chen, P., & Hub, J. S. (2015). Interpretation of solution X-ray scattering by explicit-solvent molecular dynamics. *Biophysical Journal, 108*(10), 2573–2584. https://doi.org/10.1016/j.bpj.2015.03.062.

Cho, H. S., Schotte, F., Dashdorj, N., Kyndt, J., Henning, R., & Anfinrud, P. A. (2016). Picosecond photobiology: Watching a signaling protein function in real time via time-resolved small- and wide-angle X-ray scattering. *Journal of the American Chemical Society, 138*(28), 8815–8823. https://doi.org/10.1021/JACS.6B03565.

Choa, H. S., Dashdorj, N., Schotte, F., Graber, T., Henning, R., & Anfinrud, P. (2010). Protein structural dynamics in solution unveiled via 100-ps time-resolved x-ray scattering. *Proceedings of the National Academy of Sciences of the United States of America, 107*(16), 7281–7286. https://doi.org/10.1073/PNAS.1002951107.

Da Vela, S., & Svergun, D. I. (2020). Methods, development and applications of small-angle X-ray scattering to characterize biological macromolecules in solution. *Current Research in Structural Biology, 2*, 164–170. https://doi.org/10.1016/J.CRSTBI.2020.08.004.

Doerrler, W. T., Gibbons, H. S., Christian, R., & Raetz, H. (2004). MsbA-dependent translocation of lipids across the inner membrane of Escherichia coli*. *Journal of Biological Chemistry, 279*(43), 45102–45109. https://doi.org/10.1074/JBC.M408106200.

Franke, D., Jeffries, C. M., & Svergun, D. I. (2015). Correlation map, a goodness-of-fit test for one-dimensional X-ray scattering spectra. *Nature Methods, 12*(5), 419–422. https://doi.org/10.1038/nmeth.3358.

Jahn, T. R., & Radford, S. E. (2005). The yin and Yang of protein folding. *The FEBS Journal, 272*(23), 5962–5970. https://doi.org/10.1111/J.1742-4658.2005.05021.X.

Josts, I., Gao, Y., Monteiro, D. C. F., Niebling, S., Nitsche, J., Veith, K., et al. (2021). Structural kinetics of MsbA investigated by stopped-flow time-resolved small-angle X-ray scattering. *Structure*.

Josts, I., Niebling, S., Gao, Y., Levantino, M., Tidow, H., & Monteiro, D. (2018). Photocage-initiated time-resolved solution X-ray scattering investigation of protein dimerization. *Research Letters IUCrJ, 5*, 667–672. https://doi.org/10.1107/S2052252518012149.

Konarev, P. V., Volkov, V. V., Sokolova, A. V., Koch, M. H. J., & Svergun, D. I. (2003). PRIMUS: A windows PC-based system for small-angle scattering data analysis. *Journal of applied crystallography, 36*(5), 1277–1282. https://doi.org/10.1107/S0021889803012779.

Levantino, M., Schirò, G., Lemke, H. T., Cottone, G., Glownia, J. M., Zhu, D., et al. (2015). Ultrafast myoglobin structural dynamics observed with an X-ray free-electron laser. *Nature Communications, 6*(1), 1–6. https://doi.org/10.1038/ncomms7772.

Levantino, M., Yorke, B. A., Monteiro, D. C., Cammarata, M., & Pearson, A. R. (2015). Using synchrotrons and XFELs for time-resolved X-ray crystallography and solution scattering experiments on biomolecules. *Current Opinion in Structural Biology, 35*, 41–48.

Rimmerman, D., Leshchev, D., Hsu, D. J., Hong, J., Kosheleva, I., & Chen, L. X. (2017). Direct observation of insulin association dynamics with time-resolved X-ray scattering. *The Journal of Physical Chemistry Letters, 8*(18), 4413–4418.

Roder, H., Maki, K., & Cheng, H. (2006). Early events in protein folding explored by rapid mixing methods. *Chemical Reviews, 106*(5), 1836–1861. https://doi.org/10.1021/CR040430Y.

Royer, C. A. (2006). Probing protein folding and conformational transitions with fluorescence. *Chemical Reviews, 106*(5), 1769–1784. https://doi.org/10.1021/CR0404390.

Sato, D., Ohtomo, H., Yamada, Y., Hikima, T., Kurobe, A., Fujiwara, K., et al. (2016). Ferritin assembly revisited: A time-resolved small-angle X-ray scattering study. *Biochemistry, 55*(2), 287–293. https://doi.org/10.1021/acs.biochem.5b01152.

Schreiber, G. (2002). Kinetic studies of protein–protein interactions. *Current Opinion in Structural Biology*, *12*(1), 41–47. https://doi.org/10.1016/S0959-440X(02)00287-7.

Segel, D. J., Eliezer, D., Uversky, V., Fink, A. L., Hodgson, K. O., & Doniach, S. (1999). Transient dimer in the refolding kinetics of cytochrome c characterized by small-angle X-ray scattering. *Biochemistry*, *38*(46), 15352–15359. https://doi.org/10.1021/bi991337k.

Svergun, D., Barberato, C., & Koch, M. H. (1995). CRYSOL – A program to evaluate X-ray solution scattering of biological macromolecules from atomic coordinates. *Journal of Applied Crystallography*, *28*(6), 768–773. https://doi.org/10.1107/S0021889895007047.

Syberg, F., Suveyzdis, Y., Kötting, C., Gerwert, K., & Hofmann, E. (2012). Time-resolved Fourier transform infrared spectroscopy of THE nucleotide-binding domain from THE ATP-binding cassette transporter MsbA ATP HYDROLYSIS IS THE RATE-LIMITING STEP IN THE CATALYTIC CYCLE. *Journal of Biological Chemistry*, *287*(28), 23923–23931.

Takala, H., Björling, A., Berntsson, O., Lehtivuori, H., Niebling, S., Hoernke, M., et al. (2014). Signal amplification and transduction in phytochrome photosensors. *Nature*, *509*(7499), 245–248. https://doi.org/10.1038/nature13310.

Thélot, F. A., Zhang, W., Song, K. K., Xu, C., Huang, J., & Liao, M. (2021). Distinct allosteric mechanisms of first-generation MsbA inhibitors. *Science*, *374*(6567), 580–585. https://doi.org/10.1126/SCIENCE.ABI9009.

Ward, A., Reyes, C. L., Yu, J., Roth, C. B., & Chang, G. (2007). Flexibility in the ABC transporter MsbA: Alternating access with a twist. *Proceedings of the National Academy of Sciences*, *104*(48), 19005–19010.

Zhou, Z., White, K. A., Polissi, A., Georgopoulos, C., & Raetz, C. R. H. (1998). Function of Escherichia coli MsbA, an essential ABC family transporter, in lipid a and phospholipid biosynthesis *. *Journal of Biological Chemistry*, *273*(20), 12466–12475. https://doi.org/10.1074/JBC.273.20.12466.

CHAPTER NINE

Time-resolved small-angle neutron scattering (TR-SANS) for structural biology of dynamic systems: Principles, recent developments, and practical guidelines

Anne Martel[a] and Frank Gabel[b,*]
[a]Large Scale Structures Group, Institut Laue-Langevin, Grenoble, France
[b]Univ. Grenoble Alpes, CEA, CNRS, IBS, Grenoble, France
*Corresponding author: e-mail address: frank.gabel@ibs.fr

Contents

Abstract

We present an overview of time-resolved small-angle neutron scattering (TR-SANS) applied to biological systems, with a focus on bio-macromolecules and assemblies they form, together with practical guidelines. After a brief introduction to the theory and practice of SANS, we present the general setup and specifics of time-resolved experiments, as well as an overview of diverse experimental results and applications from the past ≈25 years. Subsequently, we provide guidelines and practical instructions for the design, planning and execution for TR-SANS experiments, as a function of the time- and length-scales of the biological processes of interest, the availability of sample amount and deuterium labeling, and the structural information sought. We conclude with a discussion of the most recent instrumental and sample environment developments and perspectives for the future.

Methods in Enzymology, Volume 677
ISSN 0076-6879
https://doi.org/10.1016/bs.mie.2022.08.010

1. Introduction, principles and limitations

Small-angle neutron scattering (SANS) is increasingly used in bio-macromolecular structural biology as a complementary technique to crystallography, NMR and cryo-EM. The major strengths of neutron scattering in solution, with respect to its sister technique SAXS (small-angle X-ray scattering), are multiple: the absence of radiation damage (Jasnin, Moulin, Haertlein, Zaccai, & Tehei, 2008; Marty et al., 2013), the natural contrast between proteins, RNA/DNA and lipid/detergent compounds, the possibility to broadly cover the scattering length range of all of these biomolecules by H_2O/D_2O solvent mixture (Jacrot, 1976), and the possibility to distinguish molecules from the same family (e.g., two proteins) by specific deuteration (Haertlein et al., 2016) (see chapter "Deuteration for biological SANS: Case studies, success and challenges in chemistry and biology" by Duff et al.). In this book chapter we limit ourselves to a brief introduction of the fundamentals of SANS and focus on the specifics and practical aspects of time-resolved (TR) experiments. For a more comprehensive introduction to SANS and its comparison with SAXS, the reader is referred to recent reviews (Jeffries et al., 2021; Mahieu & Gabel, 2018) and other chapters in this book (see chapter "Planning, executing and assessing the validity of SANS contrast variation experiments" by Krueger and chapter "Technical considerations for small-angle neutron scattering from biological macromolecules in solution: Cross sections, contrasts, instrument setup and measurement" by Pietras et al.).

The present state-of-the-art of TR-SANS experiments allows to observe the evolution of biological structures on a length-scale from nm to μm and over time-scales from seconds to days (see concrete examples from literature in the next section). A schematic overview of the principle elements of a TR-SANS experiment is shown in Fig. 1A. Neutrons are generated at reactor or spallation sources (Svergun, Koch, Timmins, & May, 2013) and guided towards a sample position. (An overview of existing neutron facilities worldwide can be found on the following webpage: https://www.ncnr.nist.gov/nsources.html). The incoming neutrons are focused (collimated) and their wavelength (λ) range (typically a few Å) is either selected by a chopper device or directly measured in time-of-flight (TOF) mode (Teixeira et al., 2008). The interaction and scattering of the neutrons with the biological sample is determined by the spatial arrangements of the scattering length densities (SLD) $\rho_{molecule}$ of the solubilized biological molecules (proteins, RNA/DNA, sugars, lipids, ...) and the difference (= contrast $\Delta\rho$, see Fig. 2A) with the

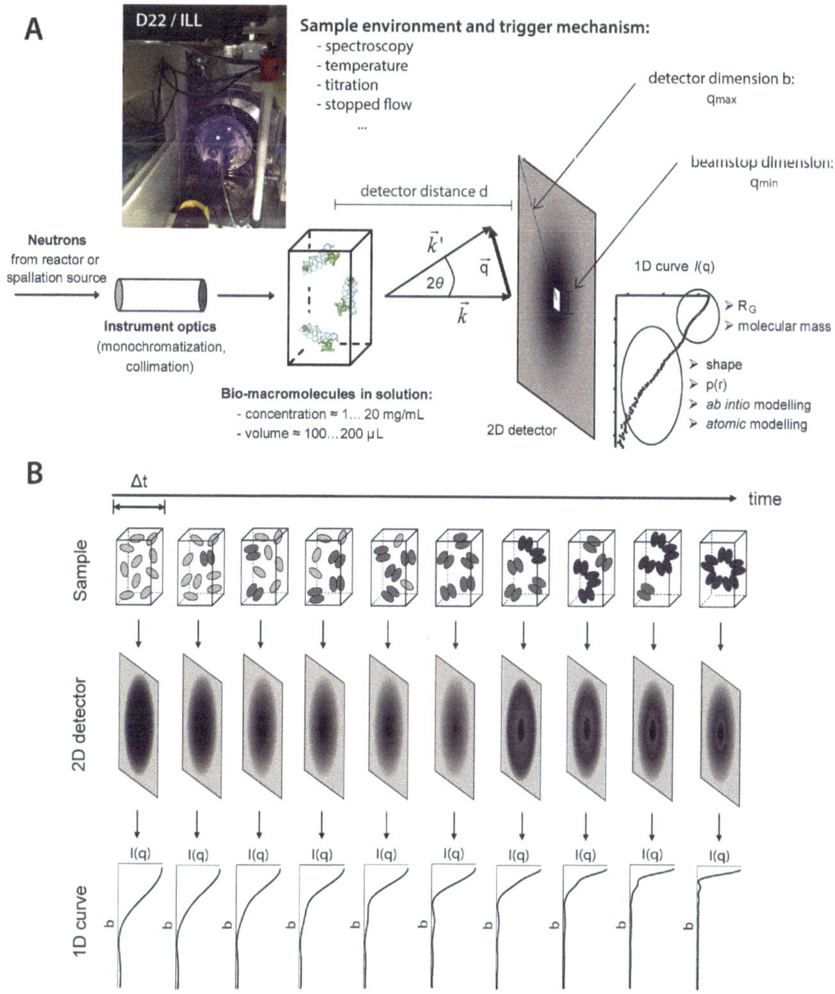

Fig. 1 Schematic overview of a TR-SANS experiment. (A) Essential elements of a typical SANS instrument and geometry, as well as information encoded in the final 1D SANS curve. **k** and **k**′ are the incident and scattered wave vectors, with $|\mathbf{k}|=|\mathbf{k}'|=2\pi/\lambda$ for elastically scattered neutrons. The inset photograph shows an online UV spectroscopic device on D22 (ILL, France). (B) Schematic work flow of a TR-SANS experiment with a changing sample state, resulting in a continuous evolution of 2D detector diffraction patterns and the associated final 1D curved I(q), which are obtained for each frame after buffer subtraction. Δt denominates the duration of a single neutron exposure frame.

solvent SLD $\rho_{solvent}$. For isotropically oriented particles in solution, the two-dimensional scattering pattern on the detector depends only on the radial distance from the center of the direct beam, and concentric rings can be integrated to yield one-dimensional intensities as a function of the scattering vector **q**, of both samples and buffers, and for each individual exposure time

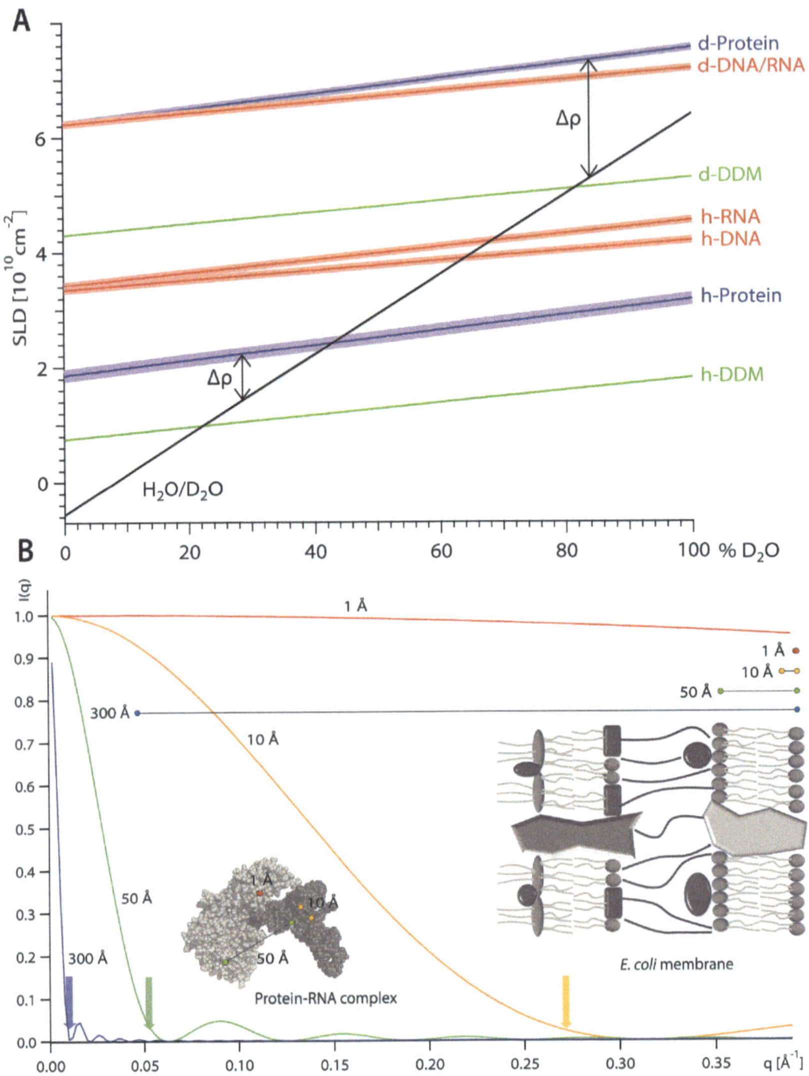

Fig. 2 (A) Scattering length densities (SLD) of protein, DNA, RNA and DDM as a function of D_2O percentage in buffer. "h" refers to hydrogenated molecules. "d" refers to per-deuterated molecules in the case of proteins, DNA and RNA, and to tail-deuterated DDM (Breyton et al., 2013). Shaded regions around the reference lines are due to a variability of SLDs due to variations of bio-macromolecular sequences (Jacrot, 1976). (B) Theoretical SANS curves of two point-like scattering elements at different distances (1, 10, 50 and 300 Å), calculated from the Debye equation (Eq. 3). The arrows indicate the approximate upper limit of the range ($q > 0.00\,Å^{-1}$) up to which the SANS curves contain statistically useful information for each pair of scattering elements. This q-range is defined by two essential criteria of the respective SANS curves:

period ("frame") Δt (Fig. 1B). The final experimental data are the buffer-subtracted sample 1D intensities as a function of time, $I(q, t_j)$, where $t_{j+1} = t_j + \Delta t$:

$$I(q, t_j) = \iiint \Delta\rho(t_j) \cdot e^{i\vec{q}\cdot\vec{r}}\, dV \tag{1}$$

where

$$q = (4\pi/\lambda)\cdot\sin(\theta) \tag{2}$$

with 2θ being the scattering angle, q being the modulus of the scattering vector **q**, **r** denoting the position of a small volume element dV within the particle. (Strictly speaking, the integral in Eq. 1 runs over the whole volume including the solubilized particle and regions of solvent in its vicinity that display different SLD than the bulk solvent).

Two important model-free parameters are the forward scattered intensity I(0), which is proportional to the square of the product of particle volume V and contrast Δρ (Eq. 1, using $q=0\,\text{Å}^{-1}$), and the radius of gyration R_G, which is a measure for the size of the solubilized object. Both parameters can be extracted using the Guinier approximation (Guinier, 1939) at very low angles. Alternatively, and neglecting subtle effects like hydration shells (Kim & Gabel, 2015; Svergun et al., 1998) Eq. (1) can be written as a squared sum of distances r_{ij} between small scattering elements dV_j, weighed by their respective contrast $\Delta\rho_j$ with respect to the solvent (Debye, 1915):

$$I(q) = \sum_{i,j} \Delta_{\rho i} dV_i \Delta_{\rho j} dV_j \frac{\sin(q\cdot r_{ij})}{q\cdot r_{ij}} \tag{3}$$

(i) The I(q) intensity in the range is significantly greater than zero. (ii) The intensity I(q) displays a significant *relative* change in this range. While criterion (i) is fulfilled for a distance of 1 Å (uppermost curve), the intensity of the respective SANS curve varies only very slightly (criterion ii). Due to the latter criterion, two scattering elements in a biological object, separated by 1 Å, contribute only little to the SANS curve. In other words, in the q-range shown here (up to $0.4\,\text{Å}^{-1}$) they cannot be separated and appear as a single element. The insets in part B show a protein-RNA complex (Schmitt et al., 2012) and a schematic cross-section of an *Escherichia coli* membrane, both drawn to scale with the four pair distances.

Eq. 3 allows to better visualize and discuss the respective contributions of scattering elements separated by a given distance r_{ij} (see insets of biological structures in Fig. 2B).

The minimum possible duration of an individual time frame Δt (and therefore the **time-resolution** of a TR-SANS experiment) depends on the incoming neutron flux, the instrumental setup (in particular the collimation length), the sample volume illuminated (i.e., the beam surface area times the sample thickness), the sample concentration and contrast, the incoherent background, and the detector efficiency. Generally, at SANS instruments at high neutron flux facilities ($\approx 10^6$–10^8 n·cm^{-2}·s^{-1}), SANS curves with good signal/noise ratio can be obtained within a few seconds for biological samples, and in favorable cases within $\approx$100 ms for strongly scattering standards such as Plexiglas (Radulescu, Pipich, Frielinghaus, & Appavou, 2012), while at medium- and low-flux facilities ($<10^6$ n·cm^{-2}·s^{-1}), exposure times are of the order of minutes to hours for weakly scattering biological samples. Importantly, the ancillary sample environment needs to be adapted to trigger the biochemical or biological process, in parallel to the start of neutron acquisition in the TR-SANS experiment, and should be in phase and fine-tuned to the duration Δt of the individual time frames. Triggering events that have been used include fast mixing (stopped flow (Grillo, 2009)), temperature jumps (Ibrahim et al., 2017; Mahieu et al., 2020) and photo-spectroscopic methods (Chaudhari, Semanat, Martel, Schneider, & Fuertes, 2021). These practical aspects are discussed in more detail in the section on guidelines for TR-SANS experiments further below, and a schematic flow chart for the design of a TR-SANS experiment is provided in Fig. 3.

The **length-scales of biological processes** observable and accessible in a TR-SANS experiment are defined by both experimental and theoretical constraints:

1) First, the geometry of the experimental setup defines the available q-range: the maximum (q_{max}) value is determined by the size, dimensions and relative position of the detector with respect to the direct beam, and its distance d from the sample. It can be calculated using Eq. 2, with $2\theta = \tan(b/d)$, where b is the maximum distance of a detector element from the incident direct beam (Fig. 1A). In the same way, the minimum value (q_{min}) can be calculated from the dimensions of the beam stop. q_{min} defines the maximum size of structural features (in particular the maximum dimension D_{max}) that can be reliably determined from SANS data, and can be estimated from the relationships

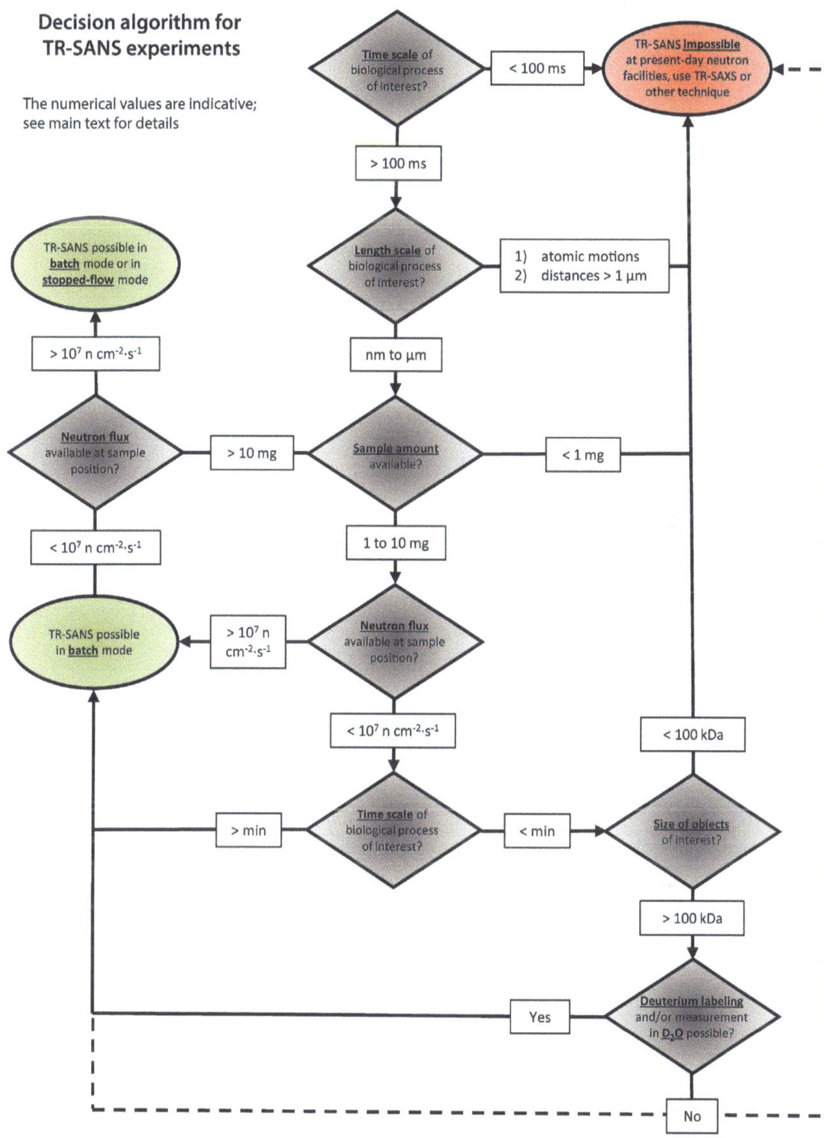

Fig. 3 Flow chart for the design and execution of TR-SANS experiments. The numerical values (times, length-scales, neutron fluxes, sample amounts and sizes of objects) are guidelines and indicate roughly the order of magnitude of the parameters involved in the respective choices. For a more detailed discussion, see the main text. The broken arrows indicate that both possibilities exist. We recommend strongly to discuss the feasibility of TR-SANS projects with the instrument scientists and local contacts at an early stage of the project (i.e., prior to proposal writing and submission).

$R_G \leq 0.5/q_{min}$ and $D_{max} \leq \pi/q_{min}$ (Koch, Vachette, & Svergun, 2003). With a $q_{min} \approx 0.001\,\text{Å}^{-1}$, a maximum dimension of $\approx 3000\,\text{Å}$ can be reliably determined. q_{max}, on the other hand, determines the minimum size of structural features that can be observed. In the case of periodic structures (e.g., stacked membrane lamella, or repetitive distances in protein α-helices or β-sheets), a minimum measureable distance L can be estimated. Indeed, a periodicity of distance L will give rise to a Bragg peak $q_{Bragg} = 2\pi/L$ in a diffraction pattern. For $L \approx 50\,\text{Å}$ (thickness of a membrane bilayer) $q_{Bragg} \approx 0.13\,\text{Å}^{-1}$ is expected, while for $L \approx 5.4\,\text{Å}$ (α-helix pitch) $q_{Bragg} \approx 1.2\,\text{Å}^{-1}$. While the former peak position is readily accessible in standard SANS q-ranges, the latter is beyond the q_{max} of usual instrumental setups. (Note that the numerical values indicated here are strictly valid only for *infinite periodic* structures and differ by a factor of ≈ 2 from the values discussed in the next paragraph and shown in Fig. 2B. The latter values are obtained from the position of the first minimum of the Debye function (Eq. 3) for two point-like objects. If the first subsidiary maximum of the Debye function is considered, the values are in better agreement with those obtained the Bragg law).

2) Generally, two isolated, point-like scattering elements, possessing contrast with the buffer, and being separated by a distance r, will give rise to a SANS pattern according to the Debye equation (Debye, 1915) (see examples in Fig. 2B). Neglecting subtle effects such as hydration shells (Kim & Gabel, 2015; Svergun et al., 1998), a SANS pattern from a complex biological sample in solution can be approximated by the square of the sum over all pairs of scattering centers (atomic nuclei in the case of neutrons), weighted by a $\sin(q \cdot r)/q \cdot r$ term and their respective contrasts (Eq. 3). The four Debye functions shown in Fig. 2B for isolated pairs of point-like scattering elements, separated by distances of 1, 10, 50 and 300 Å, respectively, illustrate where structural information from such distances is encoded in the SANS curves: the q-ranges from $0\,\text{Å}^{-1}$ to the shaded arrows correspond to regions where the scattered intensities display a significant *relative* variation, and are at the same time sufficiently strong to lie above the typical background levels from incoherent scattering (Gabel et al., 2002). The superposition of these distances with two representative biological structures (a protein-RNA complex and a schematic cross-section of an *Escherichia coli* membrane, Fig. 2B) illustrates the contributions expected for typical structural features/distances in the two biological systems. Importantly, structural features separated by very short distances ($\approx 1\,\text{Å}$), although they scatter over a wide q-range, do not yield significant relative variations of the SANS curves in the q-range shown and can therefore not be "resolved" by this technique.

3) Apart from the general experimental and theoretical limitations discussed in points (1) and (2), another practical requirement needs to be fulfilled in order for SANS data to be exploitable: the difference intensity between sample and buffer, measured in a given q-range during an exposure time Δt, must be strong enough, and above the incoherent background level, in order to be structurally interpretable. Forward scattered intensities, I(0), can be calculated for certain biomacromolecular systems such as proteins and RNA/DNA, as a function of their sequences, labeling, concentration, and contrast, by using software (Whitten, Cai, & Trewhella, 2008) and/or online tools (https://smb-research.smb.usyd.edu.au/NCVWeb/, http://psldc.isis.rl.ac.uk/Psldc/). If an atomic model is available, the entire SANS curve can be calculated using software programs, including CRYSON (Svergun et al., 1998) and Pepsi-SANS (https://pepsi.app.ill.fr).
4) Usually, a coherent I(0) intensity higher than $\approx 0.01\,\text{cm}^{-1}$ is required for an accurate extraction of basic structural parameters such as R_G. For low scattering intensities recorded at high time-resolution (i.e., short Δt), the kinetics of the experiment may have to be repeated and measured several times (Roessle & May, 2006) and the sample amount available can become a limitation

When interpreting results from TR-SANS experiments, it is important to acknowledge that the structural information is averaged on two levels: first, experimental SANS curves are obtained from typical sample volumes of some 100 μL, corresponding to $>10^{16}$ macromolecules (e.g., for proteins at ≈mg/mL concentrations), and are therefore conformationally averaged over a large thermodynamic ensemble. Second, even for relative short exposure times (≈seconds), the structural information is averaged over the duration Δt of each individual time frame. While both processes lead to a certain loss of information from individual particles, TR-SANS can provide a wealth of information from a diverse range of biological systems, as illustrated by the examples presented in the following section.

2. Examples from diverse classes of biological systems

Over the past 25 years, a broad variety of biological systems has been investigated by TR-SANS. The kinetics of **protein aggregation and fibrillation** is particularly suited for time-resolved SANS studies due to the strong scattering signal from extended and voluminous structures (Eq. 1), as well as the large changes of the aggregation numbers involved in these processes. Both processes yield readily measureable changes of both I(0) and R_G. Moreover, the absence of neutron radiation damage in biological samples (Jasnin et al., 2008; Marty et al., 2013) allows to expose the same sample over several hours and monitor the evolution of structural features.

Earliest work includes studies of the nucleation and aggregation processes of supersaturated (10–60 mg/mL) lysozyme solutions with a time-resolution of 1–2 h (Boué, Lefaucheux, Robert, & Rosenman, 1993; Niimura, Minezaki, Ataka, & Katsura, 1995). The elevated protein concentrations in D_2O buffers yielded excellent signal/noise ratios of the experiments and the lower Q-limits of 0.01–0.002 $Å^{-1}$ allowed to cover an extended range of length-scales. The effect of salt (up to 2 M NaCl) on amyloid formation was equally studied by TR-SANS (time resolution 1 h) for lower concentrations of lysozyme in solution (<10 mg/mL) in deuterated ethanol-water buffers (Fujiwara, Matsumoto, & Yonezawa, 2003).

More recently, β-lactoglobulin fibrillation has been studied in a number of buffer conditions, usually in D_2O, with best time-resolutions of 15 min (Arnaudov, de Vries, & Cohen Stuart, 2006; Bolisetty, Adamcik, & Mezzenga, 2011; Jones et al., 2011; Yoshida et al., 2019). Other systems include studies on insulin (Groenning, Frokjaer, & Vestergaard, 2009; Nayak, Sorci, Krueger, & Belfort, 2009) and α-synuclein (Bhak et al., 2018; Rekas, Lo, Gadd, Cappai, & Yun, 2009). A more exhaustive overview of fibrillation processes studied by TR-SANS (and SAXS) can be found elsewhere (Marasini & Vestergaard, 2017).

Association and aggregation phenomena other than fibrillation can also be observed by TR-SANS. Examples include $CaCO_3$ nucleation and growth, mediated by lysozyme (Pipich, Balz, Wolf, Tremel, & Schwahn, 2008), and liquid-liquid phase separation of proteins at very small angles (VSANS), as illustrated by a recent study of concentrated BSA solutions in YCl_3 salts (Da Vela et al., 2016).

A second class of systems, at the interface of biology and soft matter, that are routinely studied by TR-SANS are **membrane structures and dynamics**. Examples include lipid exchange processes in vesicles (Nakano et al., 2009; Nakano, Fukuda, Kudo, Endo, & Handa, 2007), the effect of phospholipidase on liposomes (Ferguson, De Luca, Heenan, King, & Griffiths, 2010), liposome formation in bile salts (Nawroth, Buch, Buch, Langguth, & Schweins, 2011), lipid transfer between solid and free-standing membranes (Wah et al., 2017), lipid exchange between human lipoproteins and naturally derived membranes (Maric et al., 2019) or *co*-polymers (Cuevas Arenas et al., 2017) and the dynamic interaction of peptides with *Escherichia coli* mimicking membranes (Nielsen, Prévost, Jenssen, & Lund, 2021). More exhaustive reviews on TR-SANS on lipid systems and other systems at the interface with soft matter are available elsewhere (Mata, Hamilton, & Gilbert, 2012; Perez-Salas, Garg, Gerelli, & Porcar, 2021; Urban, Heller, Katsaras, & Bras, 2021).

A third class of systems that can be readily studied by TR-SANS are **cellular ultra-structures**. Biological structures at the mesoscale (nm to μm), situated between the size of individual bio-macromolecules and whole cells, are perfectly compatible with length-scales accessible to SANS, and their temporal evolution, often as a function of external stimuli, has been studied by this technique. Several SANS studies on thylakoids report on the changes of orientation and evolution of inter-membrane distances as a function of external stimuli (e.g., light illumination), on the second to minute time-scale (Nagy et al., 2011, 2013; Ünnep et al., 2014). Another SANS study has investigated the dynamics of plasma-membrane heterogeneities in living *Bacillus subtilis* bacteria (Nickels et al., 2017). The impact of chemical and thermal treatment on **cellulose and lignin biomasses** has been characterized successfully using TR-SANS and specialized sample environment (Nishiyama, Langan, O'Neill, Pingali, & Harton, 2014; Pingali et al., 2014; Pingali Sai et al., 2020; Yang et al., 2022).

Finally, it is possible to study the **conformational kinetics of individual bio-macromolecular structures and assemblies** by TR-SANS. This approach is more challenging than the three previous ones, mainly due to the smaller sizes of bio-macromolecules (molecular weights typically $\approx$10 to a few 1000 kDa), the absence of periodic structural features and lower concentrations (typically a few mg/ml) with respect to fibrils, membrane structures and cellular ultra-structures. Both factors result in an important reduction of the I(0) intensity by two (or more) orders of magnitude and require the use of high-flux neutron sources if biological processes on short ($<$ 10 min) time-scales are investigated (see Fig. 3). Moreover, if a sophisticated and accurate interpretation of the structural dynamics of individual bio-macromolecules is sought, sample purity and monodispersity is a paramount issue (Jacques & Trewhella, 2010; Jeffries et al., 2016). Importantly, the biological process/reaction to be observed needs to be triggered and be in phase with the time-scale of the SANS experiment.

A pioneering TR-SANS study of the assembling process of an individual bio-macromolecular complex was carried out on the ≈800 kDa GroEL-GroES chaperonin from the archaeon *Thermoplasma acidophilum* on the D22 instrument at ILL Grenoble, France (Gutsche et al., 2000; Gutsche, Holzinger, Rauh, Baumeister, & May, 2001). The experimental setup used a stopped-flow device with filling, rinsing and emptying steps of a 1 mm thick quartz cell which contained the sample (Roessle & May, 2006). Individual 1 s exposure frames were recorded during 1 h. Subsequently, the same experiment was repeated 10 times under identical conditions, and respective 1 s time frames were binned and averaged. This procedure yielded very satisfactory statistics of the SANS curves (3×10^7 $n \cdot s^{-1} \cdot cm^{-2}$ at $\lambda = 8$ Å and 4 m collimation with a $55 \times 40\,mm^2$ cross section). Measuring hydrogenated GroES and per-deuterated GroEL at concentrations 2–5 mg/mL in 100% D_2O buffers in the presence of 30 mM AMP-PNP at 50 °C, the binding kinetics of two GroES units to GroEL was monitored via the evolution of the radius of gyration, which displayed a double exponential behavior.

More recently, TR-SANS was used to monitor the structural dynamics of the archaeal PAN-proteasome complex from *Methanocaldococcus jannaschii* and a GFP substrate during the active unfolding and degradation process (Ibrahim et al., 2017; Mahieu, Covès, et al., 2020). By measuring deuterated (d) GFP (28 kDa, 2 mg/mL) and hydrogenated (h) PAN unfoldase (298 kDa, 10 mg/mL) in the presence of ATP in a 42% D_2O buffer at 55 °C, the authors could follow the unfolding and aggregation of the deuterated substrate at a time-resolution of 30 s, with the PAN signal masked. Likewise, and applying an inverse deuteration scheme (h-GFP, d-PAN), the signal from the GFP aggregates could be masked in 42% D_2O, and a reversible conformational contraction of PAN, during the active unfolding process, could be characterized in terms of changes of R_G and the pair-distance distribution function p(r) (Glatter, 2018), at a time-resolution of 30 s (Ibrahim et al., 2017). When h-PAN and d-GFP were measured at 55 °C in 42% D_2O the presence of the hydrogenated proteasome core particle (751 kDa, 16 mg/mL) and ATP, no substrate aggregation was detected. Rather, the time-course of different GFP substrate populations (native state and digested oligo-peptide products) could be followed at a time resolution of 30 s while masking the signal from the much larger PAN-proteasome complex. A putative, partially unfolded substrate intermediate state as evidenced by Cryo-EM studies (de la Peña, Goodall, Gates, Lander, & Martin, 2018), could not be detected as an appreciable (>5%) population at any time during the reaction. Together with the absence of even minor amounts of aggregated GFP, the authors could propose a mechanistic model of GFP degradation in solution by the PAN-proteasome complex (Mahieu, Covès, et al., 2020).

It is important to stress that the sophisticated analysis of a small product (GFP) in the presence of a large macromolecular machine (PAN-20S), or the large machine in the presence of a substrate aggregate would have been completely impossible by SAXS since the signal of the part of interest would have been dominated by the other components (Mahieu et al., 2020). SANS, combined with deuteration and contrast variation, therefore allows to "hide the elephant in the room" (Gupta, 2020). The PAN-proteasome work, as well as future perspectives, are presented and discussed in detail in a separate chapter in volume B (see chapter "Observing protein degradation in solution by the PAN-20S proteasome complex: A state-of-the-art example of bio-macromolecular TR-SANS" by Gabel in Methods in Enzymology, vol. 678).

The PAN-proteasome project demonstrated that good quality SANS curves can realistically be obtained at high-flux neutron sources for relatively small ($\approx$ 30 kDa) and low concentration ($\approx$2 mg/mL) per-deuterated proteins in the sub-minute range, even at moderate contrast (i.e., in 42% D_2O). At the other extreme, SANS, deuteration and contrast variation can also be used to monitor biological processes with characteristic times of hours or even days, like slow structural changes in large macromolecular complexes involved in circadian clock processes (Murakami et al., 2019; Sugiyama et al., 2016; Yunoki et al., 2022) or subunit exchange in oligomeric αB crystalline complexes from the eye (Inoue et al., 2016, 2021).

3. Considerations on sample state and practical guidelines for the design of TR-SANS experiments

SANS instruments are not standardized industrial products but each beamline is created and evolves over the years as a function of the local environment at the respective neutron facility, its designated mission, and the requirements of the user community. While a few beamlines are exclusively dedicated to biological experiments (Heller et al., 2014), the majority share their beam-time with other disciplines, including chemistry, material sciences, magnetism, and soft matter. As a consequence, there exists a broad range of instruments with versatile characteristics (https://www.ill.eu/users/scientific-groups/large-scale-structures/more) and it is difficult to indicate a single, standardized protocol for TR-SANS experiments. Moreover, the instrumental setup and the sample environment will depend, among others, on the time- and length-scales of the biological process to be studied (see sections above), the availability of sample amount and deuterium labeling, the performance of the beamline, and the total beam-time allocated for the scientific project. We therefore propose to present general guidelines for the design and setup of TR-SANS experiments in the form of a flow chart (Fig. 3), including check points and decision algorithms as a function of the above-mentioned parameters. In all cases, it is highly recommendable that the neutron users discuss the feasibility and practical details of their experiment with the instrument scientists and/or local contacts at an early stage of the project.

1) *Expected* ***time-scale*** *of the biological process.* SANS instruments feature significantly smaller fluxes with respect to synchrotron SAXS beamlines (e.g., 10^7 n·s^{-1}·cm^{-2} at D22, ILL, vs 10^{13} photons·s^{-1}·cm^{-2} at ID02, ESRF). While certain mechanical restraints (e.g., shutter motions) limit individual exposure times, the major limitation of time resolution is determined by the relatively broad wavelength resolution of the incident neutrons ($\approx$10%) (Bee, 1988; Svergun et al., 2013). This distribution results in different neutron flight times and consecutive frames need to be separated by a minimum of $\approx$10 ms (for $\lambda \approx 6$ Å and sample-detactor distance of 1.4 m) to clearly separate slow neutrons scattered during frame i from fast neutrons scattered during frame i + 1. At larger sample-detector distances (8 m) and better resolution (2%), this time can be shorter (2 ms), at the expense of flux (Radulescu et al., 2016). For longer exposure times, the wavelength spread can be released (to $\approx$20%) to increase flux. In practice, the best time resolution reachable by current TR-SANS is around 100 ms. Biological processes that occur on shorter time-scales are not accessible to TR-SANS, and other techniques are required to study them, including TR-SAXS (Blanchet et al., 2015) (see chapter "Stopped-flow-time-resolved SAXS for studies of ligand-driven protein dimerization" by Tidow and Josts). On the opposite end, the duration of a SANS experiment of a single sample is typically limited to 24 h (or a few days), and slower biological processes are not accessible. However, under particular circumstances, it is in principle conceivable to schedule a series of individual SANS beam-time slots over several weeks or months, depending on availability at the respective neutron facility. In such a setup, a biological sample could be left to evolve "offline," and occasionally be exposed to neutrons at regular intervals throughout this period. A prerequisite of this approach is of course the sample stability over such long periods.

2) *Expected* ***length-scale*** *of the biological process.* As discussed in the *Introduction, principles and limitations* section, there is a limited range of distances to which SANS is sensitive. Maximum distances >3000 Å and periodic features <5 Å are not accessible to SANS, due to limitations of the q-range by the beam-stop and the detector size, respectively. For structural features larger than several thousand Å, microscopic approaches can be used, while atomic resolution techniques (crystallography Cryo-EM and NMR) are available for very short distances of a few Å. It should be noted, however, that SANS curves are an integration over all distances found in a sample (Eq. 3). Thus, even though the maximum distances in objects larger than several thousand Å are not accessible, partial information on structural features at shorter distances within such extended objects can still be obtained, if they differ in contrast (Fig. 2B). Likewise, while Bragg peaks originating from periodic features of a few Å are outside the accessible q-ranges, small conformational changes in non-periodic structures (e.g., proteins) can still be accessed in SANS. A more comprehensive discussion of these aspects and a comparison with crystallography can be found elsewhere (Putnam, Hammel, Hura, & Tainer, 2007).
3) *Minimum* ***sample amount*** *required.* The sample size and volume in SANS experiments are mainly determined by two parameters: on the one hand, the shape and area of the focused neutron beam, which is typically ≈0.1 to 3 cm^2 (Glatter, 2018; Svergun et al., 2013), and on the other hand the thickness of SANS samples, which is usually limited to ≈2 mm due to weak transmission and multiple scattering, in particular for samples containing H_2O in the buffer (Carsughi, May, Plenteda, & Saroun, 2000). Both factors result in typical sample volumes of a few hundred μL. Depending on the size of the biological objects of interest and their contrast (see next point), concentrations of at least ≈1 mg/mL are required to obtain good quality SANS curves in some seconds (at the best high-flux neutron facilities) or several minutes (at medium- and low-flux neutron facilities). Thus, a **minimum sample amount of ≈1 mg of biomaterial** is in general required for carrying out a TR-SANS experiment successfully in a static batch mode (i.e., by illuminating the same sample continuously over a longer period) (Fig. 3). Exposed sample volumes in TR-SANS are about 2 orders of magnitude larger than those exposed during individual frames in SAXS experiments at high-brilliance synchrotron beamlines (Blanchet et al., 2015) (see chapter "Stopped-flow-time-resolved SAXS for studies of ligand-driven

protein dimerization" by Tidow and Josts). However, the absence of radiation damage in neutron scattering (Jasnin et al., 2008; Marty et al., 2013) allows to expose the same sample over several hours, or even days. This specific advantage of SANS compensates for the smaller sample volumes exposed in a single SAXS frame, but which has to be replaced by a new sample volume for each frame in a TR-SAXS series.

4) *Importance of sample* ***contrast***. Biological structures can only be studied by SANS if their neutron scattering length density (SLD) is different from the one of the solvent, i.e., if they have a contrast $\Delta\rho$. Luckily, SLDs of most biomolecules in their natural hydrogenated (i.e., non-deuterated) state differ and, in addition, can be matched by adjusting the H_2O:D_2O ratio of the buffer solution (Fig. 2A). Thus, it is possible to distinguish structural features of different types of biomolecules within large and complex assemblies (Jacrot, 1976; Mahieu & Gabel, 2018; Neylon, 2008; Zaccai & Jacrot, 1983). In the case of TR-SANS experiments, it is important to be aware that the scattered intensity of a given sample volume of a biological molecule is proportional to its contrast *squared*, i.e., $(\Delta\rho)^2$ (Eq. 1). According to this relationship, a hydrogenated protein scatters four times as strongly in a 0% D_2O buffer than in 20% D_2O, but only half as strongly as in 100% D_2O (see Fig. 2A). (In this context it is important to note that both positive and negative contrast yield a positive scattering intensity). Moreover, the incoherent background signal from the buffer is also reduced by over an order of magnitude in D_2O with respect to H_2O (Gabel et al., 2002), and the ratio from the coherently scattered intensities from biomolecules with respect to the background significantly improved. It is thus clear that changing the solvent contrast can either result in a significant improvement of signal/noise of the SANS curves or in a reduction of exposure times by an order of magnitude, without loss of quality of the TR-SANS experiments

5) *Need and impact of* ***deuterium labeling***. Biomolecular deuteration offers another possibility to vary contrast and signal from biological samples in SANS experiments, in addition to solvent contrast variation (i.e., H_2O/D_2O buffer exchange). First, biomolecular deuteration can increase the coherent intensity by a factor of up to 7–8 (in the case of proteins in H_2O, Fig. 2A), and therefore improves the statistics and time-resolution of TR-SANS experiments considerably. More importantly, however, biomolecular deuteration allows to **distinguish molecules of similar chemical composition**, e.g., two different proteins. Such a distinction is crucial to obtain insight into some biological processes involving two,

or more, protein partners. A recent illustration of this approach is specific protein degradation (Ibrahim et al., 2017; Mahieu, Covès, et al., 2020), and is discussed in more detail in a chapter in volume B (see chapter "Observing protein degradation in solution by the PAN-20S proteasome complex: A state-of-the-art example of bio-macromolecular TR-SANS" by Gabel in Methods in Enzymology, vol. 678). The possibility to distinguish between different biomolecules of similar (average) composition by deuteration is a *unique* strength of neutrons with respect to X-rays. Indeed, while electron-rich labels have been added to different kinds of biomolecules in SAXS experiments (Grishaev, Anthis, & Clore, 2012; Zettl et al., 2016) and allow to obtain specific distance restraints, it is hard to conceive a technique that modifies the electron densities (the equivalent of neutron SLD) of bio-macromolecules *homogeneously*, and without impacting their function. Whether or not deuteration is required to obtain specific structural information should be evaluated by the experimenters when planning and designing a TR-SANS experiment (Fig. 3). Subsequently, the practical feasibility of preparing deuterated material in sufficient amounts and quality needs to be evaluated. While several small molecules (e.g., detergents and lipids) are commercially available in deuterated form, biological macromolecules (e.g., proteins and RNA) are usually deuterated by adapted protocols using recombinant expression or transcription in vitro. An overview of available protocols can be found elsewhere (Haertlein et al., 2016) (see chapter "Deuteration for biological SANS: Case studies, success and challenges in chemistry and biology" by Duff et al.). It is important to check the yield, purity and monodispersity of the deuterated samples as compared to standard protocols of the hydrogenated samples by laboratory techniques (SEC-MALLS, AUC, etc.) (Jacques & Trewhella, 2010; Jeffries et al., 2016), if possible also in D_2O buffers which are known to favor oligomerization. Several neutron centers propose deuteration services in associated facilities ("deuteration laboratories") and should be contacted when planning TR-SANS experiments.

6) *Choice of* ***triggering mechanism*** *and experimental setup.* TR-SANS experiments can essentially be carried out in two different ways: either by using mixing devices such as stopped-flow and exposing a sample volume at a specific position, corresponding to a specific time of the reaction of interest, or by exposing a static batch sample volume continuously during specific time periods (frames). While the first approach is common with TR-SAXS (Grillo, 2009; Jain et al., 2013), the second one is unique to TR-SANS due to the absence of notable radiation damage in biological samples. Both approaches require the triggering of the biological reaction (and expected concomitant structural changes) by a specific process compatible and in phase with the time-scale of the structural changes and the duration of individual neutron exposure frames. Manual mixing of reactants directly in SANS-measurement cell is the most convenient way of starting a kinetics because it is adapted to a wide range of reactions. However, it is limited to slow kinetics as the "dead time" related to closing the instrument zone is >10 s. Certain experiments can be triggered by mixing reactants via a stopped-flow equipment (Grillo, 2009). This method is appropriate to the analysis of the first 100 ms to 1 s of an experiment but very often, the very good time-resolution is balanced with poor data statistics. To improve data statistics, if sample amount allows it, the kinetics is repeated a large number of times and the corresponding exposures are binned (Rössle et al., 2000).

The triggering mode used to start a kinetics is obviously dependent upon the system studied, but it is important to think about it in advance to be technically ready and work in the best possible conditions. Examples of triggering processes already used include the following:

i) *Temperature*: the in situ unfolding process described in detail in a complementary book chapter (see chapter "Observing protein degradation in solution by the PAN-20S proteasome complex: A state-of-the-art example of bio-macromolecular TR-SANS" by Gabel in Methods in Enzymology, vol. 678) could be triggered by temperature thanks to the use of an enzyme originating from a hyperthermophile organism. At room temperature and on ice, the enzyme was inactive but as soon as the reaction cell was placed in pre-heated sample holder ($\approx$55–60 °C), the reaction started within seconds. In very specific samples temperature jumps can be triggered by magnetic fields (Reimhult, Schroffenegger, & Lassenberger, 2019).

ii) *Pressure*: the specific effect of high pressure on the conformational changes of biological samples has been shown in a SANS study on myoglobin (Loupiac, Bonetti, Pin, & Calmettes, 2002). The advantages of this technique with respect to others, in particular SAXS, has been reviewed recently in the special context of food studies (Teixeira, 2019).

iii) *Light and other spectroscopy*: certain reactions can be triggered by light and have been measured by SANS. For instance, the dimerization of the transcription factor EL222 (Chaudhari et al., 2021) or thylakoid membrane alignment (Ünnep et al., 2014, 2020). This triggering is very system-specific and can be tuned in wavelength and intensity. It can also be coupled with other complex sample environments, as in situ size-exclusion chromatography or magnetic fields in the two examples cited above.

iv) *Microfluidics*: mixing using microfluidics systems is very convenient as it translates mixing time into distance along the mixing channel. But due to the small exposed sample volume, this approach for the moment limited to strongly scattering systems (Adamo et al., 2017). Microfluidics can also be used to apply high shear rate to a solution and study the relaxation time of the structures upon exiting the high shear region (Lopez et al., 2018).

It is important to recall that the structural information obtained from a single SANS frame is integrated and averaged both over a thermodynamic ensemble of biological molecules and the duration of the neutron exposure. A triggering signal should therefore affect the entire sample volume homogeneously and completely, and its duration should typically be chosen shorter than a single neutron frame. Finally, and as stated further above, the characteristic time-scale of structural changes to be studied should at least be an order of magnitude longer than the duration of a single neutron frame.

4. Conclusions and future perspectives

The recent advent of accurate and high-throughput prediction of atomic structures in silico is revolutionizing the field of structural biology (Baek et al., 2021; Jumper et al., 2021), and impacts particularly the high-resolution techniques crystallography, NMR and Cryo-EM. While in silico structure predictions still has some shortcomings and the majority of applications are focused on isolated, compact, structures (Perrakis & Sixma, 2021), it can be expected that this approach will be applied increasingly to describe interactions between different macromolecules in the near future.

In the context of this structural biology revolution, small-angle scattering (SAXS/SANS), and in particular time-resolved studies, have a role to play. Rather than perceiving in silico structure prediction as a competitor, SAXS/SANS should focus on the benefits of being able to use the atomic models as an input to analyze complex experimental situations, e.g., mixtures of several protein species, oligomeric states and conformations.

In complement to high-resolution experimental techniques (crystallography, Cryo-EM and NMR) and in silico model predictions, TR-SANS (and SAXS) studies can contribute to understand the dynamics and complexity of transiently interacting systems but also explore the "neglected" scale of cellular supramolecular and ultra-structures (Martínez-Sanz, Gidley, & Gilbert, 2015; Semeraro, Marx, Frewein, & Pabst, 2021). The state-of-the-art of current experimental TR-SANS capacities to study transient, bio-macromolecular interactions, are presented and discussed in terms of time-resolution and details of structural analysis in a complementary book chapter in volume B (see chapter "Observing protein degradation in solution by the PAN-20S proteasome complex: A state-of-the-art example of bio-macromolecular TR-SANS" by Gabel in Methods in Enzymology, vol. 678).

Several encouraging developments in biological TR-SANS experiments are currently underway and progress can be expected in the following areas over the next years:

i) *Instrumentation and new sources*: New SANS instruments are being developed at the European Spallation Source (ESS) in Lund, Sweden (Kynde et al., 2014), and being designed for the second target station at the Spallation Neutron Source (SNS) in Oak Ridge, Tennessee (https://neutrons.ornl.gov/sts/sts-instrument-systems). These instrument developments aim at reducing the sample amount required and the range of time-scales accessible. Recent developments include also the availability of fast, multi-megahertz neutron detectors (Houston et al., 2018). For a more comprehensive review, focused on soft matter applications, see (Urban et al., 2021). Besides existing large neutron facilities, there are initiatives for the construction of small, compact high-brilliance neutron sources which should assure a broader availability and geographical spread of this technique (Gutberlet, Rücker, & Brückel, 2017; Rücker et al., 2016).

ii) *Sample environment and data analysis*: Recently, several new sample environments have been developed. They include electric or magnetic field devices allowing to induce temperature jumps in samples containing ions in situ (Hayward et al., 2021; Reimhult et al., 2019) and continuous stepped-flow device allowing to trigger and resolve chemical reactions down to 1 s (Hayward, Chiappisi, Prévost, Schweins, & Gradzielski, 2018) (both developed D11at ILL, France). In situ DLS and FTIR are available at the KWS-2 instrument (JCNS, Germany) (Balacescu, Brandl, Kaneko, Schrader, & Radulescu, 2021). The D22 instrument at ILL features now the possibility to measure SAXS data in parallel to neutrons, designed for recording kinetics in a single experiment (Metwalli et al., 2020, 2021). While first results on gold nanoparticles have been obtained (Zech et al., 2022) the current X-ray flux is unfortunately still too low to measure weakly scattering biological samples. Finally, the "event-mode" data processing (Peterson, Campbell, Reuter, Taylor, & Zikovsky, 2015) facilitates significantly the analysis of fast kinetics of strongly scattering objects (e.g., metal alloy and nanoparticles), by enabling to adapt the experimental time resolution to the actual kinetics after the data acquisition (Ioannidou et al., 2021; Kelley, Nguyen, Marquardt, Maranville, & Murphy, 2021). The development of data treatment capabilities is still required to enable its use for longer kinetics and the lower scattering power of biological molecules.

Acknowledgment

Financial support from the "Agence Nationale de la Recherche" (Grant "PROTstretch" ANR-15-CE11-0026) is acknowledged.

References

Adamo, M., Poulos, A. S., Miller, R. M., Lopez, C. G., Martel, A., Porcar, L., et al. (2017). Rapid contrast matching by microfluidic SANS. *Lab on a Chip*, *17*(9), 1559–1569. https://doi.org/10.1039/C7LC00179G.

Arnaudov, L. N., de Vries, R., & Cohen Stuart, M. A. (2006). Time-resolved small-angle neutron scattering during heat-induced fibril formation from bovine β-lactoglobulin. *The Journal of Chemical Physics*, *124*(8), 084701. https://doi.org/10.1063/1.2171418.

Baek, M., DiMaio, F., Anishchenko, I., Dauparas, J., Ovchinnikov, S., Lee Gyu, R., et al. (2021). Accurate prediction of protein structures and interactions using a three-track neural network. *Science*, *373*(6557), 871–876. https://doi.org/10.1126/science.abj8754.

Balacescu, L., Brandl, G., Kaneko, F., Schrader, T. E., & Radulescu, A. (2021). Light scattering and absorption complementarities to neutron scattering: In situ FTIR and DLS techniques at the high-intensity and extended Q-range SANS diffractometer KWS-2. *Applied Sciences*, *11*(11). https://doi.org/10.3390/app11115135.

Bee, M. (1988). Quasielastic neutron scattering. In *Principles and applications in solid state chemistry, biology and materials science*. United Kingdom: Adam Hilger,Bristol, UK.

Bhak, G., Lee, S., Kim, T.-H., Lee, J.-H., Yang, J. E., Joo, K., et al. (2018). Morphological evaluation of Meta-stable oligomers of α-synuclein with small-angle neutron scattering. *Scientific Reports*, *8*(1), 14295. https://doi.org/10.1038/s41598-018-32655-0.

Blanchet, C. E., Spilotros, A., Schwemmer, F., Graewert, M. A., Kikhney, A., Jeffries, C. M., et al. (2015). Versatile sample environments and automation for biological solution X-ray scattering experiments at the P12 beamline (PETRA III, DESY). *Journal of Applied Crystallography*, *48*(2), 431–443. https://doi.org/10.1107/S160057671500254X.

Bolisetty, S., Adamcik, J., & Mezzenga, R. (2011). Snapshots of fibrillation and aggregation kinetics in multistranded amyloid β-lactoglobulin fibrils. *Soft Matter*, 7(2), 493–499. https://doi.org/10.1039/C0SM00502A.

Boué, F., Lefaucheux, F., Robert, M. C., & Rosenman, I. (1993). Small angle neutron scattering study of lysozyme solutions. *Journal of Crystal Growth*, *133*(3), 246–254. https://doi.org/10.1016/0022-0248(93)90161-O.

Breyton, C., Gabel, F., Lethier, M., Flayhan, A., Durand, G., Jault, J.-M., et al. (2013). Small angle neutron scattering for the study of solubilised membrane proteins. *The European Physical Journal E*, *36*(7), 71. https://doi.org/10.1140/epje/i2013-13071-6.

Carsughi, F., May, R. P., Plenteda, R., & Saroun, J. (2000). Sample geometry effects on incoherent small-angle scattering of light water. *Journal of Applied Crystallography*, *33*(1), 112–117. https://doi.org/10.1107/S0021889899013643.

Chaudhari, A. S., Semanat, E. C., Martel, A., Schneider, B., & Fuertes, G. (2021). Light-induced oligomerization of the transcription factor EL222. *Acta Crystallographica Section A*, 77(a2), C868. https://doi.org/10.1107/S0108767321088309.

Cuevas Arenas, R., Danielczak, B., Martel, A., Porcar, L., Breyton, C., Ebel, C., et al. (2017). Fast collisional lipid transfer among polymer-bounded nanodiscs. *Scientific Reports*, 7(1), 45875. https://doi.org/10.1038/srep45875.

Da Vela, S., Braun, M. K., Dörr, A., Greco, A., Möller, J., Fu, Z., et al. (2016). Kinetics of liquid–liquid phase separation in protein solutions exhibiting LCST phase behavior studied by time-resolved USAXS and VSANS. *Soft Matter*, *12*(46), 9334–9341. https://doi.org/10.1039/C6SM01837H.

de la Peña, A. H., Goodall, E. A., Gates, S. N., Lander, G. C., & Martin, A. (2018). Substrate-engaged 26S proteasome structures reveal mechanisms for ATP-hydrolysis–driven translocation. *Science*, *362*(6418), eaav0725. https://doi.org/10.1126/science.aav0725.

Debye, P. (1915). Zerstreuung von Roentgenstrahlen. *Annalen der Physik (Leipzig)*, *28*, 809–823.

Ferguson, E. L., De Luca, E., Heenan, R. K., King, S. M., & Griffiths, P. C. (2010). Time-resolved small-angle neutron scattering as a tool for studying controlled release from liposomes using polymer-enzyme conjugates. *Macromolecular Rapid Communications*, *31*(19), 1685–1690. https://doi.org/10.1002/marc.201000241.

Fujiwara, S., Matsumoto, F., & Yonezawa, Y. (2003). Effects of salt concentration on Association of the Amyloid Protofilaments of hen egg white lysozyme studied by time-resolved neutron scattering. *Journal of Molecular Biology*, *331*(1), 21–28. https://doi.org/10.1016/S0022-2836(03)00722-8.

Gabel, F., Bicout, D., Lehnert, U., Tehei, M., Weik, M., & Zaccai, G. (2002). Protein dynamics studied by neutron scattering. *Quarterly Reviews of Biophysics*, *35*(4), 327–367.

Glatter, O. (2018). *Scattering methods and their application in colloid and Interface sciences*. Elsevier.

Grillo, I. (2009). Applications of stopped-flow in SAXS and SANS. *Current Opinion in Colloid & Interface Science*, *14*(6), 402–408. https://doi.org/10.1016/j.cocis.2009.04.005.

Grishaev, A., Anthis, N. J., & Clore, G. M. (2012). Contrast-matched small-angle X-ray scattering from a heavy-atom-labeled protein in structure determination: Application to a lead-substituted calmodulin-peptide complex. *Journal of the American Chemical Society*, *134*(36), 14686–14689. https://doi.org/10.1021/ja306359z.

Groenning, M., Frokjaer, S., & Vestergaard, B. (2009). Formation mechanism of insulin fibrils and structural aspects of the insulin fibrillation process. *Current Protein and Peptide Science*, *10*(5), 509–528. https://doi.org/10.2174/138920309789352038.

Guinier, A. (1939). La diffraction des Rayons X aux Tres Faibles Angles: Applications a l'Etude des Phenomenes Ultra-microscopiques. *Annales de Physique (Paris)*, *12*, 161–236.

Gupta, K. (2020). Hiding the elephant in the room with experimental neutrons. *Biophysical Journal*, *119*(2), 234–235. https://doi.org/10.1016/j.bpj.2020.05.038.

Gutberlet, T., Rücker, U., & Brückel, T. (2017). Towards compact accelerator driven Neutronsources for Europe. *Neutron News*, *28*(3), 20–25. https://doi.org/10.1080/10448632.2017.1342486.

Gutsche, I., Holzinger, J., Rauh, N., Baumeister, W., & May, R. P. (2001). ATP-induced structural change of the Thermosome is temperature-dependent. *Journal of Structural Biology*, *135*(2), 139–146. https://doi.org/10.1006/jsbi.2001.4373.

Gutsche, I., Holzinger, J., Rößle, M., Heumann, H., Baumeister, W., & May, R. P. (2000). Conformational rearrangements of an archaeal chaperonin upon ATPase cycling. *Current Biology*, *10*(7), 405–408. https://doi.org/10.1016/S0960-9822(00)00421-8.

Haertlein, M., Moulin, M., Devos, J. M., Laux, V., Dunne, O., & Trevor Forsyth, V. (2016). Chapter five - biomolecular Deuteration for neutron structural biology and dynamics. In Z. Kelman (Ed.), *Vol. 566, Methods in enzymology* (pp. 113–157). Academic Press.

Hayward, D. W., Chiappisi, L., Prévost, S., Schweins, R., & Gradzielski, M. (2018). A small-angle neutron scattering environment for in-situ observation of chemical processes. *Scientific Reports*, *8*(1), 7299. https://doi.org/10.1038/s41598-018-24718-z.

Hayward, D. W., Magro, G., Hörmann, A., Prévost, S., Schweins, R., Richardson, R. M., et al. (2021). A temperature-controlled electric field sample environment for small-angle neutron scattering experiments. *Review of Scientific Instruments*, *92*(3), 033903. https://doi.org/10.1063/5.0040675.

Heller, W. T., Urban, V. S., Lynn, G. W., Weiss, K. L., O'Neill, H. M., Pingali, S. V., et al. (2014). The bio-SANS instrument at the high flux isotope reactor of oak Ridge National Laboratory. *Journal of Applied Crystallography*, *47*(4), 1238–1246. https://doi.org/10.1107/S1600576714011285.

Houston, J. E., Brandl, G., Drochner, M., Kemmerling, G., Engels, R., Papagiannopoulos, A., et al. (2018). The high-intensity option of the SANS diffractometer KWS-2 at JCNS - characterization and performance of the new multi-megahertz detection systemThis article will form part of a virtual special issue on advanced neutron scattering instrumentation, marking the 50th anniversary of the journal. *Journal of Applied Crystallography*, *51*(2), 323–336. https://doi.org/10.1107/S1600576718004132.

Ibrahim, Z., Martel, A., Moulin, M., Kim, H. S., Härtlein, M., Franzetti, B., et al. (2017). Time-resolved neutron scattering provides new insight into protein substrate processing by a AAA+ unfoldase. *Scientific Reports*, *7*, 40948. https://doi.org/10.1038/srep40948.

Inoue, R., Sakamaki, Y., Takata, T., Wood, K., Morishima, K., Sato, N., et al. (2021). Elucidation of the mechanism of subunit exchange in αB crystallin oligomers. *Scientific Reports*, *11*(1), 2555. https://doi.org/10.1038/s41598-021-82250-z.

Inoue, R., Takata, T., Fujii, N., Ishii, K., Uchiyama, S., Sato, N., et al. (2016). New insight into the dynamical system of αB-crystallin oligomers. *Scientific Reports*, *6*, 29208. https://doi.org/10.1038/srep29208. https://www.nature.com/articles/srep29208#supplementary-information.

Ioannidou, C., Navarro-López, A., Dalgliesh, R. M., Rijkenberg, A., Zhang, X., Kooi, B., et al. (2021). Phase-transformation and precipitation kinetics in vanadium micro-alloyed steels by in-situ, simultaneous neutron diffraction and SANS. *Acta Materialia*, *220*, 117317. https://doi.org/10.1016/j.actamat.2021.117317.

Jacques, D. A., & Trewhella, J. (2010). Small-angle scattering for structural biology—Expanding the frontier while avoiding the pitfalls. *Prot Science*, *19*(4), 642–657. https://doi.org/10.1002/pro.351.

Jacrot, B. (1976). The study of biological structures by neutron scattering from solution. *Reports on Progress in Physics*, *39*(10), 911–953.

Jain, R., Petri, M., Kirschbaum, S., Feindt, H., Steltenkamp, S., Sonnenkalb, S., et al. (2013). X-ray scattering experiments with high-flux X-ray source coupled rapid mixing microchannel device and their potential for high-flux neutron scattering investigations. *The European Physical Journal E*, *36*(9), 109. https://doi.org/10.1140/epje/i2013-13109-9.

Jasnin, M., Moulin, M., Haertlein, M., Zaccai, G., & Tehei, M. (2008). Down to atomic-scale intracellular water dynamics. *EMBO Reports*, *9*, 543–547.

Jeffries, C. M., Graewert, M. A., Blanchet, C. E., Langley, D. B., Whitten, A. E., & Svergun, D. I. (2016). Preparing monodisperse macromolecular samples for successful biological small-angle X-ray and neutron-scattering experiments. *Nature Protocols, 11*(11), 2122–2153. https://doi.org/10.1038/nprot.2016.113.

Jeffries, C. M., Ilavsky, J., Martel, A., Hinrichs, S., Meyer, A., Pedersen, J. S., et al. (2021). Small-angle X-ray and neutron scattering. *Nature Reviews Methods Primers, 1*(1), 70. https://doi.org/10.1038/s43586-021-00064-9.

Jones, O. G., Handschin, S., Adamcik, J., Harnau, L., Bolisetty, S., & Mezzenga, R. (2011). Complexation of β-Lactoglobulin fibrils and sulfated polysaccharides. *Biomacromolecules, 12*(8), 3056–3065. https://doi.org/10.1021/bm200686r.

Jumper, J., Evans, R., Pritzel, A., Green, T., Figurnov, M., Ronneberger, O., et al. (2021). Highly accurate protein structure prediction with AlphaFold. *Nature, 596*(7873), 583–589. https://doi.org/10.1038/s41586-021-03819-2.

Kelley, E. G., Nguyen, M. H. L., Marquardt, D., Maranville, B. B., & Murphy, R. P. (2021). Measuring the time-evolution of nanoscale materials with stopped-flow and small-angle neutron scattering. *Journal of visualized experiments : JoVE, 174*. https://doi.org/10.3791/62873 (Accession No. 34424249).

Kim, H. S., & Gabel, F. (2015). Uniqueness of models from small-angle scattering data: The impact of a hydration shell and complementary NMR restraints. *Acta Crystallographica Section D: Biological Crystallography, 71*(Pt 1), 57–66. https://doi.org/10.1107/s1399004714013923.

Koch, M. H., Vachette, P., & Svergun, D. I. (2003). Small-angle scattering: A view on the properties, structures and structural changes of biological macromolecules in solution. *Quarterly Reviews of Biophysics, 36*(2), 147–227.

Kynde, S., Hewitt Klenø, K., Nagy, G., Mortensen, K., Lefmann, K., Kohlbrecher, J., et al. (2014). A compact time-of-flight SANS instrument optimised for measurements of small sample volumes at the European spallation source. *Nuclear Instruments and Methods in Physics Research A, 764*, 133–141.

Lopez, C. G., Watanabe, T., Adamo, M., Martel, A., Porcar, L., & Cabral, J. T. (2018). Microfluidic devices for small-angle neutron scatteringThis article will form part of a virtual special issue on advanced neutron scattering instrumentation, marking the 50th anniversary of the journal. *Journal of Applied Crystallography, 51*(3), 570–583. https://doi.org/10.1107/S1600576718007264.

Loupiac, C., Bonetti, M., Pin, S., & Calmettes, P. (2002). High-pressure effects on horse heart metmyoglobin studied by small-angle neutron scattering. *European Journal of Biochemistry, 269*(19), 4731–4737. https://doi.org/10.1046/j.1432-1033.2002.03126.x.

Mahieu, E., Covès, J., Krüger, G., Martel, A., Moulin, M., Carl, N., et al. (2020). Observing protein degradation by the PAN-20S proteasome by time-resolved neutron scattering. *Biophysical Journal, 119*(2), 375–388. https://doi.org/10.1016/j.bpj.2020.06.015.

Mahieu, E., & Gabel, F. (2018). Biological small-angle neutron scattering: Recent results and development. *Acta Crystallographica Section D, 74*(8), 715–726. https://doi.org/10.1107/S2059798318005016.

Mahieu, E., Ibrahim, Z., Moulin, M., Härtlein, M., Franzetti, B., Martel, A., et al. (2020). The power of SANS, combined with deuteration and contrast variation, for structural studies of functional and dynamic biomacromolecular systems in solution. *EPJ web Conf., 236*. Retrieved from https://doi.org/10.1051/epjconf/202023603002.

Marasini, C., & Vestergaard, B. (2017). SAS-based studies of protein fibrillation. In B. Chaudhuri, I. G. Muñoz, S. Qian, & V. S. Urban (Eds.), *Biological small angle scattering: Techniques, strategies and tips* (pp. 149–165). Singapore: Springer Singapore.

Maric, S., Lind, T. K., Raida, M. R., Bengtsson, E., Fredrikson, G. N., Rogers, S., et al. (2019). Time-resolved small-angle neutron scattering as a probe for the dynamics of lipid exchange between human lipoproteins and naturally derived membranes. *Scientific Reports*, *9*(1), 7591. https://doi.org/10.1038/s41598-019-43713-6.

Martínez-Sanz, M., Gidley, M. J., & Gilbert, E. P. (2015). Application of X-ray and neutron small angle scattering techniques to study the hierarchical structure of plant cell walls: A review. *Carbohydrate Polymers*, *125*, 120–134. https://doi.org/10.1016/j.carbpol.2015.02.010.

Marty, V., Jasnin, M., Fabiani, E., Vauclare, P., Gabel, F., Trapp, M., et al. (2013). Neutron scattering: A tool to detect in vivo thermal stress effects at the molecular dynamics level in micro-organisms. *J R Soc Interface*, *10*(82), 20130003. doi: rsif.2013.0003 [pii] https://doi.org/10.1098/rsif.2013.0003.

Mata, J. P., Hamilton, W. A., & Gilbert, E. P. (2012). Application of time-resolved small angle neutron scattering to non-equilibrium kinetic studies. In V. García Sakai, C. Alba-Simionesco, & S.-H. Chen (Eds.), *Dynamics of soft matter: Neutron applications* (pp. 289–318). New York, NY: Springer US.

Metwalli, E., Gotz, K., Lages, S., Bar, C., Zech, T., Noll, D. M., et al. (2020). A novel experimental approach for nanostructure analysis: Simultaneous small-angle X-ray and neutron scattering. *Journal of Applied Crystallography*, *53*(3), 722–733. https://doi.org/10.1107/S1600576720005208.

Metwalli, E., Götz, K., Zech, T., Bär, C., Schuldes, I., Martel, A., et al. (2021). Simultaneous SAXS/SANS method at D22 of ILL: Instrument upgrade. *Applied Sciences*, *11*(13). https://doi.org/10.3390/app11135925.

Murakami, R., Yunoki, Y., Ishii, K., Terauchi, K., Uchiyama, S., Yagi, H., et al. (2019). Cooperative binding of KaiB to the KaiC hexamer ensures accurate circadian clock oscillation in Cyanobacteria. *International Journal of Molecular Sciences*, *20*(18). https://doi.org/10.3390/ijms20184550.

Nagy, G., Kovács, L., Ünnep, R., Zsiros, O., Almásy, L., Rosta, L., et al. (2013). Kinetics of structural reorganizations in multilamellar photosynthetic membranes monitored by small-angle neutron scattering. *The European Physical Journal E*, *36*(7), 69. https://doi.org/10.1140/epje/i2013-13069-0.

Nagy, G., Posselt, D., Kovács, L., Holm, J. K., Szabó, M., Ughy, B., et al. (2011). Reversible membrane reorganizations during photosynthesis in vivo: Revealed by small-angle neutron scattering. *Biochemical Journal*, *436*(2), 225–230. https://doi.org/10.1042/BJ20110180.

Nakano, M., Fukuda, M., Kudo, T., Endo, H., & Handa, T. (2007). Determination of Interbilayer and Transbilayer lipid transfers by time-resolved small-angle neutron scattering. *Physical Review Letters*, *98*(23), 238101. https://doi.org/10.1103/PhysRevLett.98.238101.

Nakano, M., Fukuda, M., Kudo, T., Matsuzaki, N., Azuma, T., Sekine, K., et al. (2009). Flip-flop of phospholipids in vesicles: Kinetic analysis with time-resolved small-angle neutron scattering. *The Journal of Physical Chemistry B*, *113*(19), 6745–6748. https://doi.org/10.1021/jp900913w.

Nawroth, T., Buch, P., Buch, K., Langguth, P., & Schweins, R. (2011). Liposome formation from bile salt–lipid micelles in the digestion and drug delivery model FaSSIFmod estimated by combined time-resolved neutron and dynamic light scattering. *Molecular Pharmaceutics*, *8*(6), 2162–2172. https://doi.org/10.1021/mp100296w.

Nayak, A., Sorci, M., Krueger, S., & Belfort, G. (2009). A universal pathway for amyloid nucleus and precursor formation for insulin. *Proteins: Structure, Function, and Bioinformatics*, *74*(3), 556–565. https://doi.org/10.1002/prot.22169.

Neylon, C. (2008). Small angle neutron and X-ray scattering in structural biology: Recent examples from the literature. *European Biophysics Journal*, *37*, 531–541.

Nickels, J. D., Chatterjee, S., Stanley, C. B., Qian, S., Cheng, X., Myles, D. A. A., et al. (2017). The in vivo structure of biological membranes and evidence for lipid domains. *PLoS Biology*, *15*(5), e2002214. https://doi.org/10.1371/journal.pbio.2002214.

Nielsen, J. E., Prévost, S. F., Jenssen, H., & Lund, R. (2021). Impact of antimicrobial peptides on E. coli-mimicking lipid model membranes: Correlating structural and dynamic effects using scattering methods. *Faraday Discussions*, *232*(0), 203–217. https://doi.org/10.1039/D0FD00046A.

Niimura, N., Minezaki, Y., Ataka, M., & Katsura, T. (1995). Aggregation in supersaturated lysozyme solution studied by time-resolved small angle neutron scattering. *Journal of Crystal Growth*, *154*(1), 136–144. https://doi.org/10.1016/0022-0248(95)00164-6.

Nishiyama, Y., Langan, P., O'Neill, H., Pingali, S. V., & Harton, S. (2014). Structural coarsening of aspen wood by hydrothermal pretreatment monitored by small- and wide-angle scattering of X-rays and neutrons on oriented specimens. *Cellulose*, *21*(2), 1015–1024. https://doi.org/10.1007/s10570-013-0069-2.

Perez-Salas, U., Garg, S., Gerelli, Y., & Porcar, L. (2021). Chapter eleven - deciphering lipid transfer between and within membranes with time-resolved small-angle neutron scattering. In M. A. Model, & I. Levitan (Eds.), *Vol. 88, Current topics in membranes* (pp. 359–412). Academic Press.

Perrakis, A., & Sixma, T. K. (2021). AI revolutions in biology. *EMBO Reports*, *22*(11), e54046. https://doi.org/10.15252/embr.202154046.

Peterson, P. F., Campbell, S. I., Reuter, M. A., Taylor, R. J., & Zikovsky, J. (2015). Event-based processing of neutron scattering data. *Nuclear Instruments and Methods in Physics Research Section A: Accelerators, Spectrometers, Detectors and Associated Equipment*, *803*, 24–28. https://doi.org/10.1016/j.nima.2015.09.016.

Pingali, S. V., O'Neill, H. M., Nishiyama, Y., He, L., Melnichenko, Y. B., Urban, V., et al. (2014). Morphological changes in the cellulose and lignin components of biomass occur at different stages during steam pretreatment. *Cellulose*, *21*(2), 873–878. https://doi.org/10.1007/s10570-013-0162-6.

Pingali Sai, V., Smith Micholas, D., Liu, S.-H., Rawal Takat, B., Pu, Y., Shah, R., et al. (2020). Deconstruction of biomass enabled by local demixing of cosolvents at cellulose and lignin surfaces. *Proceedings of the National Academy of Sciences*, *117*(29), 16776–16781. https://doi.org/10.1073/pnas.1922883117.

Pipich, V., Balz, M., Wolf, S. E., Tremel, W., & Schwahn, D. (2008). Nucleation and growth of CaCO3 mediated by the egg-white protein ovalbumin: A time-resolved in situ study using small-angle neutron scattering. *Journal of the American Chemical Society*, *130*(21), 6879–6892. https://doi.org/10.1021/ja801798h.

Putnam, C. D., Hammel, M., Hura, G. L., & Tainer, J. A. (2007). X-ray solution scattering (SAXS) combined with crystallography and computation: Defining accurate macromolecular structures, conformations and assemblies in solution. *Quarterly Reviews of Biophysics*, *40*(3), 191–285.

Radulescu, A., Pipich, V., Frielinghaus, H., & Appavou, M. S. (2012). KWS-2, the high intensity / wideQ-range small-angle neutron diffractometer for soft-matter and biology at FRM II. *Journal of Physics: Conference Series*, *351*, 012026. https://doi.org/10.1088/1742-6596/351/1/012026.

Radulescu, A., Szekely, N. K., Appavou, M.-S., Pipich, V., Kohnke, T., Ossovyi, V., et al. (2016). Studying soft-matter and biological systems over a wide length-scale from nanometer and micrometer sizes at the small-angle neutron diffractometer KWS-2. *JoVE*, *118*, e54639. https://doi.org/10.3791/54639.

Reimhult, E., Schroffenegger, M., & Lassenberger, A. (2019). Design principles for Thermoresponsive Core–Shell nanoparticles: Controlling thermal transitions by brush morphology. *Langmuir*, *35*(22), 7092–7104. https://doi.org/10.1021/acs.langmuir.9b00665.

Rekas, A., Lo, V., Gadd, G. E., Cappai, R., & Yun, S. I. (2009). PAMAM dendrimers as potential agents against fibrillation of α-Synuclein, a Parkinson's disease-related protein. *Macromolecular Bioscience*, *9*(3), 230–238. https://doi.org/10.1002/mabi.200800242.

Roessle, M. W., & May, R. P. (2006). Structure and kinetics of proteins observed by small angle neutron scattering. In J. Fitter, T. Gutberlet, & J. Katsaras (Eds.), *Neutron scattering in biology: Techniques and applications* (pp. 187–204). Berlin, Heidelberg: Springer Berlin Heidelberg.

Rössle, M., Manakova, E., Holzinger, J., Vanatalu, K., May, R. P., & Heumann, H. (2000). Time-resolved small-angle neutron scattering of proteins in solution. *Physica B: Condensed Matter*, *276-278*, 532–533. https://doi.org/10.1016/S0921-4526(99)01732-9.

Rücker, U., Cronert, T., Voigt, J., Dabruck, J. P., Doege, P. E., Ulrich, J., et al. (2016). The Jülich high-brilliance neutron source project. *The European Physical Journal Plus*, *131*(1), 19. https://doi.org/10.1140/epjp/i2016-16019-5.

Schmitt, E., Panvert, M., Lazennec-Schurdevin, C., Coureux, P.-D., Perez, J., Thompson, A., et al. (2012). Structure of the ternary initiation complex aIF2–GDPNP–methionylated initiator tRNA. *Nature Structural & Molecular Biology*, *19*(4), 450–454. https://doi.org/10.1038/nsmb.2259.

Semeraro, E. F., Marx, L., Frewein, M. P. K., & Pabst, G. (2021). Increasing complexity in small-angle X-ray and neutron scattering experiments: From biological membrane mimics to live cells. *Soft Matter*, *17*(2), 222–232. https://doi.org/10.1039/C9SM02352F.

Sugiyama, M., Yagi, H., Ishii, K., Porcar, L., Martel, A., Oyama, K., et al. (2016). Structural characterization of the circadian clock protein complex composed of KaiB and KaiC by inverse contrast-matching small-angle neutron scattering. *Scientific Reports*, *6*, 35567. https://doi.org/10.1038/srep35567. https://www.nature.com/articles/srep35567#supplementary-information.

Svergun, D. I., Koch, M. H. J., Timmins, P. A., & May, R. P. (2013). *Small angle X-ray and neutron scattering from solutions of biological macromolecules*. Oxford: Oxford University Press.

Svergun, D. I., Richard, S., Koch, M. H. J., Sayers, Z., Kuprin, S., & Zaccai, G. (1998). Protein hydration in solution: Experimental observation by x-ray and neutron scattering. *Proceedings of the National Acadamy of Sciences USA*, *95*(5), 2267–2272.

Teixeira, S. C. M. (2019). High-pressure small-angle neutron scattering for food studies. *Current Opinion in Colloid & Interface Science*, *42*, 99–109. https://doi.org/10.1016/j.cocis.2019.05.001.

Teixeira, S. C. M., Zaccai, G., Ankner, J., Bellissent-Funel, M. C., Bewley, R., Blakeley, M. P., et al. (2008). New sources and instrumentation for neutrons in biology. *Chemical Physics*, *345*(2), 133–151. https://doi.org/10.1016/j.chemphys.2008.02.030.

Ünnep, R., Paul, S., Zsiros, O., Kovács, L., Székely, N. K., Steinbach, G., et al. (2020). Thylakoid membrane reorganizations revealed by small-angle neutron scattering of Monstera deliciosa leaves associated with non-photochemical quenching. *Open Biology*, *10*(9), 200144. https://doi.org/10.1098/rsob.200144.

Ünnep, R., Zsiros, O., Solymosi, K., Kovács, L., Lambrev, P. H., Tóth, T., et al. (2014). The ultrastructure and flexibility of thylakoid membranes in leaves and isolated chloroplasts as revealed by small-angle neutron scattering. *Biochimica et Biophysica Acta (BBA) - Bioenergetics*, *1837*(9), 1572–1580. https://doi.org/10.1016/j.bbabio.2014.01.017.

Urban, V. S., Heller, W. T., Katsaras, J., & Bras, W. (2021). Soft matter sample environments for time-resolved small angle neutron scattering experiments: A review. *Applied Sciences, 11*(12). https://doi.org/10.3390/app11125566.

Wah, B., Breidigan, J. M., Adams, J., Horbal, P., Garg, S., Porcar, L., et al. (2017). Reconciling differences between lipid transfer in free-standing and solid supported membranes: A time-resolved small-angle neutron scattering study. *Langmuir, 33*(14), 3384–3394. https://doi.org/10.1021/acs.langmuir.6b04013.

Whitten, A. E., Cai, S., & Trewhella, J. (2008). MULCh: Modules for the analysis of small-angle neutron contrast variation data from biomolecular assemblies. *Journal of Applied Crystallography, 41*(1), 222–226. https://doi.org/10.1107/S0021889807055136.

Yang, Z., Foston, M. B., O'Neill, H., Urban, V. S., Ragauskas, A., Evans, B. R., et al. (2022). Structural reorganization of noncellulosic polymers observed in situ during dilute acid pretreatment by small-angle neutron scattering. *ACS Sustainable Chemistry & Engineering, 10*(1), 314–322. https://doi.org/10.1021/acssuschemeng.1c06276.

Yoshida, K., Zenin, T., Fujiyoshi, A., Sanada, Y., Yamaguchi, T., Murata, K., et al. (2019). The effect of alkyl ammonium ionic liquids on thermal denaturation aggregation of β-lactoglobulin. *Journal of Molecular Liquids, 293*, 111477. https://doi.org/10.1016/j.molliq.2019.111477.

Yunoki, Y., Matsumoto, A., Morishima, K., Martel, A., Porcar, L., Sato, N., et al. (2022). Overall structure of fully assembled cyanobacterial KaiABC circadian clock complex by an integrated experimental-computational approach. *Communications Biology, 5*(1), 184. https://doi.org/10.1038/s42003-022-03143-z.

Zaccai, G., & Jacrot, B. (1983). Small Angle Neutron Scattering. *Annual Review of Biophysics and Biomolecular Structure, 12*, 139–157.

Zech, T., Metwalli, E., Götz, K., Schuldes, I., Porcar, L., & Unruh, T. (2022). Investigating growth of gold Nanorods by simultaneous small-angle X-ray and Neutron scattering. *Particle & Particle Systems Characterization, n/a*(n/a), 2100172. https://doi.org/10.1002/ppsc.202100172.

Zettl, T., Mathew, R. S., Seifert, S., Doniach, S., Harbury, P. A. B., & Lipfert, J. (2016). Absolute intramolecular distance measurements with angstrom-resolution using anomalous small-angle X-ray scattering. *Nano Letters, 16*(9), 5353–5357. https://doi.org/10.1021/acs.nanolett.6b01160.

CHAPTER TEN

Protein fibrillation from another small angle: Sample preparation and SAXS data collection

Bente Vestergaard and Annette Eva Langkilde*
Department of Drug Design and Pharmacology, University of Copenhagen, Universitetsparken 2, Copenhagen, Denmark
*Corresponding author: e-mail address: annette.langkilde@sund.ku.dk

Contents

Abstract

Protein fibrillation associates with several chronic, progressive, and fatal disorders, counting well-known maladies as Parkinson's, Alzheimer's, and Huntington's disease. The fibrillation process includes structural changes and aggregation of the disease specific protein, resulting in a mixture of different structural states covering nm to μm scale in varying volume fractions. SAXS uniquely enables structural investigations of such evolving mixtures but requires that the underlying main data collection experiment is carefully prepared. In this chapter, we provide very detailed instructions on how to plan and perform such protein fibrillation experiments, both before and during the SAXS data

Methods in Enzymology, Volume 677
ISSN 0076-6879
https://doi.org/10.1016/bs.mie.2022.08.041

collection. The chapter is based on our own experience mainly using high-end synchrotron radiation facilities for the data collection but can equally well be applied on state-of-the-art laboratory based SAXS instruments. We accumulate the know-how from our group, established via the study of different amyloid-like proteins, applying fibrillation either in batch or in plate reader, with or without known process quenching conditions.

1. Introduction

Several severe pathological conditions associate with protein misfolding and fibrillation (Ankarcrona et al., 2016). Proteins, which under nonpathological conditions assign with a biological function, form abnormal, inherently aggregation prone and dysfunctional structures ultimately resulting in the deposition of amyloid fibrils. Amyloidosis may have systemic impact (Merlini et al., 2018; Morris et al., 2019; Papa & Lachmann, 2018; Si et al., 2021); alternatively, deposits of the amyloid fibrils are observed in disease specific organs and tissues (Hoppener & Lips, 2006; Lim, 2019; McAlary et al., 2019; Srinivasan et al., 2021; Trejo-Lopez et al., 2022). Indeed, the finding of protein fibril deposits in cerebral amyloid plaques is the hallmark of several neurodegenerative disorders, prevalently amyotrophic lateral sclerosis (ALS) and Parkinson's, Alzheimer's and Huntington's diseases (Lim, 2019). Combined crystal structures and SAXS were used to characterize aggregation propensities of superoxide dismutase ALS mutants and show that they mirror clinical phenotypes (Pratt et al., 2014). Neurodegenerative diseases are frequent in the elderly population, and in an aging world these chronic, progressive diseases pose an almost unmanageable socio-economic burden (Zahra et al., 2020). The association of protein fibril formation with such grave diseases certainly primes significant interest for the investigation of the structural changes and aggregation processes underlying fibril formation. Additional interest is sparked by the understanding that some proteins form non-toxic fibrils (AKA native or functional amyloids (recent reviews in Levkovich et al., 2021; Otzen & Riek, 2019; Rubel et al., 2020) playing diverse roles ranging from hormone storage (Maji et al., 2009) and sperm maturation (Cornwall et al., 2019) in mammals to fungal coating (Mackay et al., 2001) or bacterial biofilm formation (Khambhati et al., 2021). Interesting examples of functional amyloids include more recent findings of cross-α fibril structures, i.e., fibrils where α-helical fragments, rather than β-strands, form the amyloid core. Examples include a bacterial cytotoxin (Tayeb-Fligelman et al., 2017) and an amphibian antimicrobial peptide, the latter capable of switching between cross-α and cross-βbstructures (Salinas et al., 2021).

Spider silk composition is also reminiscent of fibril structure (Salinas et al., 2021) resulting in a nano-material with the strength of steel (Wang & Schniepp, 2018). Correspondingly, the use of fibril structural principles for the development of self-assembly functional biomaterials receives notable interest (Li & Zhang, 2021), and designed amyloids suggestibly can be used as drug depots (Maji et al., 2008), nanosensors (Hauser et al., 2014) and nanocatalysts (Huang et al., 2013) to mention a few.

There is thus a tremendous interest in the structure of amyloid-like fibrils, and indeed fibrils present impressive structural features. Fibrils grow to μm length and (with the few former mentioned exceptions) exhibit cross-β structure, named by the characteristic cross visible in fiber diffraction images from amyloid-like material (Eanes & Glenner, 1968; Jahn et al., 2010). The cross-β signal arises from several extensive, highly ordered β-strands twisting around the fibril axis (resulting in a meridian signal at 4.7–4.8 Å and equatorial signals approximately at 10 Å in fiber diffraction images) and is common to all amyloid-like fibril material. This said, there is significant variation over this central structural theme in the core structure of the fibrils that have been investigated to high resolution, using cryo-electron microscopy (cryo-EM). The advances in our understanding of the structure of the fibril core are tremendous, and although a review is beyond the scope of the current chapter, we highlight impressive recent examples of high resolution cryo-EM fibril structures, including a tour de force structure determination of no less than 76 different recombinant tau fibril structures (Lövestam et al., 2022), and several structures based on amyloid material isolated directly from patient tissue (e.g., tau filaments from Alzheimer's brains (Fitzpatrick et al., 2017; Shi et al., 2021); amyloid light chain fibrils, extracted from the explanted heart of a patient (Radamaker et al., 2021); α-synuclein (aSN) from the brains of diseased MSA and DLB patients (Schweighauser et al., 2020) and Aβ amyloid fibrils from the brain tissue of diseased Alzheimer's patients (Kollmer et al., 2019; Yang et al., 2022)). It has become evident that several polymorphic fibril structures of the same protein can be isolated, and each such structure is seemingly disease specific. Another common result is that fibrils consist of a highly structured core, surrounded by a less structured halo of protein tails, the latter not visible in cryo-EM structures.

While these high-resolution structures of fibrils are very important, complementary methods are still needed as large fractions of the proteins are often not observed due to a high degree of flexibility. It also remains highly relevant to investigate the complex process leading to fibril formation, not least since the high-resolution structures of the end-product so far have not paved the way to control of the self-assembly processes, not to mention therapeutic intervention. The process can be initiated in vitro, enabling the study of certain aspects of the fibrillation process principles using fibril specific fluorescent dyes such as thioflavin T (ThT) (Gade Malmos et al., 2017). The process can be described as a nucleated polymerization process, with characteristic lag-, elongation- and steady state phases (Fig. 1) (Buell, 2019), although it should immediately be highlighted here, that the latter phase is much less steady, from a structural perspective, than the name may suggest (Nielsen et al., 2013; Skamris et al., 2019). It is of evident therapeutic interest to understand the structural basis of both nucleation and expansion of the fibril material, the latter often dominated by secondary nucleation events (Zimmermann et al., 2021). Of therapeutic importance

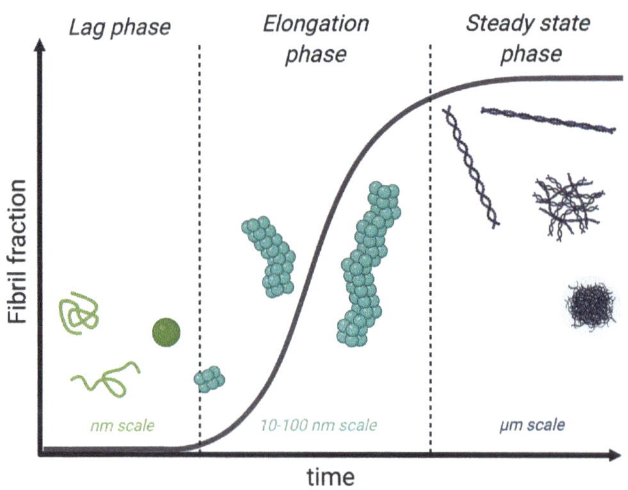

Fig. 1 A schematic view of the fibrillation process. The fibril fraction (gray curve) is estimated by fluorescence spectroscopy using a fibril specific fluorescent probe, such as Thioflavin T. The three classic phases in the fibrillation reaction are shown, and the approximate scale of the dominant species within each phase is indicated. Under fibrillation conditions the native state is in equilibrium with a re-/un-folded amyloidogenic state, which nucleates during the lagphase (light green colors). These are primary nucleations. After nucleation fibrils grow by elongation (teal colors). Fibrils are the source of secondary nucleation, resulting in an exponential increase in fibril material, during the elongation phase. In the steady state phase, the maximum fluorescence signal has been reached, where the sample is either fully fibrillated or rests in an equilibrium with other (monomeric or oligomeric) species. Despite the name, the steady state phase does not represent steady fibril structures, as maturation and further packing at both micro- and macroscopic levels may occur as indicated with different fibril species (dark blue). Note that all the described structural changes may take place throughout the fibrillation reaction, although with very different probabilities. *Created with BioRender.com*

are the structural investigation of the initial structural conversion from native (nonpathological) to amyloidogenic conformations, the molecular basis for secondary nucleation events, and, finally, the structural elucidation of potential intermediate oligomeric structures, formed along the fibrillation pathway.

In cases where it is possible to identify stabilizing principles, which will prevent the initial structural conversion to fibrillogenic conformations, this can be used as a treatment strategy. An example is the transthyretin case, where so-called kinetic stabilizers (first example is the drug tafamidis) delay disease progression (Bulawa et al., 2012). Biologically active transthyretin is (predominantly) a folded tetramer, and the formation of a highly unfolded transthyretin monomer is a central step in the amyloid cascade (Groenning et al., 2015). Tafamidis binds and stabilizes the dimer-dimer interface, thereby increasing the kinetic barrier of the initial and rate-limiting tetramer dissociation (Bulawa et al., 2012). This is one way of inhibiting the primary nucleation event in fibril formation. The molecular basis of secondary nucleation events is poorly understood, but includes fibril fragmentation, template induced unfolding and altered local environment at the fibril surface (Kumari et al., 2021; Zimmermann et al., 2021). Finally, on-pathway and intermediately formed oligomeric species have been suggested to play a directly cytotoxic role (Rochet et al., 2000; Winner et al., 2011) in contrast to the mature fibrils, which in some cases can be considered as non-toxic reservoirs of such oligomeric species (Franco et al., 2021; Madhu & Mukhopadhyay, 2021; Maity et al., 2020). The molecular mechanism behind the allegedly toxic action of these species, and whether such potentially toxic oligomers are on- or off-pathway remains an open question. On-pathway species may establish an equilibrium with the fibril state (Bigi et al., 2021; Skamris et al., 2019), and as such, targeting of the fibril structure itself, driving the structural state toward a soluble one could have unwanted side-effects. Indeed, it has been demonstrated that oligomeric species appear intracellularly in neurons upon fibril docking (Bigi et al., 2021).

As it appears, several structural aspects of the fibrillation process are very important to investigate further. A central point to understand is the fact that the fibrillation process is a multi-step process, and that structures at each of such steps are in equilibrium with other structures, and hence co-exist in solution. Under certain conditions the native species may change to one or more non-native conformations, sparking a nucleation event after which (potentially several) oligomeric species will be present in concert with the growing fibrils, and likewise fibrils may be present in several different populations (Fig. 1). If attempting to isolate one such species the structural equilibrium is perturbed, and hence whichever species isolated will not represent an in-process species. It is this a prerequisite for structural studies, to leave the structural equilibrium undisturbed. Most structural methods require very high sample purity, and this is also the case for biological solution small angle X-ray scattering (BioSAXS). In a common BioSAXS experiment it is normally recommended to have sample purity of at least 95% and sample polydispersity of maximally 0.1% (Graewert & Jeffries, 2017). In some cases data from monodisperse solutions, freshly separated from a mixture, can be obtained by inline chromatography:SAXS experiments (see chapters "Size exclusion coupled small angle X-ray scattering with tandem multiangle light scattering at the SIBYLS beamline" by Rosenberg et al. and "SEC-SAXS: Experimental set-up and software developments build up a powerful tool" by Pérez et al., in this issue). In relatively simple mixtures, data from a concentration series of the sample may also enable analysis of the individual components and indeed there are standardized analysis methods for this purpose (Konarev et al., 2003). The fibrillation mixtures, however, are far more complex.

However, as we have previously argued, a central quality of BioSAXS experiments is the fact that the method—with care—can be applied also to complex mixtures such as the fibrillation process (Herranz-Trillo et al., 2017; Langkilde et al., 2018; Sagar et al., 2021; Vestergaard, 2016). When properly performed, the experiments are non-invasive, and all structural states that are present at any given timepoint of measurement will contribute to the scattering data. If structure factors are negligible SAXS data are additive. Ultimately, this enables the study of the developing fibrillating solutions, even if they consist of species that differ with orders of magnitude in length, and existing at very different time-scales. Although the information obtained is low-resolution it is highly complementary to, e.g., the cryo-EM data available in the field. Indeed, the BioSAXS experiments are unique in character, providing direct, in-solution quantitative and qualitative structural information from the un-disturbed fibrillation equilibriums, ultimately resulting in information about the number of species present at different timepoints, together with the volume fractions and model-free (ab initio) 3D low-resolution structural information from individual species. To our knowledge, no other method available will provide similar information. In this chapter, we aim to provide our best advice on how to prepare such an experiment, with emphasis on sample preparation and data collection. In a later chapter (Langkilde & Vestergaard, 2022), we describe the subsequent data analysis. We recommend that the data analysis chapter is also studied carefully before preparing for the experiment, since an understanding of the pitfalls and challenges associated with data evaluation will improve the understanding of the sample preparation demands.

2. General method and safety notes

2.1 Safety when handling amyloid and amyloid-like material

Caution must be taken when working with amyloid and amyloid-like material. Tragic cases of prion-specialized scientists developing Creutzfeldt-Jakob disease are currently investigated for the potential link to accidents during laboratory experiments (Casassus, 2021). Accumulating evidence appoints a risk of such contamination also for other amyloid-like material; examples from the literature covering alpha-synuclein fibrillation is included here (Holmqvist et al., 2014; Peelaerts et al., 2015) and the reader is urged to search the literature further for any given amyloid-like protein of interest. We recommend to follow safety procedures, as described in Bousset et al. (2016) and with particular caution if handling protein in powder form. For increased reproducibility (see below) we originally used sonication of the fibrillated (and potentially gel-like) samples during data collection (Langkilde et al., 2018; Langkilde & Vestergaard, 2012). However, for safety reasons, instead we recommend the use of freeze-thaw cycles as this method avoids formation of fibril containing aerosols.

2.2 Fibrillation "in batch" or in plate reader?

We have experience with conducting the main fibrillation experiment both "in batch" (i.e., all sample in one sample vial (e.g., in Groenning et al., 2015; Vestergaard et al., 2007) and in a fluorescence plate reader (e.g., in Giehm et al., 2011; Herranz-Trillo et al., 2017; Langkilde et al., 2015; Nielsen et al., 2013; Nors Pedersen et al., 2015)). While both types of experiments are valid, the plate reader experiments have the advantage of a generally higher reproducibility and, not least, the optional simultaneous and automated collection of the developing fluorescence signal from fibril-sensitive probes. When monitoring of the fluorescence signal it is possible to collect complementary biophysical data, which can be analyzed in concert with the SAXS data, collected at the same timepoint, i.e., using the fluorescence trace as the timer between different experiments. If collecting from batch mode, it is highly recommended to either do parallel fluorescence measurements on extracted samples or use, e.g., a Probedum setup (Palmadottir et al., 2021) allowing for continuous observation of the developing fluorescence signal. Keeping the entire sample in batch ensures that there is only one onset of fibrillation and hence that the data from extracted samples are chronologically developing, which may not be the case in a plate reader. However, the continuous sampling from one vial may also represent repeated interference and hence have unwanted side effects. The final choice between batch or plate reader will depend on the users experience with the system. The procedures described in this chapter are valid for both types of experiments.

2.3 Synchrotron radiation or laboratory instrument SAXS data: In situ or ex situ experiments?

While our published studies so far have been conducted at synchrotron beamlines, we also highlight the option to collect data on an in-house SAXS instrument. Elegant fibrillation studies have been published by other groups (e.g., in Oliveira et al., 2009; Rasmussen et al., 2021) based on data from a laboratory SAXS instrument where the fibrillating solution remains in the sample cell throughout the experiment. Here we call this experiment type an in situ fibrillation experiment, while ex situ experiments denote the approach where individual samples are extracted from batch or plate readers and mounted one after the other in the sample cell. Such an approach has obvious advantages and conveniences and eliminates the risk of diverging fibrillation pathways in different vials in a plate reader setup, while ensuring parallelization of the sample and data collection progresses. There may be fibrillation experiments which would be very difficult to conduct at a synchrotron facility, but which are possible at home (whether working in situ or ex situ), and the reader is urged to check the feasibility of a planned experiment with the beam line responsible at a large-scale facility well in advance of applying for beam time (ask directly whether you can bring your plate reader to the beamline, whether they allow handling of amyloid samples at or in the vicinity of the beamline, etc.). A few potential pitfalls in concert of in situ experiments can be mentioned: First, if radiation damage is induced during an in situ experiment, all subsequent measurements will be affected. We do not recommend in situ experiments at synchrotron radiation facilities. The risk of radiation damage is however considerably smaller when working in-house, while even ex situ data from synchrotron experiments must be carefully checked for signs of radiation damage (see below). Lower X-ray intensity, however, also means that longer exposure times are necessary, which must be considered when estimating the general feasibility of the experiment. Second, in an in situ experiment there is a risk of partial sedimentation of fibril material during the data collection (when using a horizontal sample cell), and/or aggregates may accumulate or stick at a particular area in the sample cell (using either horizontal or periscope cells). The procedures described in this chapter are based on our own experiences with ex situ experiments and applying synchrotron radiation.

3. Feasibility, design and optimization

The main experiment (MainExp, Fig. 2 and Section 4) is the optimized fibrillation experiment conducted while collecting BioSAXS data. Prior to MainExp you must conduct a series of experiments to ensure that the conditions are optimized with regards to timing, reproducibility, sample concentration, etc. Even before this, however, feasibility should firstly be assessed by two aspects: sample quality and sample quantity. Here, the general guidelines for sample quality at beamlines applies (Graewert & Jeffries, 2017). It must be possible to purify the sample to high purity, and the protein must remain stable in a SAXS-suitable buffer long enough to allow for MainExp. Although this sounds like a very basic assumption, this demand can be difficult to fulfill when it comes to amyloid-like material, due to the often high aggregation propensity. We have, e.g., published an analysis of a fibrillating short peptide, which immediately initiated aggregation when transferred to an aqueous buffer (Langkilde et al., 2015) in which case a special protocol had to be developed for the data collection to ensure measurements of the initial monomeric state. Do not assume that you can rely solely on published purification protocols for your protein of interest, often additional purification steps are necessary to improve the final SAXS data quality or ensure reproducibility (van Maarschalkerweerd et al., 2015). Also, consider that BioSAXS analysis of fibrillation is conducted at concentrations in the mg/ml range, and hence the protein must be handled at these concentrations for the duration of the experiment. Given that several data points are required to perform the subsequent data analysis (Langkilde & Vestergaard, 2022), sample quantity becomes an issue, even if sample consumption has been minimized at many beamlines (Blanchet et al., 2015; Dyer et al., 2014; Round et al., 2015). You must thus consider whether it is possible for you to purify the necessary quantity for both optimization experiments and MainExp (Section 4). Once the basic feasibility of the experiment has been evaluated, the actual optimization of the experiment can begin, as described below.

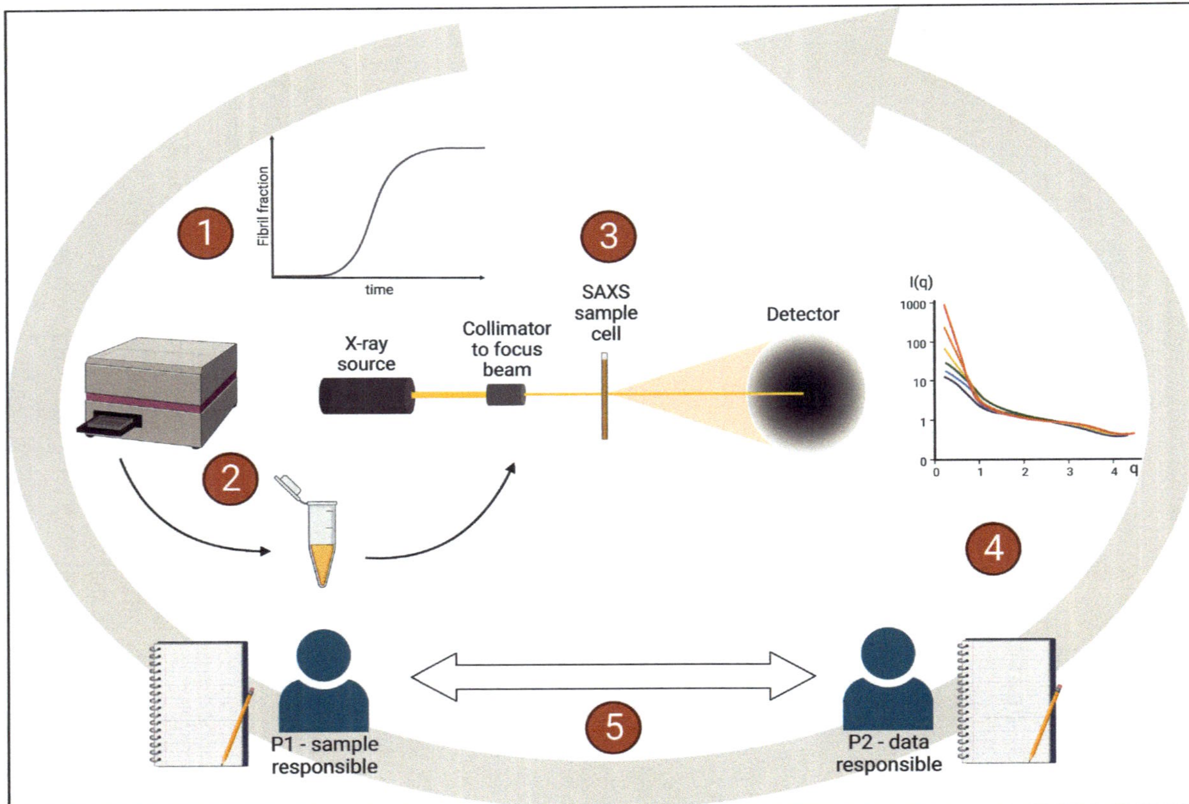

Fig. 2 See figure legend on opposite page.

3.1 Equipment

- Spectrophotometer, e.g., NanoDrop for accurate protein concentration determination
- Fluorescence plate reader with black 96-well plates with clear bottom, sealing tape, potentially glass beads (note 1)
- Heat block for preheating samples/solvents (note 2) prior to initiating fibrillation and as basis for the plate when extracting samples
- Vessel for liquid N_2

3.2 Reagents

- Highly purified protein or peptide for fibrillation
- Buffers to test and optimize the fibrillation conditions, filtered with 0.22 μm filters
- Stock solution of a fibril specific fluorescent probe (note 3)
- Liquid N_2

3.3 Procedure

1. Produce a list of known fibrillation conditions for your protein
2. Compare and modify these conditions to the general demands for MainExp:
   - Protein concentration must be >1 mg/mL (note 4)
   - Concentration of any buffer reagent must be minimized (note 4)
   - Samples must contain a fibril-specific fluorescent probe (note 3)
   - Samples should preferably include a radical scavenger (note 5)
   - Temperature during the fibrillation reaction is preferably >20C
   - The total fibrillation time is optimally between 6 and 12 h (note 6)
   - The lag-phase is at least 1 h (note 6)
   - The elongation phase is at least 3 h (note 6)
   - The stationary phase remains stationary for at least 2 h (note 7)

   For these experiments, we strongly recommend the use of a fluorescence plate reader, such that many fibrillation conditions can be tested.

Fig. 2 The main experiment illustrated with person 1 (P1) responsible for the sample side and person 2 (P2) overseeing the data collection and initial analysis. Attention points are marked with red numbers: (1) Monitor fibril specific fluorescence signal to ensure sampling from the entire fibrillation process. (2) Extract samples for transfer to the SAXS sample cell. (3) Monitor sample loading, data collection and sample cell cleaning; ensure collection of buffer data. (4) Keep track of the development in the data series. (5) Feedback and communication, e.g., adapt extraction times, restart fibrillation experiments. Keep extensive notes from all parts of the experiment. The sequence of experiments is repeated multiple times (lightgray arrow). *Created with BioRender.com.*

3. Plan the number of fibrillation experiments running during MainExp
During beamtime, you may initiate the fibrillation reaction several times, in order to cover a particular part of the process more thoroughly than what it possible if initiating the process only once. Hence, you should consider how many wells in the 96-well plate that will be filled and initiated at different time points. The volume for each initiated reaction should be adequate for the planned data acquisition.
4. Plan complementary experiments
Inclusion of complementary biophysical data, using, e.g., DLS, SLS, UV, CD is strongly recommended. In some alternative settings collection of complementary data is possible simultaneously with collection of SAXS (or SANS) data (Dicko et al., 2020; Falke et al., 2018; Haas et al., 2014; Lafleur et al., 2011) but these settings are not necessarily compatible with the collection of SAXS data during fibrillation. Instead, the fibril-specific fluorescence signal is used to align the timing of the MainExp with the timing of complementary data collection (Nielsen et al., 2013; Rasmussen et al., 2021; Skamris et al., 2019). If collecting data at a laboratory-based facility, collection of complementary data may be possible from the exact same samples as the SAXS data collection. If collecting data at a synchrotron radiation facility, complementary data should be collected before and after the beamtime, unless you have the means to quench the fibrillation reaction very thoroughly (see point 7 below). Consider your options, and plan accordingly.
5. Optimize reproducibility
   - Use preheated solutions (note 2)
   - Stabilize the starting point, such that your onset of fibrillation is reproducible (note 8)
   - Vary sample volume in each well of the 96-well plate, with focus on reproducibility. Also test how partial extraction of the sample volume from the wells influences the fibrillation process (note 9).
6. Optimize the handling of fibrillated samples
The fibrillated samples are most often highly viscous. This makes it difficult to pipette the samples at later time points in the fibrillation reaction from the plate reader/sample vial to the SAXS sample cell or robotic sample changer (see below). Make sure to use pipetting tips with an adequately large diameter. Potentially use freeze-thaw cycles in liquid N_2 to lower the viscosity of the sample (Note 10). In order to avoid that you miss the opportunity to collect data from your fibril samples, this must be tested at home, prior to beamtime.

7. **Develop quenching mechanism (optional)**
In principle, if it is not possible to develop a fibrillation protocol with a timing that can be aligned with the timing of MainExp, a quenching mechanism (i.e., halting the fibrillation reaction at the point of extraction) can be developed. If the sample is fibrillated at high temperatures and/or using mechanical stress, it may be enough to extract the sample and keep it at quiescent conditions at lower temperatures. This, however, is by far not the typical picture, and very careful tests must be conducted to determine the efficiency of quenching over time if this strategy is applied (note 11).There is significant risk to induce structural changes in the samples during quenching, hence, we strongly recommend that the SAXS data collection is directly aligned with the fibrillation reaction. We also stress that although mail-in SAXS data collection is a convenient option at several high-end synchrotron SAXS beamlines, this should not be applied even when efficient quenching is possible, since several very careful measures must be taken during data collection, and hence at least one dedicated scientist with specialized knowledge about the fibrillation reaction is an absolute prerequisite during data collection (Section 4).
8. **Save fully fibrillated samples for MainExp**
You need fully fibrillated samples at the beamline to test how the sample loads into the sample cell (see Section 4). These samples must be produced at home in advance of MainExp. Preferably, save a sample volume corresponding to at least 5 times the SAXS sample volume, such that you can repeat the loading test if necessary. This sample is also used (along with a native state sample) to test for radiation damage before starting MainExp.
9. **Which beamline?**
Not all beamlines are suitable for a fibrillation experiment. First, check whether the facility allows entry of amyloid-like material (see safety notes above). Second, check if the beamline offers access to a wet lab or similar, such that you can run the fibrillation experiment on site. And third, consult the beamline responsible, to request if the data-range on this instrument is suitable for a fibrillation experiment. It is important that the q-range covers low angles, such that the longest dimension of the repeating unit in the fibril structure is covered in reciprocal space. These aspects are covered in more detail in a following chapter (Langkilde & Vestergaard, 2022). Fourth, ask the beamline scientist how the samples are loaded into the sample cell (manual loading or sample robot).

If a sample robot is used, ask whether you can control the sample temperature in the sample storage rack, and whether you can potentially load the samples with low flow rates. If your samples will be stored temporarily in a sample rack, consider this aspect in the light of your experience with (a) quenching and (b) sample stability. For (a): optimally, the fibrillation reaction is quenched at the given storage temperature. Alternatively, discuss with the beamline scientist how often you can enter the hutch of the beamline to place samples in the sample rack, such that you minimize the storage time. In any case: note down when you place the sample in the rack, and when the sample is loaded. For (b): check from home whether fibrils precipitate in a test tube when stored. If yes, this could result in only partial loading of the fibrillated sample into the sample cell. This can be solved if the robot can perform a gentle mixing of the sample before loading. If not, robotic sample loading is not an option, or you must change the buffer conditions of the fibrillation reaction such that fibrils do not precipitate within the time frame of your operations—i.e., back to optimization. Regardless of the approach consider the full turnover time including buffer measurements and cleaning procedures in order to sample all phases of the fibrillation process.

10. Plan transportation
It may be challenging to transport equipment, amyloid-like samples, and reagents across large distances and national borders. Plan well in advance.

11. Concrete example of optimized sample conditions
As an example of how the above-mentioned points have been addressed in a particular case by us, here we include the detailed and commented experimental conditions, used for SAXS data collection during the fibrillation of aSN. In this particular case we investigated the reversibility of aSN fibrillation by temperature variation, and compared the species formed after reversing fibrillation to the species that are formed on-pathway to fibrillation. Reversibility was investigated by placing pre-fibrillated samples (fibrillated on-site, 24 h prior to beamtime) in a temperature controlled SAXS sample changer (Round et al., 2015) while the fibrillating samples were loaded during the fibrillation, as described in detail below. The resulting SAXS data and corresponding analyses are unpublished. Recombinant human aSN was purified, as described in van Maarschalkerweerd et al. (2015), in order to avoid the unwanted formation of aSN dimers. By following this purification procedure,

reproducibility of the fibrillation kinetics greatly improved. The purified protein was lyophilized and stored at −20 °C until use. The lyophilization protocol has been optimized, using native gel electrophoresis and the reproducibility of fibrillation kinetics as tests. In order to have freshly fibrillated samples available for the cooling experiments, samples were prepared for fibrillation using wet lab facilities at the beamline 1 day prior to the start of beamtime. Lyophilized powder was dissolved in PBS Buffer, filtrated through a pre-washed 0.22 μm filter, diluted to 3.5 mg/mL including Thioflavin T (ThT) dye at a final concentration of 20 μM. The fibrillation was performed in a FLUOstar Optima platereader (BMG Labtech) where the ThT fluorescence intensity was followed by the emission at 480 ± 5 nm upon excitation at 450 ± 5 nm by orbital shaking, and $10\times$ flashes per well were recorded every 6th min. Samples were distributed in the 96-well optical bottom plate (Thermo Fischer) with 150 μL sample in each well. Previous experience has shown that this fibrillation volume results in reproducible fibrillation and enables up to three repeated sample extractions during the beamtime. An autoclaved 3 mm glass bead (Sigma Aldrich) was added to each well for agitation and increased reproducibility. The plate was sealed using sealing tape (Thermo Fisher) and the protein was fibrillated by incubation and orbital shaking at 37 °C. The 96-well plate was filled in all wells, except the outer rim of the plate, to avoid any potential effects of diverging temperature at these positions. 1 h after the fluorescence trace reached a plateau, samples were extracted, pooled into 300 mL fractions, and stored at room temperature, until the start of the beamtime. We have used SAXS data to check for the best possible quenching conditions for aSN fibrillating samples, and room temperature provides the optimal conditions, while, e.g., storage on ice results in on-going fibrillation. Human aSN fibril samples, extracted 1 h after reaching plateau, are not very viscous, and hence further handling of the samples, prior to loading into the SAXS sample cell, was not necessary. This is in contrast to fibril samples, extracted, e.g., 12 h after reaching the plateau, where we use freeze-thaw cycles in order to handle the viscous samples.

After the start of the beamtime, the sample preparation procedure was repeated, and the fibrillation reaction was initiated with freshly dissolved protein in a 96-well plate. These samples were used for sample extraction, aligned with the on-going data collection, as described further below in Section 4.

3.4 Notes

1. As mentioned in Section 2.2. the fibrillation experiments may be carried out in batch or in a fluorescence plate reader. We recommend the use of a fluorescence plate reader when optimizing the fibrillation conditions, since this enables the testing of many conditions, and in general increases the reproducibility of the fibrillation experiment. Please consider, however, that if using a fluorescence plate reader for the fibrillation, it must also be possible to transport this equipment to the synchrotron beamline or laboratory based SAXS instrument, when performing MainExp (Section 3). Shaking with a glass bead in the sample vial is a common method to initialize fibrillation, as done for aSN (Giehm & Otzen, 2010).
2. After optimizing the fibrillation conditions the fibrillation procedure is repeated to ensure reproducibility of the fibrillation experiment during MainExp (Section 4). For this, all solutions are stabilized to a given starting temperature prior to initiating the fibrillation experiment. If your plate reader has a cooler, you can apply fibrillation temperatures below ambient temperature. In this case you must consider how to keep the 96-well plate as close as possible to the applied temperature while extracting the sample. This is very important, both to avoid influencing the fibrillation reaction (note that temperature changes have a very strong influence on the fibrillation state, see, e.g., Skamris et al., 2019) and to avoid temperature dependent variation in your fluorescent signal. If your plate reader does not have a cooler, we recommend using temperatures from ambient +5 °C and above. Remember to consult your synchrotron contact scientist/SAXS laboratory contact to check the expected ambient temperature on site. When extracting samples, it is advised to keep the plate on a temperature-controlled block to minimize temperature variations during the process. Note that the extracted sample should also be kept at the desired temperature while transferring the sample to the sample changer (Section 4) or sample cell (or the fibrillation reaction should be thoroughly quenched, see Section 3.3, point 5).
3. E.g., ~1 mM ThT in H_2O. The concentration is determined spectroscopically at 412 nm after filtering the solution with 0.22 μm filters using an absorption coefficient of $\varepsilon_{412} = 36{,}000\,M^{-1}\,cm^{-1}$. A ThT stock solution should be protected from light and stored at 4 °C, and the exact concentration must be re-determined after storage, immediately prior to use, due to the limited stability of the solution (Fodera et al., 2008). The fibrillation process can of course be followed without

a fluorescent probe, but in such a case other means of tracking the fibrillation process should be applied.

4. To ensure that the signal:noise ratio in the SAXS data is suitable for the data analysis a relatively high concentration of the protein is necessary. We recommend that the concentration is in the mg/mL range, and that the beamline/SAXS laboratory responsible is consulted, such that the final protein concentration will be high enough for the specific SAXS instrument used. The concentration of any other buffer component must be minimized since any buffer component will contribute to the background scattering and hence reduce the contrast and thus the signal:noise ratio, also see Graewert & Jeffries, 2017. The latter demand, however, must be balanced against the need to control the fibrillation conditions properly, such as to ensure reproducibility and to fulfill the other general demands for MainExp.
5. A radical scavenger will reduce the risk of radiation damage to the sample during data collection (Jeffries et al., 2015). Radiation damage must be avoided since radiation damage may cause non-fibrillation related aggregation in the sample, and likely disturb the fibrillation reaction. The radical scavenger could be DTT or low concentrations of glycerol, and this compound must be included, also during the optimization of the fibrillation experiment since all compounds will influence the fibrillation procedure. If the inclusion of a radical scavenger hampers the fibrillation reaction, alternatives such as beam attenuation can be considered keeping in mind the requirements for suitable signal:noise levels.
6. This recommended timing is intended for a synchrotron radiation experiment with a limited beamtime allocation of, e.g., 24 h. In a laboratory based SAXS setting the ideal timing is likely longer, since both the recommended exposure time and the time of access to the equipment is typically significantly longer than at a synchrotron beamline. We recommend that each measurement is finalized before you extract sample for the next measurement (or the fibrillation process must be completely quenched in the extracted sample until the measurement is performed, see Section 3.3, point 5). See Section 4 for an evaluation of the time necessary for each measurement. The goal is to extract and measure data from an adequate number of samples, thereby enabling proper data analysis. The necessary number of samples partially depends on the data analysis method (see (Langkilde & Vestergaard, 2022)). When using manual data decomposition *at least* 6, 8 and 6 measurements are necessary during the lag-, elongation and steady state phases,

respectively. For COSMiCS based data decomposition we recommend at least twice the number of measurements, preferably more. In general, we recommend collecting as much data as possible during each phase, only limited by the available protein quantity and the available beamtime.

7. In many cases a so-called maturation of the fibril structural state will happen during the steady state phase, which may not result in detectable changes in the fluorescence signal. Most likely, the intermediate structural states in the fibrillation reaction are on-pathway to the early fibril structural state, but not the later fibril structure. It is thus preferable that several measurements are made during the steady state, where the early fibril structure is the dominant structural state, since this is preferable for the data decomposition, especially when using manual decomposition analysis. The full maturation could be far beyond the timeframe of MainExp but complementing the measurements with samples prepared ahead may add to the understanding of such maturation. See, e.g., Skamris et al. (2019, 2022) for advice. Likewise, be aware that initial structural changes happen during the lagphase before changes are detectable in the fluorescence signal. That is, be aware that the fluorescent signal is an important tracer of your fibrillation progress, but that by far, not all structural changes are detectable by the probe.
8. A stable starting point means that the fibrillation will not initiate until you trigger the onset. The onset may be triggered, e.g., by a temperature shift, by shaking or by addition of a specific chemical. A stable starting point will increase reproducibility of the fibrillation reaction. A stable starting point will also enable SAXS data collection of the sample at different concentrations, providing further information about your system (Langkilde & Vestergaard, 2022).
9. You can balance sample consumption, reproducibility, and available space for several simultaneous fibrillation reactions on the 96-well plate by varying sample volume and the number of repeated extractions of sample for SAXS exposure from each well. Be aware that partial extraction will influence the fluorescence signal and consequences of repeated extractions should be tested. Make very careful notes during MainExp, also about repeated extraction from a well.
10. Freeze-thaw cycles will primarily influence the fibril macroscopic assembly (lateral association, plate formation, etc.), rather than the actual fibril formation. We recommend the use of TEM or AFM prior to the SAXS beamtime to test the effect of freeze-thawing cycles on the specific system. An alternative homogenization procedure is to vortex samples in the presence of a seed bead, but this may be difficult to do

reproducibly with low sample volumes. Previously we have recommended the use of a rod sonicator (Langkilde et al., 2018; Langkilde & Vestergaard, 2012) which we no longer recommend for safety reasons (see Section 2).

11. As already mentioned, a fibrillated sample will likely mature continuously, also after ending your actual fibrillation experiment, unless quenching conditions are carefully developed. In many cases such a matured sample is still useful to test the loading conditions, but the SAXS data cannot be included in the data pool for decomposition. Please consider carefully how you store such preformed samples, since temperature changes may influence the fibril state dramatically (Skamris et al., 2019).

4. The main experiment: SAXS data collection

Brief intro

The main experiment (MainExp) is the culmination of your hard work. Here, you collect the central batch of SAXS data, into which you will subsequently invest significant time and efforts in your data analysis, and to which you will refer with your complementary biophysical data. It cannot be stressed enough how important this experiment is for the final outcome of your investigations. We thus recommend that you invest everything you possibly can into this procedure. If possible, travel one or more days in advance, such that you have ample time to setup equipment, possibly start an experiment (to extract late time points), and not least get a good night sleep before your beamtime. Bring enough manpower, such that you can work around the clock handling the plate reader, loading the samples, and performing an initial analysis of your incoming SAXS data. That is, if possible, plan to have two to three people at the beamline/in the lab at all times throughout the experiment. Schedule work shifts and distribute the responsibilities before the experiment is running and ensure that there is time planned for hand-over of experience when there is a shift, e.g., in the middle of the night. MainExp is a fascinating event.

4.1 Equipment

- Spectrophotometer, e.g., NanoDrop for accurate protein concentration determination
- Fluorescence plate reader with black 96-well plates with clear bottom, sealing tape, potentially glass beads

- Heat block for preheating samples/solvents prior to initiating fibrillation and as basis for the plate when extracting samples
- Vessel for liquid N_2
- Synchrotron bioSAXS beamline or state-of-the-art in-house BioSAXS instrument
- Software for initial data evaluation (note 1)

Bring everything from home. Do not rely on availability of anything at the beamline, unless specifically assured by the beamtime responsible. E.g., some filter types may induce aggregation or proper sample extraction may rely on specific pipette tips

4.2 Reagents

- Highly purified protein or peptide for fibrillation
- Optimized buffer, filtered with 0.22 μm filters. Buffer must exactly match the protein buffer (note 2)
- Stock solution of a fibril specific fluorescent probe
- Liquid N_2
- Protein standard, e.g., bovine serum albumin to be dissolved in a recommended buffer (typically 50 mM HEPES, pH 7.5). The protein standard is used for relative scaling of the molecular mass (see details elsewhere in this volume). Most beamlines provide a suitable protein standard and recommended buffers—ask your beamline responsible
- Cleaning reagent for sample cell (note 4, often provided by beamline, but note that extra cleaning may be necessary)
- Cleaning reagent for wetlab space and equipment (see Section 2)
- Bring everything from home, including buffer stocks for backup experiments

4.3 Procedure

1. Upon arrival, install and test your equipment in the wet lab well in advance of the beamtime. Prepare the fibrillation experiment, but do not initiate the fibrillation reaction yet (unless you plan to produce late stage samples for your measurements).
2. When beamtime commences, collect SAXS data from the recommended protein standard and buffers. Carefully determine the protein concentration of both the protein standard, the non-fibrillated protein (point 3) and the protein stock for your main fibrillation experiment (point 5).

3. Collect SAXS data from the non-fibrillated amyloid-like protein at different concentrations, preferably in the fibrillation buffer (unless this will immediately initiate the fibrillation reaction, in which case we recommend that you use a buffer where the protein remains stable). Check carefully for any signs of radiation damage or structure factors (note 3). If you observe radiation damage, you must modify your fibrillation buffer by including a higher concentration of the radical scavenger or even a different protectant (Graewert & Jeffries, 2017; Jeffries et al., 2015). Ideally the sample concentration will be high enough to collect data with good signal:noise and low enough to avoid structure factor effects
4. Now test the loading procedure for a fully fibrillated sample, produced at home. This test has dual reason: first, you establish whether the viscous sample is moving nicely in and out of the sample cell, which hence also includes a test of the cleaning procedure. Second, you establish whether the sample has propensity to align in the sample cell

   The first part you can control as follows: (a) if there is a real-time camera on the sample cell, watch the sample while it is loading. Does the sample load at all, or is it left behind in the sample rack? Do you see bubbles (=trouble)? It may be necessary to modify the sample loading procedure, if any of these conditions apply (also see Section 3, point 9); (b) once the sample is loaded and you have collected the data, the signal:noise ratio at higher angles should compare to the signal:noise of the non-fibrillated sample (at similar concentration)? (c) if the sample nicely leaves the sample cell (rather than sticking to the windows) the SAXS data from the buffer before is measured before and after the fibril sample, and the two should perfectly superimpose. Alternatively, you must improve the cleaning procedure of the sample cell.

   To check for unwanted alignment in the sample cell, inspect the 2D images of the scattering data, before radial averaging. The scattering should be isotropic, indicating random orientation of the sample. Non-isotropic scattering indicates that the fibrils are aligned (by flow) in the sample cell (for example images, please refer to Langkilde & Vestergaard, 2012). If you detect alignment, you must either homogenize the sample or establish an alternative loading procedure (e.g., lower flow rates during loading, but always consult your beamline scientist before changing such procedures). Make sure that any potential problems with loading of the fibril samples are solved before commencing the fibrillation reaction.

5. From now on make notes about virtually everything. Note down the time for any action point (onset of fibrillation, extraction time, halting time, exposure time, temperatures, changes at the beamline, etc.). Also check the synchronicity of the different time stamps (in the various data files) that are made by different computers, to avoid later confusion
6. Initiate the fibrillation reaction. Note down the time. Ensure that the temperature in the plate reader is stabilized before you continue
7. Note down the time. Pause the plate reader. Take out your 96-well plate from the plate reader and place it on the tempered heating block to minimize temperature variation. Extract a sample from a well (note down the well number), using a pipette tip with adequately large diameter (to make sure that fibrillated material can enter the tip). Reinsert the plate as quickly as possible in the plate reader. Note down the time when the plate reader is restarted; this is often tracked by the reader control program, but thorough notes facilitate on the fly evaluations of outliers or similar
8. If possible, one may choose to quench the fibrillation reaction. Quenching conditions must be established prior to the beamtime, and the conditions are system specific. For instance, we have established that fibrillating aSN remains stable for several minutes at ambient temperature under quiescent conditions, when using the experimental conditions described, e.g., in Giehm et al. (2011). If you use quenching, don't forget to confirm that the fibrillation reaction is indeed quenched. You do this by performing at least two (preferably more) SAXS exposures from a sample placed under quenching conditions in a considerable amount of time
9. If the sample is viscous (i.e., at later time points in the fibrillation reaction) it must be homogenized, such that the fibril material can be loaded into the sample cell. We recommend the use of freeze-thaw cycles, but the choice of homogenization method is system-specific. Please note safety recommendations
10. The sample is now
    a. Loaded directly into the sample cell or
    b. Placed in the sample changer (note down the position in the sample changer)
11. Collect SAXS data from the sample
12. Rinse the sample cell thoroughly. Likely there are beamline standards for maintaining a clean sample cell, which must be followed as a minimum. In some cases, it is necessary to perform additional cleaning steps

13. Load the buffer into the sample cell and collect SAXS data
14. Repeat from point 7
15. Plot the last recorded fluorescence data point before each extraction/measurement in a separate sheet. These points should ideally end up covering the entire fibrillation reaction
16. Aim to leave a few samples untouched in the 96-well plate, such that an undisturbed fluorescence trace can be directly compared to the trace created in point 15

As mentioned above, we recommend that one person has the responsibility to control the incoming data. This person can hence react if something unforeseen is happening, which will give you the chance to extract samples from a particular time window of the fibrillation reaction, or even alter the wet lab experiments accordingly. We recommend that the data-responsible person does the following, while the data is collected:

17. Keep an eye on the camera on the sample cell, while the sample is loading. Watch out for bubbles (i.e., if the sample changer does not pipette the fibrillated sample correctly).
18. Repeatedly check that the data from buffers remain completely stable, i.e., that buffer-data superimpose perfectly over time. If this is not the case, perform additional cleaning steps, and re-measure the data from a buffer, until the superimposition is optimal
19. Repeatedly check some of the 2D images to look for signs of non-isotropic scattering. See point 4 above
20. Plot the radially averaged data from your protein samples, before background subtraction. Are spikes or other unusual features visible in the data? If yes, this may be due to fibril alignment and/or improper cleaning. See point 4 above
21. Some beamlines use repeated exposures on individual samples and use automated software to subsequently average data-files that do not reveal radiation damage, or other differences (Blanchet et al., 2015). This is a very convenient and user-friendly setup, but in the case of fibrillation data, it is important to perform an analysis of the individual, non-averaged data files. We describe this in detail in a later chapter (Langkilde & Vestergaard, 2022). Here, during data collection, we recommend that you make spot checks of your data: React if you see that the signal:noise in our averaged data-files suddenly drops. Read the header in such a datafile, to see how many of the individual exposures (frames) that are included in the averaged data. If the number is lower than normal, look further into the situation (note 5).

22. Plot the background subtracted data together, as they are collected from the developing fibrillation series. You will (almost certainly) be able to see evolvement in the samples before the changes are visible by fluorescence spectroscopy, and we recommend intensifying sample extraction and data collection in this time window. The appearance of fibrils in the samples will be clearly visible in the data, by a strong increase in scattering intensity at lower angles. Again, once you see such an increase, intensify data collection. Compare your observations to the fluorescence spectroscopy traces in the plate reader. If you can see that you miss data in a particular time interval, try to identify wells in the 96-well plate that seem to correspond to this time point (i.e., judged by the fluorescence trace). If you do not have any such samples available, consider if you can initiate new fibrillation reactions in empty wells, such that you can deliberately extract and measure once these samples reach the given time point
23. Remember to make clear notes about any observations made during MainExp. Such notes may save you a lot of effort later during data evaluation

For data analysis, we refer to our chapter in the subsequent volume of this publication (Langkilde & Vestergaard, 2022).

4.4 Notes

1. At most beamlines software will already be installed enabling direct viewing of your raw datafiles (i.e., 2D intensity plots) as well as the radially averaged files. Alternatively you can download and install freely available packages such as SASview (www.sasview.org), RAW (Hopkins et al., 2017; Nielsen et al., 2009) or SAXSview (https://github.com/emblsaxs/saxsview) or Fit2D (Hammersley, 1997). Most dedicated bioSAXS beamlines have advanced and automated pipelines for data reduction and analysis installed (Blanchet et al., 2015), which is very convenient for non-fibrillated samples. Among many other things, such a pipeline will typically provide radially averaged, frame-averaged, background-subtracted datafiles, which conveniently can be plotted such that you can follow the changes in the data from the evolving fibrillation reaction.
2. As for any SAXS experiment, the background buffer should exactly match the solvent in the protein sample. In a standard SAXS experiment it is recommended to use dialysis as a final step before collecting your

SAXS data, such that you can use the dialysate as the perfectly matching buffer. This is very often not possible for a fibrillation experiment since many fibrillation-prone proteins will not remain in the native state during the dialysis. In some cases, it is necessary to dissolve a freeze-dried powder directly into the fibrillation buffer, after which the fibrillation reaction will immediately commence. In such cases, it is important to investigate (at home, well in advance of the beamtime) how much salt is stored in your freeze-dried powder, since this will modify the buffer, when dissolving the powder. Your background buffer must thus include the corresponding amount of salt. In other cases the fibrillation reaction is initiated by the addition of a chemical or a change in pH. Remember to include the exact same chemicals in your background buffer.

3. Ideally, if you are a returning user at the beamline, or if you have access to other SAXS equipment at home, you have already checked the solution state of your native starting point at an earlier beamtime. Any changes in the fibrillation buffer at this stage will be suboptimal, since changes to the optimized conditions include the risk to change the timing of your fibrillation reaction, or even (worst case scenario) induce a different fibrillation reaction. It is, however, the least of two evils since radiation damage will induce non-fibril aggregation, which will ruin your data quality and hence hamper your subsequent data analysis.
4. At most beamlines a very thorough cleaning procedure is standard, and in most cases it is sufficient to follow the guidelines at the beamline. Always keep an eye out, however, for signs of contamination of the SAXS sample cell. This is most easily done by continuously superimposing the most recent buffer measurement with the earliest buffer measurements. The frames must superimpose perfectly. If not, perform additional cleaning, preferably with the cleaning reagent recommended by the beamline staff. Subsequently include longer cleaning procedures after all protein measurements. Never start cleaning the SAXS sample cell with harsh chemicals, without first consulting the beamline staff.
5. When the beamline uses repeated exposures and data averaging you must independently evaluate the variability of the individual radially averaged files. This is particularly important before commencing the data analysis process, which is covered in (Langkilde & Vestergaard, 2022), but also highly relevant to be aware off during the on-site evaluation. During data collection, if, e.g., the signal:noise lowers in the background subtracted data, this may be because the automated software has detected discrepancies between individual exposures, which normally would signify

radiation damage (Graewert et al., 2015). In our specialized case, there may however be different explanations, and thus the data are not necessarily damaged. We distinguish between the two following scenarios:

i. The heterogeneous sample (containing, e.g., native protein, oligomers and fibrils) distributes unevenly in the SAXS sample cell while flushed through during X-ray radiation. This results in varying scattering intensities in the individual frames, but all frames are relevant to characterize the entire sample and should be averaged. The user will typically see random variations in the different frames. In this case data collection can continue but keep an eye on the individual frames in subsequent measurements.

ii. X-rays induce aggregation. In this case the first few frames may be identical after which a systematic increase in scattering intensity will be observed, and all subsequent files show the same trend. Only the first files are useful (if they are identical and thus unaffected) and the situation is non-optimal. Make sure to inspect the individual files from your next exposure of a protein sample. If radiation damage remains a problem, you must alter your strategy (consult the beamline scientist!). Potential solutions include shorter individual exposure times; larger sample volumes in combination with higher flowrates during exposure; attenuation of the beam; inclusion of higher concentrations of radical scavengers. Neither solution is optimal, if imposed in the middle of a data series, but it may be the least of two evils. If you do not alter your conditions at this point, likely all your subsequent data will suffer from radiation damage and hence you risk to not have any useful data from the latter part of your fibrillation series.

5. Summary and conclusions

In this chapter, we aim to provide a point-by-point protocol for the planning, optimization, and execution of a protein fibrillation experiment, optimized for and ultimately performed ex situ in parallel with SAXS data collection. The format of this volume enables the inclusion of numerous detailed tips and tricks that should prepare the reader for the most typical types of obstacles that the authors have experienced over the years while developing the know-how described here. They are based on our very positive experience with using high-end synchrotron radiation SAXS facilities but are directly transferable also to a setting with a laboratory based SAXS instrument, given that this instrument meets adequate standards. Protein fibrillation is, at least from the authors point of view, one of the most complex and fascinating structural biology phenomena known, involving transient structures ranging from the nm to μm scale. Several of the individual structures only exist in context of the other structures in the mixture, meaning that they cannot be isolated from the mixture without changing their structural properties. This immediately calls for the use of a structural biology method, that can be used on evolving mixtures, and indeed SAXS is an ideal method for this (Brosey & Tainer, 2019; Grawert & Svergun, 2020; Konarev & Svergun, 2018; Meisburger et al., 2021; Vestergaard, 2016). Parts of the methods described here can be used when preparing for data collection on other evolving bio- or nano-systems, such as protein filaments and designed nanoscale self-assembly materials.

Acknowledgments

The experience assembled in this chapter has been established based on the much-appreciated hard work of current and former members of the BioSAXS group at the University of Copenhagen. It has only been possible due to numerous visits to different synchrotron facilities. We extend our gratitude to the dedicated staff at X33 and P12 at EMBL-Hamburg, and the continued scientific support and interest from Dmitri I. Svergun as well as the great support at BM29 at ESRF; SWING beamline at SOLEIL; I711 and I911-4 at MAX-lab. We appreciate funding from the Independent Research Fund Denmark | Medical Sciences, the Lundbeck Foundation, the Novo Nordisk Foundation, The Carlsberg Foundation and DANSCATT.

References

Ankarcrona, M., et al. (2016). Current and future treatment of amyloid diseases. *Journal of Internal Medicine*, *280*(2), 177–202.

Bigi, A., et al. (2021). Exploring the release of toxic oligomers from alpha-synuclein fibrils with antibodies and STED microscopy. *Life (Basel)*, *11*(5), 431.

Blanchet, C. E., et al. (2015). Versatile sample environments and automation for biological solution X-ray scattering experiments at the P12 beamline (PETRA III, DESY). *Journal of Applied Crystallography*, *48*(Pt. 2), 431–443.

Bousset, L., et al. (2016). An efficient procedure for removal and inactivation of alpha-synuclein assemblies from laboratory materials. *Journal of Parkinson's Disease*, *6*(1), 143–151.

Brosey, C. A., & Tainer, J. A. (2019). Evolving SAXS versatility: Solution X-ray scattering for macromolecular architecture, functional landscapes, and integrative structural biology. *Current Opinion in Structural Biology*, *58*, 197–213.

Buell, A. K. (2019). The growth of amyloid fibrils: Rates and mechanisms. *The Biochemical Journal*, *476*(19), 2677–2703.

Bulawa, C. E., et al. (2012). Tafamidis, a potent and selective transthyretin kinetic stabilizer that inhibits the amyloid cascade. *Proceedings of the National Academy of Sciences of the United States of America*, *109*(24), 9629–9634.

Casassus, B. (2021). France halts prion research amid safety concerns. *Science*, *373*(6554), 475–476.

Cornwall, G. A., et al. (2019). The epididymal amyloid matrix: Structure and putative functions. *Andrology*, 7(5), 603–609.

Dicko, C., et al. (2020). NUrF—Optimization of in situ UV–vis and fluorescence and autonomous characterization techniques with small-angle neutron scattering instrumentation. *Review of Scientific Instruments*, *91*(7), 075111.

Dyer, K. N., et al. (2014). High-throughput SAXS for the characterization of biomolecules in solution: A practical approach. *Methods in Molecular Biology*, *1091*, 245–258.

Eanes, E. D., & Glenner, G. G. (1968). X-ray diffraction studies on amyloid filaments. *The Journal of Histochemistry and Cytochemistry*, *16*(11), 673–677.

Falke, S., et al. (2018). Multi-channel in situ dynamic light scattering instrumentation enhancing biological small-angle X-ray scattering experiments at the PETRA III beamline P12. *Journal of Synchrotron Radiation*, *25*(Pt. 2), 361–372.

Fitzpatrick, A. W. P., et al. (2017). Cryo-EM structures of tau filaments from Alzheimer's disease. *Nature*, *547*(7662), 185–190.

Fodera, V., et al. (2008). Thioflavin T hydroxylation at basic pH and its effect on amyloid fibril detection. *The Journal of Physical Chemistry B*, *112*(47), 15174–15181.

Franco, A., et al. (2021). All-or-none amyloid disassembly via chaperone-triggered fibril unzipping favors clearance of alpha-synuclein toxic species. *Proceedings of the National Academy of Sciences of the United States of America*, *118*(36), e2105548118.

Gade Malmos, K., et al. (2017). ThT 101: A primer on the use of thioflavin T to investigate amyloid formation. *Amyloid*, *24*(1), 1–16.

Giehm, L., & Otzen, D. E. (2010). Strategies to increase the reproducibility of protein fibrillization in plate reader assays. *Analytical Biochemistry*, *400*(2), 270–281.

Giehm, L., et al. (2011). Low-resolution structure of a vesicle disrupting α-synuclein oligomer that accumulates during fibrillation. *Proceedings of the National Academy of Sciences of the United States of America*, *108*(8), 3246–3251.

Graewert, M. A., et al. (2015). Automated pipeline for purification, biophysical and x-ray analysis of biomacromolecular solutions. *Scientific Reports*, *5*, 10734.

Graewert, M. A., & Jeffries, C. M. (2017). Sample and buffer preparation for SAXS. *Advances in Experimental Medicine and Biology*, *1009*, 11–30.

Grawert, T. W., & Svergun, D. I. (2020). Structural modeling using solution small-angle X-ray scattering (SAXS). *Journal of Molecular Biology*, *432*(9), 3078–3092.

Groenning, M., et al. (2015). Considerably unfolded transthyretin monomers preceed and exchange with dynamically structured amyloid protofibrils. *Scientific Reports*, *5*, 11443.

Hammersley, A. P. (1997). FIT2D: An introduction and overview. In *ESRF Internal Report, ESRF97HA02T* European Synchrotron Radiation Facility.

Hauser, C. A., et al. (2014). Amyloid-based nanosensors and nanodevices. *Chemical Society Reviews*, *43*(15), 5326–5345.

Herranz-Trillo, F., et al. (2017). Structural Analysis of multi-component amyloid systems by chemometric SAXS data decomposition. *Structure*, *25*(1), 5–15.

Holmqvist, S., et al. (2014). Direct evidence of Parkinson pathology spread from the gastrointestinal tract to the brain in rats. *Acta Neuropathologica*, *128*(6), 805–820.

Hopkins, J. B., et al. (2017). BioXTAS RAW: Improvements to a free open-source program for small-angle X-ray scattering data reduction and analysis. *Journal of Applied Crystallography*, *50*(Pt. 5), 1545–1553.

Hoppener, J. W., & Lips, C. J. (2006). Role of islet amyloid in type 2 diabetes mellitus. *The International Journal of Biochemistry & Cell Biology*, *38*(5–6), 726–736.

Huang, Z., et al. (2013). Self-assembly of amphiphilic peptides into bio-functionalized nanotubes: A novel hydrolase model. *Journal of Materials Chemistry B*, *1*(17), 2297–2304.

Haas, S., et al. (2014). Combined SAXS/UV-vis/Raman as a diagnostic and structure resolving tool in materials and life sciences applications. *The Journal of Physical Chemistry B, 118*(8), 2264–2273.
Jahn, T. R., et al. (2010). The common architecture of cross-beta amyloid. *Journal of Molecular Biology, 395*(4), 717–727.
Jeffries, C. M., et al. (2015). Limiting radiation damage for high-brilliance biological solution scattering: Practical experience at the EMBL P12 beamline PETRAIII. *Journal of Synchrotron Radiation, 22*(Pt. 2), 273–279.
Khambhati, K., et al. (2021). Gene regulation of biofilm-associated functional amyloids. *Pathogens, 10*(4), 490.
Kollmer, M., et al. (2019). Cryo-EM structure and polymorphism of Abeta amyloid fibrils purified from Alzheimer's brain tissue. *Nature Communications, 10*(1), 4760.
Konarev, P. V., & Svergun, D. I. (2018). Direct shape determination of intermediates in evolving macromolecular solutions from small-angle scattering data. *IUCrJ, 5*(Pt. 4), 402–409.
Konarev, P. V., et al. (2003). PRIMUS: A Windows PC-based system for small-angle scattering data analysis. *Journal of Applied Crystallography, 36*, 1277–1282.
Kumari, P., et al. (2021). Structural insights into alpha-synuclein monomer-fibril interactions. *Proceedings of the National Academy of Sciences of the United States of America, 118*(10), e2012171118.
Lafleur, J. P., et al. (2011). Automated microfluidic sample-preparation platform for high-throughput structural investigation of proteins by small-angle X-ray scattering. *Journal of Applied Crystallography, 44*, 1090–1099.
Langkilde, A. E., et al. (2018). Noninvasive structural analysis of intermediate species during fibrillation: An application of small-angle X-ray scattering. *Methods in Molecular Biology, 1779*, 209–239.
Langkilde, A. E., et al. (2015). The architecture of amyloid-like peptide fibrils revealed by X-ray scattering, diffraction and electron microscopy. *Acta Crystallographica. Section D, Biological Crystallography, 71*(Pt. 4), 882–895.
Langkilde, A. E., & Vestergaard, B. (2012). Structural characterization of prefibrillar intermediates and amyloid fibrils by small-angle X-ray scattering. *Methods in Molecular Biology, 849*, 137–155.
Langkilde, A. E., & Vestergaard, B. (2022). Protein fibrillation from another small angle – SAXS data analysis of developing systems. *Methods in Enzymology, 678*. In press.
Levkovich, S. A., et al. (2021). Two decades of studying functional amyloids in microorganisms. *Trends in Microbiology, 29*(3), 251–265.
Li, J., & Zhang, F. (2021). Amyloids as building blocks for macroscopic functional materials: Designs, applications and challenges. *International Journal of Molecular Sciences, 22*(19), 10698.
Lim, K. H. (2019). Diverse misfolded conformational strains and cross-seeding of misfolded proteins implicated in neurodegenerative diseases. *Frontiers in Molecular Neuroscience, 12*, 158.
Lövestam, S., et al. (2022). Assembly of recombinant tau into filaments identical to those of Alzheimer's disease and chronic traumatic encephalopathy. *Elife, 11*, e76494.
Mackay, J. P., et al. (2001). The hydrophobin EAS is largely unstructured in solution and functions by forming amyloid-like structures. *Structure, 9*(2), 83–91.
Madhu, P., & Mukhopadhyay, S. (2021). Distinct types of amyloid-beta oligomers displaying diverse neurotoxicity mechanisms in Alzheimer's disease. *Journal of Cellular Biochemistry, 122*(11), 1594–1608.
Maity, D., et al. (2020). Sub-stoichiometric inhibition of IAPP aggregation: A peptidomimetic approach to anti-amyloid agents. *RSC Chemical Biology, 1*(4), 225–232.
Maji, S. K., et al. (2008). Amyloid as a depot for the formulation of long-acting drugs. *PLoS Biology, 6*(2), e17.

Maji, S. K., et al. (2009). Functional amyloids as natural storage of peptide hormones in pituitary secretory granules. *Science, 325*(5938), 328–323.
McAlary, L., et al. (2019). Prion-like propagation of protein misfolding and aggregation in amyotrophic lateral sclerosis. *Frontiers in Molecular Neuroscience, 12*, 262.
Meisburger, S. P., et al. (2021). REGALS: A general method to deconvolve X-ray scattering data from evolving mixtures. *IUCrJ, 8*(Pt. 2), 225–237.
Merlini, G., et al. (2018). Systemic immunoglobulin light chain amyloidosis. *Nature Reviews. Disease Primers, 4*(1), 38.
Morris, A. D., et al. (2019). The pathology and changing epidemiology of dialysis-related cardiac beta-2 microglobulin amyloidosis. *Cardiovascular Pathology, 42*, 30–35.
Nielsen, S. B., et al. (2013). Wildtype and A30P mutant alpha-synuclein form different fibril structures. *PLoS One, 8*(7), e67713.
Nielsen, S. S., et al. (2009). BioXTAS RAW, a software program for high-throughput automated small-angle X-ray scattering data reduction and preliminary analysis. *Journal of Applied Crystallography, 42*, 959–964.
Nors Pedersen, M., et al. (2015). Direct correlation between ligand-induced alpha-synuclein oligomers and amyloid-like fibril growth. *Scientific Reports, 5*, 10422.
Oliveira, C. L. P., et al. (2009). A SAXS study of glucagon fibrillation. *Journal of Molecular Biology, 387*(1), 147–161.
Otzen, D., & Riek, R. (2019). Functional amyloids. *Cold Spring Harbor Perspectives in Biology, 11*(12), 1–28.
Palmadottir, T., et al. (2021). Charge regulation during amyloid formation of alpha-synuclein. *Journal of the American Chemical Society, 143*(20), 7777–7791.
Papa, R., & Lachmann, H. J. (2018). Secondary, AA, amyloidosis. *Rheumatic Diseases Clinics of North America, 44*(4), 585–603.
Peelaerts, W., et al. (2015). alpha-Synuclein strains cause distinct synucleinopathies after local and systemic administration. *Nature, 522*(7556), 340–344.
Pratt, A. J., et al. (2014). Aggregation propensities of superoxide dismutase G93 hotspot mutants mirror ALS clinical phenotypes. *Proceedings of the National Academy of Sciences of the United States of America, 111*(43), E4568–E4576.
Radamaker, L., et al. (2021). Role of mutations and post-translational modifications in systemic AL amyloidosis studied by cryo-EM. *Nature Communications, 12*(1), 6434.
Rasmussen, H. O., et al. (2021). A multimethod approach for analyzing FapC fibrillation and determining mass per length. *Biophysical Journal, 120*(11), 2262–2275.
Rochet, J. C., et al. (2000). Inhibition of fibrillization and accumulation of prefibrillar oligomers in mixtures of human and mouse alpha-synuclein. *Biochemistry, 39*(35), 10619–10626.
Round, A., et al. (2015). BioSAXS sample changer: A robotic sample changer for rapid and reliable high-throughput X-ray solution scattering experiments. *Acta Crystallographica. Section D, Biological Crystallography, 71*, 67–75.
Rubel, M. S., et al. (2020). Functional mammalian amyloids and amyloid-like proteins. *Life (Basel), 10*(9), 156.
Sagar, A., et al. (2021). Structure and thermodynamics of transient protein-protein complexes by chemometric decomposition of SAXS datasets. *Structure, 29*(9), 1074–1090.e4.
Salinas, N., et al. (2021). The amphibian antimicrobial peptide uperin 3.5 is a cross-α/cross-β chameleon functional amyloid. *Proceedings of the National Academy of Sciences of the United States of America, 118*(3), e2014442118.
Schweighauser, M., et al. (2020). Structures of alpha-synuclein filaments from multiple system atrophy. *Nature, 585*(7825), 464–469.
Shi, Y., et al. (2021). Structure-based classification of tauopathies. *Nature, 598*(7880), 359–363.

Si, J. B., et al. (2021). Transthyretin misfolding, a fatal structural pathogenesis mechanism. *International Journal of Molecular Sciences*, *22*(9), 4429.
Skamris, T., et al. (2019). Early stage alpha-synuclein amyloid fibrils are reservoirs of membrane-binding species. *Scientific Reports*, *9*(1), 1733.
Skamris, T., et al. (2022). Identifying biological and biophysical features of different maturation states of α-synuclein amyloid fibrils. *Methods in Molecular Biology*. In press.
Srinivasan, E., et al. (2021). Alpha-synuclein aggregation in Parkinson's disease. *Frontiers in Medicine (Lausanne)*, *8*, 736978.
Tayeb-Fligelman, E., et al. (2017). The cytotoxic *Staphylococcus aureus* PSMα3 reveals a cross-α amyloid-like fibril. *Science*, *355*(6327), 831–833.
Trejo-Lopez, J. A., et al. (2022). Neuropathology of Alzheimer's disease. *Neurotherapeutics*, *19*(1), 173–185.
van Maarschalkerweerd, A., et al. (2015). Formation of covalent di-tyrosine dimers in recombinant alpha-synuclein. *Intrinsically Disordered Proteins*, *3*(1), e1071302.
Vestergaard, B. (2016). Analysis of biostructural changes, dynamics, and interactions—Small-angle X-ray scattering to the rescue. *Archives of Biochemistry and Biophysics*, *602*, 69–79.
Vestergaard, B., et al. (2007). A helical structural nucleus is the primary elongating unit of insulin amyloid fibrils. *PLoS Biology*, *5*(5), e134.
Wang, Q., & Schniepp, H. C. (2018). Strength of recluse spider's silk originates from nanofibrils. *ACS Macro Letters*, 7(11), 1364–1370.
Winner, B., et al. (2011). In vivo demonstration that alpha-synuclein oligomers are toxic. *Proceedings of the National Academy of Sciences of the United States of America*, *108*(10), 4194–4199.
Yang, Y., et al. (2022). Cryo-EM structures of amyloid-β 42 filaments from human brains. *Science*, *375*(6577), 167–172.
Zahra, W., et al. (2020). The global economic impact of neurodegenerative diseases: Opportunities and challenges. In C. Keswani (Ed.), *Bioeconomy for sustainable development* (pp. 333–345). Singapore: Springer.
Zimmermann, M. R., et al. (2021). Mechanism of secondary nucleation at the single fibril level from direct observations of abeta42 aggregation. *Journal of the American Chemical Society*, *143*(40), 16621–16629.

CHAPTER ELEVEN

High-pressure SAXS, deep life, and extreme biophysics

Richard E. Gillilan*
Cornell High Energy Synchrotron Source, Cornell University, Ithaca, NY, United States
*Corresponding author: e-mail address: reg8@cornell.edu

Contents

Abstract

The biological relevance of hydrostatic pressure is becoming much more widely understood and appreciated as discoveries of new niches for extreme life continue to emerge. The unusual chemistry and physiological adaptations of organisms under extreme pressure promises to be a rich source of new insights in the years ahead if structural information can be obtained at the molecular level. Fortunately, recent advances in instrumentation are making structural biology techniques easier to perform at extreme pressures and more widely available. In addition to biological applications, hydrostatic pressure is a useful biophysical tool that can perturb systems in ways directly connected to the presence of atomic-level voids, cavities, and other volumetric properties. Under pressure, individual molecular complexes can dissociate, and monomers can unfold; transitions can occur in lipid mesophases, and liquid phases can dissolve and re-form. Small angle X-ray solution scattering (SAXS) can detect and characterize pressure-induced changes in all these situations. This chapter reviews what is known about pressure effects in a wide variety of biomolecular systems and how those effects display in X-ray scattering data. The influence of hydrostatic pressure on solution

Methods in Enzymology, Volume 677
ISSN 0076-6879
https://doi.org/10.1016/bs.mie.2022.08.025

scattering is discussed, and the most widely used data processing methods are re-examined considering pressure effects. The chapter concludes with an overview of the high-pressure SAXS instrument design followed by recommended data collection protocol.

1. Introduction

Small angle X-ray solution scattering (SAXS) measures the intensity of X-rays scattered from a sample at angles ranging from 0° to ~5°. The end result of a SAXS measurement is called the *scattering profile*, a one-dimensional curve, $I(q)$, that records the number of scattered photons of wavelength λ of as a function of the *momentum transfer* $q = 4\pi \sin(\theta)/\lambda$. 2θ is the angle between incident beam and scattered beam.

SAXS is popular for its simplicity and ability to measure structural features in solution under a wide range of conditions. For dilute solutions, it routinely yields basic size, molecular weight, shape information, and even low-resolution electron density. A particularly valuable feature of SAXS is its ability to characterize unfolding and disorder. When combined with known high-resolution structures, the method can distinguish between different models and provide information about ensembles of conformations in solution. With concentrated solutions, SAXS is also sensitive to interactions between neighboring biomolecules. For readers new to SAXS, there are a number of recent texts, reviews and protocols (including those in this volume) that have already been written on the subject (Brosey & Tainer, 2019; Jeffries, Graewert, Svergun, & Blanchet, 2015; Lattman, Grant, & Snell, 2018; Skou, Gillilan, & Ando, 2014; Svergun, Koch, Timmins, & May, 2013; Tuukkanen, Spilotros, & Svergun, 2017). This discussion will focus exclusively on SAXS as a tool for understanding how biomolecules respond to high hydrostatic pressure.

The International System of Units (SI) specifies pressure in Pascals (Pa). For the range of pressures studied in biological systems, the megapascal ($1\,\text{MPa} = 10^6\,\text{Pa}$) is the unit of more useful magnitude. The *bar*, a popular non-SI metric unit very close to one atmosphere of pressure, is defined as 0.1 MPa. Pressure effects in biological studies are small enough that there is usually no harm in rounding pressures to the nearest MPa. Consequently, atmospheric pressure will be referred to simply as 0 MPa or *ambient pressure* throughout this work.

For a sense of scale, hydrostatic pressure in the ocean increases by approximately 1 MPa for every 100 m of depth. Approximately 88% of the world's ocean volume is above 10 MPa with the deepest trenches reaching 110 MPa. In deep subduction zones, where the temperature approaches what is thought to be the limit for life, 122 °C, the pressure may be as high as 340 MPa (Plümper et al., 2017). High pressure food processing (HPP) can reach 600 MPa where it is effective at killing pathogens.

Effects of pressure on biological materials have long been known (Bridgman, 1914), but the significance of pressure to biology itself was not widely recognized until late in the 20th century. Starting with the surprise discovery of life centered around ocean hydrothermal vents in the late 1970s, it became increasingly apparent that life could exist in some of the most extreme environments on Earth. Speculations about vast subsurface microbial communities supported by chemical sources of energy in the "deep hot biosphere" in the 1990s are now firmly supported by evidence (Colman, Poudel, Stamps, Boyd, & Spear, 2017). Microbial life is known to exist not only in hydrothermal vents and deep ocean trenches, but deep subsurface coal beds, mine shafts, wells, and porous rock kilometers below the surface (Ando et al., 2021). Though sometimes very sparsely distributed and living on a severely energy-restricted diet, *deep life* constitutes a major portion of the Earth's biomass and has significant impact on the environment (Bar-On, Rob, & Ron, 2018; Worman, Pratson, Karson, & Schlesinger, 2020). The unique chemical and physiological adaptation strategies of these organisms promises to be a rich source of discovery and of valuable tools for molecular biology (Hatti-Kaul & Abouhmad, 2020). While pressure is a fundamental variable necessary for any complete thermodynamic characterization of a molecular system, it has important implications for how life functions and the course that evolution has taken in many organisms.

This chapter begins with an overview of biologically relevant pressure-induced structural changes that are observable with SAXS. We then proceed to describe how the various terms contributing to the SAXS scattering profile respond to pressure, including implications for basic model-independent processing and analysis. Essential elements of design and operation for our high-pressure SAXS system are then presented, concluding with recommended protocols for good practice in data collection.

2. Structural effects of pressure on biomolecular systems

The response of whole microbes to high hydrostatic pressure is a complex and still poorly understood process that involves seemingly every aspect of cellular function, including the properties of lipids as well as proteins and other biomolecules (Akasaka, 2015). Based on the similarity between pressure-temperature viability diagrams observed for microorganisms and the pressure-temperature stability curves for proteins, it has been suggested that the disruption of protein-protein interactions may play a major role in the pressure sensitivity of organisms (Meersman, Smeller, & Heremans, 2006). Although protein folds are relatively resistant to pressure, macromolecular complexes are much more susceptible to disruption (Fig. 1; molecular complexes).

SAXS can readily be applied to follow changes in oligomeric state under pressure (Fujisawa & Kato, 2002). The scattering profile for yeast enolase at ambient pressure, for example, fits the predicted scattering for the crystallographic dimer (PDB ID: 3ENZ), but cannot be made to match the scattering

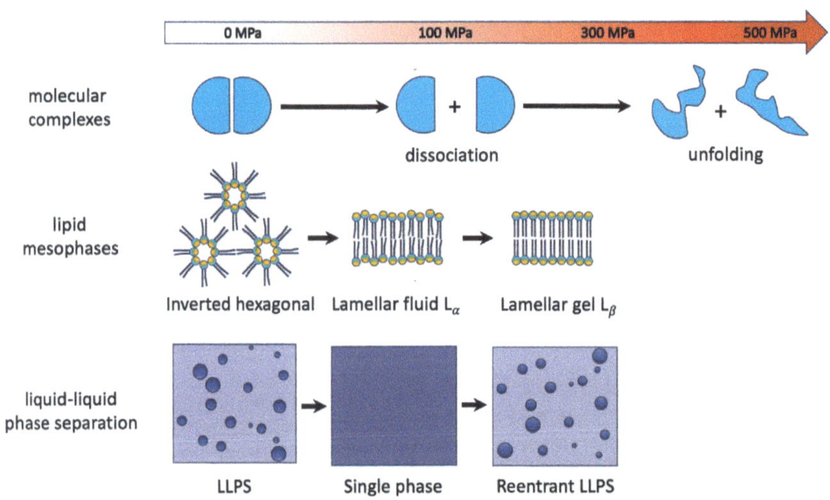

Fig. 1 Categories of pressure-induced transformations in biological systems characterizable with small-angle X-ray scattering (SAXS). As hydrostatic pressure climbs to 100 MPa, molecular complexes can start to dissociate. Individual monomer units largely retain their internal structure until pressures exceed 300 MPa or more where unfolding is observed. Changes in oligomeric state and folding are observable with standard small angle scattering techniques. Lipid mesophases are far more sensitive to pressure. Less efficiently packed non-lamellar phases, such as the inverted hexagonal form, give way to lamellar phases at pressures well below 100 MPa. Because mesophases are extended structures with well-defined atomic separations, X-ray scattering patterns display distinctive peaks. Like lipidic phases, the biologically important phenomenon of liquid-liquid phase separation (LLPS) is also pressure sensitive. Pressure can induce the dissolution and re-formation of heterogenous states (represented here). Though less studied than the other two phenomena, X-ray scattering techniques have been applied.

predicted for the monomer (Fig. 2A). At 100 MPa and 4 °C, however, neither monomer nor dimer structures fit the experimental profile. The program OLIGOMER (Konarev, Volkov, Sokolova, Koch, & Svergun, 2003) can achieve a profile match only by combining 51% monomer with 49% dimer (Fig. 2B). This result is consistent with early spectroscopic observations of enolase under pressure (Paladini & Weber, 1981).

Important foundations for our current understanding of protein-protein interactions under pressure were laid some time ago by Gregorio Weber using ground-breaking techniques in spectroscopy (Weber, 1992). For pressure p and temperature T, the dissociation constant K_p can be described as an Arrhenius-type equation that depends upon the dissociation constant at atmospheric pressure K_{atm}, the gas constant R, and the change in volume ΔV that occurs during *dissociation*:

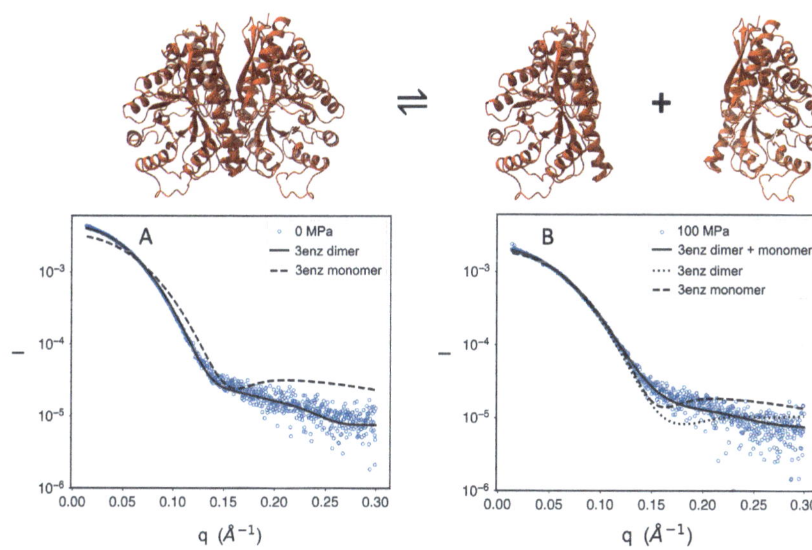

Fig. 2 Dissociation of enolase under pressure. The 93 kDa homodimer yeast enolase exists purely in dimeric form at ambient pressure. (A) The scattering profile predicted from the dimer crystal structure (red ribbon models; PDB ID: 3ENZ) fits the data (blue circles) whereas the monomer structure does not (dashed line). (B) Application of 100 MPa of hydrostatic pressure at 4°C, however, shifts equilibrium in favor of dissociation. Neither the monomer (dashed line) nor the dimer (dotted line) scattering profiles can be fit to the data. Best fit is obtained with a scattering profile that is a mix of 51% monomer and 49% dimer (solid line). Data were collected at CHESS beamline ID7A (HP-Bio): enolase (Millipore Sigma E6126 lot #SLCC98590) pH 6.8 in 25 mM BisTRIS pH 6.8 in 10 mM NaCl and 5 mM DTT.

$$K_p = K_{atm}\, exp\left(\frac{-p\Delta V}{RT}\right). \tag{1}$$

In general, hydrostatic pressure favors changes in biomolecular systems that reduce the overall volume. On first thought, biomolecules would merely compress into spherical blobs in response to pressure, but the story is much richer and more complex. A considerable body of work is now starting to accumulate on the structural changes induced in biomolecules by pressure (Silva et al., 2014). Though atoms themselves can be thought of as hard incompressible spheres, folded biopolymers like proteins are never perfectly packed collections of spheres; they contain voids and cavities (Chen & Makhatadze, 2017). *Voids* are defined here as spaces inside the biomolecule that are not occupied by atoms of the molecule itself. *Cavities*, are voids large enough to contain solvent. Cavities, are often important for function (think catalytic sites), which is what makes pressure effects so interesting and biologically relevant. The processes of dissociation and unfolding exposes trapped voids to the bulk solvent thereby reducing the total volume. So, a paradoxical consequence is that complexes tend to dissociate under hydrostatic pressure and individual proteins tend to unfold. Apparently, unpacking is a simple and accessible way for nature to reduce the volume of an imperfectly packed molecule.

The cavities may or may not be filled with water, and water itself has its own complex volume behavior which must be considered part of the system (Akasaka, 2015). Hydrogen bonds and salt bridges are broken, and hydrophobic patches exposed when a molecule unfolds or when a complex dissociates. This can induce changes in total volume as water solvates the exposed surfaces (Silva et al., 2014). Contraction of bulk water volume in the presence of an electric field, for example, is a well-known phenomenon called the *electrostriction effect* (Hamann, 1980). Despite the complexity of the changes in a biomolecule and solvent upon exposure to pressure, it has now been clearly demonstrated, in protein folding at least, that voids are the dominant contribution to the net volume change (Roche & Royer, 2018). *The occurrence of voids and packing defects in a structure is therefore an important indicator for potential pressure effects.*

Most protein folds are disrupted only by pressures above 300 MPa. The leucine-rich repeat protein pp32 is an exception. Used as a model system for protein folding, the L60A mutant of pp32 begins to unfold at the modest pressure of 100 MPa (Jenkins et al., 2018). Changes in folding are typically detected using the Kratky plot ($q^2I(q)$ as a function of q). The distinctive signature of unfolding or disorder is when the Kratky plot does not fall to zero at high q. The effect in pp32 is shown in Fig. 3 where the pressurized structure (dashed line) rises above the ambient structure (solid line) for $q > 0.1\,\text{Å}^{-1}$.

Lipids comprise another very important class of molecules that are affected by pressure and that have a strong, distinctive SAXS signature (Fig. 1; lipid mesophases). Arguably, lipids are more sensitive than proteins to pressure and have been proposed as major drivers of adaptation in deep sea environments (Somero, 1992). By virtue of the fact that phospholipid molecules have a polar head group and a non-polar tail (two acyl chains), they self-assemble into a variety of extended structures or *mesophases*. Unlike protein solutions, lipid mesophases are partially ordered and therefore can display strong scattering peaks. Since the average electron density of lipid is significantly different than that of water, the scattering contrast of such samples is also naturally stronger than that of proteins in water.

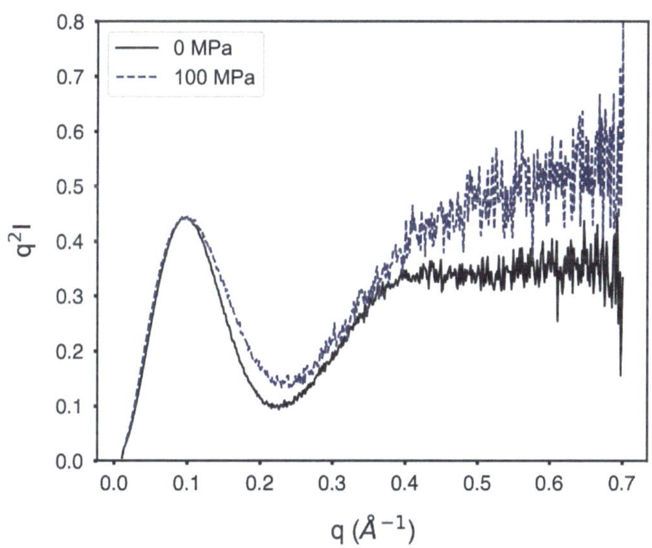

Fig. 3 Pressure-induced unfolding of leucine-rich repeat protein pp32. While most proteins unfold at pressures above 300 MPa, the L60A mutant of pp32 starts to show effects at the modest pressure of 100 MPa. As a Kratky plot ($q^2I(q)$ vs q) the scattering profile for a fully folded protein should fall to zero at large q. Here, the 100 MPa profile (dashed line) rises above the ambient pressure sample (solid line) for $q > 0.1\,\text{Å}^{-1}$. L60A pp32 sample provided courtesy of S. Zhang and C. Royer; data collected at CHESS beamline ID7A (HP-Bio).

The composition of real biological membranes can be complex, consisting of sometimes hundreds of different types of lipids along with other embedded small molecules and proteins. A wide range of biologically-relevant lipidic phases have now been identified which are readily-observable using X-ray scattering techniques (Rappolt, 2006). Represented in Fig. 1 are phases from the neutral phospholipid dioleoylphosphatidylethanolamine (DOPE). At ambient pressure, the inverted hexagonal phase places the polar head groups (spheres in Fig. 1) in the interior of the structure with the hydrophobic tails splayed outwards. This *non-bilayer* structure is not particularly well packed and is therefore pressure sensitive. Application of modest pressure (a few 10's of MPa) leads to the formation of the fluid lamellar phase (L_α). L_α is the bilayer structure fundamental to all biological membranes and its fluidity is essential for function. Application of still more pressure leads to a rigidification of the membrane and the formation of the lamellar gel phase L_β. To maintain a functional fluid membrane in cold high-pressure environments, deep-sea organisms can counter the rigidification by incorporating more cis-unsaturated lipids in their tails, an example of *homeoviscous* adaptation (Winter, 2015). The presence of non-bilayer forming lipids *in vivo* has been implicated in important membrane functions such as fusion and fission events (Jouhet, 2013). The various lipidic phases distinguishable via X-ray scattering may therefore serve as an important guide in understanding key cellular responses and adaptations to hydrostatic pressure. As an example, Fig. 4 shows the scattering profile for DOPE. At ambient pressure the three main peaks represent the repeat distances seen in the inverted hexagonal phase (solid lines). Upon application of only 80 MPa, the hexagonal phase has transformed into the standard fluid lamellar phase with its regular peak spacing (dashed line).

A final phenomenon of growing importance and lively debate in biology is *liquid-liquid phase separation* (LLPS), also known as *biomolecular condensation*. Since the discovery that P granules are liquid-like in nature and that their cellular localization is based on condensation/dissolution (Brangwynne et al., 2009), a wide range of other subcellular structures have come under investigation as possible examples of phase separation (de Oliveira, Cordeiro, Silva, & Vieira, 2019; Wang et al., 2021). The last decade has seen enormous growth of interest in this area, particularly with implications for human disease (Bouchard et al., 2018). That thermodynamically-driven phase transitions could be responsible for assembly processes and localization in cells is a significant paradigm shift for the field (Falahati & Haji-Akbari, 2019).

Proteins with intrinsically disordered regions (IDRs) have been found to play a special role in mediating the process of LLPS (Kato et al., 2012; Molliex et al., 2015). Established theory of polymer-mediated phase

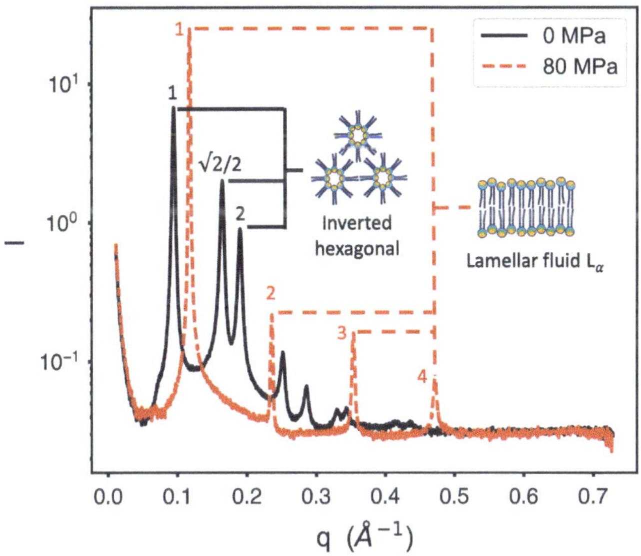

Fig. 4 Pressure-induced phase transition in the phospholipid dioleoylphosphatidylethanolamine (DOPE). At ambient pressure DOPE exists in the non-lamellar inverted hexagonal phase (black line) where the hexagonal symmetry gives rise to a characteristic unevenly spaced pattern of peaks. By 80 MPa, the hexagonal phase has fully converted to the simpler Lamellar fluid phase L_α having regularly spaced peaks. DOPE sample courtesy of I. Budin and J. Winnikoff. Data collected at CHESS beamline ID7A (HP-Bio).

transitions led to the insight that LLPS formation in concentrated IDRs solutions could be understood by examining conformational transitions of the same IDRs in dilute solution (Martin et al., 2020). While IDRs pose a challenge to conventional SAXS due to their low solubility, tendency to aggregate, and degree of disorder, size-exclusion coupled SAXS (SEC-SAXS) has now been used with considerable success to characterize the behavior of monomeric IDR's (Martin, Hopkins, & Mittag, 2021).

The effect of pressure on LLPS has begun to be examined in recent years (Cinar et al., 2019). As in the case of lipid mesophases, LLPS appears to be much more sensitive to pressure than protein fold (Cinar et al., 2020; Cinar, Cinar, Chan, & Winter, 2019). Liquid-liquid phase separation can be induced in a wide variety of biomolecular systems and need not contain intrinsically disordered molecules. Lysozyme, for example, is a famously rigid model system in protein crystallography that shows LLPS at sufficiently-high concentrations and low temperatures (Muschol & Rosenberger, 1997). Winter and co-workers have extensively studied the concentrated solution phases of lysozyme under pressure and temperature with various conditions and osmolytes

(Möller et al., 2012; Schroer et al., 2011). Lysozyme solutions near 20wt% exist in a phase-separated state at ambient pressure but become a single homogeneous phase above 100MPa. Remarkably, the single phase gives way to a *reentrant* phase-separated state above 200MPa (represented schematically in Fig. 1; Liquid-liquid phase separation).

Rather than use conventional dilute SAXS methodology, Winter and coworkers divided out the scattering of the lysozyme molecule from the scattering profiles of concentrated solutions to yield the *structure factor*, a component that contains information about protein-protein interactions (Möller et al., 2014). Well-established techniques can extract from the structure factor an averaged potential of interaction $V(r)$, where r represents the distance between proteins (Winter, 2015). The form of $V(r)$ can thus be examined as a function of pressure, temperature, and solution conditions including additives such as osmolytes. The technique has already shed light on how deep-sea osmolytes control water-mediated protein-protein interactions at high pressure (Julius, Al-Ayoubi, Paulus, Tolan, & Winter, 2018). A detailed discussion of this method is involved and beyond the scope of this chapter. A readable account of the practical details can be found in (Winter, 2015).

3. Pressure dependence of SAXS

Most routine BioSAXS seeks to characterize the changes in structural features of individual biomolecules in the context of their physiological function. What is the overall shape? How are the various domains organized? What is the oligomeric state in solution? Is the molecule compact or unfolded? What states are involved with function? To answer these questions, comparisons are made between different sample preparations having potentially different buffer conditions and additives. Pressure and temperature effects in SAXS are potentially much more subtle and usually involve measurements of the same sample with no variation in concentration or buffer composition. While the absolute errors in the various structural parameters determined by SAXS may be high, the relative errors can potentially be low given the reduction in systematic errors due to working with a single sample. The absolute error of a molecular mass estimation in SAXS, for example, is roughly 10% when comparing different proteins and algorithms (Hajizadeh, Franke, Jeffries, & Svergun, 2018). The standard error of a single forward scattering measurement used to compute molecular mass, on the other hand, can easily be less than 1%. It is therefore worth re-examining some of the underlying assumptions of the common processing algorithms to check for subtle pressure dependency. To understand these effects, we need to first examine the basic pressure dependence of the scattering profile itself.

The scattering profile $I(q)$ is almost universally introduced as a product of terms: a prefactor, K, which is independent of q, a *form factor*, $P(q)$, which represents scattering from pairs of atoms in a single biomolecule, and an *effective structure factor*, $S_{eff}(q)$, which represents scattering from pairs of atoms in which each atom is in a different molecule (Kotlarchyk & Chen, 1983):

$$I(q) = K\, P(q)\, S_{eff}(q). \tag{2}$$

Traditionally, structural biology has concerned itself largely with form factors, working with dilute monodisperse solutions where $S_{eff}(q)$ is constant. The much broader field of soft matter which includes colloid and polymer science, on the other hand, has frequently required modeling of structure factors as well as $P(q)$. In this chapter we focus primarily on how pressure affects K and $P(q)$, because those components are better understood and more widely used by structural biologists. Our brief discussion is necessarily limited to basic *model-free* methods.

3.1 Prefactor and contrast

Overall scattering intensity scales according to instrument parameters and composition of sample as

$$K = Td\frac{f}{r^2}\,\frac{\nu^2\Delta\rho^2}{N_a}\,Mc. \tag{3}$$

The pressure-independent terms are thickness of the sample d (cm^2), incident flux f (particles/s), sample-to-detector distance r (cm), molecular mass M (g/mol), and Avogadro's number N_a. Transmission T through the sample varies with pressure due to changes in sample density, nonetheless, the effect can be ignored since scattering profiles are typically normalized at the beamline using beam intensity transmitted through the sample. Concentration also varies simply with pressure according to buffer density. If ρ_0 is buffer density (m/v), c(0) is sample concentration at atmospheric pressure, and ρ_P density at pressure P, then concentration under pressure is $c(P) = c(0)\ \rho_P/\rho_0$.

Specific volumes, ν, for biomolecules have been measured by a variety of techniques (Mylonas & Svergun, 2007). In principle these values can change with hydrostatic pressure, but measurements of the intrinsic compressibility of folded proteins tend to yield values that are much smaller than the compressibility of water itself (Kharakoz, 2000; Taulier & Chalikian, 2002). The possibility of significant changes in specific volume with pressure shouldn't be entirely ignored, especially for unfolding. Though changes are expected to be quite small, the value is used in several common calculations involving SAS data and further research is warranted. For the purposes of the present discussion, we will assume that protein specific volume is constant ($\sim 0.72\,\text{cm}^3\,\text{g}^{-1}$) (Wang, Lin, Bougie, & Gillilan, 2018).

The *excess scattering length density*, $\Delta\rho$, is derived from the average *electron* density difference between buffer and sample molecules. The classical electron radius, r_0, customarily multiplies the density term. The squared term $\Delta\rho^2$ appearing in Eq. (3) is called *contrast*; it is a small number in SAXS since the intrinsic density of protein is not much different than water. Contrast has been used to great advantage in neutron scattering due to large differences in hydrogen and deuterium scattering. In this way, isotopic labelling can help identify different components in a biomolecule that would otherwise be indistinguishable. Buffer additives such as sucrose have also been used with success to manipulate buffer-sample contrast in SAXS, but the effect is smaller.

Contrast changes with pressure according to differences in compressibility between buffer and sample. For SAXS on proteins, excess length density is expressed as

$$\Delta\rho = \left[\rho_{e,protein} - \rho_{e,buffer}\right] r_0 \tag{4}$$

where $\rho_{e,buffer}$ is the average length density of the buffer and $\rho_{e,protein}$ the average (intrinsic) length density of protein. Typically, $\rho_{e,protein}$ is estimated as $\varepsilon_{protein}/v$, where $\varepsilon_{protein}$ is electrons per unit dry mass and assumed for this discussion to be constant with pressure. Let $\alpha = \rho_P/\rho_0$, the ratio of buffer density at the two different pressures, then the overall prefactor for scattering profiles becomes

$$\frac{K(P)}{K(0)} = \alpha \frac{\left(\rho_{e,protein} - \alpha\rho_{e,buffer}\right)^2}{\left(\rho_{e,protein} - \rho_{e,buffer}\right)^2}. \tag{5}$$

The formula is cubic in solvent density ratio α and could potentially reach zero when the electron densities of buffer and protein are close in value.

In routine scattering experiments, buffer is usually a dilute solution close to water in density. Using typical values $\varepsilon_{protein} = 3.22 \times 10^{23}\,e\,g^{-1}$, $\nu = 0.72\,cm^3\,g^{-1}$ and $\rho_{buffer} = 3.3 \times 10^{23}\,e\,g^{-1}$ (Orthaber, Bergmann, & Glatter, 2000) in combination with tabulated water density at various pressures (Grindley & Lind, 1971) one can estimate the degree to which scattering intensity is reduced under hydrostatic pressure conditions (Table 1).

For hydrostatic pressures in the 300–400 MPa range, where many proteins begin to do interesting things, the simple contrast analysis shown in Table 1 tells us that we need to be prepared for scattering signals that are weakened in intensity by ~50%. This is not a dramatic reduction factor, but it does mean that samples of low molecular weight, low solubility, or strong concentration effects may pose an extra challenge for HP-SAXS.

3.2 Form factor

Most common SAS analysis tools in structural biology rely upon the assumption of dilute monodisperse samples. In this ideal case, the form factor, $P(q)$, represents the rotationally averaged scattering from an average isolated biomolecule in solution. Since pressure induces contrast changes between the sample and buffer, it is important when using these standard tools to consider whether any subtle changes observed represent true conformational changes in the sample or simply contrast effects.

Table 1 Effect of hydrostatic pressure on X-ray scattering contrast of protein solutions.

MPa	ε (25 °C)[a]	K(P)/K(0)
100	1.04	0.81
200	1.08	0.68
300	1.10	0.55
400	1.13	0.45
500	1.15	0.37
600	1.17	0.30
700	1.19	0.25
800	1.21	0.20

[a] $\varepsilon = \rho_P/\rho_0$ from tabulated water densities in (Grindley & Lind, 1971).

3.2.1 Radius of gyration

Standard biological SAXS experiments most often begin with determination of the radius of gyration R_g and the forward scattering intensity $I(0)$ by means of the Guinier analysis. Since Andre Guinier developed his now famous small-angle approximation (Guinier, 1939), the method has become as much a quality-control check as it has a source of structural information.

$$I(q) \approx I(0)e^{-R_g^2 q^2/3} \tag{6}$$

$$R_g^2 \equiv \int \rho(\vec{r}) r^2 \mathrm{d}\vec{r} \Big/ \int \rho(\vec{r}) \mathrm{d}\vec{r} \tag{7}$$

For SAXS, R_g is related to electron density $\rho(\vec{r})$ by Eq. (7). Ideally, a plot of $\ln(I(q))$ as a function of q^2 in the small-angle region of the scattering profile yields a straight line where R_g and $I(0)$ are computed from as slope and intercept. Alternatively both parameters can be calculated indirectly from the pair-distance distribution function computed using the indirect Fourier method (Svergun, 1992).

Radius of gyration is not impervious to contrast change. As SAXS and SANS contrast variation experiments began to be conducted on a variety of proteins, it was noticed that the radius of gyration was contrast-dependent (Stuhrmann & Kirste, 1967). This dependency arises from the fact that proteins are not perfect shapes of uniform internal scattering density. Following the development laid out by Stuhrmann (Koch & Stuhrmann, 1979), the total scattering length density of a biomolecule in solution can be decomposed into a shape of uniform internal density, $\rho_F(\vec{r})$, and a term that includes deviations from the average $\rho_S(\vec{r})$:

$$\rho(\vec{r}) = \Delta\rho\, \rho_F(\vec{r}) + \rho_S(\vec{r}). \tag{8}$$

In this equation, $\Delta\rho = \overline{\rho}_{solute} - \overline{\rho}_{solvent}$ is the average excess scattering length density of the biomolecule above the solvent background. Substituting Eq. (8) into a relation similar to Eq. (7), and using the fact that the integral of $\rho_S(\vec{r})$ vanishes and that $\rho_F(\vec{r})$ is constant inside the molecule and zero outside, Stuhrmann developed an expression for the contrast dependency of R_g:

$$R_g^2 = R_c^2 + \frac{\alpha}{\Delta\rho} - \frac{\beta}{\Delta\rho^2}. \tag{9}$$

The radius of gyration of the idealized molecular shape with constant internal density R_c is thus corrected by coefficients α and β, which reflect non-constant distribution of mass within the molecule (Koch & Stuhrmann, 1979). Svergun et al., through comparisons of SAXS and SANS data have shown that the solvent hydration layer around the biomolecule is a major contributor to the corrections (Svergun et al., 1998). The effect of pressure on R_g from contrast change is likely small, nonetheless pressure pushes systems significantly closer to zero contrast. No systematic studies of this phenomenon to date have been done.

3.2.2 Forward scattering and molecular mass

The forward scattering *I(0)* obtained from the Guinier analysis is frequently used to estimate molecular mass (Mylonas & Svergun, 2007). For a protein standard of known molecular mass MM_{st} and concentration c_{st}, the molecular mass of an unknown under identical conditions is

$$MM_{unk} = \frac{I(0)_{unk}}{c_{unk}} \frac{c_{st}\, MM_{st}}{I(0)_{st}}. \tag{10}$$

Eq. (10) at ambient pressure can easily be verified by substituting Eq. (2) for *I(0)*. The presumption is that unknown and standard specific volumes and contrasts are nearly the same, so the terms in Eq. (3) effectively cancel. This is no longer true for standards taken at a different pressure than the sample. In principle, a series of standards could be measured at various pressures, but high pressures can potentially dissociate and unfold even stable standards. It may be permissible to employ Eq. (5) to correct Eq. (10), but clearly care should be taken in interpreting molecular mass changes computed by this means.

It is good practice in SAXS to place the scattering on an absolute scale of cm^{-1} using a water as a standard (Orthaber et al., 2000). This practice allows meaningful comparisons of scattering profiles between SAXS and SANS taken at different facilities. Absolute scale also enables direct computation of molecular mass from *I(0)*. The scattering of water itself is a constant in the small-angle region that depends upon scattering-length density ρ, and isothermal compressibility of water χ at temperature T, pressure P:

$$I(0; T, P)_{water} = \rho(T, P)^2 kT\chi(T, P). \tag{11}$$

Using tabulated values of density and compressibility (Fine & Millero, 1973), the scattering intensity of water falls by 3% in going from 0 °C to 20 °C at 0 MPa. On the other hand, pressurizing from 0 MPa to 400 MPa at 20 °C, the intensity falls by 6%. This characteristic drop in buffer baseline can be seen in pressure series of SAXS profiles that are normalized for transmitted X-ray intensity.

To perform a water calibration, it is necessary to measure and subtract from water the scattering of the empty (air) sample cell: $I_{water} - I_{empty}$. With the present HP-SAXS cell design, it is not possible to measure an empty cell containing air under pressure due to the thin X-ray transparent windows of the internal sample cell. This use of glassy carbon as a standard is potentially interesting here due to its structural and thermal stability (Zhang et al., 2009). Fortunately, it should be sufficient to calibrate scattering intensity using water and empty cell at ambient pressure:

$$I(q)_{absolute} = \frac{I(q)}{I(0;T,0)_{water}} \rho^2 kT\chi_{T,0}. \tag{12}$$

The correct calculation of molecular weight from $I(0)$ on the absolute scale, however, requires knowing concentration c and contrast $\Delta\rho$ at the appropriate temperature and pressure:

$$MM = \frac{N_a I(0)}{c\Delta\rho^2}. \tag{13}$$

Once again. molecular mass estimates by this method could be corrected using Eq. (5), but such correction rests on the accurate calculation of contrast.

Molecular mass estimates based on the Porod volume are popular because they do not require knowledge of the concentration. In its simplest form, Eq. (14), the molecular volume (square brackets) is multiplied by the mass density of protein ρ_m to yield the molecular mass.

$$MM = \rho_m \left[2\pi^2 \frac{I(0)}{\int I(q)q^2 dq} \right]. \tag{14}$$

Substituting Eq. (2) into the Eq. (14), the contrast dependency K completely factors out of the ratio, so that the only pressure dependence resides within ρ_m. As discussed in relation to Eq. (5), protein is thought to be much less compressible than water and consequently the intrinsic specific volume can be treated as a constant. The class of concentration-independent molecular mass calculations based on integral invariants is therefore contrast-independent and likely more resistant to spurious pressure-dependency (Piiadov, Ares de Araújo, Oliveira Neto, Craievich, & Polikarpov, 2019; Rambo & Tainer, 2013).

3.2.3 Unfolding and disorder

Beyond changes in contrast between solvent and solute, hydrostatic pressure alters the structure of biomolecules. At pressures above 300 MPa, proteins begin to unfold. One of the advantages of SAXS is its ability to provide structural information on unfolded and disordered systems. Early in the development of SAXS it was recognized that the scattering profile for a molecule having a well-defined smooth boundary decayed rapidly in the wide-angle region: $I(q) \propto q^{-4}$ (Porod, 1982). Porod's Law, as it is called now, has been generalized to treat polymers in various states of disorder by introducing a general exponent p that can be interpreted as a fractal dimension (Johansen, Trewhella, & Goldenberg, 2011):

$$I(q) \propto q^{-p}. \tag{15}$$

For proteins, p is a measure of disorder that ranges from nearly 4.0 for compact folded states, down to 1.0 for fully unraveled, extended chains. Practical computation of p and other related parameters for disordered biomolecules has been described elsewhere (Rambo & Tainer, 2011).

A more common assessment of flexibility is the Kratky plot (Fig. 3): $q^2 I(q)$ as a function of q. For folded molecules, where $p > 2$, the Kratky plot falls to zero at wide angle. In dimensionless units, $(q\,R_g)^2 I(q)/I(0)$ vs $q\,R_g$, the plot can be used to compare degree of flexibility between different molecules (Receveur-Brechot & Durand, 2012). Reliable interpretation of these plots requires accurate buffer subtraction as well as good quality Guinier analysis (see Protocol).

Guinier analysis for unfolded molecules is not without problems. Unfolded systems represent ensembles of very different structures with relatively large radii of gyration and narrow Guinier regions. The recent surge of interest in intrinsically disordered proteins (IDPs) has stimulated the development of new tools specifically designed to handle these types of problems. The molecular form factor (MFF) method uses a pre-computed collection of molecular ensembles to deduce R_g (Riback et al., 2017). The extended Guinier analysis uses a higher-order empirical correction term (Zheng & Best, 2018). Ensemble and atomistic methods will likely continue to improve and yield important insights for flexible systems that are ultimately applicable to pressure-induced unfolding.

4. Methods

The problem of designing a sample cell that is thin and transparent to X-rays, yet strong enough to withstand enormous hydrostatic pressure has been approached by a number of groups over the years (Ando, Chenevier, Novak, Tate, & Gruner, 2008; Brooks et al., 2010; Kato & Fujisawa, 1998; Krywka et al., 2008; Woenckhaus, Köhling, Winter, Thiyagarajan, & Finet, 2000). Neutron (SANS) scattering cells have been designed too; the unique features and data collection protocols are described elsewhere (Teixeira, Leão, Gagnon, & McHugh, 2018). The term *static* SAXS refers to experiments in which the sample is neither flowing nor oscillating during data collection. The introduction of sample flow to regular SAXS was an important innovation that greatly reduced the problem of radiation damage. Except for the high-pressure chromatography-coupled SAXS system under development by us, sample flow is not yet available in current high-pressure SAXS instruments. While this places some restrictions on how experiments need to be conducted, non-flowing samples are currently the only way to reach the highest hydrostatic pressure ranges for SAXS.

4.1 Design

The HP-SAXS sample cell described here and in the literature (Rai et al., 2021) was built in collaboration with Pressure BioSciences Inc (South Easton, MA) and is an improvement based upon an earlier design by Ando (Ando et al., 2008). The cell utilizes water as a pressurizing medium which effectively seals the diamond windows against optically-polished metal flats in what is referred to as a Poulter-style design (Poulter, 1932). Hydrostatic pressure is supplied by a computer-controlled high-pressure generator (HUB440, Pressure BioSciences Inc., South Easton, MA). The hardened metal cell is designed to tolerate hydrostatic pressures up to 700 MPa yet samples can be accessed easily through a piston that is held in place by two sliding metal pins that act as a stop (Fig. 5). The entire cell is enclosed by a temperature-controlled water jacket allowing measurements in the 0–70 °C range. Vacuum fittings also allow the cell to be connected directly to a synchrotron flight path without the need for additional windows.

In conventional BioSAXS, thin windows isolate sample from vacuum or air. Smooth low-Z materials such as glass, mica, or polymer films are typically used in thicknesses <25 μm. Silicon nitride windows have pushed

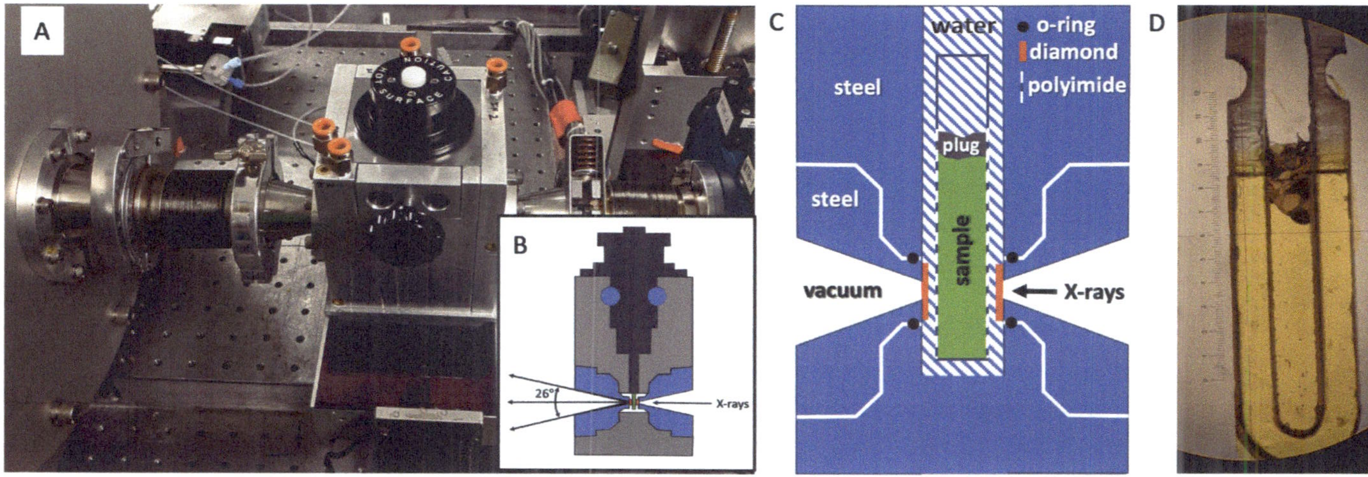

Fig. 5 High-pressure SAXS cell design with rapid sample change mechanism mounted on CHESS beamline ID7A (HP-Bio). Pressure cell (A) with cooling jacket connects to beamline vacuum flightpath system via flexible bellows. Inset (B) shows the path of the x-rays through the sample cell. Closeup side view of sample cell inside pressure cell (C) shows Poulter seal of diamond windows (red) against polished steel surface (blue). A deformable grease plug seals the sample against the surrounding pressurizing water medium while X-rays travel through thin polyimide film (dashed line) to produce scattering. Front view photo of actual sample cell (D) shows plastic cuvette with polyimide windows and plug seal.

thinness of windows down even to the submicron range. Of course, none of these options are possible for high pressure work. Single crystal diamonds are a low-Z material of extremely high strength, nonetheless, window thicknesses of 500 μm or greater are needed for the hydrostatic pressure ranges considered here. X-ray absorption by these thick windows is significant at lower X-ray energies, but not totally prohibitive (transmission of two such windows at 10 keV is ~20%). Nonetheless, higher energy X-rays allow greater sample pathlengths, so that window scattering constitutes a smaller fraction of total signal thereby reducing background noise. Reduced radiation damage may also be a benefit of working at higher energies, despite the slight loss of low-angle resolution. With non-flowing samples, radiation damage is an important limiting factor. The experiments run with this system have consequently used 3–3.5 mm sample cell thicknesses at 14 keV.

4.2 Protocol

The difficulty inherent in creating an X-ray compatible environment for high hydrostatic pressure places some important limitations on how scattering data can be collected and interpreted. Much like the early days of SAXS, contemporary HP-SAXS instruments are limited to non-flowing sample. Sample cells must also be physically removed from the beam path to be re-filled. These two circumstances require careful mitigation if SAXS profiles are to be reliably interpreted when pressure-induced structural changes are small.

Photons arriving at the detector that were not scattered by the biomolecule itself are referred to as *background* scattering in SAXS. Solvent molecules are a principal source of background, but instrumental sources such as X-ray windows and slits can also be significant. Due to the low level of contrast between biomolecules and their water solvent, careful background subtraction has always been necessary to pull out the faint signal. This requires that an identical sample of *matched* buffer be measured as a *blank* for use in subtraction. In introductory SAXS courses and materials, great emphasis is frequently placed on preparing properly matched buffer since very small differences (<1%) in salts and other additives can result in detectable mismatch (Jacques & Trewhella, 2010). The primary impact of buffer mismatch is in the WAXS region where Kratky plots and Porod analysis can give potentially misleading measures of folding and disorder. *Since pressure effects on biomolecules can involve unfolding, it is particularly important that background subtraction be carefully performed.*

Radiation damage in BioSAXS occurs primarily in the form of induced aggregation which is thought to be caused by chemical activation of the biomolecular surfaces by the radiolytic products of water. The total absorbed energy per unit mass in units of gray, $1\,\text{Gy} = 1\,\text{J}\,\text{kg}^{-1}$, can be estimated simply as

$$D = \frac{PE}{\rho V}(1 - T_s)t. \tag{16}$$

Here P is photon flux, E is energy (wavelength $\lambda = 12.4/\text{E}$), L is sample thickness, t is exposure time, ρ the density of water, and V is the volume of sample in the beam (Meisburger et al., 2013; Wang et al., 2018). Sample transmission (mostly due to water), can be conveniently calculated as $T_s = \exp(-3\text{L}\lambda^3)$ (Wang et al., 2018).

For a 3rd generation synchrotron at 10^{13} photons per second, a $0.25\,\text{mm} \times 0.25\,\text{mm}$ beam at 14keV through a 3.5mm cell will deposit 53kGy in 1s. This is significantly higher than the reported damage limits for proteins: from 400Gy to as high as 10kGy (Cho, Yang, & Song, 1999; Jeffries et al., 2015; Kuwamoto, Akiyama, & Fujisawa, 2004; Meisburger et al., 2013). HP-SAXS experiments with non-flowing cells must therefore require beam attenuation and other strategies to manage X-ray damage.

One strategy that has been employed to manage damage is the introduction of time delays (Ando et al., 2008). In practice the sample volume required for SAXS is generally much larger than the volume illuminated by X-rays. An X-ray beam with a full-width at half maximum (FWHM) of 0.25mm (standard deviation $\sigma = 0.25/2.355\,\text{mm}$) will create an equivalent-sized column of damaged material that will broaden according to the basic principle of diffusion. The timescale for such a process depends upon the diffusion constant, D, of the damaged material in buffer and can be estimated as $\text{t} = \sigma^2/(2\text{D})$. Even a small protein like lysozyme ($\text{D} = 1.1 \times 10^{-6}\,\text{cm}^2/\text{s}$) requires a time on the order of 5min. Bovine serum albumin, at $\text{D} = 6.8 \times 10^{-7}\,\text{cm}^2/\text{s}$, would take closer to 8min. Damaged material likely would take much longer due to its high molecular weight. Time delays are observed to reduce damage, but the strategy is not necessarily reliable or consistent (Rai et al., 2021).

Remarkably, high hydrostatic pressure is known to dissociate aggregates. This effect has been observed in HP-SAXS, and may indeed help mitigate radiation damage effects (Harish et al., 2021). To date, no systematic studies of the effect of pressure on radiation damage in SAXS have been conducted.

Addition of small amounts of glycerol, and reducing agents such as dithiothreitol (DTT) and tris(2-carboxyethyl)phosphine (TCEP) have long been known to protect against radiation damage in SAXS samples (Kuwamoto et al., 2004). Recent systematic experiments in the addition of radical scavengers have shown promise and will likely be valuable to non-flowing HP-SAXS experiments as more effective additives are identified (Castellví, Pascual-Izarra, Crosas, Malfois, & Juanhuix, 2020; Stachowski, Snell, & Snell, 2021).

As in standard flow-cell SAXS, HP-SAXS profiles should be collected as a series of short (usually 1 s) exposures and inspected for radiation damage. Modern processing software should include methods to determine if statistically significant changes in the profile have taken place during data collection (Franke, Jeffries, & Svergun, 2015). Caution should be used with these methods when comparing sequential sets of exposures on the same non-flowing sample. Just because exposures 1–5 and 6–10 merge separately without detectable damage, does not mean the frames 1–10 will merge. A pressure series in which all frames at each pressure have merged separately without difficulty may still carry slow cumulative radiation damage. The gold standard for damage assessment is to compare initial and final ambient pressure results. This strategy, however, is not applicable when pressure-induced conformational changes are irreversible. Sticking to a dose budget and spot-checking individual pressures by collecting them first remains the best way to insure undamaged profiles.

4.3 Sample preparation

Except for buffer choice, sample preparation for HP-SAXS follows the same protocols as normal ambient-pressure BioSAXS. A number of detailed protocols for sample preparation can already be found in the literature (Graewert & Jeffries, 2017; Jeffries et al., 2016; Skou et al., 2014).

Since high-pressure SAXS measurements are more difficult to perform and sometimes involve subtle structural changes, it is essential that samples be pre-screened and fully characterized with normal BioSAXS prior to high-pressure work. Due to the thick sample cell windows required by high-pressure work and the natural reduction in sample contrast at high pressures (Table 1), higher concentrations of sample may be necessary to obtain adequate signal for analysis. Exactly what constitutes a high-enough concentration depends upon how the scattering data are to be used (Wang et al., 2018). Since scattering intensity $I(0) \propto M\,c$, where M is molecular weight and c is concentration, the typical guidance for normal SAXS is to use $c \geq 60/M$ (M in kDa and c in mg/mL). This choice usually results in sufficient signal to obtain a low-noise Guinier plot on modern synchrotron sources but dilute enough sample to avoid concentration effects. Given the more than two-fold reduction in intensity at 400 MPa due to contrast, $c \geq 120/M$ mg/mL is a more realistic *lower bound* for sample concentration in a single HP-SAXS measurement. Higher concentrations should be produced to allow for dilution series, if needed. High concentrations also allow for shorter exposures and less radiation damage per profile if they do not result in significant concentration effects. Low molecular weight samples (<10 kDa) are most affected by the limitations of HP-SAXS due to their inherently weak scattering and the high concentrations necessary. Proper pre-screening of samples using normal SAXS should establish the limits for concentration effects and radiation damage. The HP-SAXS cell described here holds 60 μL of sample. Precisely how many exposures can be obtained from a single sample depends upon radiation sensitivity which can vary by an order of magnitude or more for different samples.

4.4 Baroresistant buffers

The behavior of biomolecules in solution can be heavily dependent upon pH, consequently buffers are universally used in biological experiments. This is particularly pertinent to high-pressure biological work where small changes are common. Early measurements of pH under pressure found that while common buffers like acetate and phosphate are susceptible to pH shifts, buffers like tris(hydroxymethyl)aminomethane (Tris), which produced no change in the number of charged species during ionization, are relatively resistant (Neuman, Kauzmann, & Zipp, 1973):

$$HPO_4^{2-} + H_3O^+ \rightarrow H_2PO_4^- + H_2O \qquad \Delta V = +24 \text{ mL/mol}$$

$$\text{Tris} + H_3O^+ \rightarrow \text{TrisH}^+ + H_2O \qquad \Delta V = -1.1 \text{ mL/mol.}$$

Like other reactions, an ionization reaction with acid dissociation constant K_a can have a non-zero reaction volume ΔV which makes it susceptible to pressure (Eq. (1)). Contraction of bulk water volume in the presence of an electric field, such as when charges are created, is called the *electrostriction effect* (Hamann, 1980). For this reason, pressure tends to favor ionization.

By example, phosphate buffer ($H_2PO_4^-$) *decreases* by 0.4 pH units at 100 MPa, and up to 1.7 MPa units at 600 MPa. In contrast, Tris buffer *increases* in pH by only 0.2 at 600 MPa according to original indicator-based measurements (Neuman et al., 1973). Table 2 lists selected biological buffers having small reaction volumes (adapted from (Kitamura & Itoh, 1987)).

Table 2 Selected pressure resistant buffers.

Buffer	pK_a	Range	ΔV	T_{coeff}
Maleic	1.92	2.0–3.0	−5.1	−0.0006
MES	6.10	5.5–6.7	3.9	−0.011
Bis-tris	6.48	5.8–7.2	3.1	−0.017
ACES	6.75	6.1–7.5	4.0	−0.02
MOPSO	6.90	6.2–7.6	3.9	−0.015
Imidazole	6.95	6.2–7.8	1.8	−0.020
BES	7.16	6.4–7.8	2.6	−0.016
MOPS	7.15	6.5–7.9	4.7	0.011
HEPES (2nd)	7.50	6.8–8.2	4.8	−0.014
TES	7.48	6.8–8.2	−1.5	−0.02
TAPSO	7.64	7.0–8.2	1.5	−0.023
Tricine	7.92	7.4–8.8	−2.8	−0.021
Tris	8.10	7.0–9.0	4.3	−0.028
TAPS	8.30	7.7–9.1	0.5	0.018
AMP	8.78	9.0–10.5	3.5	−0.032
Bis-tris-propane (2nd)	9.10	6–9.5[a]	−3.1	−0.03
CHES	9.43	8.6–10.0	1.2	−0.023
CAPS	10.6	9.7–11.1	0.4	−0.009

[a]Buffer range of first and second K_a's overlap, but only second shows pressure resistance.
Selected entries from Kitamura and Itoh (1987); $T_{coeff} = d(pK_a)/dt$ from Good et al. (1966) and Goldberg, Kishore, and Lennen (2002).

Though all buffers show some pressure dependency, the observation that carboxylates produce a volume change opposite in sign to common cationic buffers suggests that mixtures can be used to achieve true baroresistance (Quinlan & Reinhart, 2005). Mixtures may be particularly valuable at low pH where good single buffers are lacking (Table 2).

A study using a specially-designed *in situ* high-pressure pH sensor suggests that reaction volumes may over-estimate actual pH changes (Samaranayake & Sastry, 2010). Kitamura has also cautioned that control of pH does not necessarily imply control of pOH under pressure (Kitamura & Itoh, 1987).

While ionization reaction volumes are an important guide in choosing appropriate buffers for routine pressure experiments, care should still be taken to rule out pH-sensitive effects that might result from pressure-induced changes in buffer

4.5 Recommended steps

Careful characterization of the sample using normal BioSAXS is an important pre-requisite for successful HP-SAXS experiments. The Pre-HP screening protocol identifies potential sample problems, provides an undamaged dilute reference profile, and determines the acceptable dose budget for the HP experiments.

4.5.1 Pre-HP screening protocol

1. Prepare a protein sample in an appropriate buffer at concentration approximately 300/MW mg/mL, where MW is molecular weight in kilodaltons (kDa).
2. Equilibrate a SEC-SAXS column with an appropriate pressure-resistant buffer (Table 2) containing ~5% glycerol, TCEP or other protectives (>250 mL of buffer is typically required).
3. Run SEC-SAXS by injecting 100 μL of sample.
4. Evaluate the SAXS chromatogram for sample purity and save the peak SAXS profile, noting radius of gyration as an example of undamaged, dilute protein.
5. Prepare >120 μL of 60/MW mg/mL sample in the SEC buffer using SEC, dialysis, centrifugal concentrator, or other buffer-exchange method to insure a good match between sample and buffer.
6. Create a sample dilution series of at least three concentrations (>60 μL each): for example, 30/MW, 60/MW, 120/MW mg/mL using matched buffer.
7. Perform standard ambient flow-cell SAXS, examining radius of gyration to evaluate potential concentration effects.
8. Note the flux, energy, and sample cell thickness of the SAXS instrument.
9. Perform a long exposure series without any sample flow.
10. Analyze the exposure series to determine the radiation damage limit. Calculate the acceptable dose in kGy.

4.5.2 HP-SAXS protocol

1. Note the energy, flux, and sample cell thickness of the instrument.

2. Compute the maximum exposure time allowed by the previously-determined dose limit for the sample. This is the *dose budget* for the experiment.
3. Adjust beamline attenuation so that the total dose budget will not be exceeded during the pressure series measurements. Typically, a measurement at a single pressure would consist of five or more exposures statistically compared for agreement.
4. Load a sample cell with 60 μL of buffer, pre-pressurize the buffer to 20 MPa to allow cell hardware to settle, then proceed to collect buffer profiles for the whole range of pressures of interest. Take at least as much buffer exposure time at each pressure as will be tolerated by the sample according to the dose budget. To assure reproducibility, it is advisable to collect a rising pressure series followed by a falling pressure series, making careful note to match gauge pressures on the way back down. (Example: 0 MPa, 100 MPa, 200 MPa, 300 MPa, 400 MPa, 300 MPa, 200 MPa, 100 MPa, 0 MPa).
5. Using the same plastic sample cell, replace the buffer with 60 μL of sample and re-seal.
6. Pre-pressurize to 20 MPa to allow cell hardware to settle, then proceed with pressure series data collection, leaving appropriate time for sample conformational equilibration at each pressure. Delay time between pressures may also help reduce potential damage. Make sure to precisely match the noted gauge pressures used in previous buffer collection. Take the last exposure at 0 MPa for comparison with the undamaged ambient pressure sample.
7. Using the same cell, flush with fresh buffer and re-collect a buffer pressure series for comparison with the initial buffer (optional, but recommended).

5. Summary and conclusions

Small angle X-ray solution scattering is widely used in structural biology to provide insight into the behavior of molecular biological systems in solution under conditions more closely approximating "physiological" than techniques requiring cryogenic treatment or crystallization. But in recent years, the term "physiological" has taken on a whole new meaning with the realization that so much life on Earth thrives under conditions of extreme pressure, temperature, and chemistry. This chapter has focused on pressure-induced transformations in biological materials that can be followed and characterized with SAXS.

SAXS is capable of quantifying oligomeric state changes in biomolecules, an important phenomenon that can occur at biologically relevant pressures encountered in the environment. The ability of SAXS to detect and quantify disorder is also of interest to fundamental biophysics since high hydrostatic pressure ultimately induces unfolding. The growing assortment of algorithms specifically tailored for analysis of SAXS data from intrinsically disordered molecules will likely prove useful for pressure studies as well.

Strong contrast combined with extended, partially ordered structure makes lipid mesophases a natural target for HP-SAXS studies. Phase transitions of this type take place at relatively low biologically relevant pressures. Scattering signatures for various mesophases are well characterized under ambient pressure conditions and the fundamental differences in packing of those phases may provide insights into how organisms respond and adapt to pressure.

The limited studies of liquid-liquid phase separation under pressure suggest that it too is a relatively pressure sensitive phenomenon. Though LLPS systems governed by conformational changes in intrinsically disordered solutes have been studied using conventional SAXS, their pressure behavior has yet to be investigated. Rigid model systems, such as lysozyme, have been studied in concentrated form under pressure by extracting intermolecular interaction potentials from the structure-factor component of the scattering profile. As LLPS continues to gain importance in biology, this less-frequently used form of SAXS analysis seems likely to yield important insights.

Due to the additional difficulty of performing SAXS under extreme pressure, special protocol and sample handling is necessary to ensure reliable data collection and interpretation. Until flow-based instruments are developed, radiation damage will continue to be the dominant concern. Nonetheless, effective strategies exist to mitigate the problems of HP-SAXS.

Because the compressibility of biomolecules can differ from the compressibility of water, pressure unavoidably alters X-ray scattering contrast. This fact has implications for the most widely used data processing methods: radius of gyration may subtly shift in ways that are unrelated to conformational changes, and molecular weight estimates based on standards can be biased due to contrast. Fortunately, integral methods for molecular weight estimation are likely more reliable due to the cancellation of contrast effects. Contrast variation by itself is an important tool in both conventional SAXS and SANS. Whether pressure-induced contrast can be leveraged in the same way has yet to be investigated.

Small angle solution scattering has played a vital role in structural biology over the last several decades. Rather than being supplanted by higher resolution methods such as crystallography and cryo-electron microscopy, the ready availability of high-resolution structures has further empowered SAXS and SANS to make important statements about the behavior of biomolecules in solution. With appropriate care in sample preparation, data collection, and processing, the widely recognized advantages of SAXS are fully available for the study of biomolecular systems under extreme pressure.

Acknowledgments

Many thanks to Itay Budin and Jacob Winnikoff for providing DOPE scattering data and for valuable advice in constructing Figs. 1 and 4. Thanks also to Siwen Zhang and Cathy Royer for making available the pp32 sample used in Fig. 3. Durgesh Rai was instrumental in the development and testing of the HP-SAXS system described in this work. Thanks to Robert Miller, Nozomi Ando, Ben Myers and Sol Gruner for many useful discussions. Qingqiu Huang provided much-needed support in experimental setup, sample preparation, and data collection. This work is based upon research conducted at the Center for High Energy X-ray Sciences (CHEXS), which is supported by the National Science Foundation under award DMR-1829070, and the Macromolecular Diffraction at CHESS (MacCHESS) facility, which is supported by award 1-P30-GM124166-01A1 from the National Institute of General Medical Sciences, National Institutes of Health, and by New York State's Empire State Development Corporation (NYSTAR).

References

Akasaka, K. (Ed.). (2015). *Vol. 72. High pressure bioscience basic concepts, applications and frontiers. Subcellular biochemistry*. New York: Springer.

Ando, N., Barquera, B., Bartlett, D. H., Boyd, E., Burnim, A. A., Byer, A. S., et al. (2021). The molecular basis for life in extreme environments. *Annual Review of Biophysics*, *50*(1), 343–372.

Ando, N., Chenevier, P., Novak, M., Tate, M. W., & Gruner, S. M. (2008). High hydrostatic pressure small-angle X-ray scattering cell for protein solution studies featuring diamond windows and disposable sample cells. *Journal of Applied Crystallography*, *41*(1), 167–175.

Bar-On, Y. M., Rob, P., & Ron, M. (2018). The biomass distribution on earth. *Proceedings of the National Academy of Sciences of the United States of America*, *115*(25), 6506–6511.

Bouchard, J. J., Otero, J. H., Scott, D. C., Szulc, E., Martin, E. W., Sabri, N., et al. (2018). Cancer mutations of the tumor suppressor SPOP disrupt the formation of active, phase-separated compartments. *Molecular Cell*, *72*(1), 19–36.e8.

Brangwynne, C. P., Eckmann, C. R., Courson, D. S., Agata, R., Carsten, H., Jöbin, G., et al. (2009). Germline P granules are liquid droplets that localize by controlled dissolution/condensation. *Science*, *324*(5935), 1729–1732.

Bridgman, P. W. (1914). The coagulation of albumin by pressure. *Journal of Biological Chemistry*, *19*, 511.

Brooks, N. J., Gauthe, B. L. L. E., Terrill, N. J., Rogers, S. E., Templer, R. H., Ces, O., et al. (2010). Automated high pressure cell for pressure jump X-ray diffraction. *Review of Scientific Instruments*, *81*(6), 064103.

Brosey, C. A., & Tainer, J. A. (2019). Evolving SAXS versatility: Solution X-ray scattering for macromolecular architecture, functional landscapes, and integrative structural biology. *Current Opinion in Structural Biology*, *58*, 197–213.

Castellví, A., Pascual-Izarra, C., Crosas, E., Malfois, M., & Juanhuix, J. (2020). Improving data quality and expanding BioSAXS experiments to low-molecular-weight and low-concentration protein samples. *Acta Crystallographica Section D Structural Biology*, *76*(10), 971–981.

Chen, C. R., & Makhatadze, G. I. (2017). Molecular determinant of the effects of hydrostatic pressure on protein folding stability. *Nature Communications*, *8*(1), 14561.

Cho, Y., Yang, J. S., & Song, K. B. (1999). Effect of ascorbic acid and protein concentration on the molecular weight profile of bovine serum albumin and β-lactoglobulin by γ-irradiation. *Food Research International*, *32*(7), 515–519.

Cinar, S., Cinar, H., Chan, H. S., & Winter, R. (2019). Pressure-sensitive and osmolyte-modulated liquid–liquid phase separation of eye-lens γ-crystallins. *Journal of the American Chemical Society*, *141*(18), 7347–7354.

Cinar, H., Fetahaj, Z., Cinar, S., Vernon, R., Chan, H., & Winter, R. (2019). Temperature, hydrostatic pressure, and osmolyte effects on liquid-liquid phase separation in protein condensates: Physical chemistry and biological implications. *Chemistry-A European Journal*, *25*(57), 13049–13069.

Cinar, H., Oliva, R., Chen, X., Zhang, M., Chan, H., Winter, R., et al. (2020). Pressure sensitivity of SynGAP/PSD-95 condensates as a model for postsynaptic densities and its biophysical and neurological ramifications. *Chemistry-A European Journal*, *26*(48), 11024–11031.

Colman, D. R., Poudel, S., Stamps, B. W., Boyd, E. S., & Spear, J. R. (2017). The deep, hot biosphere: Twenty-five years of retrospection. *Proceedings of the National Academy of Sciences of the United States of America*, *114*(27), 6895.

de Oliveira, G. A. P., Cordeiro, Y., Silva, J. L., & Vieira, T. C. R. G. (2019). Chapter Nine—Liquid-liquid phase transitions and amyloid aggregation in proteins related to cancer and neurodegenerative diseases. In R. Donev (Ed.), *Vol. 118. Advances in protein chemistry and structural biology* (pp. 289–331). Academic Press.

Falahati, H., & Haji-Akbari, A. (2019). Thermodynamically driven assemblies and liquid–liquid phase separations in biology. *Soft Matter*, *15*(6), 1135–1154.

Fine, R. A., & Millero, F. J. (1973). Compressibility of water as a function of temperature and pressure. *The Journal of Chemical Physics*, *59*(10), 5529–5536.

Franke, D., Jeffries, C. M., & Svergun, D. I. (2015). Correlation map, a goodness-of-fit test for one-dimensional X-ray scattering spectra. *Nature Methods*, *12*(5), 419–422.

Fujisawa, T., & Kato, M. (2002). The small-angle X-ray scattering from proteins under pressure. In Y. Taniguchi, H. E. Stanley, & H. Ludwig (Eds.), *Biological systems under extreme conditions: Structure and function, biological and medical physics series* (pp. 121–138). Berlin, Heidelberg: Springer.

Goldberg, R. N., Kishore, N., & Lennen, R. M. (2002). Thermodynamic quantities for the ionization reactions of buffers. *Journal of Physical and Chemical Reference Data*, *31*(2), 231–370.

Good, N. E., Winget, G. D., Winter, W., Connolly, T. N., Izawa, S., & Singh, R. M. M. (1966). Hydrogen ion buffers for biological research. *Biochemistry*, *5*(2), 467–477.

Graewert, M. A., & Jeffries, C. M. (2017). Sample and buffer preparation for SAXS. In B. Chaudhuri, I. G. Muñoz, S. Qian, & V. S. Urban (Eds.), *Biological small angle scattering: Techniques, strategies and tips* (pp. 11–30). Singapore: Springer Singapore.

Grindley, T., & Lind, J. E. (1971). PVT properties of water and mercury. *The Journal of Chemical Physics*, *54*(9), 3983–3989.

Guinier, A. (1939). La diffraction des rayons X aux très petits angles: Application à l'étude de phénomènes ultramicroscopiques. *Annales de Physique*, *11*(12), 161–237.

Hajizadeh, N. R., Franke, D., Jeffries, C. M., & Svergun, D. I. (2018). Consensus Bayesian assessment of protein molecular mass from solution X-ray scattering data. *Scientific Reports*, *8*(1), 7204.

Hamann, S. D. (1980). The role of electrostriction in high pressure chemistry. *The Review of Physical Chemistry of Japan*, *50*, 147–168.

Harish, B., Gillilan, R. E., Zou, J., Wang, J., Raleigh, D. P., & Royer, C. A. (2021). Protein unfolded states populated at high and ambient pressure are similarly compact. *Biophysical Journal*, *120*(12), 2592–2598.

Hatti-Kaul, R., & Abouhmad, A. (2020). Biotechnological applications of extremophilic microorganisms. In N. M. Lee (Ed.), *Vol. 6. Life in extreme environments* (pp. 1–42). De Gruyter.

Jacques, D. A., & Trewhella, J. (2010). Small-angle scattering for structural biology—Expanding the frontier while avoiding the pitfalls. *Protein Science: A Publication of the Protein Society*, *19*(4), 642–657.

Jeffries, C. M., Graewert, M. A., Blanchet, C. E., Langley, D. B., Whitten, A. E., & Svergun, D. I. (2016). Preparing monodisperse macromolecular samples for successful biological small-angle X-ray and neutron-scattering experiments. *Nature Protocols*, *11*(11), 2122–2153.

Jeffries, C. M., Graewert, M. A., Svergun, D. I., & Blanchet, C. E. (2015). Limiting radiation damage for high-brilliance biological solution scattering: Practical experience at the EMBL P12 beamline PETRAIII. *Journal of Synchrotron Radiation*, *22*(2), 273–279.

Jenkins, K. A., Fossat, M. J., Zhang, S., Rai, D. K., Klein, S., Gillilan, R., et al. (2018). The consequences of cavity creation on the folding landscape of a repeat protein depend upon context. *Proceedings of the National Academy of Sciences of the United States of America*, *115*(35), E8153–E8161.

Johansen, D., Trewhella, J., & Goldenberg, D. P. (2011). Fractal dimension of an intrinsically disordered protein: Small-angle X-ray scattering and computational study of the bacteriophage λ N protein. *Protein Science: A Publication of the Protein Society*, *20*(12), 1955–1970.

Jouhet, J. (2013). Importance of the hexagonal lipid phase in biological membrane organization. *Frontiers in Plant Science*, *4*, 1–5.

Julius, K., Al-Ayoubi, S. R., Paulus, M., Tolan, M., & Winter, R. (2018). The effects of osmolytes and crowding on the pressure-induced dissociation and inactivation of dimeric LADH. *Physical Chemistry Chemical Physics*, *20*(10), 7093–7104.

Kato, M., & Fujisawa, T. (1998). High-pressure solution X-ray scattering of protein using a hydrostatic cell with diamond windows. *Journal of Synchrotron Radiation*, *5*(5), 1282–1286.

Kato, M., Han, T. W., Xie, S., Shi, K., Du, X., Wu, L. C., et al. (2012). Cell-free formation of RNA granules: Low complexity sequence domains form dynamic fibers within hydrogels. *Cell*, *149*(4), 753–767.

Kharakoz, D. P. (2000). Protein compressibility, dynamics, and pressure. *Biophysical Journal*, *79*(1), 511–525.

Kitamura, Y., & Itoh, T. (1987). Reaction volume of protonic ionization for buffering agents. Prediction of pressure dependence of pH and pOH. *Journal of Solution Chemistry*, *16*(9), 715–725.

Koch, M. H. J., & Stuhrmann, H. B. (1979). [52] Neutron-scattering studies of ribosomes. *Methods in Enzymology*, *59*, 670–706. Academic Press.

Konarev, P. V., Volkov, V. V., Sokolova, A. V., Koch, M. H. J., & Svergun, D. I. (2003). PRIMUS: A windows PC-based system for small-angle scattering data analysis. *Journal of Applied Crystallography*, *36*(5), 1277–1282.

Kotlarchyk, M., & Chen, S. (1983). Analysis of small angle neutron scattering spectra from polydisperse interacting colloids. *The Journal of Chemical Physics*, *79*(5), 2461–2469.

Krywka, C., Sternemann, C., Paulus, M., Tolan, M., Royer, C., & Winter, R. (2008). Effect of osmolytes on pressure-induced unfolding of proteins: A high-pressure SAXS study. *ChemPhysChem*, *9*(18), 2809–2815.

Kuwamoto, S., Akiyama, S., & Fujisawa, T. (2004). Radiation damage to a protein solution, detected by synchrotron X-ray small-angle scattering: Dose-related considerations and suppression by cryoprotectants. *Journal of Synchrotron Radiation*, *11*(6), 462–468.

Lattman, E. E., Grant, T. D., & Snell, E. H. (2018). *Biological small angle scattering: Theory and practice. International union of crystallography monographs on crystallography*. Oxford: Oxford University Press.

Martin, E. W., Holehouse, A. S., Peran, I., Farag, M., Incicco, J. J., Bremer, A., et al. (2020). Valence and patterning of aromatic residues determine the phase behavior of prion-like domains. *Science*, *367*(6478), 694.

Martin, E. W., Hopkins, J. B., & Mittag, T. (2021). Chapter Seven—Small-angle X-ray scattering experiments of monodisperse intrinsically disordered protein samples close to the solubility limit. In C. D. Keating (Ed.), *Vol. 646. Methods in enzymology* (pp. 185–222). Academic Press.

Meersman, F., Smeller, L., & Heremans, K. (2006). Protein stability and dynamics in the pressure–temperature plane. *Proteins Under High Pressure*, *1764*(3), 346–354.

Meisburger, S. P., Warkentin, M., Chen, H., Hopkins, J. B., Gillilan, R. E., Pollack, L., et al. (2013). Breaking the radiation damage limit with cryo-SAXS. *Biophysical Journal*, *104*(1), 227–236.

Möller, J., Grobelny, S., Schulze, J., Bieder, S., Steffen, A., Erlkamp, M., et al. (2014). Reentrant liquid-liquid phase separation in protein solutions at elevated hydrostatic pressures. *Physical Review Letters*, *112*(2), 028101.

Möller, J., Schroer, M. A., Erlkamp, M., Grobelny, S., Paulus, M., Tiemeyer, S., et al. (2012). The effect of ionic strength, temperature, and pressure on the interaction potential of dense protein solutions: From nonlinear pressure response to protein crystallization. *Biophysical Journal*, *102*(11), 2641–2648.

Molliex, A., Temirov, J., Lee, J., Coughlin, M., Kanagaraj, A. P., Kim, H. J., et al. (2015). Phase separation by low complexity domains promotes stress granule assembly and drives pathological fibrillization. *Cell*, *163*(1), 123–133.

Muschol, M., & Rosenberger, F. (1997). Liquid–liquid phase separation in supersaturated lysozyme solutions and associated precipitate formation/crystallization. *The Journal of Chemical Physics*, *107*(6), 1953–1962.

Mylonas, E., & Svergun, D. I. (2007). Accuracy of molecular mass determination of proteins in solution by small-angle X-ray scattering. *Journal of Applied Crystallography*, *40*(s1), s245–s249.

Neuman, R. C., Kauzmann, W., & Zipp, A. (1973). Pressure dependence of weak acid ionization in aqueous buffers. *The Journal of Physical Chemistry*, 77(22), 2687–2691.

Orthaber, D., Bergmann, A., & Glatter, O. (2000). SAXS experiments on absolute scale with Kratky systems using water as a secondary standard. *Journal of Applied Crystallography*, *33*(2), 218–225.

Paladini, A. A., & Weber, G. (1981). Pressure-induced reversible dissociation of enolase. *Biochemistry*, *20*(9), 2587–2593.

Piiadov, V., Ares de Araújo, E., Oliveira Neto, M., Craievich, A. F., & Polikarpov, I. (2019). SAXSMoW 2.0: Online calculator of the molecular weight of proteins in dilute solution from experimental SAXS data measured on a relative scale. *Protein Science*, *28*(2), 454–463.

Plümper, O., King, H. E., Geisler, T., Liu, Y., Pabst, S., Savov, I. P., et al. (2017). Subduction zone forearc serpentinites as incubators for deep microbial life. *Proceedings of the National Academy of Sciences of the United States of America*, *114*(17), 4324–4329.

Porod, G. (1982). *General theory. Small-angle X-ray scattering* (pp. 17–51). London: Academic Press.
Poulter, T. C. (1932). Apparatus for optical studies at high pressure. *Physical Review, 40*(5), 860–871.
Quinlan, R. J., & Reinhart, G. D. (2005). Baroresistant buffer mixtures for biochemical analyses. *Analytical Biochemistry, 341*(1), 69–76.
Rai, D. K., Gillilan, R. E., Huang, Q., Miller, R., Ting, E., Lazarev, A., et al. (2021). High-pressure small-angle X-ray scattering cell for biological solutions and soft materials. *Journal of Applied Crystallography, 54*(1), 111–122.
Rambo, R. P., & Tainer, J. A. (2011). Characterizing flexible and intrinsically unstructured biological macromolecules by SAS using the Porod-Debye law. *Biopolymers, 95*(8), 559–571.
Rambo, R. P., & Tainer, J. A. (2013). Accurate assessment of mass, models and resolution by small-angle scattering. *Nature, 496*(7446), 477–481.
Rappolt, M. (2006). Chapter 9. The biologically relevant lipid mesophases as "Seen" by X-rays. In A. Leitmannova Liu (Ed.), *Vol. 5. Advances in planar lipid bilayers and liposomes* (pp. 253–283). Academic Press.
Receveur-Brechot, V., & Durand, D. (2012). How random are intrinsically disordered proteins? A small angle scattering perspective. *Current Protein & Peptide Science, 13*(1), 55–75.
Riback, J. A., Bowman, M. A., Zmyslowski, A. M., Knoverek, C. R., Jumper, J. M., Hinshaw, J. R., et al. (2017). Innovative scattering analysis shows that hydrophobic disordered proteins are expanded in water. *Science, 358*(6360), 238–241.
Roche, J., & Royer, C. A. (2018). Lessons from pressure denaturation of proteins. *Journal of The Royal Society Interface, 15*(147), 20180244.
Samaranayake, C. P., & Sastry, S. K. (2010). In situ measurement of pH under high pressure. *The Journal of Physical Chemistry B, 114*(42), 13326–13332.
Schroer, M. A., Markgraf, J., Wieland, D. C. F., Sahle, C. J., Möller, J., Paulus, M., et al. (2011). Nonlinear pressure dependence of the interaction potential of dense protein solutions. *Physical Review Letters, 106*(17), 178102.
Silva, J. L., Oliveira, A. C., Vieira, T. C. R. G., de Oliveira, G. A. P., Suarez, M. C., & Foguel, D. (2014). High-pressure chemical biology and biotechnology. *Chemical Reviews, 114*(14), 7239–7267.
Skou, S., Gillilan, R. E., & Ando, N. (2014). Synchrotron-based small-angle X-ray scattering of proteins in solution. *Nature Protocols, 9*(7), 1727–1739.
Somero, G. N. (1992). Adaptations to high hydrostatic pressure. *Annual Review of Physiology, 54*, 557–577.
Stachowski, T. R., Snell, M. E., & Snell, E. H. (2021). A SAXS-based approach to rationally evaluate radical scavengers—Toward eliminating radiation damage in solution and crystallographic studies. *Journal of Synchrotron Radiation, 28*(5), 1309–1320.
Stuhrmann, H. B., & Kirste, R. G. (1967). Elimination der intrapartikulären Untergrundstreuung bei der Röntgenkleinwinkelstreuung an kompakten Teilchen. II. *Zeitschrift für Physikalische Chemie, 56*(5–6), 334–337.
Svergun, D. I. (1992). Determination of the regularization parameter in indirect-transform methods using perceptual criteria. *Journal of Applied Crystallography, 25*(4), 495–503.
Svergun, D. I., Koch, M. H. J., Timmins, P. A., & May, R. P. (2013). *Small angle X-ray and neutron scattering from solutions of biological macromolecules. International Union of Crystallography Texts on Crystallography*. Oxford: Oxford University Press.
Svergun, D. I., Richard, S., Koch, M. H. J., Sayers, Z., Kuprin, S., & Zaccai, G. (1998). Protein hydration in solution: Experimental observation by x-ray and neutron scattering. *Proceedings of the National Academy of Sciences of the United States of America, 95*(5), 2267.

Taulier, N., & Chalikian, T. V. (2002). Compressibility of protein transitions. *Biochimica Biophysica Acta, 1595*(1–2), 48–70.

Teixeira, S. C. M., Leão, J. B., Gagnon, C., & McHugh, M. A. (2018). High pressure cell for Bio-SANS studies under sub-zero temperatures or heat denaturing conditions. *Journal of Neutron Research, 20*(1/2), 11–21.

Tuukkanen, A. T., Spilotros, A., & Svergun, D. I. (2017). Progress in small-angle scattering from biological solutions at high-brilliance synchrotrons. *IUCrJ, 4*(5), 518–528.

Wang, C., Lin, Y., Bougie, D., & Gillilan, R. E. (2018). Predicting data quality in biological X-ray solution scattering. *Acta Crystallographica Section D, 74*(8), 727–738.

Wang, B., Zhang, L., Dai, T., Qin, Z., Lu, H., Zhang, L., et al. (2021). Liquid–liquid phase separation in human health and diseases. *Signal Transduction and Targeted Therapy, 6*(1), 290.

Weber, G. (1992). *Protein interactions.* New York: Chapman and Hall.

Winter, R. (2015). Pressure effects on the intermolecular interaction potential of condensed protein solutions. In K. Akasaka, & H. Matsuki (Eds.), *High pressure bioscience: Basic concepts, applications and frontiers* (pp. 151–176). Dordrecht: Springer Netherlands. Retrieved from https://doi.org/10.1007/978-94-017-9918-8_8.

Woenckhaus, J., Köhling, R., Winter, R., Thiyagarajan, P., & Finet, S. (2000). High pressure-jump apparatus for kinetic studies of protein folding reactions using the small-angle synchrotron X-ray scattering technique. *Review of Scientific Instruments, 71*(10), 3895–3899.

Worman, S. L., Pratson, L. F., Karson, J. A., & Schlesinger, W. H. (2020). Abiotic hydrogen (H_2) sources and sinks near the mid-ocean ridge (MOR) with implications for the subseafloor biosphere. *Proceedings of the National Academy of Sciences of the United States of America, 117*(24), 13283.

Zhang, F., Ilavsky, J., Long, G., Quintana, J., Allen, A., & Jemian, P. (2009). Glassy carbon as an absolute intensity calibration standard for small-angle scattering. *Metallurgical and Materials Transactions A, 41*, 1151–1158.

Zheng, W., & Best, R. B. (2018). An extended guinier analysis for intrinsically disordered proteins. *Intrinsically Disordered Proteins, 430*(16), 2540–2553.

CHAPTER TWELVE

Characterization of biological materials with soft X-ray scattering

Joshua T. Del Mundo[a], Sintu Rongpipi[b], Enrique D. Gomez[a,c,*], and Esther W. Gomez[a,d,*]

[a]Department of Chemical Engineering, The Pennsylvania State University, University Park, PA, United States
[b]Advanced Light Source and Materials Science Division, Lawrence Berkeley National Laboratory, Berkeley, CA, United States
[c]Department of Materials Science and Engineering and Materials Research Institute, The Pennsylvania State University, University Park, PA, United States
[d]Department of Biomedical Engineering, The Pennsylvania State University, University Park, PA, United States
*Corresponding authors: e-mail address: edg12@psu.edu; ewg10@psu.edu

Contents

Methods in Enzymology, Volume 677
ISSN 0076-6879
https://doi.org/10.1016/bs.mie.2022.08.042

Abstract

The complex structure of biological assemblies is crucial for function yet challenging to discern given the chemical similarities between constituent components. Hard X-ray techniques, for example, rely on small density differences between domains that lead to modest scattering intensities. Resonant soft X-ray scattering (RSoXS) uses X-rays below 2 keV to access absorption edges of low-Z elements. In this way, RSoXS can enhance scattering contrast between domains of different chemical compositions or bonding motifs, thus providing structural information about specific chemical motifs. RSoXS is emerging as a technique applicable for biological systems, having been used to characterize protein structure in solution and polysaccharide organization in plant cell walls. Sample environment instrumentation, however, is challenging in the current state of the art, particularly with liquid samples. This chapter contains a brief introduction to RSoXS and current beamline capabilities, and provides methods to prepare, store, and mount biological samples for RSoXS characterization. Furthermore, key details during RSoXS and X-ray absorption data acquisition are highlighted and some future opportunities in RSoXS instrumentation for biological systems are discussed.

1. Introduction

X-ray scattering is a valuable tool for the characterization of biological materials. One such technique, small-angle X-ray scattering (SAXS), can determine the size and shape of nanometer scale structures in macromolecules and biological assemblies. In a SAXS experiment, a focused beam of X-rays that is often hundreds of microns in diameter passes through a sample and the signal of the scattered X-rays is detected using an image (2D) detector. SAXS is typically conducted in the "hard" X-ray energy regime (generally >8 keV), and determines structural features based on the differences in electron densities of components in a material. The resulting structural information obtained over an average of hundreds of microns complements local information obtained from microscopy techniques, such as transmission electron microscopy (TEM) and atomic force microscopy (AFM).

Synchrotron light sources provide the necessary X-ray flux to make SAXS of biological samples with modest scattering achievable. SAXS has been used to reveal structural information about many biological systems, including protein oligomers (Korasick et al., 2017; Korasick & Tanner, 2018), liposomes for drug delivery (Di Cola, Grillo, & Ristori, 2016; Ristori, Oberdisse, Grillo, Donati, & Spalla, 2005), and plant cell walls (Lopez-Sanchez et al., 2020; Martínez-Sanz, Gidley, & Gilbert, 2015). Despite the utility of SAXS for structural studies, elucidating the arrangement of the various components within biological assemblies can be challenging, in part due to small scattering contrast between domains with similar density.

When an X-ray photon interacts with a material, it will undergo either transmission, absorption, or scattering. These interactions are summarized in the complex index of refraction, which is composed of a dispersive and absorptive component. The complex refractive index is defined as

$$n(\lambda) = 1 - \delta(\lambda) + i\beta(\lambda) \tag{1}$$

where λ is the X-ray wavelength, which is related to the photon energy, and δ and β are the dispersive and absorptive optical constants, respectively, which are related to the complex atomic scattering factor (Rongpipi, Del Mundo, Gomez, & Gomez, 2021). Scattering contrast arises from the difference in how X-rays interact with two or more materials. This is proportional to the sum of the squares of the differences between the optical constants as defined by

$$\frac{\Delta n \Delta n^*}{\lambda^4} = \frac{\left(\delta_i(\lambda) - \delta_j(\lambda)\right)^2 + \left(\beta_i(\lambda) - \beta_j(\lambda)\right)^2}{\lambda^4} \tag{2}$$

where $\Delta n \Delta n^*/\lambda^4$ is the contrast between materials i and j (Virgili, Tao, Kortright, Balsara, & Segalman, 2007; Ye et al., 2018). In the hard X-ray regime, absorption by low-Z elements is minimal and differences between chemical-specific components are often negligible. Small differences in density yield small differences in the dispersive component of scattering contrast. This limits the application of SAXS for the study of biological samples, which are often composed of low-Z elements.

Resonant soft X-ray scattering (RSoXS) combines small-angle scattering with X-ray absorption spectroscopy to increase contrast between low-Z species. Using X-rays in the "soft" regime (generally $<2\,\text{keV}$) allows for access to the absorption edges of low-Z elements. When the energy of incident X-rays is near the absorption edge of an atom, the scattering factor is altered due to anomalous dispersion. This, in turn, alters the scattering contrast between different components in a material. Thus, by tuning the X-ray energy, the scattering contrast between components of different chemical composition or bonding environment can be perturbed. As such, contrast can be enhanced without chemical modification of the sample, unlike other approaches such as deuteration for small-angle neutron scattering or heavy metal staining for TEM experiments. Increasing the scattering contrast near atomic absorption edges also results in smaller required sample volumes when compared to hard X-rays (Ye et al., 2018). Furthermore, the low X-ray energy of RSoXS has been shown to mitigate radiation damage for proteins in solution when compared to hard X-ray scattering experiments (Ye, Le, et al., 2018) or electron microscopy techniques. RSoXS is often complemented by near-edge X-ray absorption fine structure (NEXAFS) spectroscopy, which can be used to extract real and complex optical constants for pure components, and thereby lead to predictions of the scattering contrast of a multi-component system as a function of X-ray energy. Table 1 lists atomic absorption energies of elements commonly found in biological materials.

RSoXS is an emerging technique for the investigation of the structure of biological systems. Carbon edge RSoXS has revealed the structure of bovine serum albumin (BSA) in solution based on scattering from different carbon bonding motifs (Ye, Le, et al., 2018). RSoXS at 283.3 eV highlights the portion of the protein structure comprised of C=C bonds, while RSoXS at 288.3 eV is sensitive to CONH peptide bonds. This insight refines the structural model of BSA by revealing the shape and size of envelopes defined by

Table 1 X-ray absorption energies in eV of some common elements found in biological molecules and biological samples (Bearden & Burr, 1967).

Element	Z	K-edge	L1-edge	L2-edge	L3-edge
C	6	283.8 ± 0.4	–	6.4 ± 1.9	6.4 ± 1.9
N	7	401.6 ± 0.4	–	9.2 ± 0.6	9.2 ± 0.6
O	8	532.0 ± 0.4	23.7 ± 0.4	7.1 ± 0.8	7.1 ± 0.8
F	9	685.4 ± 0.4	31.0	8.6 ± 0.8	8.6 ± 0.8
Na	11	1072.1 ± 0.4	63.3 ± 0.4	31.1 ± 0.4	31.1 ± 0.4
P	15	2145.5 ± 0.4	189.3 ± 0.4	132.2 ± 0.5	132.2 ± 0.5
S	16	2472.0 ± 0.4	229.2 ± 0.4	164.8 ± 0.7	164.8 ± 0.7
K	19	3607.4 ± 0.4	377.1 ± 0.4	296.3 ± 0.4	293.6 ± 0.4
Ca	20	4038.1 ± 0.4	437.8 ± 0.4	350.0 ± 0.4	346.4 ± 0.4

these specific moieties. Calcium edge RSoXS, combined with SAXS, was able to identify the spacing and size of colloidal calcium phosphate (CCP) particles in milk casein micelles (Ingham et al., 2015; Ingham et al., 2016). Later RSoXS studies compared the CCP nanocluster distribution, which was similar in cow and goat milk, even though structures at larger scales differ (Ingham, Smialowska, Kirby, Wang, & Carr, 2018). Further RSoXS studies at the iron-edge, combined with anomalous small-angle X-ray scattering (ASAXS), of iron-fortified milk revealed that both iron and calcium are co-located in the same structures in casein micelles (Ingham, Kirby, Wang, Brady, & Carr, 2018). In another study, NEXAFS and calcium-edge RSoXS of onion primary cell wall revealed enhanced scattering at the calcium edge, and a peak corresponding to a 20 nm spacing between domains, as shown in Fig. 1 (Ye, Kiemle, et al., 2018). The increase in scattering at the calcium edge was determined to originate from contrast between cellulose and calcium-crosslinked pectin. Thus, the 20 nm feature was assigned to the spacing between cellulose microfibril bundles or between individual microfibrils.

Despite the tremendous potential of RSoXS, there are several experimental challenges that arise when using soft X-rays. Samples can have attenuation lengths that are sub-micron at soft X-ray energies due to the significant absorption present when working near elemental edges.

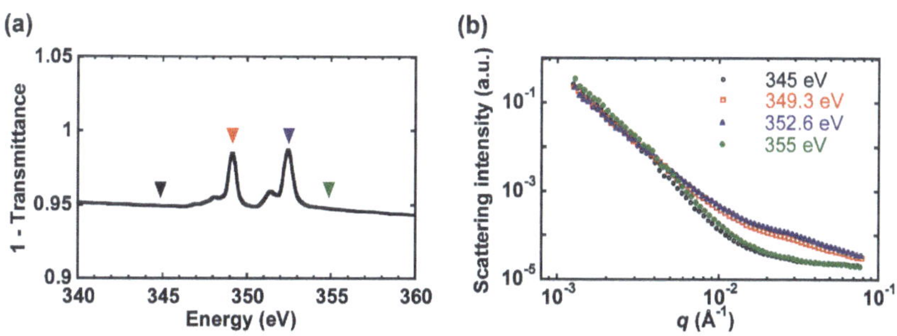

Fig. 1 NEXAFS and RSoXS data for onion outer epidermal peel collected at energies around the Ca-edge. (A) Transmission Ca-edge NEXAFS spectrum and (B) azimuthally integrated RSoXS profiles. Scattering profiles at the on-edge energies (349.3 and 352.6 eV) show a peak at approximately $q = 0.03\,\text{Å}^{-1}$, or 20 nm, due to enhanced contrast that is not present in scattering profiles at off-edge energies (345 and 355 eV). Arrowheads in panel (A) indicate energies at which the RSoXS data in panel (B) were collected. *Data are from Ye, D., Kiemle, S. N., Ron, S., Wang, X., Wang, C., Cosgrove, D. J., et al. (2018). Resonant soft X-ray scattering reveals cellulose microfibril spacing in plant primary cell walls.* Scientific Reports, *8, 12449.*

Therefore, samples must be thin, for example below 1 μm at the carbon edge (ca. 284 eV); however, multiple scattering is seldom a problem with thin samples. The longer wavelength of soft X-rays reduces the upper limit in scattering vector, such that scattering from structures below 1 nm is not accessible, whereas hard X-rays can probe Angstrom sized features (crystal lattice planes). Additionally, scattering in air or even helium is high in the soft X-ray regime. Therefore, measurements must occur in a high vacuum environment (10^{-7} Torr), which can limit the viability of certain types of samples. Current RSoXS capabilities are best suited for dry, solid samples, such as plant epidermal peels (Ye, Kiemle, et al., 2018).

Although there are a few reports of solution RSoXS, it remains a technical challenge. Many biological systems must be characterized in solution or in a hydrated state for an accurate structural model. One of the greatest challenges of RSoXS is preparing solution or hydrated solid samples that can withstand high vacuum. For hydrated biological systems, sample preparation and measurement conditions are analogous to liquid TEM, which is another technique that requires sample encapsulation within a high vacuum environment. Recently, a microfluidic flow cell was used for in-situ RSoXS characterization of the structure and dynamics of a copolymer micelle drug delivery platform (McAfee et al., 2021). A more common and simple approach to liquid RSoXS is to sandwich liquid between two silicon nitride films that are supported by a silicon frame, and that are separated by a 1 μm spacer. The edges are sealed using vacuum-tight epoxy. This approach has been used to examine milk casein (Ingham et al., 2015; Ingham et al., 2016; Ingham, Kirby, et al., 2018; Ingham, Smialowska, et al., 2018) and BSA (Ye, Le, et al., 2018).

RSoXS can be performed at beamline 11.0.1.2 at the Advanced Light Source (ALS) (Gann et al., 2012; Liu, Brady, & Wang, 2016) and the RSoXS end station at the SST-1 beamline suite at the National Synchrotron Light Source II (NSLS-II), which is operated by the National Institute of Standards and Technology (NIST) (Gann et al., 2021). The ALS is scheduled to undergo a large upgrade, dubbed "ALS-U." We will discuss the capabilities of ALS beamline 11.0.1.2 before this upgrade, although beamline specifications and setup are likely to change in the near future. The energy ranges of the ALS and NSLS-II beamlines are 165–1800 eV and 100–2500 eV, respectively. The photon flux of the beamlines varies depending on the experimental configuration and energy. While both beamlines can be adjusted for both automatic and manual data collection, the ALS workstation is slightly more suited for manual operation, where users manually select scans and sample spots. The NSLS-II workstation on the other hand is well suited for automatic data collection; pre-set energy range scans are available to users, with up to 0.1 eV resolution. Both beamlines are equipped to conduct RSoXS and transmission NEXAFS experiments, can hold a large quantity of samples (more than 50 square RSoXS substrates sized at 5 mm on a side), and can accommodate biological samples using the methods described in this text. Here, we present methods for sample preparation, handling, and transportation to beamlines, and approaches to overcome experimental challenges of RSoXS. We also discuss new instrumental opportunities for soft X-ray scattering of biological systems, which include unexplored sample environments and beamline modalities.

2. Sample preparation for dry or dehydrated materials

Scattering from air or even helium is high in the soft X-ray regime, such that vacuum from the sample chamber to the detector is crucial for RSoXS experiments. As such, samples must be compatible with high vacuum environments. Flat, film-like samples are typically used. Dried samples are well suited for this environment, though drying of some biological samples may not be representative of their native state. In this section, we will discuss sample requirements for NEXAFS and RSoXS and we will describe protocols to mount solid, dried samples onto X-ray transparent silicon nitride RSoXS substrates. As described later, wet cells can be used to isolate hydrated samples from the high vacuum environment.

As with any transmission scattering experiment, the thickness of the sample should ideally be close to the attenuation length of the sample. This is defined as the thickness where a 1/e fraction of photons is transmitted and is dependent on the sample density, elemental composition, and photon energy. The lower attenuation lengths in the soft X-ray regime when compared to the hard X-ray regime limits the sample thickness for RSoXS. For example, the attenuation length is typically a few hundreds of nanometers for polymeric materials at the carbon edge (284 eV) and is still well below 5 μm at energies up to 1 keV (Henke, Gullikson, & Davis, 1993). The low thickness of samples typically makes multiple scattering negligible, as seen for proteins in solution (Ye, Le, et al., 2018). A grazing-incidence geometry is also a possibility for surface sensitive measurements. In this chapter, we will focus on transmission RSoXS as this approach provides information relevant to the bulk volume of solid and liquid biological samples.

To support the sample and keep it upright under high vacuum, thin samples should be mounted onto "RSoXS substrates." The RSoXS substrate is composed of a thin, X-ray transparent film that is supported by a frame (often silicon) with a square "window" in the middle. Silicon nitride (Si_3N_4) is the most common film material, but aluminum oxide windows can also be fabricated for experiments that focus on examination of scattering at the silicon or nitrogen edges (Ye et al., 2019). The film thickness is typically 50–100 nm. Thicker films are stronger and therefore easier to handle but also attenuate more X-rays.

Here, we will describe RSoXS preparation of onion outer epidermal peel, which has been investigated as a model for plant primary cell wall (Zhang & Cosgrove, 2017). Previous calcium-edge RSoXS studies of onion outer epidermal peel have revealed the spacing between cellulose microfibril bundles (Ye, Kiemle, et al., 2018). This general protocol can be adapted to mount any dried biological sample that can be sectioned or cast as a thin film.

2.1 Materials

- White onion bulb (*Allium cepa*, cv: cometa), can be purchased from a local grocery store
- HEPES-Tween buffer: 20 mM HEPES (Sigma catalog # H3537), 0.1% Tween 20 (Sigma catalog # P1379), pH 6.8
- Razor blade
- Si_3N_4 RSoXS substrate: Typical dimensions are 50 nm Si_3N_4 film thickness, 1 mm window width, 5 mm frame width, 0.2 mm frame thickness (Norcada, catalog # NX5100A)
- Carbon-tipped tweezers
- Milli-Q water
- Petri dish or well plate
- Disposable plastic pipette

2.2 Procedure

2.2.1 Preparation of onion outer epidermal peel

The procedure of extracting onion epidermal peel is described as previously reported (Zhang & Cosgrove, 2017), and here adapted for transmission X-ray scattering experiments.

1. Remove dry, papery layers from the onion.
2. Select onion scale (layer) of interest. The outermost fleshy layer is scale number 1. The innermost layers are younger and thinner (see Note 1). Scales are labeled in Fig. 2A.

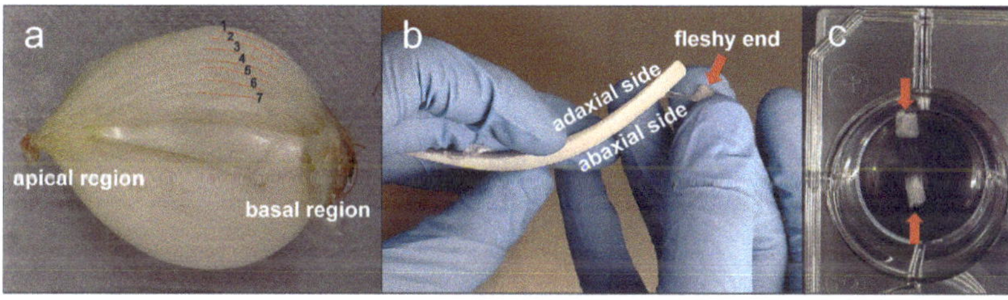

Fig. 2 Preparation of onion outer epidermal peel for NEXAFS and RSoXS experiments. (A) White onion bulb, with apical region, basal region, and scales 1 through 7 labeled. (B) Removal of outer epidermal peel. (C) Onion outer epidermal peel sample during buffer treatment. The fleshy ends (arrowheads) are used as handles for transferring the peel.

3. Using a razor blade, cut a ~5 mm wide strip along the height of the onion.
4. Make a shallow cut into the adaxial (inner) side of the onion scale. The cut should be as deep as possible before excising the abaxial (outer) epidermal layer (Fig. 2B).
5. Using a razor blade, trim off the top part of the onion strip, ~3 mm above the location of the shallow cut, to create a "fleshy end" (Fig. 2B).
6. Peel off the outer epidermis, from the apical to basal region (top to bottom) until at least 5 mm of the peel is exposed (Fig. 2B). This process ruptures the outermost layer of cells in the onion scale.
7. Using a razor blade, trim off the inner flesh to expose the epidermal peel.
8. Using a razor blade, trim off the bottom part of the onion strip ~3 mm from where peeling ceased to create another fleshy end. The onion peel should have pieces of flesh on either end. This allows for easier handling when transferring between buffers (Fig. 2C).
9. Place the onion peel into a petri dish or well plate with HEPES-Tween buffer and let sit for 1 h at room temperature. This washes the cell membranes and proteins from the sample, leaving behind the cell wall.
10. Use a disposable plastic pipette to remove the HEPES-Tween buffer while perturbing the peel as little as possible.
11. Wash the sample peel with water for 2 min. Fill the petri dish with Milli-Q water, swirl gently at 50 rpm for ~2 min, and use a disposable pipette to remove the water. Repeat this process five more times.
12. Before mounting the sample onto the Si_3N_4 RSoXS substrate, trim off the fleshy ends using a razor blade.

2.2.2 Mounting sample onto silicon nitride RSoXS substrate

1. Float the onion epidermal peel in a petri dish filled with Milli-Q water. For onion peels, the hydrophobic cuticle side will be facing up (Fig. 3; see Note 2).
2. Using carbon-tipped tweezers, submerge the Si_3N_4 RSoXS substrate, Si_3N_4 film side up, under the water (see Note 3).
3. Carefully lift the RSoXS substrate out from under the water to "scoop" the sample onto the window (see Note 4). A droplet will typically form on the window after successfully catching the sample. Make sure that the sample film is flat and is not folded on the droplet.
4. Allow the sample to dry fully in a dust-free environment. Drying typically takes 1–2h for small amounts of sample material (see Note 5).

2.3 Notes

1. The thickness of the 11th scale, which is ~1 μm, is suitable for RSoXS (Ye, Kiemle, et al., 2018).
2. This procedure can be applied to any solid sample that floats in water. If the material does not float, it can be allowed to settle onto the top of the Si_3N_4 RSoXS substrate. The water can then be slowly and carefully

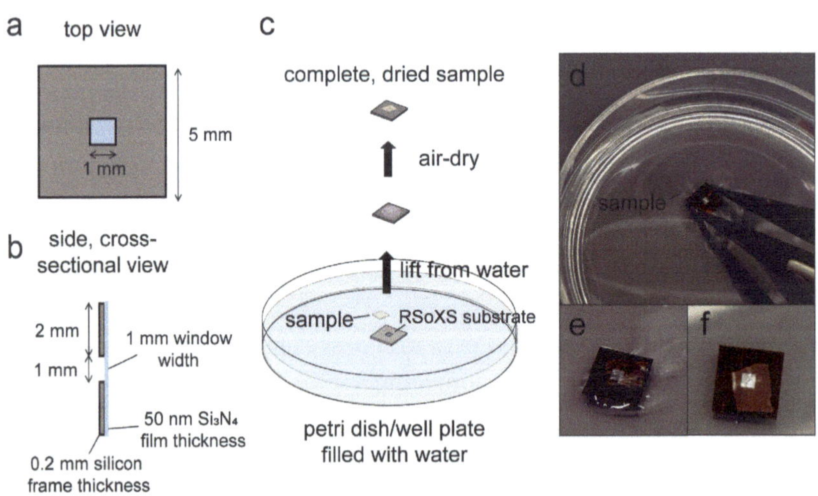

Fig. 3 Mounting onion outer epidermal peel onto a silicon nitride window. (A) Top view dimensions of silicon nitride RSoXS substrate. (B) Side, cross-sectional view of silicon nitride RSoXS substrate with dimensions. (C) Schematic of the sample mounting process. (D) Photograph of the process of lifting the RSoXS substrate out from water. (E) Onion outer epidermal peel on a silicon nitride RSoXS substrate immediately after mounting. (F) Sample after air-drying.

removed from the petri dish using a fine pipette. More robust solid samples, such as *Arabidopsis* seedlings, can be placed directly onto the Si_3N_4 RSoXS substrate.

3. The Si_3N_4 RSoXS substrate is fragile and should be handled with carbon-tipped tweezers. Use of metal tweezers will risk breaking both the frame and the Si_3N_4 film.
4. This step requires a large amount of precision and practice as the goal is to have the sample positioned such that it is centered over the window.
5. Avoid using Kimwipes to blot the RSoXS substrate, as the fibers from the Kimwipe easily transfer to the film surface.

3. Sample preparation for liquids and hydrated solids

Though challenging, extending RSoXS to liquid samples provides a unique opportunity to examine biomolecules such as proteins, lipid vesicles, or biological assemblies in solution. An initial approach has relied on encapsulating liquid in between silicon nitride RSoXS substrates that are separated by a spacer. These sample cells have been used to obtain chemical specific structures of casein micelles and BSA in solution (Ingham et al., 2015; Ingham et al., 2016; Ingham, Kirby, et al., 2018; Ingham, Smialowska, et al., 2018; Ye, Le, et al., 2018). In this section, a general approach for preparation of samples for solution RSoXS experiments will be described.

3.1 Materials

- Solution of protein or biological assembly of interest, typically at a concentration of 5–10 mg/mL (see Note 1)
- Buffer solvent for background measurement
- 0.2 μm syringe filter, cellulose acetate (VWR) or polyethersulfonate (EMD Millipore) filter
- Syringe
- Micropipette (1–10 μL)
- Carbon-tipped tweezers
- Si_3N_4 RSoXS substrates (see Note 2)
  - Bottom: 100 nm Si_3N_4 film thickness, 1 mm window width, 7.5 mm frame width, 0.2 mm frame thickness (Norcada, catalog # NX 7100C)
  - Top: 100 nm Si_3N_4 film thickness, 1 mm window width, 5 mm frame width, 0.2 mm frame thickness with 1 μm SU8 spacer (Silson, reference number 11608113)

- Vacuum epoxy (LOCTITE® Hysol®1C™)
- Toothpicks or fine-tipped tool

3.2 Procedure

1. Filter the buffer and protein solution using a 0.2 μm syringe filter and syringe to remove contaminates such as dust and aggregates.
2. Pipette a 4 μL droplet of the biological sample solution (see Note 1) onto the surface of the bottom RSoXS substrate, with the Si_3N_4 film side facing up (see Fig. 4).
3. Place the top RSoXS substrate with the Si_3N_4 film and spacer side facing down, onto the bottom RSoXS substrate, making sure that the windows from the top and bottom overlap (Fig. 4A; see Notes 3 and 4).
4. Use toothpicks or another fine-tipped tool to apply epoxy to cover the outer perimeter of the top RSoXS substrate.
5. Cure the epoxied sample cell at 4 °C for 24 h (Fig. 4B).
6. Repeat this procedure for the buffer solution to be used for background measurements.

3.3 Notes

1. The concentration should be chosen such that particle aggregation is minimized. Approximately 4 μL of solution is needed per sample. Do not overfill the cell, as to not cause it to rupture.
2. RSoXS substrates with 100 nm thick silicon nitride films are used for the liquid cell experiments to make it less likely for the films to break when the liquid cells are placed under vacuum.
3. It is challenging to verify that the liquid is properly encased within the sample cell, and that the cell will survive under vacuum. The success rate for the sample cell to not rupture under vacuum is about one in five (Ye, Le, et al., 2018). We recommend preparing 5–10 cells for both sample and buffer.

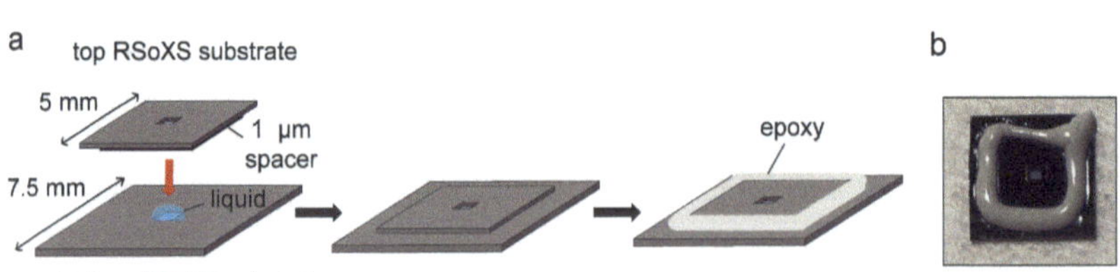

Fig. 4 (A) Schematic of liquid cell assembly. (B) Photograph of liquid cell following assembly.

4. This method can also be used to encapsulate hydrated solid samples, such as plant cell wall or polymer membranes. In this case, a spacer may not be needed, as the material acts as its own spacer. However, extra care must be taken so that the sample will not be too thick for sufficient transmission, especially at the carbon edge.

4. Storing and transporting samples

We recommend that liquid and hydrated samples be prepared on-site at the beamline or at user chemistry labs. Dried samples should be prepared in advance. Samples that are prepared away from the beamline must be transported long-distance to be mounted at the beamline workstation. Precautions must be taken to store and transport RSoXS substrates as they are fragile. In this section, we present a recommended method for storing and transporting samples.

4.1 Materials

- Samples, mounted onto RSoXS substrates
- Wide, shallow plastic container (e.g., TedPella catalog # 139-206)
- Double sided carbon tape (SPI catalog # 05081-AB)
- Carbon-tipped tweezers

4.2 Procedure

1. On the inside of a wide, shallow plastic container, adhere double-sided carbon tape in strips that are spaced at a distance that is slightly less than the diagonal length of the RSoXS substrates.
2. With a gloved finger, press the carbon tape a few times to weaken the adhesive. While samples should be securely adhered to the tape, a bond that is too strong may increase the risk of breaking the sample when trying to remove it from the tape (see Note 1).
3. Using carbon-tipped tweezers, place the sample mounted on a RSoXS substrate in between the rows of carbon tape so that opposite diagonal sides of the frame are touching the tape (Fig. 5A). Tap the corners with carbon-tipped tweezers gently to secure.
4. Label the samples as they are placed within the container.
5. When all samples are adhered to the inside of the container and labeled, seal the container with tape. The container is now ready to be stored or transported (see Note 2).

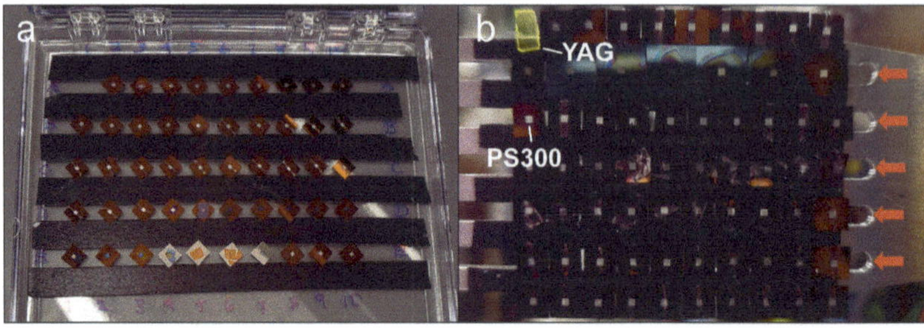

Fig. 5 (A) Inside surface of sample storage box to hold samples mounted on silicon nitride RSoXS substrates for storage and transportation. (B) A mounted RSoXS bar for transmission RSoXS/NEXAFS, with samples that include cellulose films, *Arabidopsis thaliana* hypocotyls, and onion epidermal peel. YAG crystal and PS300 calibration samples are labeled. Arrows indicate slits in bar where X-ray beam will pass through.

6. To remove the samples, use carbon-tipped tweezers to separate the RSoXS substrate frame from the tape, one corner at a time.

4.3 Notes

1. It is critical to secure the samples without the use of strong adhesives. Adhering two corners of the frame onto a weak adhesive tape is typically sufficient. Removing a RSoXS substrate from an adhesive that is strong will risk breaking it, because the frame and window are very fragile.
2. Samples can alternatively be stored in gel capsules, which are often provided by RSoXS substrate suppliers.

5. Mounting samples onto a sample bar

For NEXAFS and RSoXS experiments, the RSoXS substrates are first adhered onto a sample bar, which is then placed in the fixed path of the X-ray beam. A motor controls the position of the bar so that the beam can probe it at different spots. Bars are custom designed for the instrument and can typically hold more than 50 RSoXS substrates ($L \times W = 5\,mm \times 5\,mm$). Mounting samples onto the bar can be done at the beamline. If possible, samples may also be mounted onto the bar at the home institution and then shipped to the beamline.

The bar is then loaded into the sample chamber and the chamber is purged to 10^{-7} Torr. Pumping down the chamber can take up to several hours. For ex-situ samples, sample bars can be loaded well in advance.

RSoXS sample chambers do not require a secure hutch that is present in hard X-ray beamlines because of the low penetration depth of soft X-rays.

5.1 Materials

- Samples (mounted onto RSoXS substrates or encapsulated inside liquid cells)
- Double sided carbon tape (SPI catalog # 05081-AB)
- Carbon-tipped tweezers
- Sample bar, usually provided by the beamline
- Calibration standard samples (polystyrene spheres): 100 nm (TedPella catalog # 610-58) or 300 nm polystyrene spheres (TedPella catalog # 610-20), drop casted onto Si_3N_4 RSoXS substrate (see Notes 1 and 2)
- Blank sample (empty RSoXS substrate or liquid cell filled with buffer), for background subtraction

5.2 Procedure

1. Cover the sample bar with double-sided carbon tape in the regions where the RSoXS substrates will be mounted.
2. With a gloved finger, press the carbon tape a few times to weaken the adhesive.
3. Using carbon-tipped tweezers, transfer the samples onto the bar. The window area of the RSoXS substrate should be positioned within the bar slits (Fig. 5B).
4. Tap the RSoXS substrate frame with tweezers gently to secure to the adhesive carbon tape.
5. RSoXS substrates should be placed adjacent to each other so that the frames are touching (Fig. 5B), to maximize the number of samples that can be loaded at one time and to prevent shadows from scattering earlier in the beam path.
6. Along with the samples, mount calibration standard samples and a blank sample (see Notes 1 and 2). These are needed to calibrate and normalize both NEXAFS and RSoXS. If needed, also mount an yttrium aluminum garnet (YAG) crystal (see Note 3).

5.3 Notes

1. We recommend using PS300 to calibrate the beam energy for NEXAFS and PS100 to calibrate for beam center and sample detector distance for RSoXS. PS300 can also be used to calibrate the sample detector distance,

though PS100 has a preferable size range for RSoXS q-ranges. Other common calibration standard samples for detector distances include silver behenate (AgB) and highly oriented pyrolytic graphite (HOPG). Calibration standards are used to determine (1) the X-ray energy offset from NEXAFS, (2) the sample-to-detector distance, and (3) the beam center pixel.

2. To prepare polystyrene calibration samples, pipette 4 μL of PS100 or PS300 solution onto a 50 nm or 100 nm thick silicon nitride RSoXS substrate and let air dry for 2 h.
3. The sample bar is used to determine the location of the beam. There are several methods of accomplishing this. One method is to mount a YAG crystal to the bar (Fig. 5B). YAG glows yellow when exposed to the X-ray beam. The beam spot location can also be determined by scanning across fiducials on the sample bar.

6. Performing NEXAFS experiments

Data from NEXAFS, or X-ray absorption near edge structure (XANES), complements RSoXS measurements by allowing for the prediction of the scattering contrast between two components. As such, analyses of NEXAFS spectra can be used to identify energy ranges of interest for RSoXS experiments.

There are three common methods to obtain a NEXAFS spectrum. X-ray radiation can excite a core electron to a higher energy anti-bonding orbital and will subsequently produce a characteristic Auger electron or a fluorescence X-ray (Stöhr, 1992). As such, both the electron yield and fluorescent X-ray intensity are proportional to X-ray absorbance (Nahid, Gann, Thomsen, & McNeill, 2016). NEXAFS measurements in the transmission geometry probe the forward-scattered X-ray flux as a function of energy. Sample conditions are similar to RSoXS, where samples must be mounted on soft X-ray transparent RSoXS substrates such as silicon nitride. This approach measures the total absorption of the film. Measurements of the total electron yield (TEY) gives the number of electrons that are emitted from the sample due to X-ray radiation. This is taken as the drain current flowing into the sample as it becomes electrically neutral. Because the electric current is measured, samples must have an electrical connection to the bar. To achieve this, the sample should be mounted on a silicon wafer substrate and adhered to the bar with fresh carbon tape. Silicon nitride substrates will also work, but silicon wafers are preferred. The wafer should be doped and not have a thermal oxide layer. TEY NEXAFS is surface sensitive in the soft X-ray regime. The short electron escape depth limits sensitivity to the first few nanometers at the sample surface. Lastly, the total fluorescence yield (TFY) measures the intensity of fluoresced X-rays, typically using a 2D detector. The depth sensitivity depends on the X-ray energy; for example, near the carbon edge fluorescence emission is limited to the top 100 nm of the film. Reviews of NEXAFS, mostly focused on measurements of TEY or TFY, can be found elsewhere (Nahid et al., 2016; Stöhr, 1992). In this chapter, we focus on transmission because it depends on the composition throughout the entire sample thickness.

One approach to measure the forward scattered X-ray flux is to use a photodiode. This is often integrated into the beamstop in a scattering instrument; in this configuration X-rays scattered beyond the beamstop are ignored. The transmitted intensity (I) is taken from the sample and the initial intensity (I0) is taken from the blank sample. NEXAFS spectra can be plotted as absorbance ($A = -Log(T)$) versus the beam energy, where the transmittance T is I/I0. If absorbance is very high, NEXAFS can alternatively be plotted as 1-T. Peaks in NEXAFS spectra correspond to different electron transitions, mainly $1s \rightarrow \pi^*$ and $1s \rightarrow \sigma^*$. The energies of these transitions will vary depending on the bonding environment. Such differences were seen in X-ray absorption spectra of amino acids (Le Guillou, Bernard, De La Pena, & Le Brech, 2018). A NEXAFS spectrum of an onion outer epidermal peel collected near the calcium L-edge is shown in Fig. 1A. The peaks in the NEXAFS spectrum correspond to the calcium L_2 and L_3 edge (349.3 and 352.6 eV, respectively).

Because time is limited, planning of the experiment in advance and communication with the beamline scientists is necessary for a successful experimental run. Beamline staff and/or manuals should be consulted for instructions on operating all equipment, including the sample chambers, stages, optical devices, and detectors, as these factors may vary over time and at different beamlines. The following is a general procedure outlining the necessary measurements needed to produce a NEXAFS spectra at a RSoXS beamline. Raw data is acquired as photodiode current as a function of beam energy. Tools such as QANT are available to facilitate analysis.

6.1 Materials

- PS300 calibration standard sample, mounted on bar in beam path
- Blank sample (empty RSoXS substrate or liquid cell filled with buffer), mounted on bar in beam path

- Biological sample of interest (mounted onto RSoXS substrate or encapsulated inside liquid cell), mounted on bar in beam path

6.2 Procedure

1. Acquire carbon photodiode scan (270 to 320 eV with 0.1 eV steps) of the PS300 calibrant (see Note 1). The polystyrene absorbance peak is at 285.1 eV (Dhez, Ade, & Urquhart, 2003). If this is not the value, the offset in beam energy should be corrected for steps 2–4.
2. Select energy values for the NEXAFS spectrum. The energy range should encompass the entire atomic absorption edge of interest. At energies near peaks in the absorption edges, the energy resolution of the scan should be high (every 0.1 eV).
3. Select a sample spot with minimal scattering artifacts. This must be done using RSoXS (see Section 7.2.1).
4. Acquire photodiode scan of the sample (I) at selected energies.
5. Acquire photodiode scan of the blank sample (I0) at selected energies (see Note 2). If the beam intensity is known to fluctuate throughout the day, we recommend retaking the blank photodiode scan every 3 h. Sample photodiode scans should be normalized with the blank photodiode scan taken within the closest span of time.

6.3 Notes

1. A "photodiode scan" refers to measurements of the photodiode intensity at every beam energy for the NEXAFS spectrum.
2. Steps 4 and 5 do not necessarily need to be completed in this order, but both I and I0 photodiode scans are required to obtain a normalized NEXAFS spectrum.

7. Performing RSoXS experiments

RSoXS data can be collected with either point detectors or 2D detectors, the latter of which is the focus of this chapter. 2D detectors enable collection of data over a wide q range simultaneously. This can identify anisotropy in a sample if present. The scattering vector is denoted as q, which is related to the scattering angle θ by $q = 4\pi \sin\theta / \lambda$, where λ is the X-ray wavelength. Real space length is related to q by the relationship $d = 2\pi / q$. 2D scattering data can be azimuthally integrated and reduced into scattering intensity versus scattering vector q profiles. The achievable q range is limited by the sample to detector distance and X-ray wavelength. The maximum q is limited by the maximum scattering angle of the instrument. The minimum q is limited by the collimation of the beam. To achieve a large range in q, RSoXS data are typically taken at multiple detector positions with different sample to detector distances. A common practice is to take data at two detector positions: (1) position where the detector is more than 100 mm away from the sample, corresponding to low values of q ("wide-angle" scattering), and (2) where the detector is on the order of tens of millimeters away, corresponding to high values in q ("small-angle" scattering). When available, the high-q detector can be shifted to capture even higher values of q if necessary. RSoXS curves can be "stitched" to map the full q-range, at a specified point scaled to a constant local second derivative. At 284 eV, (near the carbon edge) a range of $0.1° < 2\theta < 70°$ corresponds to a q-range of $0.00025\,\text{Å}^{-1} < q < 0.165\,\text{Å}^{-1}$ and a real space range of approximately 4 nm to 2.5 μm. Soft X-ray detectors will produce a significant amount of electronic noise that is proportional to the exposure time. Therefore, "dark" images, where the beam shutter is closed, must be taken to subtract detector noise. In addition to the detector background, incoherent fluorescence emission can also add a background intensity to scattering data, particularly at the carbon edge. This can be subtracted using a previously described method (Ferron et al., 2017).

Ensuring minimal artifacts is important in RSoXS experiments and can require some fine adjustment of the instrumentation. Heterogeneity is often present in biological samples. Wrinkles in the sample and dust particles can cause unwanted flare-like scattering artifacts at certain regions (Fig. 6A). Therefore, RSoXS and NEXAFS data should ideally be taken at the same spot. We first present a short method to find ideal regions to probe on a sample, which should be completed before both NEXAFS and RSoXS measurements. Then in the succeeding section, we present a protocol for RSoXS data acquisition. For data reduction, the RSoXS beamlines use modified versions of Nika on Igor Pro that are available as open source software (Ilavsky, 2012).

7.1 Materials

- PS100 calibration standard sample, mounted on bar in beam path
- Blank sample (empty RSoXS substrate or liquid cell filled with buffer), mounted on bar in beam path
- Biological sample of interest (mounted onto RSoXS substrate or encapsulated inside liquid cell), mounted on bar in beam path

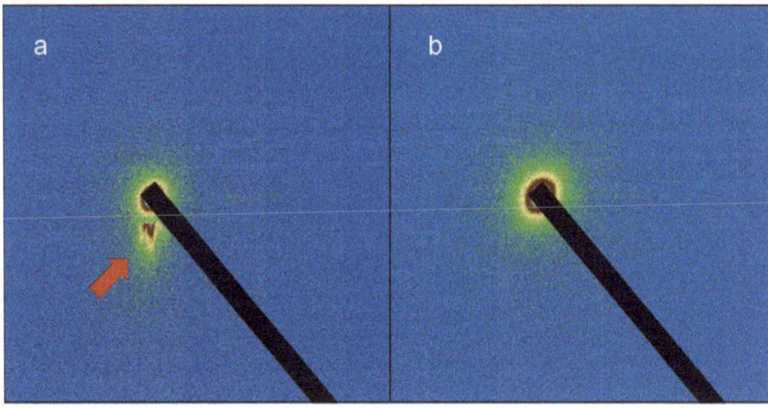

Fig. 6 (A) 2D RSOXS image of an *Arabidopsis thaliana* sample with undesirable streaks (arrow) and (B) 2D RSOXS image of the same sample probed at a different location with minimal artifacts. Black bars indicate the beam stop.

7.2 Procedure

7.2.1 Selecting sample spot

1. Change the X-ray beam energy to an off-edge value near the absorption edge of interest (see Note 1). For example, 270 eV is a suitable energy for the carbon edge.
2. Set the detector position so that the beam center is near the center of the detector, as is the case in Fig. 6. The detector position can be in either a "low q" or "high q" range, as potential artifacts can appear in both.
3. Collect RSoXS data with a minimal exposure time at several different spots on the sample window. The beam, which is a few hundred μm in diameter, can hit multiple locations within a 1 mm by 1 mm window. If available, the process of scanning multiple spots can be done automatically (see Note 2).
4. Select the sample spot with the least artifacts and record the bar motor position. Sharp streaks that extend from the beam center are likely artifacts that arise from wrinkles in the films or dust particles (Fig. 6A). Ideal scattering data has no streaks, minimal shadows, and continuous features in the scattering (Fig. 6B). If there are streaks that cannot be avoided by selecting a different spot, they should be masked out before data reduction.

7.2.2 Acquiring RSoXS data

1. Select size range of interest. This will determine the q-range that should be acquired for RSoXS measurements, and thus the detector positions required.

2. Acquire RSoXS data of PS100 standard at a minimum of one energy for every detector and detector position needed to achieve the desired q-range. For example, if performing RSoXS experiments at two detector positions: high-q (shorter detector position) and low-q (farther detector position) ranges, RSoXS data of PS100 must be taken at both detector positions to calibrate for the beam centers and sample to detector distances. Short, <1 s exposures at 270 eV are typically sufficient for scattering from polystyrene.
3. Select X-ray beam energies of interest (see Note 1). We recommend selecting a wide range of energies for RSoXS experiments, though not necessarily as many as for a NEXAFS measurement, as this is not always doable in a limited amount of beamtime. A high energy resolution scan includes collecting measurements usually about every 0.1 eV throughout the absorption edge. The beamline at NSLS-II is particularly well equipped for this, with pre-set energy scans available to users for automatic data collection. If time is scarce or samples are too plentiful for full elemental scans, then energies can be selected using the NEXAFS spectrum. A good practice can be to select a few energies on and around X-ray absorption peaks of interest. For example, Fig. 1B shows RSoXS profiles for onion outer epidermal peels at on-edge energies (349.3 and 352.6 eV) and off-edge energies (345 and 355 eV) around the calcium L-edge.
4. Select a sample spot with minimal scattering artifacts (see Section 7.2.1).
5. Determine optimal sample exposure times. The scattering intensity can vary greatly at different beam energies around an absorption edge, so the exposure time at each energy should be adjusted to accommodate for energies with low scattering (see Note 3). Exposure times can vary from <1 s to multiple minutes depending on the scattering from the sample. Ideally, the maximum intensity in the detector image should reach 50–70% of the saturation limit. Intensity must be normalized by exposure time.
6. When detector positions, energies, sample spot, and exposure times are determined, collect RSoXS data of the sample (see Notes 4 and 5).
7. Collect dark detector images. This is the signal from the detector where the beam shutter is closed. For every unique exposure time from which RSoXS is collected, there should be an accompanying dark image with the same exposure time. This is needed to subtract the electronic noise of the detector from the RSoXS data.
8. Repeat steps 4–7 for the blank sample. The scattering from the blank should be subtracted from the sample as a background. Sometimes

acquiring data with no sample or blank (empty chamber) can be useful for background subtraction.

9. Acquire a photodiode scan (see Section 6.2) of the direct beam, where the beam transmits through only vacuum. This is used to normalize the beam intensity at different energies (see Note 6). If the beam intensity is known to fluctuate throughout the day, we recommend retaking the direct beam photodiode scan every 3 h. RSoXS measurements should be normalized with the direct beam photodiode scan taken within the closest span of time (see Note 7).

7.3 Notes

1. Users should communicate with beamline scientists about energies of interest to properly accommodate the optics needed to perform the desired experiments.
2. The beamline at NSLS-II has developed a "spiral scan" macro that takes short RSoXS exposures in a spiral pattern around a specified position. Spiral scans can take several hours to complete but can sometimes take less time than manually attempting to find an appropriate spot for every sample.
3. High-q images generally need more exposure time than low-q images. This is because a 2D detector will capture less solid angle per pixel at higher q, leading to less intensity per pixel.
4. RSoXS users should pay close attention to the quality of their data during the beamtime. Often, biological samples do not display strong discernable diffraction that is visible during an experiment. Therefore, it is difficult to discern the data quality from the 2D image alone. On the fly data reduction of the RSoXS images is highly recommended. Having multiple personnel at the beamline at a time, one for instrumental control and one for analysis, can be helpful here.
5. For automated data acquisition reduction, it is important to monitor the data throughout the beamtime in the event of sample or instrumental complications.
6. At RSoXS beamlines, the beam intensity may vary with energy. For NEXAFS spectra, this is corrected when the photodiode scan of the sample is normalized with the blank. For RSoXS analysis, the variation of the beam intensity with energy must be normalized with a photodiode scan of the direct beam.
7. At beamlines where the beam intensity varies throughout the day, it can be convenient to acquire photodiode scans of both the blank sample

(NEXAFS normalization) and the direct beam (RSoXS normalization) regularly; for example, every 3 h throughout the beamtime.

8. Opportunities in instrumentation for resonant soft X-ray scattering of biological systems

In terms of hydrated samples, there is much room for improvement from the previously described encapsulated liquid cell method, especially given the high failure rate of the sample cells. The precise thickness of the liquid film can be difficult to control using spacers. X-ray transparent films such as silicon nitride also tend to bow under high vacuum, further altering the liquid film thickness. The lack of an accurate estimate of the film thickness confounds estimates of the fluorescence intensity (Ingham, Kirby, et al., 2018). At ALS, a microfluidic instrument made from a commercial TEM holder has been used for characterization of micelles in solution (McAfee et al., 2021). This microfluidics system could be used for other systems including proteins and lipid vesicles in solutions; however, larger particles are at higher risk of clogging the microfluidic channels. Commissioning for the first liquid environment RSoXS at NSLS-II was started in early 2020, but has since halted significantly due to the pandemic (Collins & Gann, 2021). One underexplored approach is the application of commercial liquid cells for RSoXS instrumentation. For example, Silson currently has liquid film cells available for X-ray microscopy down to 1 μm in thickness that may prove beneficial for solution RSoXS experiments. Cryogenic stages, similar to stages available for cryogenic TEM experiments, may further alleviate the effects of radiation damage in RSoXS experiments.

Several beam modalities and related resonant X-ray techniques are applicable to solid samples with domains of specific molecular orientation. While many biological samples are isotropic in solution, there are several opportunities where these techniques could be applied to elucidate anisotropic structural information about solid biological systems. The absorption edge energy of soft X-rays is sensitive to molecular orientations in a material. Thus, polarized RSoXS can be used to determine preferred molecular orientation at higher resolution than ellipsometry (Collins et al., 2012; Litofsky et al., 2019; Mukherjee et al., 2021). Resonant effects can also be incorporated into X-ray reflectivity, which is known as resonant soft X-ray reflectivity (RSoXR). This has been applied widely to characterize the compositional depth profiles of multilayer organic films (Ade et al., 2009). RSoXS in the grazing-incidence geometry may provide additional information along the depth of biological films with chemical sensitivity (Schaffer, Wang, Hexemer, & Müller-Buschbaum, 2016). Finally, scanning soft X-ray transmission microscopy (STXM) has been used for the last several decades in polymer science and relies on the same photon-sample interactions as RSoXS. In this approach, the molecular orientations and domain compositions of soft matter structure can be imaged with resolution up to tens of nanometers (Watts, Schuettfort, & McNeill, 2011). Extension of these modalities to biological systems may reveal new insights into assembly and structure of biomolecules.

9. Conclusion and outlook

RSoXS has advanced our understanding of the structure of polymer thin films, including materials used for organic electronics. Furthermore, recent efforts have demonstrated that RSoXS is a promising technique for the structural characterization of biological molecules, films, and assemblies. We have discussed how RSoXS at energies around an atomic absorption edge can boost scattering contrast of materials with less heterogenous electron densities, without the need for tagging. RSoXS is a complementary tool to current characterization techniques, which can be used to refine models for the structure of biological systems. Currently, the greatest challenges for biological RSoXS are the preparation of the sample and sample environment due to the requirements of low sample thickness and of high vacuum environment of the RSoXS chamber. Dry, solid samples that can be in contact with high vacuum are the easiest samples to investigate, although biological molecules in solution have been characterized with the use of liquid cells. Liquid and hydrated environments for RSoXS measurements, which are technically more challenging, will allow studies of the native structure of many biological assemblies. While we have discussed the most common method for liquid sample preparation, we expect that soon there will be newer capabilities that can better accommodate sample requirements, such as liquid flow cells.

Acknowledgments

This work was supported as part of the Center for Lignocellulose Structure and Formation, an Energy Frontier Research Center funded by the US Department of Energy, Office of Science, Basic Energy Sciences under award no. DE-SC0001090. This work is based upon research conducted at the 1) Advanced Light Source, supported by the Director,

Office of Science, Office of Basic Energy Sciences, of the U.S. Department of Energy under Contract No. DE-AC02-05CH11231 and 2) National Synchrotron Light Source II, supported by the Office of Science, Office of Basic Energy Sciences, of the U.S. Department of Energy under Contract No. DE-SC0012704.

References

Ade, H., Wang, C., Garcia, A., Yan, H., Sohn, K. E., Hexemer, A., et al. (2009). Characterization of multicomponent polymer trilayers with resonant soft X-ray reflectivity. *Journal of Polymer Science Part B: Polymer Physics*, *47*(13), 1291–1299. https://doi.org/10.1002/polb.21730.

Bearden, J. A., & Burr, A. F. (1967). Reevaluation of X-ray atomic energy levels. *Reviews of Modern Physics*, *39*(1), 125–142. https://doi.org/10.1103/RevModPhys.39.125.

Collins, B. A., Cochran, J. E., Yan, H., Gann, E., Hub, C., Fink, R., et al. (2012). Polarized X-ray scattering reveals non-crystalline orientational ordering in organic films. *Nature Materials*, *11*(6), 536–543. https://doi.org/10.1038/nmat3310.

Collins, B. A., & Gann, E. (2021). Resonant soft X-ray scattering in polymer science. *Journal of Polymer Science*, *60*, 1199–1243. https://doi.org/10.1002/pol.20210414.

Dhez, O., Ade, H., & Urquhart, S. G. (2003). Calibrated NEXAFS spectra of some common polymers. *Journal of Electron Spectroscopy*, *128*, 85–96.

Di Cola, E., Grillo, I., & Ristori, S. (2016). Small angle X-ray and neutron scattering: Powerful tools for studying the structure of drug-loaded liposomes. *Pharmaceutics*, *8*(2), 10. https://doi.org/10.3390/pharmaceutics8020010.

Gann, E., Crofts, T., Holland, G., Beaucage, P., McAfee, T., Kline, R. J., et al. (2021). A NIST facility for resonant soft x-ray scattering measuring nano-scale soft matter structure at NSLS-II. *Journal of Physics: Condensed Matter*, *33*(164001).

Gann, E., Young, A. T., Collins, B. A., Yan, H., Nasiatka, J., Padmore, H. A., et al. (2012). Soft x-ray scattering facility at the advanced light source with real-time data processing and analysis. *Review of Scientific Instruments*, *83*(4), 045110. https://doi.org/10.1063/1.3701831.

Henke, B. L., Gullikson, E. M., & Davis, J. C. (1993). X-ray interactions: Photoabsorption, scattering, transmission, and reflection at E = 50-30,000 eV, Z = 1-92. *Atomic Data and Nuclear Data Tables*, *54*(2), 181–342. https://doi.org/10.1006/adnd.1993.1013.

Ilavsky, J. (2012). Nika: Software for two-dimensional data reduction. *Journal of Applied Crystallography*, *45*(2), 324–328. https://doi.org/10.1107/s0021889812004037.

Ingham, B., Erlangga, G. D., Smialowska, A., Kirby, N. M., Wang, C., Matia-Merino, L., et al. (2015). Solving the mystery of the internal structure of casein micelles. *Soft Matter*, *11*(14), 2723–2725. https://doi.org/10.1039/c5sm00153f.

Ingham, B., Kirby, N., Wang, C., Brady, M., & Carr, A. (2018). Elemental fingerprinting of mineral species in iron-fortified milk: Anomalous small-angle X-ray scattering and resonant soft X-ray scattering studies. *Journal of Synchrotron Radiation*, *25*(Pt 4), 1106–1112. https://doi.org/10.1107/S1600577518007774.

Ingham, B., Smialowska, A., Erlangga, G. D., Matia-Merino, L., Kirby, N. M., Wang, C., et al. (2016). Revisiting the interpretation of casein micelle SAXS data. *Soft Matter*, *12*(33), 6937–6953. https://doi.org/10.1039/c6sm01091a.

Ingham, B., Smialowska, A., Kirby, N. M., Wang, C., & Carr, A. J. (2018). A structural comparison of casein micelles in cow, goat and sheep milk using X-ray scattering. *Soft Matter*, *14*(17), 3336–3343. https://doi.org/10.1039/c8sm00458g.

Korasick, D. A., Singh, H., Pemberton, T. A., Luo, M., Dhatwalia, R., & Tanner, J. J. (2017). Biophysical investigation of type A PutAs reveals a conserved core oligomeric structure. *The FEBS Journal*, *284*(18), 3029–3049. https://doi.org/10.1111/febs.14165.

Korasick, D. A., & Tanner, J. J. (2018). Determination of protein oligomeric structure from small-angle X-ray scattering. *Protein Science, 27*(4), 814–824. https://doi.org/10.1002/pro.3376.

Le Guillou, C., Bernard, S., De La Pena, F., & Le Brech, Y. (2018). XANES-based quantification of carbon functional group concentrations. *Analytical Chemistry, 90*(14), 8379–8386. https://doi.org/10.1021/acs.analchem.8b00689.

Litofsky, J. H., Lee, Y., Aplan, M. P., Kuei, B., Hexemer, A., Wang, C., et al. (2019). Polarized soft X-ray scattering reveals chain orientation within nanoscale polymer domains. *Macromolecules, 52*(7), 2803–2813. https://doi.org/10.1021/acs.macromol.8b02198.

Liu, F., Brady, M. A., & Wang, C. (2016). Resonant soft X-ray scattering for polymer materials. *European Polymer Journal, 81*, 555–568. https://doi.org/10.1016/j.eurpolymj.2016.04.014.

Lopez-Sanchez, P., Martinez-Sanz, M., Bonilla, M. R., Sonni, F., Gilbert, E. P., & Gidley, M. J. (2020). Nanostructure and poroviscoelasticity in cell wall materials from onion, carrot and apple: Roles of pectin. *Food Hydrocolloids, 98*, 105253. https://doi.org/10.1016/j.foodhyd.2019.105253.

Martínez-Sanz, M., Gidley, M. J., & Gilbert, E. P. (2015). Application of X-ray and neutron small angle scattering techniques to study the hierarchical structure of plant cell walls: A review. *Carbohydrate Polymers, 125*, 120–134. https://doi.org/10.1016/j.carbpol.2015.02.010.

McAfee, T., Ferron, T., Cordova, I. A., Pickett, P. D., McCormick, C. L., Wang, C., et al. (2021). Label-free characterization of organic nanocarriers reveals persistent single molecule cores for hydrocarbon sequestration. *Nature Communications, 12*(1), 1–10. https://doi.org/10.1038/s41467-021-23382-8.

Mukherjee, S., Streit, J. K., Gann, E., Saurabh, K., Sunday, D. F., Krishnamurthy, A., et al. (2021). Polarized X-ray scattering measures molecular orientation in polymer-grafted nanoparticles. *Nature Communications, 12*(1), 4896. https://doi.org/10.1038/s41467-021-25176-4.

Nahid, M. M., Gann, E., Thomsen, L., & McNeill, C. R. (2016). NEXAFS spectroscopy of conjugated polymers. *European Polymer Journal, 81*, 532–554. https://doi.org/10.1016/j.eurpolymj.2016.01.017.

Ristori, S., Oberdisse, J., Grillo, I., Donati, A., & Spalla, O. (2005). Structural characterization of cationic liposomes loaded with sugar-based carboranes. *Biophysical Journal, 88*(1), 535–547. https://doi.org/10.1529/biophysj.104.049080.

Rongpipi, S., Del Mundo, J. T., Gomez, E. D., & Gomez, E. W. (2021). Resonant X-ray scattering of biological assemblies. *MRS Communications, 11*, 1–17. https://doi.org/10.1557/s43579-021-00020-4.

Schaffer, C. J., Wang, C., Hexemer, A., & Müller-Buschbaum, P. (2016). Grazing incidence resonant soft X-ray scattering for analysis of multi-component polymer-fullerene blend thin films. *Polymer, 105*, 357–367. https://doi.org/10.1016/j.polymer.2016.05.056.

Stöhr, J. (1992). *NEXAFS spectroscopy. Vol. 25.* Springer Science & Business Media.

Virgili, J. M., Tao, Y., Kortright, J. B., Balsara, N. P., & Segalman, R. A. (2007). Analysis of order formation in block copolymer thin films using resonant soft X-ray scattering. *Macromolecules, 40*(6), 2092–2099. https://doi.org/10.1021/ma061734k.

Watts, B., Schuettfort, T., & McNeill, C. R. (2011). Mapping of domain orientation and molecular order in polycrystalline semiconducting polymer films with soft X-ray microscopy. *Advanced Functional Materials, 21*(6), 1122–1131. https://doi.org/10.1002/adfm.201001918.

Ye, D., Kiemle, S. N., Ron, S., Wang, X., Wang, C., Cosgrove, D. J., et al. (2018). Resonant soft X-ray scattering reveals cellulose microfibril spacing in plant primary cell walls. *Scientific Reports, 8*, 12449. https://doi.org/10.1038/s41598-018-31024-1.

Ye, D., Le, T. P., Kuei, B., Wang, C., Gomez, E. D., Gomez, E. W., et al. (2018). Resonant soft X-ray scattering provides protein structure with chemical specificity. *Structure*, *26*(11), 1513–1521. https://doi.org/10.1016/j.str.2018.07.018.

Ye, D., Rongpipi, S., Litofsky, J. H., Lee, Y., Culp, T. E., Yoo, S. H., et al. (2019). Aluminum oxide free-standing thin films to enable nitrogen edge soft x-ray scattering. *MRS Communications*, *9*(1), 224–228. https://doi.org/10.1557/mrc.2018.195.

Zhang, T., & Cosgrove, D. (2017). Preparation of onion epidermal cell walls for imaging by atomic force microscopy (AFM). *Bio-Protocol*, 7(24), 2647. https://doi.org/10.21769/bioprotoc.2647.

CHAPTER THIRTEEN

Scattering measurements on lipid membrane structures

Lin Yang*
National Synchrotron Light Source II, Brookhaven National Laboratory, Upton, NY, United States
*Corresponding author: e-mail address: lyang@bnl.gov

Contents

Abstract

Membranes isolate, maintain, and regulate the biochemical environment within cells and organelles for essential biological processes to take place. While resolving the structures of proteins that inhabit the membranes remains a focus of current structural biology research, studying the structure of the membrane itself can help elucidate how functional components organize within the membrane and how the interplay between the lipid components and membrane proteins facilitates biological functions. X-ray and neutron scattering provides a means for characterizing the basic structural parameters of membrane structures, from which the energetics of biological processes can be inferred, and the different physical states and species of the lipid can be distinguished. This chapter describes different membrane structures suitable for scattering measurements. We will discuss the practical considerations for carrying out these measurements and outline the methods for data interpretation.

1. Introduction

The basic structure of biological membranes can be described by the fluid mosaic model as a two-dimensional liquid of oriented membrane proteins and lipids (Singer & Nicolson, 1972). The lipid components form a

Methods in Enzymology, Volume 677
ISSN 0076-6879
https://doi.org/10.1016/bs.mie.2022.08.033

bilayer structure, which membrane proteins are incorporated into internally or attached to peripherally. Membrane proteins control the material exchange and signal transduction across the bilayer, and drive remodeling and recycling of membrane structures. However, the lipid bilayers are not purely passive structures. From a biophysical point of view, biological processes are dictated by the energy landscape within the configuration space that the system can explore. All components, including the bilayer itself, contribute to the overall free energy and therefore shape the energy landscape.

The elastic energy stored in a membrane structure is determined by its shape. Each leaflet in the bilayer structure of lipid membranes exhibits its own spontaneous curvature, as determined by the lipid composition. Both leaflets could be frustrated in a given membrane configuration, leading to a finite bending energy (Helfrich, 1973). The interaction between the bilayer and membrane inclusions is also sensitive to the deformation energy resulting from the bilayer deviating from a uniform thickness (Huang, 1986). The interaction between lipid membranes is another factor that affects the energy landscape in processes that involve membrane remodeling, such as the fusion of two bilayers. In addition to electrostatic and van der Waals interactions that are common between surfaces, the organization of water in the hydration layers must be disrupted when two bilayers are in close proximity, resulting a repulsive hydration force (Gruen & Marčelja, 1983; Leneveu, Rand, & Parsegian, 1976; Lis, McAlister, Fuller, Rand, & Parsegian, 1982).

Scattering measurements reveal the membrane structure parameters that can be used for distinguishing between different lipid species (chain length) and physical state (fluid vs gel-like). More importantly, these parameters provide a basis for estimating the free energy contribution of the membrane structure in biological processes. For instance, lipid water mixtures are known to form nonlamellar structures that resemble the membrane configurations that appear in membrane remodeling and fusion. Therefore, studying the transitions between these structures can shed light on how relevant proteins may shape the energy landscape in these processes (Aeffner, Reusch, Weinhausen, & Salditt, 2012; Gruner, 1989; Siegel, 1999; Yang, Ding, & Huang, 2003). As another example, structural studies based on X-ray and neutron scattering measurements and energy analysis have also been used to understand membrane-mediated interactions between ion channels (Harroun, Heller, Weiss, Yang, & Huang, 1999) and how membrane-active antimicrobial peptides disrupt cell membranes (Huang, 2000; Lee, Sun, Hung, & Huang, 2013).

The bilayers in biological membranes contain a variety of lipid species and are not homogeneous two-dimensional liquids. In recent years, domain formation in biological membranes has attracted a lot of interest (Simons & Sampaio, 2011). The formation of large domains in lipid vesicles of ternary lipid mixtures can be observed using fluorescence microscopy (Korlach, Schwille, Webb, & Feigenson, 1999; Veatch & Keller, 2003). On the other hand, scattering methods (Marquardt, Heberle, Nickels, Pabst, & Katsaras, 2015), particularly neutron scattering using deuterated components to enhance the contrast of the domains against the average bilayer structure, are more suitable for studying nanoscale domains that are believed to be more relevant to biological processes. These measurements contribute to the overall understanding of the parameters that control domain formation.

There has been a long-standing interest in making use of lipid membranes in biotechnological applications (Cullis & Hope, 2017; Lee, Lee, & Nam, 2013). Scattering-based structural characterization can be particularly helpful in applications that involve lipid-based formulations. Notably, the mRNA vaccines (Schoenmaker et al., 2021) we rely on to combat the current COVID-19 pandemic are lipid nanoparticles (LNPs). Solution scattering can be used as an effective quality control tool to estimate the composition of the LNPs (Arteta et al., 2018) and track particle stability.

In this chapter, we review different types of biological and model membrane structures that are suitable for X-ray and neutron scattering measurements, describe relevant methods for sample preparation and data collection, and discuss the data analysis methods that can be applied to extract structural information from the data.

2. Information contents in the scattering data

The methods for data collection and analysis are specific to each type of structure to be discussed in this chapter. However, it is helpful to first discuss what information can be extracted from the scattering data. In general terms, the scattering intensity is related to the Fourier transform of the electron density (for X-ray scattering) or scattering length density (for neutron scattering) contrast in the sample being probed. This intensity varies as a function of the scattering vector, $\mathbf{q}$, defined as the wave vector difference between the scattered beam and the incident beam. The magnitude of the scattering vector is given by $q = \frac{4\pi}{\lambda}\sin\theta$, where λ is the wavelength of the probing beam and 2θ is the scattering angle. For membrane structures, the electron density distribution can be typically expressed as a function of two coordinates, $\rho(r, z)$, where r is within the plane of the bilayer (in-plane) and z is along the bilayer normal (out-of-plane), reflecting the fluid-like in-plane structure in membranes. Accordingly, the Fourier transform is also a 2D functions, $I(\mathbf{q}) = I(q_r, q_z)$. In some membrane structures, the direction of the bilayer normal is aligned, therefore this 2D intensity distribution can be directly observed in the scattering data (Fig. 1A). On the other hand, in powder or solution samples the membrane structures are randomly oriented. Therefore, the scattering pattern is isotropic (Fig. 1B). Even so, features of different structural origins could still be resolved since they appear at different reciprocal coordinates.

For data interpretation, the scattering intensity in general can be described as a product of a form factor $F(\mathbf{q})$ and a structure factor $S(\mathbf{q})$: $I(\mathbf{q}) \sim |F(\mathbf{q})|^2 S(\mathbf{q})$. In small-angle scattering of membrane structures, the form factor comes from the structure of the bilayer. In the case of uncorrelated bilayers, $S(\mathbf{q}) = 1$ and the scattering data is due to the form factor alone, showing gradual variation in the q-range of $\sim$0–0.6 Å^{-1} (Section 5.2.2). The structure factor due to the positional correlation comes into play when multiple bilayers interact with each other. $S(\mathbf{q})$ then shows peaks of different magnitudes and shapes, depending on the degree of correlation. Highly ordered structures produce Bragg diffraction peaks. The structure in the unit cell can be calculated from the peak intensities by Fourier synthesis.

The in-plane electron density variations also contribute to measurable scattering intensity. On molecular length scales, the liquid-like organization of lipid molecules gives rise to a correlation peak, located at $q_r \sim 1.45$ Å^{-1}. This is sometimes referred to as the chain-packing peak. In this context, the form factor corresponds to the structure of the lipid chain and the structure factor corresponds to their in-plane organization (see Section 5.3.2). Like conventional liquids, the membrane structures can undergo phase transitions, leading to dramatic changes in the appearance of the chain-packing peak, reflecting the structural order internal to the bilayer structure (Smith, Sirota, Safinya, & Clark, 1988; Yang & Fukuto, 2005) (Fig. 1E).

At longer length scales, membrane components can form domain structures, which have been observed in biological membranes as well in model systems. The in-plane density variations due to these domain structures produce a signature in the scattering data as well. Owing to the large size of these domains, the corresponding scattering signals appear at low q ($\ll$0.1 Å^{-1}). Other membrane inclusions, such as membrane-active peptides, induce in-plane structures that are smaller and therefore produce scattering features

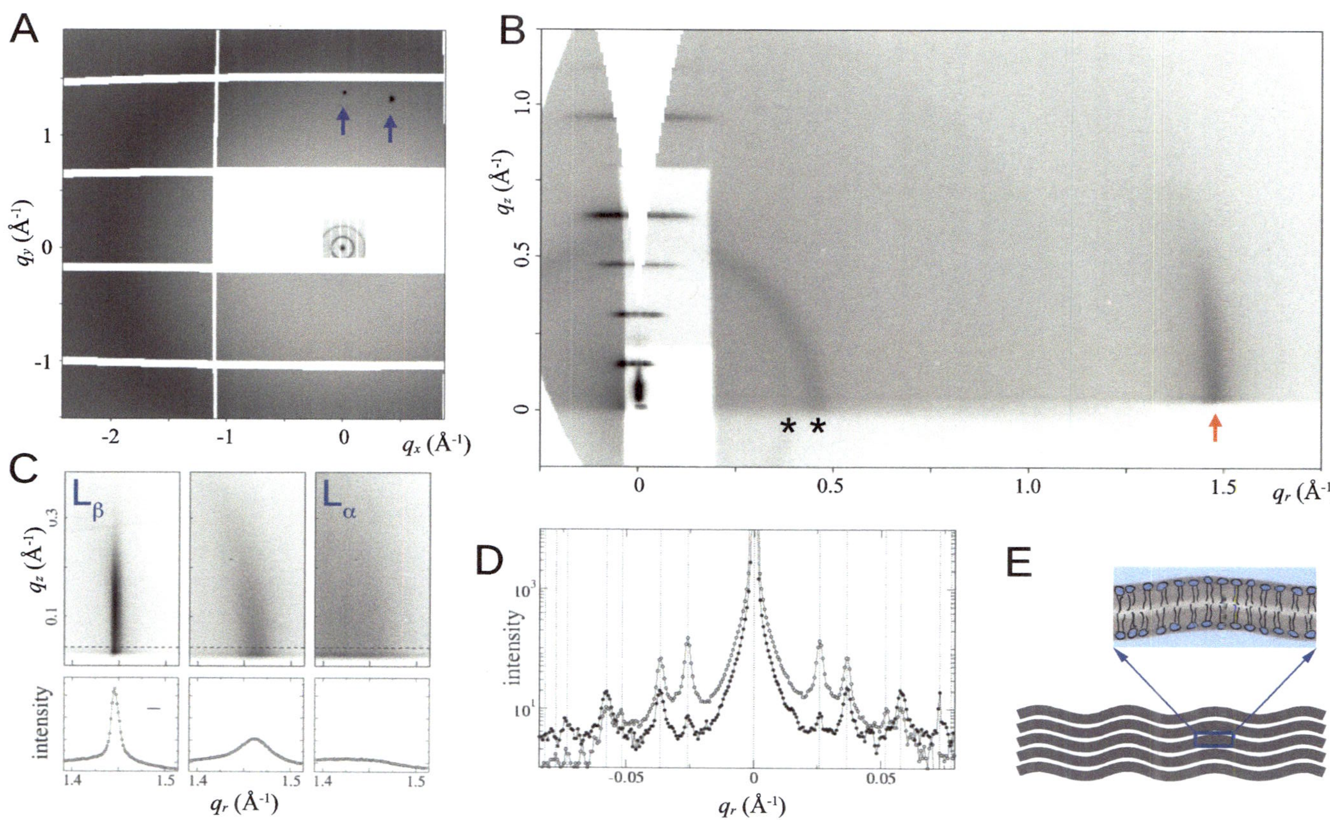

Fig. 1 See figure legend on next page.

that are more easily accessible (e.g., Yang, Harroun, Weiss, Ding, & Huang, 2001; Yang, Weiss, Harroun, Heller, & Huang, 1999). The internal structures of these membrane inclusions may give rise to additional, albeit difficult-to-observe, features in the scattering data (Salditt, Li, & Spaar, 2006).

Stacked bilayers can form in-plane modulations with either one- or two-dimensional periodicity that give rise to additional diffraction peaks near the layer peaks. Some of these modulations can exhibit very long spatial periods and therefore the resulting diffraction peaks can only be resolved with high q resolution (Yang & Fukuto, 2005) (Fig. 1C).

Fig. 1 Examples of scattering data from membrane structures. (A) Grazing-incident scattering data collected using a MarCCD detector from a substrate-support multi-bilayer sample, shown as a 2D intensity map (q_z is along the bilayer normal). The bilayer structure and the stacking of the bilayers (bottom cartoon depiction in panel (D)) give rise to the features near the q_z axis. Satellite peaks appear at the same q_z position as the layer peaks, and q_r positions that suggest an in-plane structural modulation with square symmetry and a lattice constant of $\sim 300\,Å$, as shown in the high-resolution data in panel (C). To resolve these peaks, the sample-to-detector distance was increased, and the incident angle was adjusted such that the central peak is fully visible on the detector. The peak at $q_r \sim 1.45\,Å^{-1}$ (indicated by the red arrow) is due the correlation between the lipid chains within the bilayer (top cartoon depiction in panel (D)). High-resolution chain correlation data and the corresponding 1D q_r-profiles are shown in panel (E). Notably, the appearance of this peak (middle panel) suggests that the lipid chains are much more ordered than expected in the fluid phase (L_α phase), but less so compared to the hexagonal order in the L_β phase. In addition to the scattering signal from the membrane structure itself, background scattering from the Kapton window (indicated by the asterisks) in the beam path is also visible. The ring with the higher intensity (darker) is due to the direct beam and therefore partially blocked by the substrate, while the dimmer ring is due to the reflected beam. The diffuse background due to the air and water vapor in the beam path is less obvious but must be subtracted to accurately preserve the diffused features in the scattering pattern. The rectangular, low-intensity region near the layer peaks is due to an attenuator intended for keeping the layer peaks from saturating the detector (Yang & Fukuto, 2011). (B) Solution scattering data from multilayered vesicles. The scattering patterns collected on a pair of small- and wide-angle Pilatus pixel array detectors have been combined and plotted using the same intensity scale and reciprocal coordinates (q_x and q_y, perpendicular to the beam). Even though the two scattering patterns are not contiguous and the are no data in the gaps between the detector sensing modules, the azimuthally averaged data span a continuous q-range (see data Fig. 3). The high intensity (dark color) beyond $1.0\,Å^{-1}$ is due to water and must be subtracted before data interpretation. The two high-intensity spots (blue arrows) are due to mica vacuum windows and are masked off during azimuthal averaging.

3. General experimental planning considerations

Given the discussion above, measurements on membrane structures should be designed to cover the appropriate parts of the reciprocal space that contain the structural information of interest. This may determine the configuration of the lab-based diffractometer, or the selection of a suitable synchrotron or neutron beamline, to be used for the measurement. Measurements on solution-based samples can typically take advantage of instrumentation dedicated to biomolecular solution scattering, often with automated data collection and processing (e.g., Yang et al., 2020). Data collection on other systems should be possible on any small angle scattering or general-purpose scattering or diffraction instruments. Most measurements can be done using area detector to reduce data collection time. On the other hand, reflectivity measurements on substrate-supported model membranes are typically conducted using specialized instruments based on point detectors.

It should be noted that the information along the bilayer normal is only accessible if the membrane is rotated so that the Ewald sphere intersects with the q_z axis at different locations. In samples where the bilayers are randomly oriented, this is achieved automatically (powder average). For substrate-supported membrane structure, the coverage of the reciprocal space is dependent on the incident angle of the beam with respect to the sample normal (Yang, Harroun, Heller, Weiss, & Huang, 1998). The information on the q_z axis can be collected in a reflectivity measurement, also known as a θ–2θ scan, where the sample is rotated continuously. In some experiments, curved substrates are used to create a range of incident angles over the illuminated sample to equivalently provide sample rotation (Tristram-Nagle, Liu, Legleiter, & Nagle, 2002). In grazing incident measurements, information at very low q_z is blocked by the substrate. In addition, the surface-enhanced scattering can distort the peak shape along the q_z direction. These complications can be avoided by an additional measurement in the transmission geometry, with the beam incident from the back of the substrate, either using a substrate like a silicon nitride window that has little absorption (Yang & Huang, 2003), or using a short X-ray wavelength to reduce the absorption by the substrate (Lu, Yager, Johnston, Black, & Ocko, 2013).

In scattering measurements, all materials illuminated by the probing beam contribute to the recorded scattering intensity. Proper background subtraction is needed to obtain high-quality data. A common source of scattering background is from air surrounding the sample. On dedicated small angle scattering instruments, an evacuated flight path is typically used to reduce scattering of the direct beam and absorption of scattered beam by air along the relatively long distance between the sample and the detector. A helium-filled beam path can be used when the sample-to-detector distance is short. In addition, the beam stop can be installed close to the sample, as in crystallography measurements, to reduce the scattering of the direct beam, at the cost of losing the coverage of very small scattering angles.

The resolution of the scattering data is another consideration, particularly in measurements where peaks due to periodic structures cannot be easily resolved or it is important to characterize the shape of the peak. From the definition of the scattering vector, the q-resolution can be estimated as $\frac{dq}{q} \sim \frac{1}{\lambda}\frac{d\lambda}{\lambda}\sin\theta + \cos\theta\frac{d\theta}{\theta}$. Therefore, the dominant factor in neutron scattering at wide scattering angles is the energy spread, which is typically a few percent. The angular resolution is a combination of contributions from the detector pixel size, the illuminated sample size, the sample-to-detector distance, and the scattering angle. The sample size is particularly important in grazing incident measurements, where the beam footprint can extend over the entire sample. This can be mitigated by either focusing the beam in the direction of sample normal, or extending the sample-to-detector distance proportionally (Huang & Yang, 2009).

The emergence of photon-counting pixel array detectors in recent year has been very helpful for scattering measurements on weakly scattering samples, such as vesicles in solution. Unlike CCD detectors that add to the data an accumulative dark noise, a random readout noise, and zingers from the fiberoptic taper, these new detectors are essentially free of noise. These detectors typically have gaps between sensing modules (see data in Fig. 1B). However, this limitation can be easily overcome, either by merging two scattering patterns collected using different detector positions, or by resorting to the symmetry and intensity continuity expected of the scattering patterns to fill these gaps and reconstruct complete scattering patterns.

Modern synchrotron beamlines deliver very high X-ray flux density (e.g., on the order of 10^{12} photons per second within a spot size of a few microns at the LiX beamline of NSLS-II). Consequently, radiation damage to the sample can happen very quickly. With fast-readout, photon-counting detectors that are free of readout noise, scattering data can be collected at high frame rates to detect radiation damage and exclude bad data, which typically manifest as the loss of contrast in the scattering patterns, or the rise of low-q intensity that is indicative of aggregates in solution scattering. In addition, solution samples should flow though the beam during data collection and solid samples should be scanned, to maximize the irradiated sample volume and spread out the absorbed dose. It is helpful to estimate the upper limit of measurement time based on established dose limits ($\ll$1 MGy) for similar samples (Cheng & Caffrey, 1996; Cherezov, Riedl, & Caffrey, 2002).

Neutron and X-ray scattering can be used to complement each other. Large numbers of samples can be measured quickly using X-ray scattering at synchrotron beamlines owing to the high beam intensity. These measurements typically can be performed at an acquisition rate of one second or less per scattering pattern. But the overall throughput (on the order of minutes per measurement) is limited by sample delivery in automated measurements or equilibration against varying environmental conditions (e.g., humidity). The small beam size can be used for spatially resolved measurements on heterogeneous samples or time-resolved measurements using specialized mixing devices. On the other hand, neutron scattering is determined by scattering length density contrast rather than electron density contrast. Therefore, deuteration can be used to effectively accentuate specific parts of the molecular structure (e.g., lipid chains). Radiation damage is also not a typical concern in neutron scattering.

Membrane structures are sensitive to temperature and water contents, requiring sample-specific environments for the scattering measurements. Fig. 2 show some sample holders in use at the LiX beamline of NSLS-II. In general, the sample holder should provide the necessary environmental control (temperature and hydration), low and reproducible scattering background, and access to the scattering angles needed for the structural information of interest (e.g., chain-packing peak).

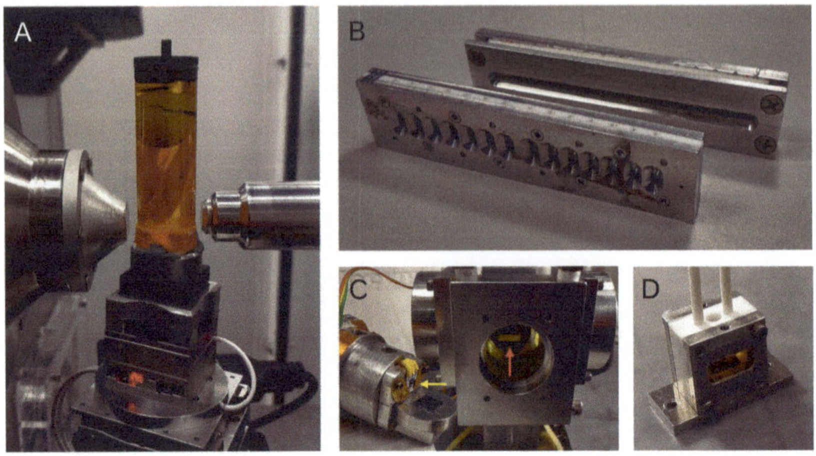

Fig. 2 Various sample holders for X-ray scattering measurements on membrane structures at the LiX beamline. (A) Brain tissues submerged under a buffer solution (Georgiadis et al., 2021). The diameter of the Kapton tube holder is ~25 mm. (B) Multiple-cell sample holders for solution samples. The holder in the background is for samples in standard thin-walled capillaries, whereas the holder in the foreground is constructed from a Teflon body and mica windows and the samples are filled from the opening on the top of each cell. These holders can be scanned during measurements to spread out the radiation dose in the sample. The Teflon holder is expected to have more reproducible sample and wall thicknesses to facilitate reliable subtraction of water scattering background. (C) An aluminum cell for substrate-supported multi-bilayers for measurements under controlled humidity, which is achieved by heating and cooling the water reservoir (red arrow). The temperatures of the cell and the sample are the same and controlled by an external water circulator. The sample support (left), which is equipped with a humidity sensor (yellow arrow) and permits in-plane rotations to improve powder average when collecting diffraction data, is inserted into the humidity cell for measurements. (D) A Teflon cell with Kapton windows for measurements on substrate-supported bilayer submerged under a buffer solution. The substrate is held at the bottom of the cell by the Teflon rods. Solutions are injected from the port on the top.

4. Biological membranes

The myelin sheath in neurons presents a rare opportunity for biological membranes to be studied by scattering methods. This multilayered structure is formed by Schwann cells or oligodendrocytes that concentrically wrap around the axon. It exhibits a periodicity that consist of two bilayers and gives rise to a series of diffraction peaks (Worthington & Blaurock, 1969). The swelling method was developed to assign the phase for each diffraction amplitude and calculate the electron density distribution (Blaurock, 1971). In this method, the water contents outside of the Schwann cell is controlled by equilibrating the specimen in sucrose solution of various concentrations. As the overall periodicity of the structure varies, the unit cell structure remains the same. Therefore, the diffraction peaks sample the form factor of the unit cell at different locations, revealing its shape. The form factor is expected to be a continuous function and thus must change its sign when the amplitude crosses zero. Finally, the overall sign of the electron density contrast is inferred based on prior knowledge of the myelin structure, namely, the membrane should exhibit higher density compared to water between the membranes.

In early X-ray scattering experiments on animal sciatic nerve fibers, the beam size was reduced by slits on lab-based X-ray diffractometers to match the sample size (tens of microns), with multiple-hour exposure times (Akers & Parsons, 1970; Blaurock & Worthington, 1969). With modern synchrotron sources and instrumentation for X-ray microdiffraction, it is possible to perform spatially resolved measurements quickly, to provides a basis for formulating structural maps that correspond to different parameters extracted from the scattering data. For instance, Inouye and coworkers were able to resolve variations in the myelin orientation and periodicity in the vicinity of a node of Ranvier (Inouye et al., 2014). However, due to the requirement of sample rotation discussed above, calculating the bilayer profile from these data is no longer feasible for all data points in the scan.

The diffraction signal from myelin has also been used as a contrast mechanism for tomographic imaging of nerve and brain tissues. Tomography is a method for resolving the spatial distribution of signal sources based on differences in the observed data upon translation and rotation of the sample. It is typically implemented based on absorption contrast, as is the case for medical Computed Tomography (CT). X-ray fluorescence-based tomography for resolving elemental distribution in the sample is also widely used at synchrotron facilities. Performing scattering-based tomography offers a unique capability of revealing the distribution of specific structural species that exhibit characteristic scattering signatures, such as myelin in nerve tissues. A recent study (Georgiadis et al., 2021) demonstrated that scattering-based tomography was able to characterize the abundance and orientation of myelin, and also distinguish the myelin species that appear at the interface between the central and peripheral nervous systems.

Unlike absorption or fluorescence emission, the scattering intensity used for tomographic reconstruction comes from diffraction due to periodic structures, and therefore is dependent on sample orientation. In general, the signal corresponding to each voxel must be represented by a vector, requiring specialized reconstruction algorithms. However, when the spatial resolution is moderate, the structure within each voxel can be considered as isotropic or exhibiting fiber symmetry, the reconstruction can then be simplified. This is particularly true for measurements for the purpose of quantifying material abundance. Under this simplification, it is even possible to recover the scattering data that correspond to each voxel in the tomogram (Jensen et al., 2011).

5. Model membrane structures

Model membranes provide a lot of flexibility for the purpose of understanding the biophysics of lipid membranes in biological processes. The composition of the bilayers can be precisely controlled, to adjust the physical properties bilayers and therefore the energy landscape, as discussed in the Introduction. Other membrane components, ranging from small molecules to membrane proteins, could be incorporated into appropriate model systems to mimic the relevant biological process more closely.

5.1 Simple lipid-water dispersions

The simplest model membrane structures can be created by simply mixing lipids with water. Due to the amphiphilic nature of lipid molecules, they self-assemble into various structures in which the hydrocarbon tails are hidden away from water to avoid the energy penalty due to the hydrophobic effect. Early studies on these model systems revealed rich phase diagrams (Tardieu, Luzzati, & Reman, 1973), as functions of temperature and water contents. In addition to stacked bilayers, nonlamellar structures that are absent of extended planar bilayers were also observed.

The most pronounced effect of temperature on planar bilayers is the main phase transition, corresponding to the melting of the hydrocarbon chains. This main transition coupled to the interaction between bilayers can qualitatively account for the behavior of stacked bilayers in the temperature-hydration phase diagram (Goldstein & Leibler, 1988; Wack & Webb, 1989). Highly hydrated membrane structures tend to fluctuate, which introduces additional repulsive interaction between bilayer and can eventually lead to the so-called unbinding transition (Lipowsky & Leibler, 1986). With excess water, non-lamellar structures tend to form by lipid species with non-zero spontaneous curvature, driven by the elastic energy (Gruner, Cullis, Hope, & Tilcock, 1985). On the other hand, the potential energy due to the repulsive force between bilayers can build up upon dehydration to promote deviation from flat bilayer structures (Yang et al., 2003; Yang & Fukuto, 2005).

Since membrane structures are sensitive to the water contents, it is important to achieve uniform hydration during sample preparation. This is typically done by repeated freeze-thaw cycles, or by slowly rehydrating lyophilized lipid powder in water vapor. The hydration level of the membrane structure can be controlled by several means: the amount of water can be controlled when preparing the sealed dispersion, the dispersion can be left equilibrating with water vapor of known relative humidity, or an external pressure can be exerted to counter the repulsive forces between membranes (Parsegian, Fuller, & Rand, 1979), either mechanically or osmotically by dialysis against a solution of macromolecules like dextran or polyethylene glycol.

Measurements on these dispersions can be done on practically any small angle scattering or powder diffraction instrument. Thin-walled ($\sim 10\,\mu m$) glass capillaries are often used as sample holders, due to their wide availability and that they are transparent to X-rays. At synchrotron beamlines, multiposition sample holders are often used to deal with the large number of samples that need to be measured at high throughput, for instance in studies searching for favorable conditions to crystallize membrane proteins (Joseph et al., 2011). These sample holders typically consist of a chemically inert polymer (silicone or Teflon) body with thin films like Kapton as windows. The scattering background and water permeability should be considered when choosing the window for such sample holders.

Lipid-water dispersions have been used for the characterization of phase diagrams and inter-bilayer interactions. The electron density of the membrane structure can be calculated from the diffraction data. As discussed in Section 4, the swelling method can be applied to phase diffraction amplitudes for lamellar structures. For non-lamellar structures, due to the finite number of diffraction peaks and the assumption of centro-symmetry, it is feasible to enumerate all possible phase combinations, and rank them based on some mathematical criteria. This can be based on the variation in the calculated electron density, as in the pattern recognition method (Mariani, Luzzati, & Delacroix, 1988), or the expectation that small changes in hydration will only result in small changes in the structure (Turner & Gruner, 1992).

Since these membrane structures are randomly oriented, the scattering data can only be expressed as a function of the magnitude of the scattering vector, leaving the components perpendicular and parallel to the average membrane normal unresolved. This can create confusion in data analysis, especially if the q-resolution of the scattering data is limited. For instance, in the initial report of the lipid ripple phase, the satellite peaks around the layer peaks were not fully resolved (Tardieu et al., 1973). Synchrotron-based powder diffraction data (Wack & Webb, 1989) permitted more accurate determination of the lattice constants and enabled later work (Sun, Tristram-Nagle, Suter, & Nagle, 1996) to resolve the electron density distribution. For the same reason, powder samples are also not ideal for line shape analysis to shed light on bilayer fluctuation and membrane elastic properties (Zhang et al., 1996) or to investigate the organization of lipid chains (Sun et al., 1994). We will return to these topics in Section 5.3.

5.2 Solution-based model systems

Lipid vesicles, bicelles, and nanodiscs can be prepared in aqueous solutions of controlled chemical conditions. These model systems are widely used to observe how the lipid bilayers respond to changes in the environmental conditions, and as they interact with each other or with other biologically relevant entities. Nanodiscs have been successful in aiding structural determination of membrane proteins by crystallography, NMR, and cryo-EM (Denisov & Sligar, 2016; Sligar & Denisov, 2021). In principle, it should be possible to extend the structural modeling methods that have been very successful with protein solution scattering to protein-loaded nanodiscs. In practice, this will be very challenging (Graziano, Miller, & Yang, 2018). Neutron scattering with contrast-matched nanodiscs (Josts et al., 2018) may be the best bet.

5.2.1 Sample preparation and data collection

Lipid vesicles can be prepared by homogenizing a lipid suspension. All components in their solution forms (e.g., in chloroform) are mixed together according to the desired composition. The solvent is removed by blow-drying in a nitrogen stream, then further extracted under vacuum. The dried mixture is then resuspended by vortexing in the buffer of choice, and further homogenized using a horn or tip sonicator (also known as cell disruptor or ultrasonic homogenizer), typically on ice as a lot of heat is generated in the process. When a tip sonicator is used, the homogenized sample will need to be centrifuged to remove the particles from the tip. Alternatively, an extruder can be used to repeatedly pass the suspension through a filter (MacDonald et al., 1991). After homogenization the sample should appear clearer compared to the original suspension, suggesting the formation of smaller structures. For the purpose of bilayer structure characterization, multilayered vesicles should be avoided to simplify the data analysis. This can be achieved by tweaking the physical parameters that promote the unbinding transition, for instance, by adding charged lipids or simply raising the sample temperature.

Bicelles and nanodiscs are both isolated small patches of lipid bilayers, intended for simulating the native environment of membrane proteins. Bicelles are created by adding to the lipid mixture a short-chain (therefore high-spontaneous-curvature) component (Ram & Prestegard, 1988; Sanders & Landis, 1995), which accumulates at the edge of the bilayer and stabilizes the overall bicelle structure. In lipid nanodiscs, the planar bilayer patch is surrounded by a belt of engineered membrane scaffolding protein (MSP) (Denisov, Grinkova, Lazarides, & Sligar, 2004; Denisov & Sligar, 2016), or other types of amphiphilic molecules (Knowles et al., 2009). With appropriate additives, both bicelles and nanodiscs can be aligned in a magnetic field.

These solution-based samples can be measured using instruments specialized in protein solution scattering, which are nowadays widely available and highly automated. It is therefore possible to quickly measure large numbers of samples and observe how the solution conditions affect the bilayer structure. As in other solution scattering measurements, subtraction of the solvent scattering is an essential step in data processing. It is important to use a matching buffer for background measurement. This can be accomplished by dialyzing the sample against the buffer, a common practice in the protein solution scattering community. Another consideration in buffer subtraction is the scaling factor. In addition to normalization to transmitted beam intensity, the volume fraction of the solvent being displaced by the solutes needs to be considered. This is not critical for low-q data, where the solvent scattering is constant. However, a precise subtraction can affect the shape of the chain-chain correlation peak at $\sim 1.45\,\text{Å}^{-1}$, which overlaps with the water peak at $\sim 2\,\text{Å}^{-1}$. The best strategy may be collecting data up to high q to include the water peak, and use the appearance of the subtracted data in the q-range where the water peak would be as an empirical criteria for preventing over- or undersubtraction of buffer scattering (Yang et al., 2021).

5.2.2 Data interpretation

The scattering data from the structures discussed in this section are typically analyzed based on geometric models, in which the structural components are represented as simple shapes (e.g., boxes for the bilayer structure, elliptical cylinders for representing the lateral dimensions of nanodiscs and bicelles). The electron density distribution of each component can be represented as a Gaussian function, or as a box function with an error-function transition at the boundary. The form factor of such a geometric model can be calculated analytically (e.g., Pedersen, Arleth, & Mortensen, 2013), so that model fitting can be implemented using standard software tools.

In general, reliable data interpretation requires the solution samples used for scattering measurements to contain monodisperse particles, to avoid additional fitting parameters needed to account for size distribution, or even shape variations. In the case of nanodiscs, the length of the belt molecule is expected to define the disc size. However, the preparation of the nanodiscs must follow stringent protocols to achieve monodispersed size distribution, as verified using size exclusion chromatography (Denisov et al., 2004). Similarly, the formation of bicelles is also sensitive to the composition. While the size of the bicelles can be calculated theoretically based on its composition, the overall lipid concentration and temperature both affect the formation of bicelles and their actual composition (Beaugrand et al., 2014; Li et al., 2013; van Dam, Karlsson, & Edwards, 2004).

The size requirement for vesicle samples is more relaxed. The vesicle sizes are in the range of a few hundred nanometers and typically there is a finite size distribution (Hallett, Watton, & Krygsman, 1991). Therefore, the high-frequency fringes that are expected for the form factor of a multi-layered core-shell structure (lipids heads and chains) are not observed in practice. Instead, the scattering intensity is mainly due to the bilayer structure. The average size of the vesicles does affect the shape of the scattering profile (Bouwstra, Gooris, Bras, & Talsma, 1993; Moody, 1975):

$$I(q) \sim \frac{1}{q^2}\left[|\overline{R}F_m(q) - iF_m{}'(q)|^2 + \Sigma^2 F_m^2(q)\right],$$

where $F_m(q) = \int \rho(z) e^{iqz} dz$ is the one-dimensional bilayer form factor, $F_m{}'(q) = \frac{d}{dq} F_m(q) = i \int z\, \rho(z) e^{iqz} dz$ is its derivative, and $\overline{R}$ and Σ are the average and standard deviation of vesicle radii. However, this size effect is most pronounced with small vesicles and is typically neglected, under the flat bilayer model:

$$I(q) \sim \frac{1}{q^2} |F_m(q)|^2.$$

As in all solution scattering data analysis, the structure factor due to the inter-particle correlation introduces another complication. In the case of nanodiscs and bicelles, this correlation can be controlled following the general solution scattering guidelines of limiting overall particle concentration. This strategy does not prevent the appearance of multilayered structures in vesicle samples, which depends on the method of sample preparation and the composition of lipids. The correlations between the layers result in peaks with long tails, owing to the fluctuation of the bilayers. The mathematical form for a structure factor that captures the underlying physics has been worked out (Zhang, Suter, & Nagle, 1994). However, the calculation of this structure factor is difficult to implement in practice. In addition, a separate population of "uncorrelated" bilayer must be included (Pabst, Rappolt, Amenitsch, & Laggner, 2000). For line shape analysis, substrate-supported samples described in Section 5.3 are more appropriate. In solution samples at low lipid concentrations, simpler structure factors may suffice to fit the data. A promising alternative is the one-dimensional sticky structure factor (Yang et al., 2009; Yuste & Santos, 1993), which takes into account a very short-ranged potential well at the edge of the hard-core repulsion of range σ, described by the stickiness parameter, α:

$$S(q) = \frac{1}{1 - \eta C(q)},$$

where η is the volume fraction of the 1D liquid and the Fourier transform of the direct correlation function is given by:

$$C(q) = \frac{2}{\eta(L_+ + L_-)} \left(-\frac{L_-^2}{L_+ - L_-} + \frac{L_+ L_-}{L_+ - L_-} \cos q\sigma \right.$$
$$\left. + \frac{1}{L_+ - L_-} \frac{\cos q\sigma - 1}{q^2\sigma^2} - \frac{\sin q\sigma}{q\sigma} \right),$$

with

$$L_\pm = \frac{1}{2\eta} \left[\sqrt{(1-\eta)(1-\eta+4\alpha\eta)} \pm (1-\eta) \right].$$

This structure factor qualitatively captures the characteristics of interactions between bilayers, and in effect implicitly accounts for the variations in the number of bilayers in multilayered vesicles.

Geometric models are adequate in accounting for low-q (up to $\sim 0.6\,\text{Å}^{-1}$) scattering features that correspond to the particle overall shape and/or the bilayer density profile, as shown in Fig. 3. A key issue in data analysis of this kind is that the fitting parameters must be chosen appropriately to include all critical structural features, and at the same time avoid overinterpreting the data. The selection of fitting parameters can affect the appearance of the best-fit model, even with comparable model fitness, indicative of the limited resolution that this kind of analysis can achieve. Absolute scattering intensity and chemical information are effective constraints in model fitting (Pedersen et al., 2013).

In the literature, the topic of asymmetric bilayers sometimes comes up. While this is certainly of biological relevance, in practice fitting the data with an asymmetric bilayer density profile (Brzustowicz & Brunger, 2005) is risky in light of the overfitting discussion above. At least in specific cases (Komorowski et al., 2018), it was shown that symmetric bilayer model could fit the data equally well, and that a finite vesicle size had a more pronounced effect on the shape of modeled scattering data.

A constant background is typically required when fitting vesicle data. While this is purely empirical, the physical origin of this background is likely the extension of the chain correlation peak at $q \sim 1.45\ \text{Å}^{-1}$ (Fig. 3). This is analogous to the scattering from bulk water, in which case the low-q scattering is constant and proportional to the compressibility (Guinier & Fournet, 1955). Under the flat bilayer model, the form factor of the bilayer contributes to the scattering data in the vicinity of the q_z axis only, while the organization of the lipid molecules results in a diffuse background that permeates the q_r–q_z plane, as has been shown based on molecular dynamics simulation (Hub, Salditt, Rheinstädter, & de Groot, 2007). A more general model therefore should include contributions due to both the bilayer structure and the chain organization.

The analysis of the density correlation function, calculated from the Fourier transform of the scattering intensity, can be used in conjunction with model fitting. In solution scattering, this correlation function is known as the distance distribution function, P(r). Since the scattering data is typically truncated at non-zero intensity at the ends of the observed q-range, specialized algorithms are needed to reliably perform the Fourier transform (Moore, 1980; Svergun, 1992). These tools can be used for the

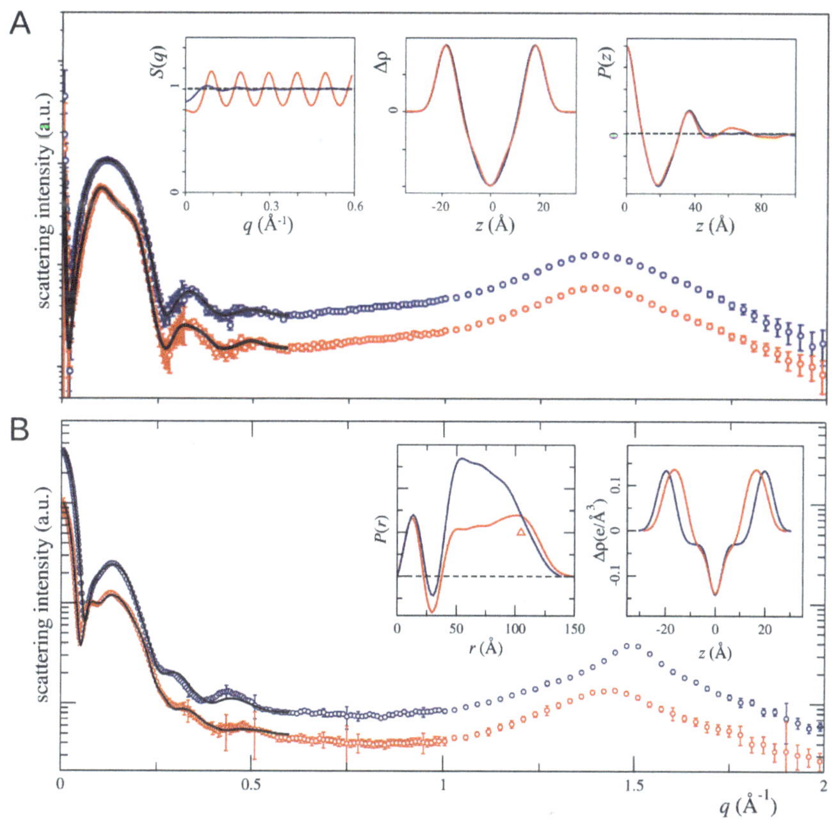

Fig. 3 Representative data from lipid vesicles and nanodiscs. (A) Representative vesicle data collected by users at the LiX beamline, with no obvious (blue symbols) and some degree of (red symbols) correlation between bilayers. The sticky structure factor (left inset) was able to produce reasonable fits (solid black lines) for both cases. The electron density profiles resulted from the fits are shown in the middle inset. They are on arbitrary scale since the sample compositions are unknown. The Patterson functions (right inset) are calculated from the data up to $0.6\,\text{Å}^{-1}$. (B) Scattering data from DMPC-nanodiscs, collected at 10 °C (blue) and 30 °C (red), below and above the main phase transition temperature of DMPC (23 °C). The best fits based on geometric models are shown as solid black lines. The insets show the P(r) function calculated from the data up to $0.6\,\text{Å}^{-1}$, and the bilayer electron density profile from the best-fit models. The triangle indicates the feature in the P(r) function that depends on the disc shape and has been attributed to the contribution from the protein belt. *Adapted from Graziano, V., Miller, L., & Yang, L. (2018). Interpretation of solution scattering data from lipid nanodiscs. Journal of Applied Crystallography, 51(1), 157–166. https://doi.org/10.1107/S1600576717018441*

P(r) calculation for nanodiscs and bicelles. For vesicles under the flat bilayer model, the scattering intensity approaches zero (more precisely a constant due to the chain correlation peak as discussed above) as q increases within the observed range. The correlation function therefore can be calculated by a simple cosine transform:

$$P(z) = \int I(q)q^2 \cos qz \, dq.$$

This model-independent transform may reveal features that are not obvious from the scattering data, such as the bilayer thickness for vesicles (see inset in Fig. 3A). It can also be used as a consistency check for the best-fit parameters. For instance, in the P(r) function of nanodiscs, a feature (indicated by the triangle in the inset of Fig. 3B) could be attributed to the diameter of the protein belt based on direct calculation from the density distribution in the geometric model density, and it is less prominent for nanodiscs whose shapes are more elliptical (Graziano et al., 2018).

5.3 Substrate-supported membrane structures

A major drawback of isotropic samples for the purpose of structural characterization is that the scattering data along the membrane normal and within the plane of the membrane are averaged together. The bilayers in substrate-supported model membranes are highly parallel to the substrate. They are therefore suitable where it is critical to resolve this ambiguity, for instance, for identifying non-lamellar membrane structures and subsequently solving the electron density distribution (Yang et al., 2003; Yang & Fukuto, 2005; Yang & Huang, 2002), for line shape analyses of the lamellar peak to extract the elastic properties of the membrane (Lyatskaya, Liu, Tristram-Nagle, Katsaras, & Nagle, 2000; Salditt, 2005), and for characterizing the organization of the lipid chains within the bilayer (Smith et al., 1988; Yang & Fukuto, 2005).

5.3.1 Sample preparation and data collection

Supported model membranes can be prepared by sandwiching the lipid-water dispersions discussed in Section 5.1 between substrates, typically silicon wafers or glass slides. Lyotropic liquid crystals contain defects that are visible by polarized microscopy and produce diffraction peaks in the direction perpendicular to the average layer normal. They can be reduced effectively by annealing, but rarely disappear completely. The drawback of this type of sample is that the hydration level is difficult to control.

Freely suspended films (Cheng, Ho, Hui, & Pindak, 1987) can also be created from lipid-water dispersions, by spreading the dispersion over an open aperture. This technique is widely used in liquid crystal research. While the film can be free of the defects discussed above and the number of layers can be controlled, they can rupture during large hydration changes.

More commonly substrate-supported multibilayers are prepared from organic solutions, by either manually spreading (e.g., Yang & Huang, 2003) or spin-coating (Perino-Gallice, Fragneto, Mennicke, Salditt, & Rieutord, 2002) over the substrate. Either way, the wetting behavior of the solvents on the substrate is the determining factor in forming uniform films. These films contain defects as well. The alignment of the bilayers can be improved by annealing in water vapor, which is observable from the scattering data. The annealed films contain micron-scale patches of stacked bilayers, resulting from the lipid film dewetting from the substrate, as shown by atomic force microscopy (Perino-Gallice et al., 2002).

These samples are typically studied in a partially dehydrated state, in equilibrium with a water vapor of known relative humidity. This can be created by a saturated salt solution, by heating a water reservoir to a temperature lower than the sample temperature, or by flushing the sample environment using an external humidity generator. While the environmental humidity can be experimentally measured, humidity sensors typically don't work well near full hydration. The sample and the sensor may also be exposed to different levels of humidity. This experimental uncertainty in the actual humidity has led to the so-called vapor pressure paradox, which was eventually recognized as an artifact (Nagle & Katsaras, 1999). It is possible to incorporate small membrane inclusions such as peptides and small proteins into multibilayers. However, the chemical condition is not well defined if the hydration is controlled by vapor equilibration. Alternatively, multibilayers can be measured under water or a buffer solution, for which the water chemical potential can be adjusted by additives (Mennicke, Constantin, & Salditt, 2006) to effectively dehydrate the bilayers.

Single bilayers can also be prepared on a substrate submerged under a buffer solution, by allowing the vesicles from the solution to spontaneously fuse on the substrate into an extended bilayer (Nováková, Giewekemeyer, & Salditt, 2006; Wang, Fukuto, & Yang, 2008). In X-ray scattering measurements on these samples, a relatively high X-ray energy is used so that the beam can penetrate the water with manageable intensity loss. As a reference, the transmission of 16 keV photons through 1 cm of water is $\sim$28%.

Water also creates a constant scattering background at low-q, which may overwhelm the scattering signals from the structures like membrane inclusions. This limits the application of this type of sample in the investigation of how the bilayer might interact with membrane proteins (Yang & Fukuto, 2011).

5.3.2 Data interpretation

As discussed in Section 3, reflectivity measurements are used to collect data along the q_z axis: $I(q_r=0, q_z)$, corresponding to the electron density averaged along the membrane normal. The reflectivity data can be analyzed by model fitting, similar to the method for solution scattering data. However, the model must include the substrate itself (Fig. 4A). The known electron density values of the substrate and water help to determine the absolute value of the electron density. For stacked bilayers, Fourier synthesis based on integrated peak intensity reveals the electron density within the unit cell (bilayer). Instead of a point-by-point reflectivity measurement, the diffraction data in this case can be collected in a single 2D diffraction pattern, with the sample rotating such that all Bragg peaks to be measured cross the Ewald sphere and therefore their full intensity captured. Before the Fourier synthesis, the Lorentz correction of $1/q$ need to be applied to the peak intensities, to account for the difference in the velocities at which the peaks cross the Ewald sphere. The variation of the illuminated sample volume at different X-ray incident angles also needs to be considered.

The fluctuations in highly hydrated bilayers lead to broadening of the Bragg peaks, along both q_r and q_z directions. The line shape analysis can be performed for each individual peak (Liu & Nagle, 2004; Salditt, Vogel, & Fenzl, 2003), circumventing the difficulty with the calculation of a structure factor that is valid throughout the full q_z-range. It should be noted that the diffraction pattern as recorded on the detector is a distorted representation of the information on the q_r–q_z plane, particularly near the q_z axis and at wide scattering angles (Yang et al., 1998). Therefore, the intensity profile taken directly from the scattering pattern may not always correctly represent the peak line shape. The "unwarping" of scattering patterns (Fig. 1A) has become routine now that grazing incident scattering is widely used in condensed matter physics research. The instrumental q-resolution is another practical issue to be considered in line shape analysis, particularly in data collected using area detectors.

As discussed in Section 5.1, upon dehydration a wide variety of membrane structures can appear. Among those that contain extended flat bilayers,

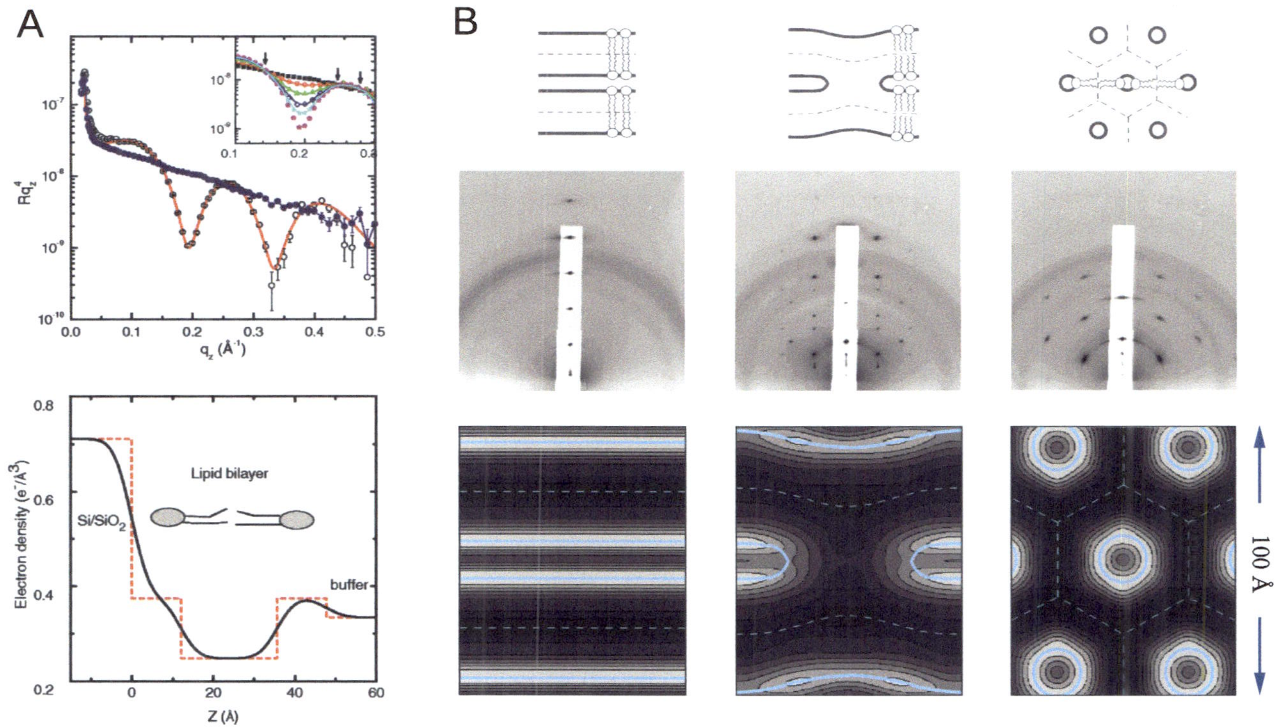

Fig. 4 Examples of data from substrate supported lipid bilayers. (A) The X-ray reflectivity data (upper panel) are recorded as the bilayer forms from vesicles on a silicon substrate submerged under a buffer solution. The time evolution of the first minimum in the data is shown in the inset. The final data can be fitted based on a box model of the bilayer structure (lower panel). (B) Structures observed in the same substrate-supported multi-bilayers under different temperatures and hydration levels. Each column shows the diffraction pattern (middle row), the electron density calculated from the data (bottom row), and a cartoon illustration of how the lipid molecules are organized in these structures (top row). *Panel (A) adapted from Wang, S. T., Fukuto, M., & Yang, L. (2008). In situ x-ray reflectivity studies on the formation of substrate-supported phospholipid bilayers and monolayers. Physical Review. E, Statistical, Nonlinear, and Soft Matter Physics, 77(3), 031909–8 and Panel (B) adapted from Yang, L., & Huang, H. W. (2002). Observation of a membrane fusion intermediate structure. Science, 297(5588), 1877–1879. https://doi.org/10.1126/science.1074354.*

the degree of structural order in the organization of the lipid chains can vary (Cheng et al., 1987; Smith et al., 1988). The analysis of the chain-packing peak profile along the q_z direction (Smith, Sirota, Safinya, Plano, & Clark, 1990) reveals that the lipid chains in the two leaflets in the same bilayer are spatially correlated and therefore contribute to the observed scattering intensity, yet the lipid heads are not. Structural modulations within the bilayer plane, with one- or two-dimensional periodicity (Sun et al., 1996; Yang & Fukuto, 2005) can also form. The chain ordering in modulated structures is more complex and the data can only be understood in the context of the shape of the bilayer modulation, such as in the ripple phase (Akabori & Nagle, 2015).

Owing to the separation of the q_r and q_z components of the scattering vector, non-lamellar structures that form in supported samples can be easily identified by the symmetry present in the diffraction pattern, from which the electron density of the unit cell can be calculated (Fig. 4B). Bragg peaks on and off the q_z axis need to be measured separately, and their integrated intensities are subject to different Lorentz corrections (Yang & Huang, 2003), due to the different path by which they "rotate" across the Ewald sphere. Phasing methods discussed in Section 5.1 also apply to these data. In addition, it is possible to take advantage of anomalous scattering from bromine labels in the lipid chain for phasing, similar to the multiwavelength anomalous dispersion (MAD) method used in protein crystallography. Since these structures never diffract to high resolution, anomalous phasing is implemented by fitting the data to the expected energy dependence to separate the contributions to the form factor (Pan, Wang, Liu, Yang, & Huang, 2006; Qian, Wang, Yang, & Huang, 2008):

$$|F(\lambda)| = \pm\left(F_0 + \frac{f'(\lambda)}{f_0}F_L\right),$$

where F_0 and F_L are the partial form factors due to non-label atoms and the labels, whose scattering amplitude is given by $f=f_0+f'(\lambda)+if''(\lambda)$, which can be derived from experimentally measured fluorescence emission. The diffraction measurements are performed at energies below the absorption edge of the anomalous scatterer to suppress the fluorescence background and therefore $f''(\lambda)\sim 0$. With the separation of partial form factors, the distribution of the bromine labels can then be calculated from the anomalous signal alone to help understand the organization of the lipid chains.

It is possible to incorporate small proteins and membrane-active peptides into substrate-supported multibilayers. Depending on the environmental conditions, these membrane inclusions can produce scattering signals that correspond to either a two-dimension liquid or a three-dimensional crystal (Yang et al., 2001). Diffraction data from the crystalline structures again can be analyzed by Fourier synthesis. The liquid-like data can be analyzed using the structure factor and form factor analysis, and splitting the structure factor into multiple terms (Yang et al., 1999):

$$S(q_r, q_z) = S_{00}(q_r) + 2 \cos q_z d \, S_{01}(q_r) + \cdots,$$

corresponding to the interaction and positional correlation between membrane inclusions in the same bilayer, or in different bilayers.

6. Concluding remarks

We have discussed different membrane structures that can be studied using scattering methods. Some of these structures are found in biological systems or engineered to accomplish a specific function (e.g., therapeutic). Other structures are created specifically for the purpose of understanding how lipid bilayers behave in biological processes. Sample-specific scattering methods provide a means to identify these structures and determine their structural parameters.

These methods measure the averaged structure in the sample volume illuminated by the probing X-ray or neutron beam. The dynamical characteristics of the membrane structures could be studied by neutron- and X-ray-based spectroscopic methods, including Inelastic X-ray or Neutron Scattering and Neutron Spin Echo (e.g., Qian, Sharma, & Clifton, 2020). There is also a vast literature from the liquid crystal community that could benefit lipid membrane research.

The discussions of elastic energy and geometric structural models implicitly consider the membrane structure as a continuum. In the context of domain formation and lipid chain organization, the individual lipid molecules may need to be considered. The complementary use with molecular dynamics simulation provides a means to correlate the scattering data with molecular interactions within the bilayer (Hub et al., 2007; Poojari, Scherer, & Hub, 2021).

References

Aeffner, S., Reusch, T., Weinhausen, B., & Salditt, T. (2012). Energetics of stalk intermediates in membrane fusion are controlled by lipid composition. *Proceedings of the National Academy of Sciences*, *109*(25), E1609–E1618. https://doi.org/10.1073/pnas.1119442109.

Akabori, K., & Nagle, J. F. (2015). Structure of the DMPC lipid bilayer ripple phase. *Soft Matter*, *11*(5), 918–926. https://doi.org/10.1039/C4SM02335H.

Akers, C. K., & Parsons, D. F. (1970). X-ray diffraction of myelin membrane: I. Optimal conditions for obtaining unmodified small-angle diffraction data from frog sciatic nerve. *Biophysical Journal*, *10*(2), 101–115. https://doi.org/10.1016/S0006-3495(70)86288-9.

Arteta, M. Y., Kjellman, T., Bartesaghi, S., Wallin, S., Wu, X., Kvist, A. J., et al. (2018). Successful reprogramming of cellular protein production through mRNA delivered by functionalized lipid nanoparticles. *Proceedings of the National Academy of Sciences*, *115*(15), E3351–E3360. https://doi.org/10.1073/pnas.1720542115.

Beaugrand, M., Arnold, A. A., Hénin, J., Warschawski, D. E., Williamson, P. T. F., & Marcotte, I. (2014). Lipid concentration and molar ratio boundaries for the use of isotropic Bicelles. *Langmuir*, *30*(21), 6162–6170. https://doi.org/10.1021/la5004353.

Blaurock, A. E. (1971). Structure of the nerve myelin membrane: Proof of the low-resolution profile. *Journal of Molecular Biology*, *56*(1), 35–52. https://doi.org/10.1016/0022-2836(71)90082-9.

Blaurock, A. E., & Worthington, C. R. (1969). Low-angle X-ray diffraction patterns from a variety of myelinated nerves. *Biochimica et Biophysica Acta (BBA) - Biomembranes*, *173*(3), 419–426. https://doi.org/10.1016/0005-2736(69)90006-6.

Bouwstra, J. A., Gooris, G. S., Bras, W., & Talsma, H. (1993). Small angle X-ray scattering: Possibilities and limitations in characterization of vesicles. *Chemistry and Physics of Lipids*, *64*(1), 83–98. https://doi.org/10.1016/0009-3084(93)90059-C.

Brzustowicz, M. R., & Brunger, A. T. (2005). X-ray scattering from unilamellar lipid vesicles. *Journal of Applied Crystallography*, *38*, 126–131.

Cheng, A., & Caffrey, M. (1996). Free radical mediated x-ray damage of model membranes. *Biophysical Journal*, *70*(5), 2212–2222. https://doi.org/10.1016/S0006-3495(96)79787-4.

Cheng, M., Ho, J. T., Hui, S. W., & Pindak, R. (1987). Electron-diffraction study of free-standing liquid-crystal films. *Physical Review Letters*, *59*(10), 1112–1115. https://doi.org/10.1103/PhysRevLett.59.1112.

Cherezov, V., Riedl, K. M., & Caffrey, M. (2002). Too hot to handle? Synchrotron X-ray damage of lipid membranes and mesophases. *Journal of Synchrotron Radiation*, *9*(6), 333–341. https://doi.org/10.1107/S0909049502014528.

Cullis, P. R., & Hope, M. J. (2017). Lipid nanoparticle Systems for Enabling Gene Therapies. *Molecular Therapy*, *25*(7), 1467–1475. https://doi.org/10.1016/j.ymthe.2017.03.013.

Denisov, I. G., Grinkova, Y. V., Lazarides, A. A., & Sligar, S. G. (2004). Directed self-assembly of monodisperse phospholipid bilayer nanodiscs with controlled size. *Journal of the American Chemical Society*, *126*(11), 3477–3487. https://doi.org/10.1021/ja0393574.

Denisov, I. G., & Sligar, S. G. (2016). Nanodiscs for structural and functional studies of membrane proteins. *Nature Structural and Molecular Biology*, *23*(6). https://doi.org/10.1038/nsmb.3195.

Georgiadis, M., Schroeter, A., Gao, Z., Guizar-Sicairos, M., Liebi, M., Leuze, C., et al. (2021). Nanostructure-specific X-ray tomography reveals myelin levels, integrity and axon orientations in mouse and human nervous tissue. *Nature Communications*, *12*(1), 2941. https://doi.org/10.1038/s41467-021-22719-7.

Goldstein, R. E., & Leibler, S. (1988). Model for lamellar phases of interacting lipid membranes. *Physical Review Letters*, *61*(19), 2213–2216. https://doi.org/10.1103/PhysRevLett.61.2213.

Graziano, V., Miller, L., & Yang, L. (2018). Interpretation of solution scattering data from lipid nanodiscs. *Journal of Applied Crystallography*, *51*(1), 157–166. https://doi.org/10.1107/S1600576717018441.

Gruen, D. W. R., & Marčelja, S. (1983). Spatially varying polarization in water. A model for the electric double layer and the hydration force. *Journal of the Chemical Society, Faraday Transactions 2: Molecular and Chemical Physics*, *79*(2), 225–242. https://doi.org/10.1039/F29837900225.

Gruner, S. M. (1989). Stability of lyotropic phases with curved interfaces. *The Journal of Physical Chemistry*, *93*(22), 7562–7570. https://doi.org/10.1021/j100359a011.

Gruner, S. M., Cullis, P. R., Hope, M. J., & Tilcock, C. P. S. (1985). Lipid polymorphism: The molecular basis of nonbilayer phases. *Annual Review of Biophysics and Biophysical Chemistry*, *14*(1), 211–238. https://doi.org/10.1146/annurev.bb.14.060185.001235.

Guinier, A., & Fournet, G. (1955). *Small-angle scattering of X-rays*. Wiley.

Hallett, F. R., Watton, J., & Krygsman, P. (1991). Vesicle sizing. *Biophysical Journal*, *59*(2), 357–362.

Harroun, T. A., Heller, W. T., Weiss, T. M., Yang, L., & Huang, H. W. (1999). Experimental evidence for hydrophobic matching and membrane-mediated interactions in lipid bilayers containing gramicidin. *Biophysical Journal*, *76*(2), 937–945. https://doi.org/10.1016/S0006-3495(99)77257-7.

Helfrich, W. (1973). Elastic properties of lipid bilayers: Theory and possible experiments. *Zeitschrift Für Naturforschung C*, *28*(11 – 12), 693–703. https://doi.org/10.1515/znc-1973-11-1209.

Huang, H. W. (1986). Deformation free energy of bilayer membrane and its effect on gramicidin channel lifetime. *Biophysical Journal*, *50*(6), 1061–1070. https://doi.org/10.1016/S0006-3495(86)83550-0.

Huang, H. W. (2000). Action of antimicrobial peptides: Two-state model. *Biochemistry*, *39*(29), 8347–8352.

Huang, H., & Yang, L. (2009). X-ray methods for investigation of structures of lipid assemblies. In *Handbook of Molecular Biophysics: Methods and applications* (pp. 457–502). Wiley-VCH.

Hub, J. S., Salditt, T., Rheinstädter, M. C., & de Groot, B. L. (2007). Short-range order and collective dynamics of DMPC bilayers: A comparison between molecular dynamics simulations, X-ray, and neutron scattering experiments. *Biophysical Journal*, *93*(9), 3156–3168. https://doi.org/10.1529/biophysj.107.104885.

Inouye, H., Liu, J., Makowski, L., Palmisano, M., Burghammer, M., Riekel, C., et al. (2014). Myelin Organization in the Nodal, Paranodal, and Juxtaparanodal regions revealed by scanning X-ray microdiffraction. *PLoS One*, *9*(7), e100592. https://doi.org/10.1371/journal.pone.0100592.

Jensen, T. H., Bech, M., Bunk, O., Menzel, A., Bouchet, A., Le Duc, G., et al. (2011). Molecular X-ray computed tomography of myelin in a rat brain. *NeuroImage*, *57*(1), 124–129. https://doi.org/10.1016/j.neuroimage.2011.04.013.

Joseph, J. S., Liu, W., Kunken, J., Weiss, T. M., Tsuruta, H., & Cherezov, V. (2011). Characterization of lipid matrices for membrane protein crystallization by high-throughput small angle X-ray scattering. *Methods*, *55*(4), 342–349. https://doi.org/10.1016/j.ymeth.2011.08.013.

Josts, I., Nitsche, J., Maric, S., Mertens, H. D., Moulin, M., Haertlein, M., et al. (2018). Conformational states of ABC transporter MsbA in a lipid environment investigated by small-angle scattering using stealth carrier Nanodiscs. *Structure*, *26*(8), 1072–1079. https://doi.org/10.1016/j.str.2018.05.007.

Knowles, T. J., Finka, R., Smith, C., Lin, Y.-P., Dafforn, T., & Overduin, M. (2009). Membrane proteins solubilized intact in lipid containing nanoparticles bounded by styrene maleic acid copolymer. *Journal of the American Chemical Society*, *131*(22), 7484–7485. https://doi.org/10.1021/ja810046q.

Komorowski, K., Salditt, A., Xu, Y., Yavuz, H., Brennich, M., Jahn, R., et al. (2018). Vesicle adhesion and fusion studied by small-angle X-ray scattering. *Biophysical Journal, 114*(8), 1908–1920. https://doi.org/10.1016/j.bpj.2018.02.040.

Korlach, J., Schwille, P., Webb, W. W., & Feigenson, G. W. (1999). Characterization of lipid bilayer phases by confocal microscopy and fluorescence correlation spectroscopy. *Proceedings of the National Academy of Sciences, 96*(15), 8461–8466. https://doi.org/10.1073/pnas.96.15.8461.

Lee, Y. K., Lee, H., & Nam, J.-M. (2013). Lipid-nanostructure hybrids and their applications in nanobiotechnology. *NPG Asia Materials, 5*(5), e48. https://doi.org/10.1038/am.2013.13.

Lee, M.-T., Sun, T.-L., Hung, W.-C., & Huang, H. W. (2013). Process of inducing pores in membranes by melittin. *Proceedings of the National Academy of Sciences, 110*(35), 14243–14248. https://doi.org/10.1073/pnas.1307010110.

Leneveu, D. M., Rand, R. P., & Parsegian, V. A. (1976). Measurement of forces between lecithin bilayers. *Nature, 259*(5544), 601–603. https://doi.org/10.1038/259601a0.

Li, M., Morales, H. H., Katsaras, J., Kučerka, N., Yang, Y., Macdonald, P. M., et al. (2013). Morphological characterization of DMPC/CHAPSO Bicellar mixtures: A combined SANS and NMR study. *Langmuir, 29*(51), 15943–15957. https://doi.org/10.1021/la402799b.

Lipowsky, R., & Leibler, S. (1986). Unbinding transitions of interacting membranes. *Physical Review Letters, 56*(23), 2541–2544. https://doi.org/10.1103/PhysRevLett.56.2541.

Lis, L. J., McAlister, M., Fuller, N., Rand, R. P., & Parsegian, V. A. (1982). Interactions between neutral phospholipid bilayer membranes. *Biophysical Journal, 37*(3), 657–665.

Liu, Y., & Nagle, J. F. (2004). Diffuse scattering provides material parameters and electron density profiles of biomembranes. *Physical Review E, 69*(4), 040901. https://doi.org/10.1103/PhysRevE.69.040901.

Lu, X., Yager, K. G., Johnston, D., Black, C. T., & Ocko, B. M. (2013). Grazing-incidence transmission X-ray scattering: Surface scattering in the born approximation. *Journal of Applied Crystallography, 46*(1), 165–172. https://doi.org/10.1107/S0021889812047887.

Lyatskaya, Y., Liu, Y., Tristram-Nagle, S., Katsaras, J., & Nagle, J. F. (2000). Method for obtaining structure and interactions from oriented lipid bilayers. *Physical Review E, 63*(1), 011907. https://doi.org/10.1103/PhysRevE.63.011907.

MacDonald, R. C., MacDonald, R. I., Menco, B. P. M., Takeshita, K., Subbarao, N. K., & Hu, L. (1991). Small-volume extrusion apparatus for preparation of large, unilamellar vesicles. *Biochimica et Biophysica Acta (BBA) - Biomembranes, 1061*(2), 297–303. https://doi.org/10.1016/0005-2736(91)90295-J.

Mariani, P., Luzzati, V., & Delacroix, H. (1988). Cubic phases of lipid-containing systems: Structure analysis and biological implications. *Journal of Molecular Biology, 204*(1), 165–189. https://doi.org/10.1016/0022-2836(88)90607-9.

Marquardt, D., Heberle, F. A., Nickels, J. D., Pabst, G., & Katsaras, J. (2015). On scattered waves and lipid domains: Detecting membrane rafts with X-rays and neutrons. *Soft Matter, 11*(47), 9055–9072. https://doi.org/10.1039/C5SM01807B.

Mennicke, U., Constantin, D., & Salditt, T. (2006). Structure and interaction potentials in solid-supported lipid membranes studied by X-ray reflectivity at varied osmotic pressure. *The European Physical Journal E, 20*(2), 221–230. https://doi.org/10.1140/epje/i2006-10014-4.

Moody, M. F. (1975). Diffraction by dispersion of spherical membrane vesicles. I. the basic equations. *Acta Crystallographica, Section A: Crystal Physics, Diffraction, Theoretical and General Crystallography, 31*(1), 8–15. https://doi.org/10.1107/S0567739475000022.

Moore, P. B. (1980). Small-angle scattering. Information content and error analysis. *Journal of Applied Crystallography, 13*(2), 168–175. https://doi.org/10.1107/S002188988001179X.

Nagle, J. F., & Katsaras, J. (1999). Absence of a vestigial vapor pressure paradox. *Physical Review E, 59*(6), 7018–7024. https://doi.org/10.1103/PhysRevE.59.7018.

Nováková, E., Giewekemeyer, K., & Salditt, T. (2006). Structure of two-component lipid membranes on solid support: An x-ray reflectivity study. *Physical Review E, 74*(5), 051911. https://doi.org/10.1103/PhysRevE.74.051911.
Pabst, G., Rappolt, M., Amenitsch, H., & Laggner, P. (2000). Structural information from multilamellar liposomes at full hydration: Full q-range fitting with high quality x-ray data. *Physical Review E, 62*(3), 4000.
Pan, D., Wang, W., Liu, W., Yang, L., & Huang, H. W. (2006). Chain packing in the inverted hexagonal phase of phospholipids: A study by X-ray anomalous diffraction on bromine-labeled chains. *Journal of the American Chemical Society, 128*(11), 3800–3807. https://doi.org/10.1021/ja058045t.
Parsegian, V. A., Fuller, N., & Rand, R. P. (1979). Measured work of deformation and repulsion of lecithin bilayers. *Proceedings of the National Academy of Sciences, 76*(6), 2750–2754. https://doi.org/10.1073/pnas.76.6.2750.
Pedersen, M. C., Arleth, L., & Mortensen, K. (2013). *WillItFit*: A framework for fitting of constrained models to small-angle scattering data. *Journal of Applied Crystallography, 46*(6), 1894–1898. https://doi.org/10.1107/S0021889813026022.
Perino-Gallice, L., Fragneto, G., Mennicke, U., Salditt, T., & Rieutord, F. (2002). Dewetting of solid-supported multilamellar lipid layers. *The European Physical Journal E, 8*(3), 275–282. https://doi.org/10.1140/epje/i2001-10101-0.
Poojari, C. S., Scherer, K. C., & Hub, J. S. (2021). Free energies of membrane stalk formation from a lipidomics perspective. *Nature Communications, 12*(1), 6594. https://doi.org/10.1038/s41467-021-26924-2.
Qian, S., Sharma, V. K., & Clifton, L. A. (2020). Understanding the structure and dynamics of complex biomembrane interactions by neutron scattering techniques. *Langmuir, 36*(50), 15189–15211. https://doi.org/10.1021/acs.langmuir.0c02516.
Qian, S., Wang, W., Yang, L., & Huang, H. W. (2008). Structure of transmembrane pore induced by Bax-derived peptide: Evidence for lipidic pores. *Proceedings of the National Academy of Sciences, 105*(45), 17379–17383. https://doi.org/10.1073/pnas.0807764105.
Ram, P., & Prestegard, J. H. (1988). Magnetic field induced ordering of bile salt/phospholipid micelles: New media for NMR structural investigations. *Biochimica et Biophysica Acta (BBA) - Biomembranes, 940*(2), 289–294. https://doi.org/10.1016/0005-2736(88)90203-9.
Salditt, T. (2005). Thermal fluctuations and stability of solid-supported lipid membranes. *Journal of Physics: Condensed Matter, 17*(6), R287–R314. https://doi.org/10.1088/0953-8984/17/6/R02.
Salditt, T., Li, C., & Spaar, A. (2006). Structure of antimicrobial peptides and lipid membranes probed by interface-sensitive X-ray scattering. *Biochimica et Biophysica Acta (BBA) - Biomembranes, 1758*(9), 1483–1498. https://doi.org/10.1016/j.bbamem.2006.08.002.
Salditt, T., Vogel, M., & Fenzl, W. (2003). Thermal fluctuations and positional correlations in oriented lipid membranes. *Physical Review Letters, 90*(17), 178101. https://doi.org/10.1103/PhysRevLett.90.178101.
Sanders, C. R., & Landis, G. C. (1995). Reconstitution of membrane proteins into lipid-rich bilayered mixed micelles for NMR studies. *Biochemistry, 34*(12), 4030–4040.
Schoenmaker, L., Witzigmann, D., Kulkarni, J. A., Verbeke, R., Kersten, G., Jiskoot, W., et al. (2021). mRNA-lipid nanoparticle COVID-19 vaccines: Structure and stability. *International Journal of Pharmaceutics, 601*, 120586. https://doi.org/10.1016/j.ijpharm.2021.120586.
Siegel, D. P. (1999). The modified stalk mechanism of lamellar/inverted phase transitions and its implications for membrane fusion. *Biophysical Journal, 76*(1), 291–313. https://doi.org/10.1016/S0006-3495(99)77197-3.
Simons, K., & Sampaio, J. L. (2011). Membrane organization and lipid rafts. *Cold Spring Harbor Perspectives in Biology, 3*(10), a004697. https://doi.org/10.1101/cshperspect.a004697.

Singer, S. J., & Nicolson, G. L. (1972). The fluid mosaic model of the structure of cell membranes. *Science, 175*(4023), 720–731. https://doi.org/10.1126/science.175.4023.720.
Sligar, S. G., & Denisov, I. G. (2021). Nanodiscs: A toolkit for membrane protein science. *Protein Science, 30*(2), 297–315. https://doi.org/10.1002/pro.3994.
Smith, G. S., Sirota, E. B., Safinya, C. R., & Clark, N. A. (1988). Structure of the L_β phases in a hydrated phosphatidylcholine multimembrane. *Physical Review Letters, 60*(9), 813.
Smith, G. S., Sirota, E. B., Safinya, C. R., Plano, R. J., & Clark, N. A. (1990). X-ray structural studies of freely suspended ordered hydrated DMPC multimembrane films. *The Journal of Chemical Physics, 92*(7), 4519–4529. https://doi.org/10.1063/1.457764.
Sun, W.-J., Suter, R. M., Knewtson, M. A., Worthington, C. R., Tristram-Nagle, S., Zhang, R., et al. (1994). Order and disorder in fully hydrated unoriented bilayers of gel-phase dipalmitoylphosphatidylcholine. *Physical Review E, 49*(5), 4665–4676. https://doi.org/10.1103/PhysRevE.49.4665.
Sun, W. J., Tristram-Nagle, S., Suter, R. M., & Nagle, J. F. (1996). Structure of the ripple phase in lecithin bilayers. *Proceedings of the National Academy of Sciences, 93*(14), 7008–7012. https://doi.org/10.1073/pnas.93.14.7008.
Svergun, D. I. (1992). Determination of the regularization parameter in indirect-transform methods using perceptual criteria. *Journal of Applied Crystallography, 25*(4), 495–503. https://doi.org/10.1107/S0021889892001663.
Tardieu, A., Luzzati, V., & Reman, F. C. (1973). Structure and polymorphism of the hydrocarbon chains of lipids: A study of lecithin-water phases. *Journal of Molecular Biology, 75*(4), 711–718.
Tristram-Nagle, S., Liu, Y., Legleiter, J., & Nagle, J. F. (2002). Structure of gel phase DMPC determined by X-ray diffraction. *Biophysical Journal, 83*(6), 3324–3335. https://doi.org/10.1016/S0006-3495(02)75333-2.
Turner, D. C., & Gruner, S. M. (1992). X-ray diffraction reconstruction of the inverted hexagonal (HII) phase in lipid-water systems. *Biochemistry, 31*(5), 1340–1355. https://doi.org/10.1021/bi00120a009.
van Dam, L., Karlsson, G., & Edwards, K. (2004). Direct observation and characterization of DMPC/DHPC aggregates under conditions relevant for biological solution NMR. *Biochimica et Biophysica Acta (BBA) - Biomembranes, 1664*(2), 241–256. https://doi.org/10.1016/j.bbamem.2004.06.005.
Veatch, S. L., & Keller, S. L. (2003). Separation of liquid phases in Giant vesicles of ternary mixtures of phospholipids and cholesterol. *Biophysical Journal, 85*(5), 3074–3083.
Wack, D. C., & Webb, W. W. (1989). Synchrotron x-ray study of the modulated lamellar phase P_β' in the lecithin-water system. *Physical Review A, 40*(5), 2712–2730. https://doi.org/10.1103/PhysRevA.40.2712.
Wang, S. T., Fukuto, M., & Yang, L. (2008). In situ x-ray reflectivity studies on the formation of substrate-supported phospholipid bilayers and monolayers. *Physical Review. E, Statistical, Nonlinear, and Soft Matter Physics*, 77(3). 031909–8.
Worthington, C. R., & Blaurock, A. E. (1969). A low-angle X-ray diffraction study of the swelling behavior of peripheral nerve myelin. *Biochimica et Biophysica Acta (BBA) - Biomembranes, 173*(3), 427–435. https://doi.org/10.1016/0005-2736(69)90007-8.
Yang, L., Antonelli, S., Chodankar, S., Byrnes, J., Lazo, E., & Qian, K. (2020). Solution scattering at the life science X-ray scattering (LiX) beamline. *Journal of Synchrotron Radiation, 27*(3), 804–812. https://doi.org/10.1107/S1600577520002362.
Yang, L., Ding, L., & Huang, H. W. (2003). New phases of phospholipids and implications to the membrane fusion problem†. *Biochemistry, 42*(22), 6631–6635. https://doi.org/10.1021/bi0344836.
Yang, L., & Fukuto, M. (2005). Modulated phase of phospholipids with a two-dimensional square lattice. *Physical Review. E, Statistical, Nonlinear, and Soft Matter Physics, 72*(1), 010901–010904.

Yang, L., & Fukuto, M. (2011). Using X-ray scattering to study the structures of membrane-associated proteins. In E. Pechkova, & C. Riekel (Eds.), *3. Synchrotron Radiation and Structural Proteomics* (pp. 357–372). Pan Sanford Publishing.

Yang, L., Harroun, T. A., Heller, W. T., Weiss, T. M., & Huang, H. W. (1998). Neutron off plane scattering of aligned membranes. I. Method of Measurement. *Biophysical Journal, 75*(2), 641–645.

Yang, L., Harroun, T. A., Weiss, T. M., Ding, L., & Huang, H. W. (2001). Barrel-stave model or toroidal model? A case study on melittin pores. *Biophysical Journal, 81*(3), 1475–1485. https://doi.org/10.1016/S0006-3495(01)75802-X.

Yang, L., & Huang, H. W. (2002). Observation of a membrane fusion intermediate structure. *Science, 297*(5588), 1877–1879. https://doi.org/10.1126/science.1074354.

Yang, L., & Huang, H. W. (2003). A rhombohedral phase of lipid containing a membrane fusion intermediate structure. *Biophysical Journal, 84*(3), 1808–1817.

Yang, L., Lazo, E., Byrnes, J., Chodankar, S., Antonelli, S., & Rakitin, M. (2021). Tools for supporting solution scattering during the COVID-19 pandemic. *Journal of Synchrotron Radiation, 28*(4), 1237–1244. https://doi.org/10.1107/S160057752100521X.

Yang, L., Wang, S., Fukuto, M., Checco, A., Niu, Z., & Wang, Q. (2009). Structure and interaction in 2D assemblies of tobacco mosaic viruses. *Soft Matter, 5*(24), 4951. https://doi.org/10.1039/b911894b.

Yang, L., Weiss, T. M., Harroun, T. A., Heller, W. T., & Huang, H. W. (1999). Supramolecular structures of peptide assemblies in membranes by neutron off-plane scattering: Method of analysis. *Biophysical Journal*, 77(5), 2648–2656.

Yuste, S. B., & Santos, A. (1993). Radial distribution function for sticky hard-core fluids. *Journal of Statistical Physics, 72*(3), 703–720. https://doi.org/10.1007/BF01048029.

Zhang, R., Suter, R. M., & Nagle, J. F. (1994). Theory of the structure factor of lipid bilayers. *Physical Review E, 50*(6), 5047–5060. https://doi.org/10.1103/PhysRevE.50.5047.

Zhang, R., Tristram-Nagle, S., Sun, W., Headrick, R. L., Irving, T. C., Suter, R. M., et al. (1996). Small-angle x-ray scattering from lipid bilayers is well described by modified Caillé theory but not by paracrystalline theory. *Biophysical Journal, 70*(1), 349–357. https://doi.org/10.1016/S0006-3495(96)79576-0.

CHAPTER FOURTEEN

Studying integral membrane protein by SANS using stealth reconstitution systems

Inokentijs Josts[a,b,*], Dominique-Maurice Kehlenbeck[a,b,c,d], Julius Nitsche[b], and Henning Tidow[a,b,*]
[a]The Hamburg Advanced Research Center for Bioorganic Chemistry (HARBOR), Hamburg, Germany
[b]Department of Chemistry, Institute for Biochemistry and Molecular Biology, University of Hamburg, Hamburg, Germany
[c]Life Sciences Group, Institut Laue–Langevin, Grenoble, France
[d]Partnership for Structural Biology (PSB), Grenoble, France
*Corresponding authors: e-mail address: josts@chemie.uni-hamburg.de; tidow@chemie.uni-hamburg.de

Contents

Abstract

Structural studies of integral membrane proteins (IMPs) are challenging as many of them require a lipid environment for full activity and stability. Reconstitution of IMPs into carrier systems such as nanodiscs or Salipro that mimic the native lipidic environment allow structural studies of membrane proteins in solution. The difficulty with this approach when applied to scattering techniques is the contribution of the carrier system to the scattering intensity and the subsequent challenging data analysis. Recently, so-called stealth carrier systems have been developed and applied to small-angle neutron scattering (SANS) studies of integral membrane proteins that become invisible to neutrons due to specific deuteration and solvent contrast-variation.

Methods in Enzymology, Volume 677
ISSN 0076-6879
https://doi.org/10.1016/bs.mie.2022.08.034

In this chapter, we describe in detail how the well-studied ATP-binding cassette (ABC) transporter protein MsbA can be reconstituted into stealth nanodiscs and subsequently be studied by SANS. This approach allows for a direct observation of the scattering signal from MsbA without the contribution of the surrounding carrier system and enables detection of different conformational states. The protocols can also be adapted to other stealth carrier systems (such as stealth Salipro).

Abbreviations

ABC ATP-binding cassette
DDM dodecylmaltoside
DIBMA diisobutylene-maleic acid
dMSP1D1 deuterated membrane scaffold protein variant D1
dPC deuterated mixed acyl phosphatidylcholine
IMP integral membrane protein
LMNG lauryl maltose neopentyl glycol
LPS lipopolysaccharide
MSP membrane scaffold protein
NBD nucleotide-binding domain
ND nanodisc
PC phosphatidylcholine
SANS small-angle neutron scattering
SAXS small-angle X-ray scattering
SLD scattering length density
sND stealth nanodiscs
TMD transmembrane domain

1. Introduction

Integral membrane proteins (IMPs) are essential for life as they transport matter and signals across cell membranes and generate electrochemical gradients. They play key roles in signal transduction, nutrient uptake and energy generation and are key targets for drug development (Li et al., 2021; Wacker, Stevens, & Roth, 2017). Structural and functional characterization of integral membrane proteins is still very challenging due to their low abundance and difficulties in purifying functional IMPs from their native membrane environment. Many IMPs require specifically-bound lipids or a lipid environment for full stability and activity (Bechara et al., 2015; Clay & Sharom, 2013; Doerrler & Raetz, 2002; Eggensperger, Fisette, Parcej, Schafer, & Tampe, 2014).

Several reconstitution systems have been developed in the last decade that allow membrane proteins to be incorporated in a native-like lipidic environment while still being amendable to structural studies (in solution). These scaffold/carrier systems include nanodiscs, Salipro and peptidiscs (Angiulli et al., 2020; Denisov & Sligar, 2016; Frauenfeld et al., 2016) that all have successfully been used for structural studies of IMPs.

Nanodiscs are discoidal protein-lipid complexes that are composed of two amphipathic membrane scaffold proteins (MSPs) encircling a lipid bilayer as a helical belt, yielding a lipidic assembly that is soluble in aqueous solution (Bayburt & Sligar, 2010; Denisov & Sligar, 2016). Nanodisc reconstitutions have been successfully used for structural studies of different IMP classes by scattering techniques (Bayburt, Grinkova, & Sligar, 2006; Choi, Rice, Stokes, & Coller, 2013; Skar-Gislinge et al., 2015). However, solution scattering studies of nanodisc-incorporated IMPs are challenging due to the complex multi-contrast scattering contribution of the scaffold disc.

To bypass this problem, so-called stealth carrier systems have been developed that become invisible to neutrons due to specific deuteration and solvent contrast-variation (Maric et al., 2014). We recently applied this stealth nanodisc (sND) technology to the ABC transporter MsbA and succeeded in observing different conformational states of MsbA in a lipid environment by SANS (Josts et al., 2018). We also used stealth nanodiscs to generate a structural model for the full-length Ca^{2+} pump ACA8 in its calmodulin-activated state (Nitsche et al., 2018).

Neutron sources such as ILL, SNS, MLZ, ISIS and ANSTO are scarce and access to SANS beamlines is limited. For this reason, an optimal sample preparation is of utmost importance for efficient use of resources and to optimize the scientific knowledge gain.

In this chapter, we describe the detailed experimental procedure to reconstitute integral membrane proteins into stealth nanodiscs using MsbA as example. MsbA is a prokaryotic ABC transporter that acts as lipid flippase, translocating lipid A, glycolipids and lipopolysaccharides (LPS) from the inner to the outer leaflet of the bacterial inner membrane (Raetz, Reynolds, Trent, & Bishop, 2007). Our method described here yields monodisperse IMP samples incorporated in stealth carrier systems, that are invisible to low-resolution neutron diffraction yet provide a favorable lipidic environment for the embedded membrane protein of interest, and thus paves the way for low-resolution structure determination of IMPs in solution using both *ab initio* and rigid body analysis approaches.

2. Contrast variation in small-angle neutron scattering (SANS) by selective deuteration of lipids and proteins for use as stealth carrier systems

A major advantage of SANS compared to SAXS is the possibility to make use of the different scattering length densities (SLD) of biological macromolecules and exploit contrast variation to distinguish between different components in the sample. This can be achieved by exploiting the natural contrast of biomolecules containing different hydrogen composition (e.g., protein/DNA or protein/lipids), by exchanging H_2O against D_2O in appropriate amounts to achieve the contrast-match point, or by controlled deuterium labeling of biomolecules leading to artificial contrast (Mahieu & Gabel, 2018).

With regard to integral membrane proteins, SANS is a particularly suitable technique due to the natural contrast between membrane protein and surrounding detergent or lipid molecules (Breyton et al., 2013). This feature has been exploited in several studies where different membrane mimetics were made "invisible" (i.e., turning them into "stealth" carriers). The detergent LMNG becomes experimentally undetectable in SANS in a solvent with 21.4% (v/v) D_2O and can be used to solubilize IMPs (Vermot et al., 2020). Mixed micelles containing a deuterated detergent analog have been described that eliminate neutron contrast between core and shell and allow the micelle scattering to be fully contrast-matched (Oliver, Pingali, & Urban, 2017). The diisobutylene maleic acid copolymer DIBMA has a contrast match point of 12% (v/v) D_2O (Guo, Sumner, & Qian, 2021). A present limitation of these contrast-matching experiments using detergents or amphipols at low D_2O content (<25%) is the incoherent scattering of water that limits acquisition of high-quality data essential for reliable structural studies (Gabel, 2017). In addition, detergent based membrane mimetics are lacking the lipid component that is often essential for the activity and structural integrity of the integral membrane protein.

An elegant alternative for the study of IMPs by SANS is the stealth nanodisc system, which combines deuterated MSP belt proteins with selectively deuterated phosphatidylcholine (PC) lipids resulting in a negligible SANS signal in 100% D_2O, thus effectively rendering the bilayer scaffold invisible (Maric et al., 2014) (Figs. 1 and 2). As the high D_2O concentration used with stealth nanodiscs minimize background from incoherent neutron

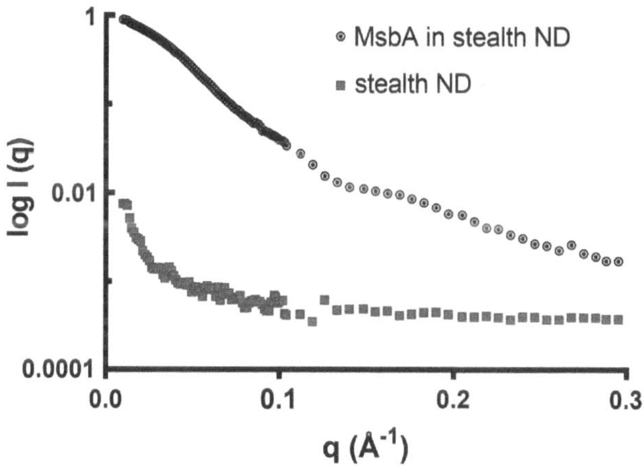

Fig. 1 Comparison of SANS data obtained from empty stealth nanodiscs (sND) (containing deuterated lipids and membrane scaffold protein only) (gray squares) with those recorded from nanodiscs that contain incorporated MsbA (black dots). The scattering signal from MsbA-sND is more than two orders of magnitude stronger than that from empty sNDs, indicating "an invisible nanodisc", i.e., that the contribution from lipids and belt protein would not influence the data analysis and structural modeling (Josts et al., 2018).

scattering and thus improves signal-to-noise ratio (Gabel et al., 2002), this method has the potential to yield high-quality data that allow to distinguish between different conformational states of incorporated membrane proteins as shown for the ABC transporter MsbA (Josts et al., 2018).

Further details regarding SANS contrast variation can be found in chapters "Contrast variation SAXS: Sample preparation protocols, experimental procedures, and data analysis" by Emeterio et al. as well as "Planning, executing, and assessing the validity of SANS contrast variation experiments" by Krueger of this edition of *Methods in Enzymology*.

For the preparation of the aforementioned stealth nanodiscs the theoretical contrast match points in 100% D_2O were calculated as 75% deuteration for the MSP1D1 belt protein (taking into account the exchangeable H atoms bound to O, N and S atoms) (Engelman & Moore, 1975) as well as 78% and 92% deuteration, respectively, for the head group atoms and acyl chain atoms of phospatidylcholine (Maric et al., 2014). These components can be produced by *Escherichia coli*-based biosynthesis. For the deuterated MSP belt protein the *E. coli* strain BL21 (DE3) was adapted and grown in

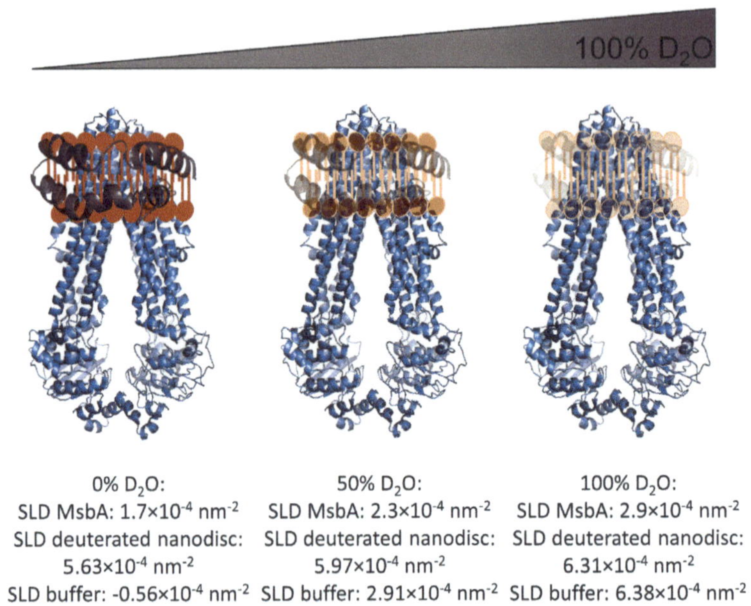

Fig. 2 Schematic representation of SANS scattering contribution from stealth nanodisc-reconstituted MsbA when measured in gradually increasing level of D_2O. At 100% H_2O based buffer (0% D_2O), all components (i.e., MsbA, lipids and belt protein) contribute to the SANS signal. With increasing D_2O concentrations, the stealth nanodiscs components (d-MSP1D1 and d-POPC, respectively) become increasingly invisible. Thus, the measured scattering data can be attributed entirely to the integrated membrane protein (MsbA). The relevant scattering length densities (SLDs) are indicated.

85% deuterated minimal medium with glycerol as carbon source (Artero, Hartlein, McSweeney, & Timmins, 2005; Rochel et al., 2011) and subsequently purified using standard procedures (Ritchie et al., 2009).

The selectively deuterated PC lipids can be produced using the genetically modified *E. coli* strain AL95 (Bogdanov, Heacock, Guan, & Dowhan, 2010) in close to 100% deuterated minimal medium supplemented with deuterated d_8-glycerol and partially deuterated choline chloride (Maric et al., 2015). Deuterated PC lipids (dPC) were then extracted and purified using silica-gel chromatography (Bligh & Dyer, 1959; Maric et al., 2015).

Further general information regarding deuteration for SANS experiments can be found in the chapter "Deuteration for biological SANS: Case studies, success and challenges in chemistry and biology" by Duff et al. of this edition of *Methods in Enzymology* as well as in several recent reviews (Abelson, Simon, Pyle, & Christianson, 2015; Haertlein et al., 2016; Pyle & Christianson, 2016).

3. Reconstitution of integral membrane proteins (IMP) into stealth carrier systems/example: Reconstitution of MsbA into stealth nanodiscs

This section describes reconstitution of the ABC transporter MsbA into stealth nanodiscs using the previously described deuterated components. The assembly process starts from detergent solubilized MsbA and dried lipids and consists of a biobead-driven assembly followed by size-exclusion chromatography (Fig. 3). The reconstitution efficiency, stability and activity are monitored by SDS-PAGE, nDSF and (Baginski) activity assays, respectively. After successful reconstitution, the samples can be directly used for SANS experiments or stored at −80 °C.

3.1 Equipment

- Glass vials (Carl Roth)
- Roller table (Cole-Parmer)
- ÄKTA pure chromatography system (Cytiva)
- Superdex200 10/300 GL size-exclusion chromatography column (Cytiva)
- Electrophoresis chamber (Bio-Rad)
- SDS-PAGE gels (Bio-Rad)
- Concentrators (Millipore)
- Centrifuge 5810 R (Eppendorf)
- Ultracentrifuge Optima XE-90 (Beckman Coulter)
- Microtiter plates (ThermoFisher)
- Tecan Infinite200 microplate reader (Tecan)
- Prometheus NT.48 nanoDSF differential scanning fluorimeter (NanoTemper Technologies)
- Prometheus NT.48 Capillaries nanoDSF Grade (NanoTemper Technologies)

3.2 Chemicals

- Tris (Carl Roth)
- NaCl (Carl Roth)
- β-mercaptoethanol (Carl Roth)
- sodium cholate (Carl Roth)
- biobeads (Bio-Rad)
- $MgCl_2$ (Carl Roth)
- ATP (Jena Biosciences)

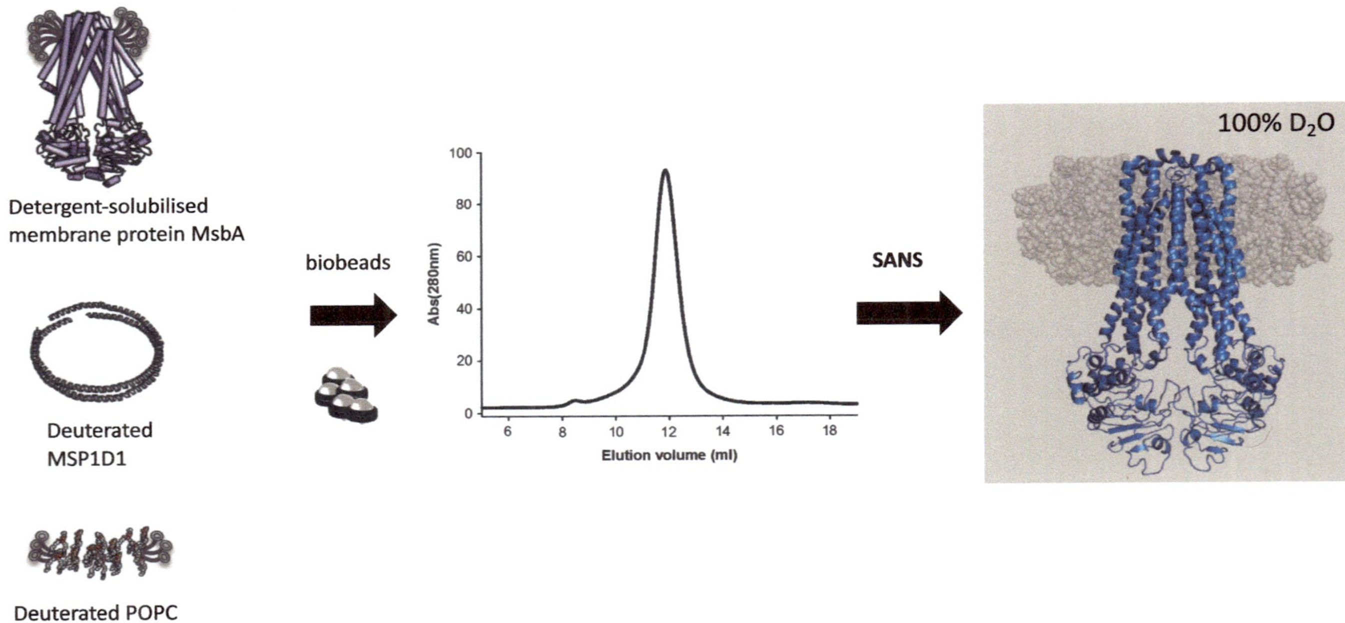

Fig. 3 Overview of stealth nanodisc assembly. The assembly process starts from detergent solubilized MsbA, dried deuterated lipids and deuterated MSP1 belt protein and consists of a biobead-driven assembly followed by size-exclusion chromatography. The resulting stealth nanodisc is invisible to neutron radiation in 100% D_2O and allows the structural investigation of the incorporated integral membrane protein by SANS.

- Ascorbic acid (Carl Roth)
- SDS (Carl Roth)
- HCl (Carl Roth)
- ammonium heptamolybdate (Sigma-Aldrich)
- sodium citrate (Carl Roth)
- sodium metaarsenite (Sigma-Aldrich)
- acetic acid (Carl Roth)
- D_2O (Sigma-Aldrich)

3.3 Protocol

Reconstitution of MsbA into stealth nanodiscs

- Resuspend selectively deuterated PC lipid (see above) from a dry lipid film in buffer containing 30 mM Tris (pH 7.5), 150 mM NaCl, 1 mM β-mercaptoethanol and 100 mM cholate to yield a 50 mM stock solution. Several mild sonication steps are necessary to fully dissolve the lipids.
- Mix detergent solubilized MsbA (in Tris/NaCl buffer with 0.03% DDM), deuterated MSP1D1 and deuterated PC (see above) in a molar ratio of 1:1:25 (i.e., two MSP1D1 belt proteins per MsbA homodimer) in buffer containing 20 mM cholate and incubate at 4 °C for 1 h.
- Initiate the nanodisc formation by adding 0.5–0.8 g/mL biobeads to the reaction and incubate overnight at 4 °C under constant agitation.
- Purify the stealth nanodisc-reconstituted MsbA by size-exclusion chromatography (SEC) on a Superdex200 10/300 GL column (GE Healthcare) in a buffer containing 30 mM Tris (pH 7.5), 150 mM NaCl and 1 mM β-mercaptoethanol (Fig. 4B).
- Analyze the elution fractions by SDS-PAGE (Fig. 4A)
- Concentrate fractions containing stealth nanodisc-embedded MsbA up to 5–8 mg/mL using a 100 kDa cut-off concentrator.
- Samples can be snap-frozen in liquid nitrogen and stored at −80 °C or directly used for stealth SANS experiments.
- On the day before to the SANS experiments, MsbA-nanodisc assemblies are dialyzed extensively against buffer prepared with 100% D_2O.

Evaluate activity of stealth nanodisc-reconstituted MsbA using activity assay

- Prepare solutions A and B of Baginski assays (Baginski, Epstein, & Zak, 1975; Chifflet, Torriglia, Chiesa, & Tolosa, 1988)
  - solution A: 140 mM ascorbic acid, 0.5 M HCl, 0.1% SDS, 5 mM ammonium heptamolybdate
  - solution B: 75 mM sodium citrate, 2% (w/v) sodium metaarsenite and 2% (v/v) acetic acid

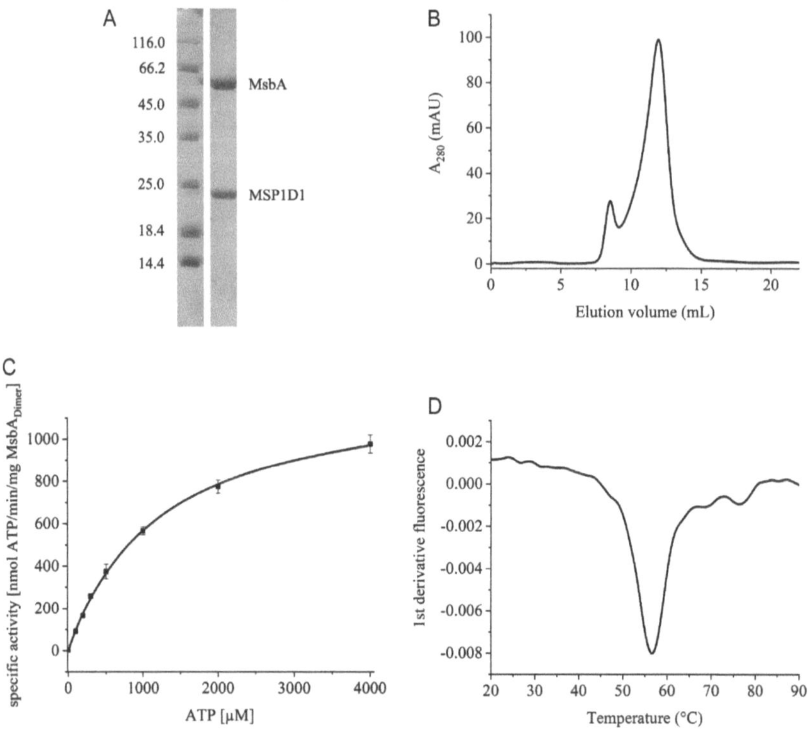

Fig. 4 Recommended quality control experiments during stealth nanodisc reconstitution. (A) SDS-PAGE sowing bands for belt protein (MSP1D1) and incorporated membrane protein (MsbA). (B) Size-exclusion chromatography (SEC) profile showing a minor void peak and a main peak corresponding to stealth nanodisc-reconstituted MsbA. (C) Activity assay based on detection of inorganic phosphate from ATP hydrolysis showing classical enzyme kinetics. (D) nDSF thermal melt profiles of stealth nanodisc-reconstituted MsbA.

- Incubate 5 µg MsbA in stealth nanodiscs in a buffer containing 30 mM Tris (pH 7.5), 150 mM NaCl, 5 mM $MgCl_2$ in 50 µL volume with different ATP concentrations (0/100/200/300/500/1000/2000/4000 µM) for 12 min at 25 °C.
- Stop the reaction by adding 50 µL ascorbic acid solution A (140 mM ascorbic acid, 0.5 M HCl, 0.1% SDS, 5 mM ammonium heptamolybdate).
- Stop the colorimetric reaction after a defined time by addition of 75 µL solution B (75 mM sodium citrate, 2% (w/v) sodium metaarsenite and 2% (v/v) acetic acid).
- Measure the absorbance at 860 nm using a Tecan Infinite200 microplate reader after 20 min.
- Calculate MsbA activity using standard curve obtained from defined phosphate concentrations (Fig. 4C).

Evaluate stability of stealth nanodisc-reconstituted MsbA using differential scanning fluorimetry (nDSF)

- Load stealth nanodisc-reconstituted MsbA ($c = 0.5$ mg/mL, $V = 10$ μL) in a buffer containing 20 mM Tris (pH 7.5) and 150 mM NaCl in a capillary (Prometheus NT.48 capillaries nanoDSF Grade) and measure the unfolding at a heating rate of 1 °C/min.
- Use the intrinsic fluorescence of tryptophan residues at 330 and 350 nm after excitation at 280 nm to monitor the fluorescence change upon heat unfolding.
- Use the first derivative of the unfolding curves (F350/F330) to determine the transition midpoint (Fig. 4D).

3.4 Notes

- It is recommended to evaluate the long-term stability of reconstituted samples before long-lasting SANS studies. For this purpose, the samples should be analyzed by analytical size-exclusion chromatography after several days using a Superdex S200 10/300 GL column (GE Healthcare):
  - o Centrifuge identical volumes of the samples and load them onto the column to allow peak area comparison
  - o Quantify the amount of intact MsbA/stealthND by peak integration
- Empty stealth nanodiscs should be prepared in a similar way as described above (using 1:40 molar ratio of dMSP1D1:dPC) in order confirm that they are really "invisible" to neutron radiation. They should show neglectable scattering when measured in 100% D_2O.
- A non-optimal ratio of IMP/MSP1D1/lipid can result in the formation of empty nanodiscs in addition to the correctly reconstituted complex. This can be detected by an additional peak in the SEC-profile as the empty nanodiscs elute significantly later (if the incorporated IMP has a reasonable size). If the IMP still contains an affinity tag (e.g., His-tag), reconstituted complexes could be re-bound to an affinity column (e.g., Ni-NTA) and empty discs could be washed off.
- For several detergents with low cmc (such as LMNG), detergent removal using biobeads is not very efficient and often results in the assembly of empty nanodiscs. For these cases, the use of cyclodextrin is recommended. Cyclodextrin is composed of a sugar ring with a hydrophobic core inside which has a higher affinity for detergents and is applicable to a wide range of detergents (Degrip, Vanoostrum, & Bovee-Geurts, 1998; Vargas, Schonbeck, Heimann, & Keller, 2018).

- It is recommended to acquire SAXS data of the reconstituted IMP samples in stealth nanodiscs before SANS measurements (using different samples from the same batch). These SAXS data contains the signal from the IMP as well as from the stealth disc and should therefore results in larger R_g and D_{max} values compared to the SANS data.
- As an additional quality control, we further recommend to measure SAXS data on the identical samples as used for SANS (after samples are released from radiation safety procedures) as soon as possible after the SANS measurements to confirm sample stability and integrity.
- Some IMPs co-purify with endogenous lipids and in the case of MsbA we introduced a 1% DDM wash (5–10 column volumes) while the protein was still bound to the Ni-NTA resin to remove the hydrogenated endogenous lipids from the sample prior to incorporation into the stealth nanodiscs.
- This methodology could also be applied (with minor adjustments) to other scaffold/carrier systems such as Salipro.

4. SANS data acquisition and analysis of IMP/stealth nanodisc samples

SANS data were measured in 100% D_2O buffer (30 mM Tris (pH 7.5), 150 mM NaCl) at 1–6 mg/mL protein concentration at 10 °C. Data were measured at the D11 beamline at ILL in two different configurations of 1.5 m/1.5 m and 8 m/8 m (collimation length/sample-detector distances) to cover a sufficiently wide q-range ($0.03 < q < 0.58\ \text{Å}^{-1}$), with a fixed neutron wavelength (λ) of 4.6 Å (Josts et al., 2018). Data reduction was performed using GRASP after measuring water reference samples, buffers, empty cell, direct beam, and the total absorber boron-cadmium.

As the previously described fractional deuteration of the components make the stealth nanodiscs invisible to neutron diffraction, the measured data only contain the signal from the incorporated membrane protein without contribution from the surrounding nanodisc. Thus, the reduced one-dimensional scattering intensities (I(q)) can be treated as homogeneous IMP scattering data and analyzed using standard small-angle scattering procedures. In the examples case of MsbA in stealth nanodiscs, the pipeline included:

- Buffer subtraction using PRIMUS (Konarev, Volkov, Sokolova, Koch, & Svergun, 2003)
- Calculation of radii of gyration (R_g) using the Guinier approximation

- Comparison with back-calculated data from atomic models using CRYSON (Svergun et al., 1998)
- Use of the MULCH server (http://smb-research.smb.usyd.edu.au/NCVWeb/) (Whitten, Cai, & Trewhella, 2008) to determine contrast and partial specific volume at 100% D_2O from the sample components and protein sequence
- Calculation of low-resolution *ab initio* models and their alignment using DAMMIF and SUPCOMP (Franke & Svergun, 2009; Petoukhov & Svergun, 2015)
- Model averaging and filtering using DAMAVER and DAMFILT (Kozin & Svergun, 2001; Petoukhov & Svergun, 2015)
- Rigid-body modeling using SASREF (Petoukhov & Svergun, 2005)

5. Summary and conclusions

The stealth nanodiscs methodology could not only be adapted to other carrier systems such as Salipro but it can also be used to study complexes between integral membrane proteins and soluble protein binding partners (Nitsche et al., 2018). In this case, contrast variation via selective deuteration of one of the proteins in the complex will make this component invisible when measured at 100% D_2O. Combining several datasets with different contrasts (deuteration/hydrogenation) will then allow more accurate modeling of the IMP complex in the stealth environment. Additionally, combining X-ray and neutron scattering of IMPs with the stealth technology allows for modeling multiple protein conformations and characterizing highly flexible regions of IMPs systems such as disordered regions that are often not visible using high-resolution techniques such as X-ray crystallography and cryo-EM. These models could then finally be combined with AlphaFold2 predictions of full-length proteins.

Acknowledgments

We are grateful to the staff at beamlines D11 and D22 (ILL, Grenoble) and P12 (EMBL, Hamburg). We thank Jens Frauenfeld (Salipro Biotech AB) for plasmids and protocols and members of the Tidow lab for helpful discussions. We acknowledge access to the Deuteration Laboratory (D-Lab) within ILL's Life Sciences Group that produced deuterated material for this study and the Sample Preparation and Characterization (SPC) Facility of EMBL Hamburg. This research was funded by an Emmy Noether grant, a PIER Seed Funding Grant and the excellence cluster "The Hamburg Centre for Ultrafast Imaging—Structure, Dynamics and Control of Matter at the Atomic Scale" of the Deutsche Forschungsgemeinschaft (DFG EXC 1074 - project ID 194651731/DFG EXC 2056 - project ID 390715994).

References

Abelson, J. N., Simon, M. I., Pyle, A. M., & Christianson, D. W. (2015). *Isotope labeling of biomolecules—Labeling methods*. New York: Academic Press.

Angiulli, G., Dhupar, H. S., Suzuki, H., Wason, I. S., Duong Van Hoa, F., & Walz, T. (2020). New approach for membrane protein reconstitution into peptidiscs and basis for their adaptability to different proteins. *eLife*, *9*, e53530. https://doi.org/10.7554/eLife.53530.

Artero, J. B., Hartlein, M., McSweeney, S., & Timmins, P. (2005). A comparison of refined X-ray structures of hydrogenated and perdeuterated rat gammaE-crystallin in H2O and D2O. *Acta Crystallographica. Section D, Biological Crystallography*, *61*(Pt. 11), 1541–1549. https://doi.org/10.1107/S0907444905028532.

Baginski, E. S., Epstein, E., & Zak, B. (1975). Review of phosphate methodologies. *Annals of Clinical and Laboratory Science*, *5*(5), 399–416.

Bayburt, T. H., Grinkova, Y. V., & Sligar, S. G. (2006). Assembly of single bacteriorhodopsin trimers in bilayer nanodiscs. *Archives of Biochemistry and Biophysics*, *450*(2), 215–222. https://doi.org/10.1016/j.abb.2006.03.013.

Bayburt, T. H., & Sligar, S. G. (2010). Membrane protein assembly into Nanodiscs. *FEBS Letters*, *584*(9), 1721–1727. https://doi.org/10.1016/j.febslet.2009.10.024.

Bechara, C., Noll, A., Morgner, N., Degiacomi, M. T., Tampe, R., & Robinson, C. V. (2015). A subset of annular lipids is linked to the flippase activity of an ABC transporter. *Nature Chemistry*, 7(3), 255–262. https://doi.org/10.1038/nchem.2172.

Bligh, E. G., & Dyer, W. J. (1959). A rapid method of total lipid extraction and purification. *Canadian Journal of Biochemistry and Physiology*, *37*(8), 911–917. https://doi.org/10.1139/o59-099.

Bogdanov, M., Heacock, P., Guan, Z., & Dowhan, W. (2010). Plasticity of lipid-protein interactions in the function and topogenesis of the membrane protein lactose permease from Escherichia coli. *Proceedings of the National Academy of Sciences of the United States of America*, *107*(34), 15057–15062. https://doi.org/10.1073/pnas.1006286107.

Breyton, C., Gabel, F., Lethier, M., Flayhan, A., Durand, G., Jault, J. M., et al. (2013). Small angle neutron scattering for the study of solubilised membrane proteins. *The European Physical Journal. E, Soft Matter*, *36*(7), 71. https://doi.org/10.1140/epje/i2013-13071-6.

Chifflet, S., Torriglia, A., Chiesa, R., & Tolosa, S. (1988). A method for the determination of inorganic phosphate in the presence of labile organic phosphate and high concentrations of protein: Application to lens ATPases. *Analytical Biochemistry*, *168*(1), 1–4.

Choi, W. S., Rice, W. J., Stokes, D. L., & Coller, B. S. (2013). Three-dimensional reconstruction of intact human integrin alphaIIbbeta3: New implications for activation-dependent ligand binding. *Blood*, *122*(26), 4165–4171. https://doi.org/10.1182/blood-2013-04-499194.

Clay, A. T., & Sharom, F. J. (2013). Lipid bilayer properties control membrane partitioning, binding, and transport of p-glycoprotein substrates. *Biochemistry*, *52*(2), 343–354. https://doi.org/10.1021/bi301532c.

Degrip, W. J., Vanoostrum, J., & Bovee-Geurts, P. H. (1998). Selective detergent-extraction from mixed detergent/lipid/protein micelles, using cyclodextrin inclusion compounds: A novel generic approach for the preparation of proteoliposomes. *The Biochemical Journal*, *330*(Pt. 2), 667–674. https://doi.org/10.1042/bj3300667.

Denisov, I. G., & Sligar, S. G. (2016). Nanodiscs for structural and functional studies of membrane proteins. *Nature Structural & Molecular Biology*, *23*(6), 481–486. https://doi.org/10.1038/nsmb.3195.

Doerrler, W. T., & Raetz, C. R. (2002). ATPase activity of the MsbA lipid flippase of Escherichia coli. *The Journal of Biological Chemistry*, *277*(39), 36697–36705. https://doi.org/10.1074/jbc.M205857200.

Eggensperger, S., Fisette, O., Parcej, D., Schafer, L. V., & Tampe, R. (2014). An annular lipid belt is essential for allosteric coupling and viral inhibition of the antigen translocation complex TAP (transporter associated with antigen processing). *The Journal of Biological Chemistry*, *289*(48), 33098–33108. https://doi.org/10.1074/jbc.M114.592832.

Engelman, D. M., & Moore, P. B. (1975). Determination of quaternary structure by small angle neutron scattering. *Annual Review of Biophysics and Bioengineering*, *4*, 219–241. https://doi.org/10.1146/annurev.bb.04.060175.001251.

Franke, D., & Svergun, D. I. (2009). DAMMIF, a program for rapid ab-initio shape determination in small-angle scattering. *Journal of Applied Crystallography*, *42*(Pt. 2), 342–346. https://doi.org/10.1107/S0021889809000338.

Frauenfeld, J., Loving, R., Armache, J. P., Sonnen, A. F., Guettou, F., Moberg, P., et al. (2016). A saposin-lipoprotein nanoparticle system for membrane proteins. *Nature Methods*, *13*(4), 345–351. https://doi.org/10.1038/nmeth.3801.

Gabel, F. (2017). Applications of SANS to study membrane protein systems. *Advances in Experimental Medicine and Biology*, *1009*, 201–214. https://doi.org/10.1007/978-981-10-6038-0_12.

Gabel, F., Bicout, D., Lehnert, U., Tehei, M., Weik, M., & Zaccai, G. (2002). Protein dynamics studied by neutron scattering. *Quarterly Reviews of Biophysics*, *35*(4), 327–367. https://doi.org/10.1017/s0033583502003840.

Guo, R., Sumner, J., & Qian, S. (2021). Structure of diisobutylene maleic acid copolymer (DIBMA) and its lipid particle as a "stealth" membrane-mimetic for membrane protein research. *ACS Applied Bio Materials*, *4*(6), 4760–4768. https://doi.org/10.1021/acsabm.0c01626.

Haertlein, M., Moulin, M., Devos, J. M., Laux, V., Dunne, O., & Forsyth, V. T. (2016). Biomolecular deuteration for neutron structural biology and dynamics. *Methods in Enzymology*, *566*, 113–157. https://doi.org/10.1016/bs.mie.2015.11.001.

Josts, I., Nitsche, J., Maric, S., Mertens, H. D., Moulin, M., Haertlein, M., et al. (2018). Conformational states of ABC transporter MsbA in a lipid environment investigated by small-angle scattering using stealth carrier nanodiscs. *Structure*, *26*(8), 1072–1079 e1074. https://doi.org/10.1016/j.str.2018.05.007.

Konarev, P. V., Volkov, V. V., Sokolova, A. V., Koch, M. H. J., & Svergun, D. I. (2003). PRIMUS: A windows PC-based system for small-angle scattering data analysis. *Journal of Applied Crystallography*, *36*, 1277–1282. <Go to ISI>://000185178600026.

Kozin, M. B., & Svergun, D. I. (2001). Automated matching of high- and low-resolution structural models. *Journal of Applied Crystallography*, *34*, 33–41.

Li, F., Egea, P. F., Vecchio, A. J., Asial, I., Gupta, M., Paulino, J., et al. (2021). Highlighting membrane protein structure and function: A celebration of the protein data Bank. *The Journal of Biological Chemistry*, *296*, 100557. https://doi.org/10.1016/j.jbc.2021.100557.

Mahieu, E., & Gabel, F. (2018). Biological small-angle neutron scattering: Recent results and development. *Acta Crystallographica Section D: Structural Biology*, *74*(Pt. 8), 715–726. https://doi.org/10.1107/S2059798318005016.

Maric, S., Skar-Gislinge, N., Midtgaard, S., Thygesen, M. B., Schiller, J., Frielinghaus, H., et al. (2014). Stealth carriers for low-resolution structure determination of membrane proteins in solution. *Acta Crystallographica. Section D, Biological Crystallography*, *70*(Pt. 2), 317–328. https://doi.org/10.1107/S1399004713027466.

Maric, S., Thygesen, M. B., Schiller, J., Marek, M., Moulin, M., Haertlein, M., et al. (2015). Biosynthetic preparation of selectively deuterated phosphatidylcholine in genetically modified Escherichia coli. *Applied Microbiology and Biotechnology*, *99*(1), 241–254. https://doi.org/10.1007/s00253-014-6082-z.

Nitsche, J., Josts, I., Heidemann, J., Mertens, H. D., Maric, S., Moulin, M., et al. (2018). Structural basis for activation of plasma-membrane ca(2+)-ATPase by calmodulin. *Communications Biology*, *1*, 206. https://doi.org/10.1038/s42003-018-0203-7.

Oliver, R. C., Pingali, S. V., & Urban, V. S. (2017). Designing mixed detergent micelles for uniform neutron contrast. *Journal of Physical Chemistry Letters*, *8*(20), 5041–5046. https://doi.org/10.1021/acs.jpclett.7b02149.

Petoukhov, M. V., & Svergun, D. I. (2005). Global rigid body modeling of macromolecular complexes against small-angle scattering data. *Biophysical Journal*, *89*(2), 1237–1250. https://doi.org/10.1529/biophysj.105.064154.

Petoukhov, M. V., & Svergun, D. I. (2015). Ambiguity assessment of small-angle scattering curves from monodisperse systems. *Acta Crystallographica. Section D, Biological Crystallography*, *71*(Pt. 5), 1051–1058. https://doi.org/10.1107/S1399004715002576.

Pyle, A. M., & Christianson, D. W. (2016). *Isotope labeling of biomolecules—Applications*. New York: Academic Press.

Raetz, C. R., Reynolds, C. M., Trent, M. S., & Bishop, R. E. (2007). Lipid a modification systems in Gram-negative bacteria. *Annual Review of Biochemistry*, *76*, 295–329. https://doi.org/10.1146/annurev.biochem.76.010307.145803.

Ritchie, T. K., Grinkova, Y. V., Bayburt, T. H., Denisov, I. G., Zolnerciks, J. K., Atkins, W. M., et al. (2009). Chapter 11—Reconstitution of membrane proteins in phospholipid bilayer nanodiscs. *Methods in Enzymology*, *464*, 211–231. https://doi.org/10.1016/S0076-6879(09)64011-8.

Rochel, N., Ciesielski, F., Godet, J., Moman, E., Roessle, M., Peluso-Iltis, C., et al. (2011). Common architecture of nuclear receptor heterodimers on DNA direct repeat elements with different spacings. *Nature Structural & Molecular Biology*, *18*(5), 564–570. https://doi.org/10.1038/nsmb.2054.

Skar-Gislinge, N., Kynde, S. A., Denisov, I. G., Ye, X., Lenov, I., Sligar, S. G., et al. (2015). Small-angle scattering determination of the shape and localization of human cytochrome P450 embedded in a phospholipid nanodisc environment. *Acta Crystallographica. Section D, Biological Crystallography*, *71*(Pt. 12), 2412–2421. https://doi.org/10.1107/S1399004715018702.

Svergun, D. I., Richard, S., Koch, M. H., Sayers, Z., Kuprin, S., & Zaccai, G. (1998). Protein hydration in solution: Experimental observation by x-ray and neutron scattering. *Proceedings of the National Academy of Sciences of the United States of America*, *95*(5), 2267–2272.

Vargas, C., Schonbeck, C., Heimann, I., & Keller, S. (2018). Extracavity effect in cyclodextrin/surfactant complexation. *Langmuir*, *34*(20), 5781–5787. https://doi.org/10.1021/acs.langmuir.8b00682.

Vermot, A., Petit-Hartlein, I., Breyton, C., Le Roy, A., Thepaut, M., Vives, C., et al. (2020). Interdomain flexibility within NADPH oxidase suggested by SANS using LMNG stealth carrier. *Biophysical Journal*, *119*(3), 605–618. https://doi.org/10.1016/j.bpj.2020.06.025.

Wacker, D., Stevens, R. C., & Roth, B. L. (2017). How ligands illuminate GPCR molecular pharmacology. *Cell*, *170*(3), 414–427. https://doi.org/10.1016/j.cell.2017.07.009.

Whitten, A. E., Cai, S., & Trewhella, J. (2008). MULCh: Modules for the analysis of small-angle neutron contrast variation data from biomolecular assemblies. *Journal of Applied Crystallography*, *41*(1), 222–226.

CHAPTER FIFTEEN

Predicting solution scattering patterns with explicit-solvent molecular simulations

Leonie Chatzimagas and Jochen S. Hub*
Theoretical Physics and Center for Biophysics, Saarland University, Saarbrücken, Germany
*Corresponding author: e-mail address: jochen.hub@uni-saarland.de

Contents

Abstract

Small-angle X-ray or neutron scattering (SAXS/SANS/SAS) is widely used to obtain structural information on biomolecules or soft-matter complexes in solution. Deriving a molecular interpretation of the scattering signals requires methods for predicting SAS patterns from a given atomistic structural model. Such SAS predictions are nontrivial because the patterns are influenced by the hydration layer of the solute, the excluded

Methods in Enzymology, Volume 677
ISSN 0076-6879
https://doi.org/10.1016/bs.mie.2022.08.035

solvent, and by thermal fluctuations. Many computationally efficient methods use simplified, implicit models for the hydration layer and excluded solvent, leading to some uncertainties and to free parameters that require fitting against experimental data. SAS predictions based on explicit-solvent molecular dynamics (MD) simulations overcome such limitations at the price of an increased computational cost. To rationalize the need for explicit-solvent methods, we first review the approximations underlying implicit-solvent methods. Next, we describe the theory behind explicit-solvent SAS predictions that are easily accessible via the WAXSiS web server. We present the workflow for computing SAS pattern from a given molecular dynamics trajectory. The calculations are available via a modified version of the GROMACS simulations software, coined GROMACS-SWAXS, which implements the WAXSiS method. Practical considerations for running routine explicit-solvent SAS predictions are discussed.

1. Introduction

Small- and wide-angle X-ray scattering (SWAXS) has originally been used to probe the overall shape of biomolecules or soft-matter complexes in solution. In recent decades, SWAXS has developed into an increasingly quantitative probe, primarily due to improved sample preparation, more brilliant light sources, and single-photon counting detectors (Graewert & Svergun, 2013; Koch, Vachette, & Svergun, 2003; Putnam, Hammel, Hura, & Tainer, 2007). Thanks to the advent of free-electron lasers, time-resolved SWAXS (TR-SWAXS) measurements are capable of tracking ultrafast dynamics of biomolecules or small molecules in solution (Arnlund et al., 2014; Brinkmann & Hub, 2016; Levantino et al., 2015). Complementary, small-angle neutron scattering (SANS) has remained popular because it allows for contrast-variation experiments by changing the D_2O concentration of the buffer (Gabel, 2015). Henceforth, we use the term SAS (small-angle scattering) when referring to both SWAXS and SANS.

Analysis of the SAS curve gives information on the overall shape and macroscopic properties of the probe such as the radius of gyration, maximum particle size, the structural order (globular vs unfolded), molecular mass, and particle volume. In contrast, an atomistic interpretation cannot be drawn from the data alone owing to the low information content of the SAS curves, but instead requires methods for theoretically predicting SAS intensities from a given atomistic structure. Such SAS curve predictions, so-called forward models, enable the validation or the refinement of structural models or ensembles against SAS data.

The physical properties that give rise to a solution scattering curve $I(q)$ are well understood. Namely, the intensities are given by

$$I(q) = \langle |F(\mathbf{q})|^2 \rangle_\Omega, \tag{1}$$

where $\langle \cdot \rangle_\Omega$ denotes the orientational average in reciprocal space, and $F(\mathbf{q})$ is the Fourier transform of the electron density contrast $\Delta\rho(\mathbf{r})$ between the solution and the pure-buffer system:

$$F(\mathbf{q}) = \int \Delta\rho(\mathbf{r})\ e^{-i\mathbf{q}\cdot\mathbf{r}}\ \mathrm{d}^3r \tag{2}$$

However, computing $I(q)$ from a given structural model such as a crystal structure or a molecular dynamics (MD) simulation of a biomolecule is complicated by several aspects:

(i) Since SAS detects the electron density contrast in solution, computing $I(q)$ requires knowledge of the volume of solvent that is displaced by the solute. For a biomolecule with internal disorder and a rough surface, the displaced volume is far from obvious. In addition, as discussed below, the volume taken by atoms of a certain chemical element depends on the chemical environment, suggesting that tabulated atomic volumes are subject to marked uncertainties.

(ii) The density of the hydration layer of biomolecules differs from the density of bulk solvent, thus contributing to the density contrast $\Delta\rho$. The hydration layer is influenced by properties such as the biomolecule's charge and geometry, the amino acid composition of the surface (anionic, cationic, hydrophobic, hydrophilic neutral), or the salt type and concentration of the buffer. For instance, highly charged biomolecules such as DNA/RNA were shown to exhibit a tight hydration layer (Pollack, 2011). For biomolecules with large surface-to-volume ratio, such as intrinsically disordered proteins, the hydration layer is expected to strongly contribute to the SAS signal. However, how such properties determine the hydration layer is not yet understood on a quantitative level (Kim et al., 2016).

(iii) SAS curves are influenced by thermal fluctuations of the biomolecule. Using analytic models, Moore (2014) showed that including Debye–Waller factors or correlated atomic fluctuations influences the calculated SAS curves at relatively small scattering angles of $q \gtrsim 0.2$ Å^{-1}. These findings are in line with earlier experimental observations (Tiede, Zhang, & Seifert, 2002) and with MD simulations, which revealed alterations of SAXS curves at $q \gtrsim 0.25\ \text{Å}^{-1}$ upon modulating atomic fluctuations during the simulations (Chen & Hub, 2014).

In the last three decades, a wide range of methods have been developed for predicting SAS intensities from atomistic structures, reflecting the increasing importance of SAS experiments for biomolecular research and the need for a physically founded interpretation of the data (Bardhan, Park, & Makowski, 2009; Chen & Hub, 2014; Grishaev, Guo, Irving, & Bax, 2010; Grudinin, Garkavenko, & Kazennov, 2017; Knight & Hub, 2015; Köfinger & Hummer, 2013; Liu, Morris, Hexemer, Grandison, & Zwart, 2012; Merzel & Smith, 2002 a, 2002 b; Nguyen, Pabit, Meisburger, Pollack, & Case, 2014; Oroguchi, Hashimoto, Shimizu, Sato, & Ikeguchi, 2009; Oroguchi & Ikeguchi, 2012; Park, Bardhan, Roux, & Makowski, 2009; Poitevin, Orland, Doniach, Koehl, & Delarue, 2011; Putnam, Weiner, Woetzel, Lowe, & Meiler, 2015; Ravikumar, Huang, & Yang, 2013; Schneidman-Duhovny, Hammel, & Sali, 2010; Schneidman-Duhovny, Hammel, Tainer, & Sali, 2013; Stovgaard, Andreetta, Ferkinghoff-Borg, & Hamelryck, 2010; Svergun, Barberato, & Koch, 1995; Tjioe & Heller, 2007; Virtanen, Makowski, Sosnick, & Freed, 2011; Yang, Park, Makowski, & Roux, 2009). The methods mainly differ by

- the method for modeling the hydration layer, involving explicit all-atom models, increased atomic form factors for solvent-exposed solute atoms, layers of uniform electron density, implicit models with internal structure, or several others;
- modeling of the excluded solvent based on tabulated atomic volumes up to explicit all-atom models;
- mathematical methods for computing the orientational average involving the Debye equation, a spherical harmonics expansion, or numerical quadrature;
- the spatial resolution, that is, whether the SAS curves are computed from individual atoms or from coarse-grained beads.

In this chapter, we focus on the calculation of SAS curves from atomistic explicit-solvent MD simulations as implemented in GROMACS-SWAXS, an extension of the GROMACS simulation software (https://gitlab.com/cbjh/gromacs-swaxs) (Chen & Hub, 2014; Chen et al., 2019). The method is easily accessible via the WAXSiS web server at https://waxsis.uni-saarland.de (Knight & Hub, 2015). As a rationale behind the explicit-solvent methods, we first review the approximations and limitations underlying widely used implicit-solvent methods, with a focus on the role of the hydration layer and excluded solvent for obtaining the density contrast. Next, we briefly describe the theory of explicit-solvent SAS curve predictions from explicit-solvent MD simulations as used by GROMACS-SWAXS and WAXSiS. Finally, the workflow of SAS calculations from an MD trajectory and practical considerations are discussed.

2. Implicit-solvent methods

Computationally efficient methods for computing SAS curves, as implemented in CRYSOL, FoXS, PepsiSAXS, or several other tools, use an implicit representation of the hydration layer and excluded solvent. To rationalize the benefits of explicit-solvent SAS calculations, we discuss the limitation of implicit-solvent methods in the following.

2.1 Displaced solvent

Many methods represent the displaced (or excluded) solvent with the help of water dummy beads, which are typically placed at the positions of the biomolecule's heavy atoms (see Fig. 1D) (Grudinin et al., 2017; Svergun et al., 1995). The electron density of the bead is essentially taken as $\rho_d(r) = \rho_s \exp\left(-\pi r^2/v^{2/3}\right)$, where ρ_s is the electron density of the solvent and v is the displaced volume such that the bead carries $\rho_s v$ electrons, as desired. The atomic form factor of the bead is given by the Fourier transform of the electron density, $f_d(q) = \rho_s v \exp\left(-v^{2/3}q^2/4\pi\right)$. In the implicit-solvent description, the buffer subtraction is carried out by reducing the atomic form factor of the biomolecule's heavy atom with the form factor of the dummy bead $f_d(q)$, leading to so-called "reduced form factors."

This method requires knowledge of the displaced volumes v for all atoms. However, estimates for the displaced volume differ substantially within the literature, leading to some uncertainty in the predicted SAS curve (Hub, 2018). Most SAXS prediction methods use and cite the displaced volumes reported by Fraser, MacRae, and Suzuki (1978). These data originally date back to a publication in 1895 by Isidor Traube, which was based on the volume change of water in response to the solvation of various organic compounds (Traube, 1895). Table 1 presents the atomic volumes reported by Fraser et al. in Å^3 as well as the original values by Traube in ccm/mol for several atoms or chemical groups. Traube and Fraser values are identical since ccm/mol = 1.66 Å^3.

More recently, atomic volumes have been obtained by Pontius et al. using Voronoi tessellation applied to the core region of high-resolution protein structures (Pontius, Richelle, & Wodak, 1996). The volumes by Traube and Fraser greatly differ from the volumes obtained from crystal structures

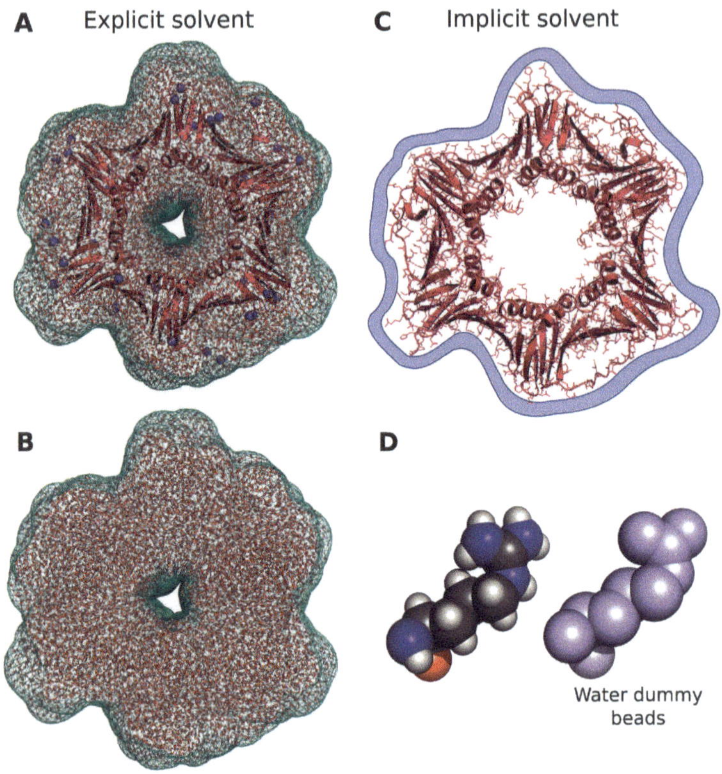

Fig. 1 (A/C) Hydration layer (B/D) displaced solvent, as represented by (A/B) explicit-solvent and (C/D) several implicit-solvent methods. (A) A spatial envelope is defined (blue mesh) that includes the biomolecule and the hydration layer, here shown for the protein PCNA (pdb code 4D2G) (De Biasio et al., 2015). (B) The same spatial envelope is used to define the displaced solvent. (C) Implicit-solvent methods use a simplified description of the hydration layer, often modeled as a layer of uniform density, while (D) the displaced solvent is typically described by water dummy beads with estimated volumes taken from a publication by Traube (1895) and fitted against the data.

(Table 1, columns 2 and 3). Compared to the Pontius values, the Fraser values greatly overestimate the volumes of CH and CH_2 groups, and they underestimate the volumes of N and O atoms and of NH or OH groups. Notably, Traube reported volumes of N or O atoms for alternative chemical environments such as nitro or (today outdated) "pentavalent" environments as well as for hydroxyl groups; however, these were not cited by Fraser, which may explain why they are not used in SAS calculations today (Table 1, column 4).

Besides the uncertainty owing to the choice of tabulated atomic volumes, additional uncertainty may arise because the atomic volumes depend

Table 1 Displaced volumes from Voronoi tessellation of crystal structure cores by Pontius et al. (1996), as well as from densitometric data by Traube (1895), as cited by Fraser et al. (1978).

Atomic group	Pontius et al. (1996) [Å³]	Fraser et al. (1978) [Å³]	Traube (1895) [ccm/mol]
H		5.15	3.1
C		16.44	9.9
N	8.8 (0.8)	2.49	1.5 (trivalent)
			10.7 (pentavalent)
			8.5–10.7 (in nitro compound)
O	22.3 (0.4)	9.13	5.5 (carbonyl oxygen)
			2.3 or 0.4 (hydroxy oxygen)
OH	23.9 (0.9)	14.28	
NH	14.1 (0.3)	7.64	
CH	11.8 (0.6)	21.59	
CH_2	20.9 (1.8)	26.74	
CH_3	33.9 (1.2)	31.89	

Pontius values represent averages and standard deviations from all amino acids. Traube and Fraser values are identical, using ccm/mol = 1.66Å^3. Fraser volumes of chemical groups represent sums of single-atom volumes.

on the chemical environment. Whereas the volumes by Pontius et al. represent the atomic packing in the core of stable proteins, it seems unlikely that the volumes also hold for more flexible protein surfaces, for detergent micelles, or for nucleotides. For n-hexadecane, for instance, the Pointius and Fraser volumes would imply a molecular volume of 360.4 and 438.1 Å^3, respectively, whereas the experimental value is 486.6 Å^3.

2.2 Hydration layer (HL)

Several tools model the hydration layer (HL) as a layer with a predefined thickness and a uniform excess density (Fig. 1C). For instance, CRYSOL constructs a homogeneous 3 Å-wide layer described by a two-dimensional angular function leading to a simplistic representation of the HL, which may not be suitable for molecules with cavities or nonglobular shapes (Svergun et al., 1995). CRYSOL3 aims to overcome this limitation by incrusting the surface (Franke et al., 2017)) similar to the grid-based density employed by PepsiSAXS (Grudinin et al., 2017). In contrast, FoXS models the HL by increasing atomic form factors of solvent-exposed atoms (Schneidman-Duhovny et al., 2010).

The density of HLs of biomolecules likely depends on the type of the water–surface interactions and, therefore, on the chemical composition of the surface. For instance, Kim et al. (2016) found that acidic amino acids (Glu/Asp) exhibit a tighter HL as compared to basic amino acids (Arg/Lys), an effect that our group recently observed in MD simulations (unpublished data). Likewise, it seems plausible that the HL of maltoside or sodium dodecyl sulfate (SDS) head groups of detergent micelles exhibit different densities as compared to the HL of charged or hydrophobic amino acids. Such chemical specificity of the HL is not captured by implicit solvent methods, explaining why they require fitting of the HL against experimental data.

2.3 Risk of overfitting solvent-related parameters

Owing to the simplified description of the HL and excluded solvent, implicit-solvent methods require two or three parameters that are adjusted upon fitting the calculated curve to the experimental curve. One parameter adjusts the excess density of the HL, while another parameter adjusts the volumes of the dummy beads that represent the excluded solvent. Some methods adjust in addition the overall displaced volume. The parameters are typically fitted by minimizing the residuals between the calculated and experimental SAS curves. Whether the fitted parameters reflect the physical situation such as the true HL density, or whether they merely absorb errors in the structural model or in the experimental data, is typically not clear. Hence, fitting solvent-related parameter may hide structurally relevant information.

A possible consequence of fitting parameters is illustrated in Fig. 2, which presents the validation of MD simulations of a protein-detergent complex against SAXS data (Berthaud et al., 2012; Chen & Hub, 2015 b). The complexes were composed of an aquaporin-0 tetramer solvated in 250 to 330 βDDM detergent molecules. Computing SAXS curves from the simulations with implicit-solvent methods and fitting the curves to experimental data leads to χ-values that hardly depend on the number of βDDM molecules, suggesting that alterations in the structural model have been absorbed into the fitting parameters (Fig. 2D). Consequently, the experimental number of βDDM molecules cannot be obtained by comparison with the data. This example illustrates that structurally relevant information may be hidden upon fitting free parameters. In contrast, when using explicit-solvent SAXS calculations described below, the χ-agreement of models with different βDDM numbers with the data can be distinguished because no

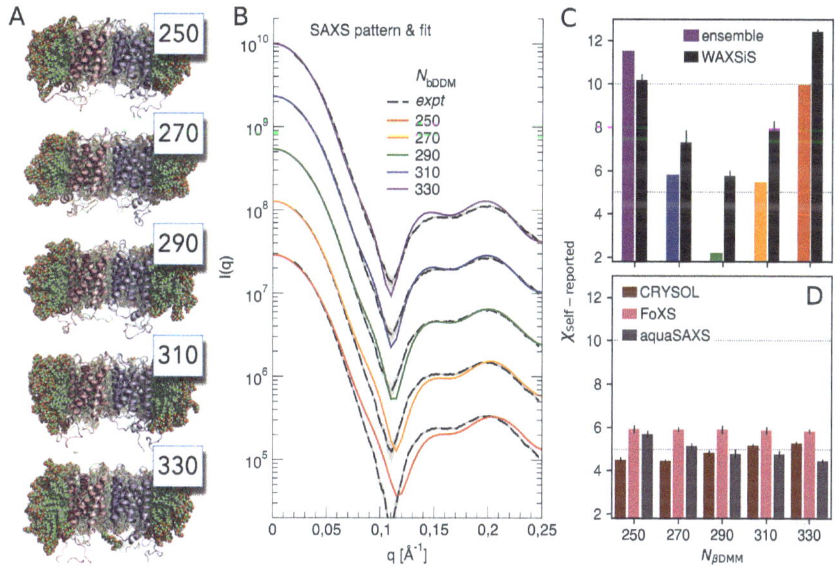

Fig. 2 SAXS curve prediction and snapshots of a protein-detergent complex. (A) MD simulation systems of aquaporin-0 solubilized in 250 to 330 βDDM detergent molecules simulated with the Charmm36 force field. (B) Explicit-solvent SAXS curve predictions for different number of βDDM using an aggregated ensemble of 90 ns to 100 ns MD trajectories (*colored lines*). Experimental data (*black dashed line*) from Berthaud, Manzi, Pérez, and Mangenot (2012). (C) χ-agreement between experimental SAXS curves and explicit-solvent SAXS calculations from free MD simulations (*colored bars*) or from the WAXSiS webserver (*black bars*). WAXSiS restrains the DDM molecules in a short simulation, rationalizing the increased χ-values. $N_{\beta DDM} = 290$ reveals the best agreement with experiment. (D) χ-agreement by CRYSOL, FoXS, and AquaSAXS, revealing nearly constant χ due to the adjustment of free parameters. *Reprinted with permission from Chen, P.-c., & Hub, J. S. (2015b). Structural properties of protein-detergent complexes from SAXS and MD simulations. The Journal of Physical Chemistry Letters, 6, 5116–5121, Copyright 2020 American Chemical Society.*

solvent-related parameters are fitted, thereby revealing that the model with 290 βDDM exhibits the best agreement with the data (Fig. 2C).

3. Explicit-solvent SAS predictions with the WAXSiS method

Several methods for predicting SAS curves have been presented that model both the excluded solvent and the HL with explicit-solvent MD simulations (Chen & Hub, 2014; Köfinger & Hummer, 2013; Oroguchi et al., 2009; Park et al., 2009). Here, we focus on the WAXSiS method, which is based on the formalism by Park et al. (2009). However, whereas the implementation by Park et al. required a constrained biomolecule, the WAXSiS method allows SAS predictions from MD simulations of flexible biomolecules by defining the SAS-contributing solvent with a spatial envelope. In addition, the WAXSiS method corrects (i) imprecise densities of water models and (ii) small density mismatches between the biomolecule and the pure-solvent simulation systems, both of which may influence the SAS prediction significantly. A careful comparison of different implicit and explicit SAXS prediction methods has been presented by Bernetti and Bussi (2021).

In the WAXSiS method, a spatial envelope is defined that encloses the biomolecule and the heterogeneous density of the HL (Figs. 1A and 3A). Hence, the HL structure is fully defined by the force field and does not require a free fitting parameter. The buffer subtraction is carried out by computing the scattering intensity difference between (a) the biomolecule with the envelope-enclosed solvent and (b) an explicit-solvent pure-buffer simulation, from which an identical volume is taken with the help of the same envelope (Fig. 1B). Thereby, the displaced solvent is modeled with atomic detail and without need of knowing atomic volumes of solute atoms.

Compared to implicit-solvent methods, the explicit-solvent calculations are computationally far more expensive as they require MD simulations, but they have several crucial advantages:

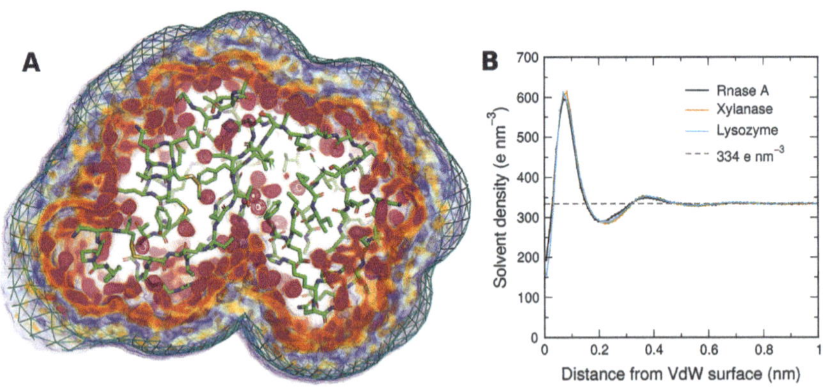

Fig. 3 (A) Solvent electron density around Rnase A, taken from an explicit-solvent MD simulations with position restraints on all heavy atoms, revealing the marked structure of the hydration layer. Only solvent density inside an envelope (*blue mesh*) is shown, where the envelope was constructed at a 7 Å distance from the protein surface. (B) Solvent density as function of the distance from the Van der Waals surface for three different proteins (see legend), averaged over the protein surface. Simulations were carried out with the Amber99sb-ildn/Tip3p force fields. Here, the solvent density was scaled with a small constant factor to match the experimental bulk density of 334 e nm^{-3}, as implemented in GROMACS-SWAXS to correct imprecise densities of some water models.

(i) The calculations do not require any free solvent-related parameters that must be fitted to experimental data. Thus, the amount of structural information, which can be extracted from the low-information SAS data, is not further reduced. By circumventing the fitting of the HL density, the radius of gyration is not adjusted against the data.

(ii) The use of "reduced form factors" is avoided and, with this, the uncertainties subject to atomic volumes. This may explain the better accuracy compared to implicit-solvent methods for predicting the SAS intensities of heterogeneous systems (such as protein-detergent complexes) and for molecules with nonglobular geometries (Ivanović, Bruetzel, Lipfert, & Hub, 2018; Ivanović, Hermann, Wójcik, Pérez, & Hub, 2020).

(iii) The explicit-solvent description remains valid at wide scattering angles ($q > 1.5$ Å), where the internal structure of water becomes relevant (Park et al., 2009).

(iv) By computing the SAS patterns from an MD trajectory, the method naturally accounts for thermal fluctuations, which influence the SAS patterns at remarkably small scattering angles of $q \gtrsim 0.25$ Å^{-1} (Chen & Hub, 2014, 2015 b; Moore, 2014; Tiede et al., 2002).

As an example, Fig. 4A compares the experimental SAXS curve of the ring-shaped protein PCNA with a curve computed with the WAXSiS

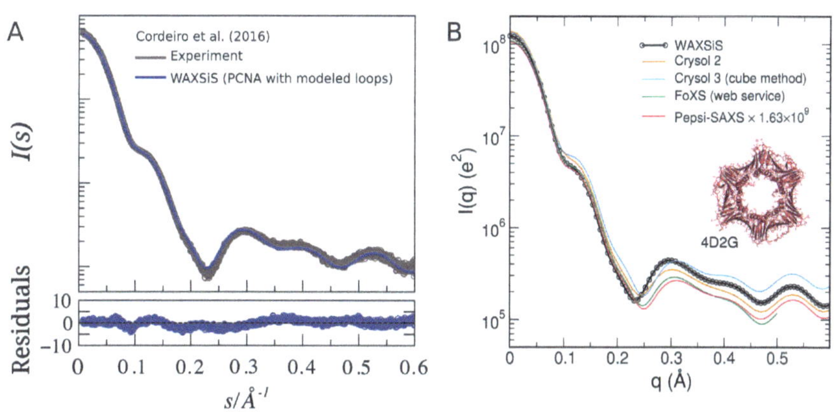

Fig. 4 (A) Experimental SAXS data of the ring-shaped protein PCNA (*gray*) and SAXS curve computed with the WAXSiS model from a PCNA model with modeled loops ($\chi^2 = 1.35$) (Cordeiro et al., 2017). (B) Comparison of SAXS curves of the PCNA crystal structure (pdb code 4D2G) computed with different methods with default settings (see legend). *Panel A: Adapted and reused with permission from Cordeiro, T. N., Chen, P.-c., De Biasio, A., Sibille, N., Blanco, F. J., Hub, J. S., …Bernadó, P. (2017). Disentangling polydispersity in the PCNA-p15PAF complex, a disordered, transient and multivalent macromolecular assembly. Nucleic Acids Research, 45, 1501–2015.*

method. The WAXSiS curve was computed from the 4D2G crystal structure of PCNA, to which missing termini were added (Cordeiro et al., 2017), and the computed curve was henceforth scaled with a constant factor ($I_{\text{scaled}}(q) = f\, I_{\text{WAXSiS}}(q)$). No other parameters were adjusted. Fig. 4B compares SAXS curves computed from the 4D2G structure (without termini) computed with WAXSiS and with several implicit-solvent methods. All methods used default settings, and the calculated curves were not fitted against the experiment. Evidently, the predicted SAXS curves differ significantly, presumably owing to different modeling of the hydration layer and excluded solvent.

4. Theory

The experimental SAS intensity is given by the difference between the scattering intensity of the solution $I_{\text{sample}}(q)$ and the solvent $I_{\text{solvent}}(q)$:

$$I(q) = I_{\text{sample}}(q) - I_{\text{solvent}}(q) \tag{3}$$

To calculate this excess intensity $I(q)$ from MD trajectories, the WAXSiS method follows the formalism by Park et al. (2009). The low-dilution limit is considered, such that correlations between different solute molecules can be neglected and the scattering experiment can be modeled by a single solute molecule in a solvent bath. A spatial envelope is constructed around the solute including the hydration layer (Fig. 1A). The instantaneous electron density of the solute system $A(\mathbf{r})$ and of the solvent system $B(\mathbf{r})$ is divided into the electron density inside and outside of the envelope, indicated by subscripts i and o, respectively,

$$A(\mathbf{r}) = A_{\text{i}}(\mathbf{r}) + A_{\text{o}}(\mathbf{r}), \tag{4}$$

$$B(\mathbf{r}) = B_{\text{i}}(\mathbf{r}) + B_{\text{o}}(\mathbf{r}). \tag{5}$$

Assuming that the envelope is large enough such that density correlations between the inside and the outside of the envelope are due to bulk water, the excess scattering intensity only requires knowledge of the Fourier transforms of the electron densities *inside* the envelope $\tilde{A}_{\text{i}}(\mathbf{q})$ and $\tilde{B}_{\text{i}}(\mathbf{q})$:

$$I(q) = \langle D(\mathbf{q}) \rangle_{\Omega}, \tag{6}$$

where

$$D(\mathbf{q}) = \left\langle |\tilde{A}_i(\mathbf{q})|^2 \right\rangle^{(\omega)} - \left\langle |\tilde{B}_i(\mathbf{q})|^2 \right\rangle^{(\omega)} + 2\,\mathrm{Re}\left[-\left\langle \tilde{B}_i^*(\mathbf{q}) \right\rangle^{(\omega)} \left\langle \tilde{A}_i(\mathbf{q}) - \tilde{B}_i(\mathbf{q}) \right\rangle^{(\omega)} \right]. \tag{7}$$

Here, $\langle \ldots \rangle_\Omega$ denotes the orientational average in $\mathbf{q}$-space, $\langle \ldots \rangle^{(\omega)}$ the average over solute and solvent fluctuations at fixed solute orientation ω, and Re is the real part. In practice, $\langle \ldots \rangle^{(\omega)}$ represents the average over MD frames after superimposing the solute onto a reference structure of orientation ω. $\tilde{A}_i(\mathbf{q})$ and $\tilde{B}_i(\mathbf{q})$ are calculated using the atomic form factors $f_j(q)$ and coordinates $\mathbf{r}_j$ of atom j,

$$\tilde{A}_i(\mathbf{q}) = \sum_{j=1}^{N_A} f_j(q) e^{-i\mathbf{q}\cdot\mathbf{r}_j}, \tag{8}$$

where N_A is the number of atoms within the envelope in the respective MD frame (Fig. 1A). $\tilde{B}_i(\mathbf{q})$ is calculated analogously over the solvent atoms within the envelope (Fig. 1B). The form factors are approximated using the Cromer–Mann parameters a_k, b_k, and c of atom j (Cromer & Mann, 1968),

$$f_j(q) = \sum_{k=1}^{4} a_k e^{-b_k (q/4\pi)^2} + c. \tag{9}$$

The formalism is identical for SANS calculations, except that the atomic form factors $f_j(q)$ are replaced by the coherent neutron scattering lengths b_j of the scattering atoms (Chen et al., 2019; Dias-Mirandela et al., 2018).

4.1 Comparing to experimental data

To compare a predicted curve with experimental data, the experimental curve I_{exp} can be fitted to the calculated curve I_c by minimizing

$$\chi^2 = N_q^{-1} \sum_{i=1}^{N_q} \frac{\left[I_c(q_i) - \left(f I_{\mathrm{exp}}(q_i) + c \right) \right]^2}{\sigma_i^2}, \tag{10}$$

where N_q denotes the number of q-points, and σ_i are the experimental errors. Besides the overall scale f, an offset c may be fitted to account for small experimental uncertainties from the buffer subtraction. Here, the experimental curves are fitted instead of the calculated curves, because the calculated curves are predicted without any free parameters. In practice, it is advisable to fit also without the offset c and, thereby, to test whether c absorbs only a small constant offset, as desired, or whether fitting the offset would hide a structurally relevant discrepancy between I_{exp} and I_c.

5. Workflow

We present a brief workflow for computing a SAS curve from an explicit-solvent MD simulation (Fig. 5). Because the SAS calculations are implemented into an extension of GROMACS, it is convenient (but not required) to run the MD simulation with GROMACS as well. This

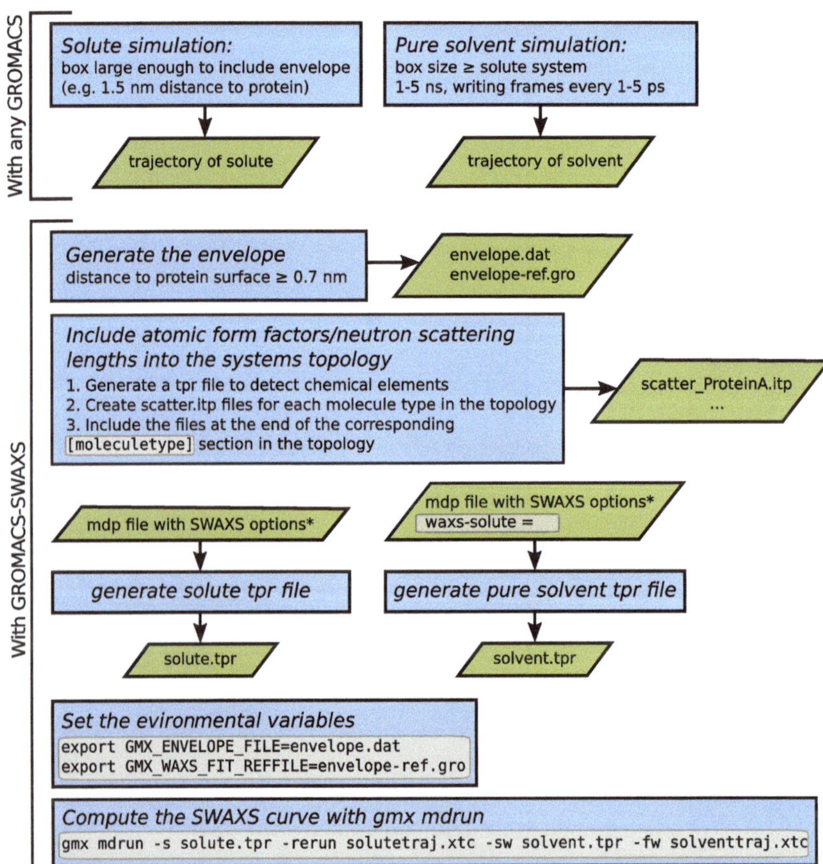

Fig. 5 Workflow for SAS curve predictions with GROMACS-SWAXS. *Suggested mdp parameters are shown in Listing 1.

workflow is automated also on the web server WAXSiS (https://waxsis.uni-saarland.de). However, the MD simulations on WAXSiS are carried out with the Yasara software which allows fully automated MD simulations even in presence of ligands or modified amino acids (Krieger & Vriend, 2015). Several additional tutorials and more detailed documentation for GROMACS-SWAXS are available at https://cbjh.gitlab.io/gromacs-swaxs-docs.

5.1 A: With any GROMACS

1. Run an MD simulation of your biomolecule in solution with either GROMACS-SWAXS or any other GROMACS version. Choose the simulation box large enough, such that the envelope (see below) will fit into the box. To generate the box use, for example,

```
gmx editconf -d 1.5 -f protein.gro -bt dodecahedron -o box.gro
```

   where for the `-d` option a distance between `1.5` and `2.0` is usually sufficient.
2. Run an MD simulation of pure solvent for the buffer subtraction. Use the same water model and MD parameter (mdp) file as used for the solute simulations to ensure that solute and solvent simulations exhibit identical bulk solvent densities, as required for an accurate buffer subtraction. The solvent simulation box should be at least as large as your solute simulation box. Produce 1 to 5 ns of simulation, writing the simulation frames every 1 to 5 ps, that is, ~1000 frames.

5.2 B: With GROMACS-SWAXS

When the simulations of solute and pure solvent have finished, we continue with preparing and running the SAS curve calculation. Once the steps below are understood, they can be scripted to automate the SAS curve calculation.

1. Download and compile GROMACS-SWAXS (https://gitlab.com/cbjh/gromacs-swaxs), and load it into your path:

```
source /path/to/gromacs-swaxs/bin/GMXRC
```

5.2.1 Specify the atomic form factors/neutron scattering lengths

2. Generate a run-input (tpr) file using `gmx grompp` of GROMACS-SWAXS to detect the chemical elements, which are used below for

assigning the Cromer–Mann parameters or neutron scattering lengths. Chemical elements are recognized using atom names and masses to ensure that a C_α carbon atom (C) is not confused with calcium (Ca), Fluor (F) not with Iron (Fe), etc. Atomic masses are reliably provided by a tpr file. An empty mdp file may be used for this step:

```
rm -f anymdp.mdp; touch anymdp.mdp
gmx grompp -f anymdp.mdp -p topol.top -o tmp.tpr
```

3. Now, the previously created tpr file is used to generate a `scatter.itp` file for each molecule, which contains the Cromer–Mann parameters or the neutron scattering lengths. If the solute contains nondefault groups, such as chromophore, a heme group, or similar, first prepare an index file with the *entire* solute. Then run

```
gmx genscatt -s tmp.tpr [-n index.ndx] [-vsites]
```

Select your *complete* solute, such as Protein, Protein_HEME or Protein_Chromophore. If you use virtual sites, use `-vistes` and select Prot-Masses (or a group with heme, the chromophore, etc.). The `-visites` option instructs `gmx genscatt` that atomic masses deviate from the physical masses, as common when using virtual sites. `gmx genscatt` writes one itp file for each molecule type. These must be added to each moleculetype definition in the topology, similar to the `#include "posres.itp"` line for defining position restraints.

5.2.2 Generating the envelope

4. Generate the envelope from the trajectory of the solute simulations

```
gmx genscatt -s tmp.tpr -f solutetraj.xtc -d 0.7
```

The output `envelope.dat` lists inner and outer radii of the envelope surface, whereas `envelope-ref.gro` is a reference structure with orientation ω used to superimpose the solute frames onto the envelope (see Theory). The envelope may be visualized with PyMol with the file `envelope.py`.

5.2.3 Generating the solute and solvent run-input files

5. Generate tpr files with `gmx grompp` and with an mdp file that contains all SAS-specific parameters, as presented in Listing 1.

```
gmx grompp -f rerun.mdp -c solutesystem.gro
\[-n index.ndx] -o solute.tpr
```

In addition, a tpr file of the pure-solvent system is required, prepared with a mdp file with an empty solute entry (`waxs-solute =`)

LISTING 1 A typical set of SAS-specific mdp parameters for calculating a SWAXS/SANS curve with GROMACS-SWAXS. The calculation of SANS curves is optional, as indicated by square brackets.

```
; read scattering info from topology
define = -DSCATTER
; turn on SAXS calc. and, optionally, multiple SANS calc.
scatt-coupl = xray [neutron neutron ...]
; solute group
waxs-solute = Protein ; or Protein-Masses
; solvent group
waxs-solvent = Water_and_ions
; rotational fit group as used with gmx genenv
waxs-rotfit = C-alpha
; or use a PBC atom near the geometric center,
; as suggested by gmx genenv
waxs-pbcatom = -1
; for uniform average over all simulation frames
waxs-tau = 0
; number of q-points
waxs-nq = 101
; qmin and qmax in nm^(-1)
waxs-startq = 0
waxs-endq = 10
; number of q-vectors for spherical average,
; use ~0.2*(D*qmax)^2
waxs-nsphere = 1500
; experimental solvent density in e/nm3, for solvent density
; correction
waxs-solvdens = 334
; Use I(sample)-I(buffer) as buffer subtraction scheme
waxs-correct-buffer = no
; D2O concentrations of SANS calculations
waxs-deuter-conc = [0.42 1]
```

```
gmx grompp -f solvent.mdp -c solventsystem.gro -o solvent.tpr
```

5.2.4 Compute the SAS curve

6. Finally, compute the SAS curves from the solute simulations with the rerun functionality of `gmx mdrun`. The envelope files are specified with environment variables, as follows:

```
export GMX_ENVELOPE_FILE=envelope.dat
export GMX_WAXS_FIT_REFFILE=envelope-ref.gro
gmx mdrun -s solute.tpr -rerun solutetraj.xtc
         \-sw solvent.tpr -fw solventtraj.xtc
```

The SAS calculations strongly benefit from the use of a GPU or from using multiple CPU cores, for instance using `gmx mdrun -ntomp 16 -gpu_id 0` .... The SAS curves are written to the files `waxs_final.xvg`.

6. Practical considerations

6.1 Convergence

The number of simulation frames required to receive a converged SAS curve depends on the contrast between solute and solvent, where a lower contrast leads to slower convergence. The contrast is lower if the hydration layer is large compared to the biomolecule, that is, if the envelope contains mostly water. Specifically, small biomolecules or intrinsically disordered proteins (IDPs) require a larger number of simulation frames as compared to larger or globular biomolecules. As a reference, for a smaller protein such as the GB3 domain, approximately 300 frames are required to obtain a well-converged SAS curve. A larger protein such as glucose isomerase requires approximately 70 frames. The statistical errors listed in `waxs_final.xvg` are computed with error propagation, assuming that each simulation frame provides an independent solvent structure, which is fulfilled when using frames with time spacing $\gtrsim 1$ ps.

For IDPs, the convergence should be carefully assessed, for instance by dividing the trajectory of *both* the biomolecule and the buffer into ten independent blocks, and then comparing the ten SAS curves computed from the blocks. The standard error over the blocks provides a rigorous measure for statistical uncertainty.

6.2 Orientational average

The parameter J (mdp option waxs-nsphere) determines the number of **q**-vectors used to take the orientational average. It should be taken as $J \approx \alpha(Dq_{max})^2$. Where D is the maximum diameter of the solute, q_{max} is the maximum momentum transfer (mdp option waxs-endq), and the constant α determines the accuracy of the orientational average. For most biomolecules, $\alpha = 0.2$ is sufficient, while for excessively elongated rod-like structures, a value of $\alpha = 0.5$ may be required.

6.3 Envelope size

The distance of the envelope from the biomolecular surface (specified with gmx genenv -d) should be at least 0.6 to 0.7 nm, thereby containing the density modulations along the hydration layers. For highly charged biomolecules, a larger envelope may be used to account for the entire counter ion cloud. We found previously that a distance of ~3 times the Debye length is required to include most effects from the counter ion cloud on the radius of gyration of glucose isomerase (Ivanović, Bruetzel, Shevchuk, Lipfert, & Hub, 2018). At a salt concentration of 100 mM, this may require an envelope distance of ~3 nm and, hence, large simulation systems.

6.4 Solvent density correction

A solvent density correction is necessary for two reasons: (i) Due to finite-size effects, the bulk solvent densities and correlations may slightly differ between the solute simulation and the pure-solvent simulation (Köfinger & Hummer, 2013). (ii) The density of certain water models, such as the popular Tip3p model, deviates from the experimental water densities. The parameter for solvent density correction should be chosen to match the experimental electron density of the solvent (mdp option `waxs-solvdens`).

6.5 Atomic fluctuations

Structural fluctuations influence the SAS profile. Depending on the scientific question, position restraints may or may not be applied to the biomolecule, for instance with the aim to obtain the SAS curve for structure with given backbone coordinates or for a flexible biomolecular ensemble, respectively. To resolve the structure of the HL, it may be useful to restrain all heavy atoms of the biomolecule to avoid that side chain fluctuations smear out the HL structure, as used to obtain the solvent densities shown in Fig. 3.

6.6 Water models

We validated previously that most popular water force fields lead to nearly identical SAS curves at moderate scattering angles, suggesting that most water models exhibit similar packing at the biomolecular surface (Chen & Hub, 2014). Only at very wide angle of $q \gtrsim 2$ Å^{-1}, where the water scattering is dominant, large effects from the water models are visible. However, we recently observed that water models with increased dispersion interactions such as the Tip4p2005s or Tip4p-D may impose tighter packing on the protein surface, leading to slightly increased radii of gyration (unpublished data) (Best, Zheng, & Mittal, 2014; Piana, Donchev, Robustelli, & Shaw, 2015). These observations open a novel route for validating protein–water interactions by modern force fields against experimental SAS data.

6.7 Computational costs

Run times of SAS predictions depend on the number of atoms, on the applied convergence criteria, and on the hardware. Typical costs may be illustrated by the execution time for different proteins on the WAXSiS webserver at https://waxsis.uni-saarland.de, which is currently (by August 2022) equipped with 16-core AMD Ryzen 9 processor and an Nvidia RTX 3070Ti graphics card. Using the convergence setting "normal," which is suitable for tests, SAS predictions of lysozyme (PDB id 1LYS), glucose isomerase (PDB id 1MNZ), and RNA polymerase II (PDB id 1TWF) required 2.7min, 4.8min, and 15.2 min, respectively. Using the convergence setting "thorough," which is recommended for publications, the execution times for the same biomolecules increased to 10.8min, 14.2 min, and 36.3 min, respectively. These numbers may change as the hardware improves or as the convergence criteria are adapted in future releases of the WAXSiS webserver.

7. Summary

Both implicit- and explicit-solvent SAS prediction methods are highly valuable for the structural interpretation of experimental SAS methods. Implicit-solvent methods are computationally efficient, thus allowing high-throughput calculations on a laptop. They are unequivocally capable of validating the overall shape of structural models against SAS data, and to reveal effects from larger conformational transitions. To harvest structural details, as encoded in high-precision SAS data collected from modern

SEC-SAXS or SEC-SANS experiments, explicit-solvent methods are needed. They are (i) based on an accurate representation of the hydration layer and the excluded solvent, thereby avoiding solvent-related fitting parameters, (ii) are accurate for inhomogeneous systems since they do not require tabulated atomic volumes, and (iii) include effects from structural fluctuations on SAS curves. The WAXSiS method presented here may be further used for structure refinement simulations by restraining an MD simulation to experimental SAS data (Chen & Hub, 2015 a; Chen et al., 2019). More details on structure refinement are provided in a chapter of Part B of this monograph.

Running a SAS prediction on the WAXSiS webserver requires between 3 and 40 min including the MD simulation. On a fast workstation or a computer cluster, tens to hundreds of explicit-solvent calculations are feasible within days, suggesting that the cost for explicit-solvent SAS calculations is negligible as compared to the costs for experimental sample preparation or for data collection at the beamline. Therefore, we expect that the methods presented here will become a routinely used tool for SAS-based structural biology.

Acknowledgments

We thank Tobias Fischbach for support with preparing Fig. 2A. This study was supported by the Deutsche Forschungsgemeinschaft (HU 1971-3/1).

References

Arnlund, D., Johansson, L. C., Wickstrand, C., Barty, A., Williams, G. J., Malmerberg, E., … Neutze, R. (2014). Visualizing a protein quake with time-resolved X-ray scattering at a free-electron laser. *Nature Methods*, *11*(9), 923–926.

Bardhan, J., Park, S., & Makowski, L. (2009). SoftWAXS: A computational tool for modeling wide-angle X-ray solution scattering from biomolecules. *Journal of Applied Crystallography*, *42*(5), 932–943.

Bernetti, M., & Bussi, G. (2021). Comparing state-of-the-art approaches to back-calculate SAXS spectra from atomistic molecular dynamics simulations. *European Physical Journal B*, *94*(9), 180. https://doi.org/10.1140/epjb/s10051-021-00186-9.

Berthaud, A., Manzi, J., Pérez, J., & Mangenot, S. (2012). Modeling detergent organization around Aquaporin-0 using small-angle X-ray scattering. *Journal of the American Chemical Society*, *134*(24), 10080–10088.

Best, R. B., Zheng, W., & Mittal, J. (2014). Balanced protein–water interactions improve properties of disordered proteins and non-specific protein association. *Journal of Chemical Theory and Computation*, *10*(11), 5113–5124. https://doi.org/10.1021/ct500569b.

Brinkmann, L. U. L., & Hub, J. S. (2016). Ultrafast anisotropic protein quake propagation after CO-photodissociation in myoglobin. *Proceedings of the National Academy of Sciences of the United States of America*, *113*, 10565–10570.

Chen, P.-c., & Hub, J. S. (2014). Validating solution ensembles from molecular dynamics simulation by wide-angle X-ray scattering data. *Biophysical Journal*, *107*, 435–447.

Chen, P.-c., & Hub, J. S. (2015a). Interpretation of solution X-ray scattering by explicit-solvent molecular dynamics. *Biophysical Journal, 108*, 2573–2584.
Chen, P.-c., & Hub, J. S. (2015b). Structural properties of protein-detergent complexes from SAXS and MD simulations. *The Journal of Physical Chemistry Letters, 6*, 5116–5121.
Chen, P.-c., Shevchuk, R., Strnad, F. M., Lorenz, C., Karge, L., Gilles, R., ... Hub, J. S. (2019). Combined small-angle X-ray and neutron scattering restraints in molecular dynamics simulations. *Journal of Chemical Theory and Computation, 15*(8), 4687–4698. https://doi.org/10.1021/acs.jctc.9b00292.
Cordeiro, T. N., Chen, P.-c., De Biasio, A., Sibille, N., Blanco, F. J., Hub, J. S., ... Bernadó, P. (2017). Disentangling polydispersity in the PCNA-p15PAF complex, a disordered, transient and multivalent macromolecular assembly. *Nucleic Acids Research, 45*, 1501–2015.
Cromer, D. T., & Mann, J. B. (1968). X-ray scattering factors computed from numerical Hartree-Fock wave functions. *Acta Crystallographica. Section A, Foundations of Crystallography, 24*(2), 321–324.
De Biasio, A., de Opakua, A. I., Mortuza, G. B., Molina, R., Cordeiro, T. N., Castillo, F., ... Blanco, F. J. (2015). Structure of p15PAF–PCNA complex and implications for clamp sliding during DNA replication and repair. *Nature Communications, 6*(1), 6439. https://doi.org/10.1038/ncomms7439.
Dias-Mirandela, G., Tamburrino, G., Ivanović, M. T., Strnad, F. M., Byron, O., Rasmussen, T., ... Javelle, A. (2018). Merging in-solution X-ray and neutron scattering data allows fine structural analysis of membrane-protein detergent complexes. *The Journal of Physical Chemistry Letters, 9*, 3910–3914.
Franke, D., Petoukhov, M. V., Konarev, P. V., Panjkovich, A., Tuukkanen, A., Mertens, H. D. T., ... Svergun, D. I. (2017). *ATSAS 2.8*: A comprehensive data analysis suite for small-angle scattering from macromolecular solutions. *Journal of Applied Crystallography, 50*(4), 1212–1225. https://doi.org/10.1107/S1600576717007786.
Fraser, R. D. B., MacRae, T. P., & Suzuki, E. (1978). An improved method for calculating the contribution of solvent to the X-ray diffraction pattern of biological molecules. *Journal of Applied Crystallography, 11*(6), 693–694.
Gabel, F. (2015). Small-angle neutron scattering for structural biology of protein–RNA complexes. In *Methods in enzymology: Vol. 558* (pp. 391–415). Elsevier.
Graewert, M. A., & Svergun, D. I. (2013). Impact and progress in small and wide angle X-ray scattering (SAXS and WAXS). *Current Opinion in Structural Biology, 23*(5), 748–754.
Grishaev, A., Guo, L., Irving, T., & Bax, A. (2010). Improved fitting of solution X-ray scattering data to macromolecular structures and structural ensembles by explicit water modeling. *Journal of the American Chemical Society, 132*, 15484–15486.
Grudinin, S., Garkavenko, M., & Kazennov, A. (2017). Pepsi-SAXS: An adaptive method for rapid and accurate computation of small-angle X-ray scattering profiles. *Acta Crystallographica. Section D, Structural Biology, 73*(5), 449–464.
Hub, J. S. (2018). Interpreting solution X-ray scattering data using molecular simulations. *Current Opinion in Structural Biology, 49*, 18–26.
Ivanović, M. T., Bruetzel, L. K., Lipfert, J., & Hub, J. S. (2018). Temperature-dependent atomic models of detergent micelles refined against small-angle X-ray scattering data. *Angewandte Chemie International Edition, 57*, 5635–5639.
Ivanović, M. T., Bruetzel, L. K., Shevchuk, R., Lipfert, J., & Hub, J. S. (2018). Quantifying the influence of the ion cloud on SAXS profiles of charged proteins. *Physical Chemistry Chemical Physics, 20*(41), 26351–26361. https://doi.org/10.1039/C8CP03080D.
Ivanović, M. T., Hermann, M. R., Wójcik, M., Pérez, J., & Hub, J. S. (2020). Small-angle X-ray scattering curves of detergent micelles: effects of asymmetry, shape fluctuations, disorder, and atomic details. *The Journal of Physical Chemistry Letters, 11*(3), 945–951. https://doi.org/10.1021/acs.jpclett.9b03154.
Kim, H. S., Martel, A., Girard, E., Moulin, M., Härtlein, M., Madern, D., ... Gabel, F. (2016). SAXS/SANS on supercharged proteins reveals residue-specific modifications of the hydration shell. *Biophysical Journal, 110*(10), 2185–2194.

Knight, C. J., & Hub, J. S. (2015). WAXSiS: A web server for the calculation of SAXS/WAXS curves based on explicit-solvent molecular dynamics. *Nucleic Acids Research*, *43*, W225–W230.

Koch, M. H. J., Vachette, P., & Svergun, D. I. (2003). Small-angle scattering: A view on the properties, structures and structural changes of biological macromolecules in solution. *Quarterly Reviews of Biophysics*, *36*(02), 147–227.

Köfinger, J., & Hummer, G. (2013). Atomic-resolution structural information from scattering experiments on macromolecules in solution. *Physical Review. E, Statistical, Nonlinear, and Soft Matter Physics*, *87*, 052712.

Krieger, E., & Vriend, G. (2015). New ways to boost molecular dynamics simulations. *Journal of Computational Chemistry*, *36*(13), 996–1007. https://doi.org/10.1002/jcc.23899.

Levantino, M., Schiro, G., Lemke, H. T., Cottone, G., Glownia, J. M., Zhu, D., … Cammarata, M. (2015). Ultrafast myoglobin structural dynamics observed with an X-ray free-electron laser. *Nature Communications*, *6*, 6772.

Liu, H., Morris, R. J., Hexemer, A., Grandison, S., & Zwart, P. H. (2012). Computation of small-angle scattering profiles with three-dimensional Zernike polynomials. *Acta Crystallographica. Section A, Foundations of Crystallography*, *68*(2), 278–285.

Merzel, F., & Smith, J. C. (2002a). Is the first hydration shell of lysozyme of higher density than bulk water? *Proceedings of the National Academy of Sciences of the United States of America*, *99*, 5378–5383.

Merzel, F., & Smith, J. C. (2002b). SASSIM: A method for calculating small-angle X-ray and neutron scattering and the associated molecular envelope from explicit-atom models of solvated proteins. *Acta Crystallographica. Section D, Biological Crystallography*, *58*, 242–249.

Moore, P. B. (2014). The effects of thermal disorder on the solution-scattering profiles of macromolecules. *Biophysical Journal*, *106*, 1489–1496.

Nguyen, H. T., Pabit, S. A., Meisburger, S. P., Pollack, L., & Case, D. A. (2014). Accurate small and wide angle X-ray scattering profiles from atomic models of proteins and nucleic acids. *The Journal of Chemical Physics*, *141*(22), 22D508.

Oroguchi, T., Hashimoto, H., Shimizu, T., Sato, M., & Ikeguchi, M. (2009). Intrinsic dynamics of restriction endonuclease EcoO109I studied by molecular dynamics simulations and X-ray scattering data analysis. *Biophysical Journal*, *96*, 2808–2822.

Oroguchi, T., & Ikeguchi, M. (2012). MD–SAXS method with nonspherical boundaries. *Chemical Physics Letters*, *541*, 117–121.

Park, S., Bardhan, J. P., Roux, B., & Makowski, L. (2009). Simulated X-ray scattering of protein solutions using explicit-solvent models. *The Journal of Chemical Physics*, *130*, 134114. https://doi.org/10.1063/1.3099611.

Piana, S., Donchev, A. G., Robustelli, P., & Shaw, D. E. (2015). Water dispersion interactions strongly influence simulated structural properties of disordered protein states. *The Journal of Physical Chemistry B*, *119*(16), 5113–5123. https://doi.org/10.1021/jp508971m.

Poitevin, F., Orland, H., Doniach, S., Koehl, P., & Delarue, M. (2011). AquaSAXS: A web server for computation and fitting of SAXS profiles with non-uniformally hydrated atomic models. *Nucleic Acids Research*, *39*, W184–W189.

Pollack, L. (2011). Saxs studies of ion–nucleic acid interactions. *Annual Review of Biophysics*, *40*, 225–242.

Pontius, J., Richelle, J., & Wodak, S. J. (1996). Deviations from standard atomic volumes as a quality measure for protein crystal structures. *Journal of Molecular Biology*, *264*(1), 121–136.

Putnam, C. D., Hammel, M., Hura, G. L., & Tainer, J. A. (2007). X-ray solution scattering (SAXS) combined with crystallography and computation: Defining accurate macromolecular structures, conformations and assemblies in solution. *Quarterly Reviews of Biophysics*, *40*(3), 191–285. https://doi.org/10.1017/S0033583507004635.

Putnam, D. K., Weiner, B. E., Woetzel, N., Lowe, E. W., & Meiler, J. (2015). BCL::SAXS: GPU accelerated Debye method for computation of small angle X-ray scattering profiles. *Proteins: Structure, Function, and Bioinformatics*, *83*(8), 1500–1512.

Ravikumar, K. M., Huang, W., & Yang, S. (2013). Fast-SAXS-pro: A unified approach to computing SAXS profiles of DNA, RNA, protein, and their complexes. *The Journal of Chemical Physics*, *138*(2), 01B609.

Schneidman-Duhovny, D., Hammel, M., & Sali, A. (2010). FoXS: A web server for rapid computation and fitting of SAXS profiles. *Nucleic Acids Research*, *38*, W540–W544.

Schneidman-Duhovny, D., Hammel, M., Tainer, J. A., & Sali, A. (2013). Accurate SAXS profile computation and its assessment by contrast variation experiments. *Biophysical Journal*, *105*(4), 962–974.

Stovgaard, K., Andreetta, C., Ferkinghoff-Borg, J., & Hamelryck, T. (2010). Calculation of accurate small angle X-ray scattering curves from coarse-grained protein models. *BMC Bioinformatics*, *11*(1), 429.

Svergun, D., Barberato, C., & Koch, M. H. J. (1995). CRYSOL–a Program to Evaluate X-ray Solution Scattering of Biological Macromolecules from Atomic Coordinates. *Journal of Applied Crystallography*, *28*, 768–773.

Tiede, D. M., Zhang, R., & Seifert, S. (2002). Protein conformations explored by difference high-angle solution x-ray scattering: Oxidation state and temperature dependent changes in cytochrome c. *Biochemistry*, *41*(21), 6605–6614.

Tjioe, E., & Heller, W. T. (2007). Ornl_sas: Software for calculation of small-angle scattering intensities of proteins and protein complexes. *Journal of Applied Crystallography*, *40*(4), 782–785.

Traube, J. (1895). Ueber Das Molekularvolumen. *Berichte der Deutschen Chemischen Gesellschaft*, *28*(3), 2722–2728. https://doi.org/10.1002/cber.18950280378.

Virtanen, J. J., Makowski, L., Sosnick, T. R., & Freed, K. F. (2011). Modeling the hydration layer around proteins: Applications to small-and wide-angle x-ray scattering. *Biophysical Journal*, *101*(8), 2061–2069.

Yang, S., Park, S., Makowski, L., & Roux, B. (2009). A rapid coarse residue-based computational method for X-ray solution scattering characterization of protein folds and multiple conformational states of large protein complexes. *Biophysical Journal*, *96*(11), 4449–4463.

CHAPTER SIXTEEN

From dilute to concentrated solutions of intrinsically disordered proteins: Sample preparation and data collection

Samuel Lenton[a,b], Mark D. Tully[c], and Marie Skepö[b,d,*]

[a]Drug Delivery and Biophysics of Biopharmaceuticals, Department of Pharmacy, University of Copenhagen, Copenhagen, Denmark
[b]Division of Theoretical Chemistry, Lund University, Lund, Sweden
[c]BioSAXS beamline, BM29, European Synchrotron Radiation Facility, ESRF, Grenoble, France
[d]LINXS—Institute of Advanced Neutron and X-ray Science, Lund, Sweden
*Corresponding author: e-mail address: marie.skepo@teokem.lu.se

Contents

Abstract

It is well-known that an increasing proportion of proteins, protein regions, and partners of globular proteins are being recognized as having an intrinsic disorder, and therefore, not adopting a single three-dimensional structure in solution. For these proteins, small-angle X-ray scattering (SAXS) has become a premier method for examination, since it can provide information about the ensemble of the structural conformations as well as the intermolecular interactions. SAXS measurements can be performed from low to high protein concentrations under different physicochemical properties of the solution. The focus of this chapter is to introduce the basics of how to use SAXS for protein samples, for new and less experienced users, in a simple and concise manner,

Methods in Enzymology, Volume 677
ISSN 0076-6879
https://doi.org/10.1016/bs.mie.2022.08.036

with emphasis on highly flexible proteins and regions. Methodological aspects in the sample preparation, experiment design, and data collection stages are raised that should be considered prior to attempting SAXS experiments. This is to ensure that high-quality SAXS data is obtained that enables accurate analysis. However, many of the points raised will also be worth considering for SAXS experiments of globular proteins.

1. Introduction

Determination of the three-dimensional structure of proteins has long been considered as an important step in understanding the mechanism through which they accomplish their physiological function. The relationship between protein structure and function can be easily understood in the case of globular proteins where, for example, specific solvent-exposed pockets of residues introduce a structurally and thermodynamically favorable region through which specific ligands can interact with the protein (Blake et al., 1965). In the case of globular proteins, there are a wide variety of experimental methods available that can be used to obtain high-resolution structural information such as X-ray/neutron crystallography, nuclear magnetic resonance (NMR), and cryo-electron microscopy (cryo-EM). The basic philosophy underpinning these studies is that the function of a protein is defined by the specific spatial organization of the amino acid side chains, and that by determining this organization, more insight can be gained into the mechanism of proteins.

In more recent years it has become clear that a significant proportion of proteins, or regions of proteins, do not adopt a well-defined three-dimensional structure in solution. This class of proteins is known as intrinsically disordered proteins, IDPs, or regions, IDRs, from now on denoted IDPs. In contrast to globular proteins, IDPs are highly flexible proteins that rapidly sample an ensemble of structural conformations (Uversky, 2021). Hence, the precise function and mechanism of the IDPs cannot be understood by a single fixed protein conformation (Schiro et al., 2015), it should rather be interpreted as an ensemble of spatially distinct conformations (Fisher & Stultz, 2011). Yet significant insight can be gained regarding the function of IDPs by determining the properties of the ensemble of structural conformations. The heterogeneric nature of IDPs makes them less amiable to traditional high-resolution structural techniques, such as X-ray crystallography and cryo-EM, due to the requirement that the protein is fixed in either a crystal surface or upon a grid. Instead, solution-based techniques such as NMR and small-angle X-ray scattering (SAXS), (Bernado & Svergun, 2012; Gomes et al., 2020; Kikhney & Svergun, 2015; Sibille & Bernado, 2012) are valuable.

In many traditional biophysical methods, the sample is probed under conditions that do not accurately represent intracellular conditions, for example, when densely packed within a crystal structure or frozen on a grid. These limitations are often not hugely problematic when studying the structure of globular proteins, however, in the case of IDPs, non-physiological conditions may result in vastly different results compared to *in vivo* conditions (Theillet et al., 2014; Yang et al., 2020). The highly flexible nature of IDPs, combined with a high content of charged amino acid side chains, results in a sensitivity of the structural ensemble by both excluded volume effects of the crowding molecules as well as charge-charge interactions between the protein and the crowding molecules (Miller, Kim, & Mittal, 2016; Qin & Zhou, 2013). Many modern experimental methods strive to represent *in vivo* cellular conditions more accurately. This may include performing *in cellular* or *in vitro* experiments in the presence of crowding molecules or other biologically relevant co-solutes (Theillet et al., 2016). More specifically, the concentration of biological macromolecules can reach as high as 400 mg/mL *in vivo* (Zimmerman & Minton, 1993; Zimmerman & Trach, 1991).

The presence of crowding molecules has been observed to modify the structural properties of IDPs, characterized by, for example, a change in radius of gyration, R_g, and induce effects such as liquid-liquid phase separation (Qin & Zhou, 2013). In recent years solution SAXS has become an increasingly popular technique to study the structure and the conformation of IDPs, under both dilute and concentrated conditions, including the effects of crowding molecules (Bonucci et al., 2021; Goldenberg & Argyle, 2014; Johansen, Jeffries, Hammouda, Trewhella, & Goldenberg, 2011; Zhou & Bilsel, 2014). However, as is the case for all experimental methods, obtaining good experimental data requires the preparation of a high-quality sample. One downside to SAXS experiments is that all constituents of the sample scatter X-rays, this includes the protein of interest, but also buffer components, oligomeric species, and unwanted aggregates (Grishaev, 2012). Hence, the quality of the sample is crucial. In this chapter, we will briefly discuss the main principles of a typical SAXS experiment, the sample preparation, characterization prior to SAXS, and quality controls that can be performed during measurements. We will focus on both dilute and concentrated samples of IDPs, and many of the points raised will also be worth considering for SAXS experiments of globular proteins.

2. What is measured during an experiment?

During a biological SAXS measurement, a sample is exposed to X-rays, that are scattered by the sample, and the scattered intensity (I(q)) is measured as a function of the scattering angle (θ). For biological SAXS experiments, the measured solution typically consists of protein molecules suspended in a water-based buffer. When exposed to X-rays, all constituents of the solution contribute to the measured scattering pattern. The intensity of the scattered X-rays is measured as a function of the momentum transfer, denoted by q:

$$q = \frac{4\pi \sin\theta}{\lambda} \tag{1}$$

where λ is the wavelength of the incident X-rays and 2θ is the angle between the incident and scattered X-rays, see Fig. 1.

If we now consider a solution of IDPs containing N particles orientated randomly, then I(q) can be described as follows:

$$I(q) = S(q) \sum_{i}^{N} \left[(\Delta p_i V_i)^2 P(q) \right] \tag{2}$$

where i is each particle in the solution, V_i is the volume of the particle i, Δp_i is the difference between the contrast of the particle and the solvent, P(q) is the form factor, and S(q) the structure factor. Each individual component of the solution contributing toward the total I(q) is therefore dependent upon its volume and the contrast. At synchrotron sources, incident X-rays typically have a wavelength of 1 Å, which is ~12.4 KeV, though this can vary between beamlines and can be adjusted if required. During a SAXS measurement, the detector is placed close to the incident beam, at small θ, with the typical biological SAXS experiment probing the q-range of $0.004 < q < 0.5\,Å^{-1}$. The distances, D, can be calculated by the relation $D = 2\pi q^{-1}$. Applying the typical q-range for SAXS experiments, this yields distances corresponding to 1500–12.5 Å. However, these figures do not represent the resolution of the experiment in the same way as for high-resolution structural biology techniques due to the contrast. The intensity at which X-rays are scattered by an atom is dependent upon its scattering length density (SLD) which determines the scattering power of a molecule. Since X-rays interact with electrons, the SLD is related to the electron

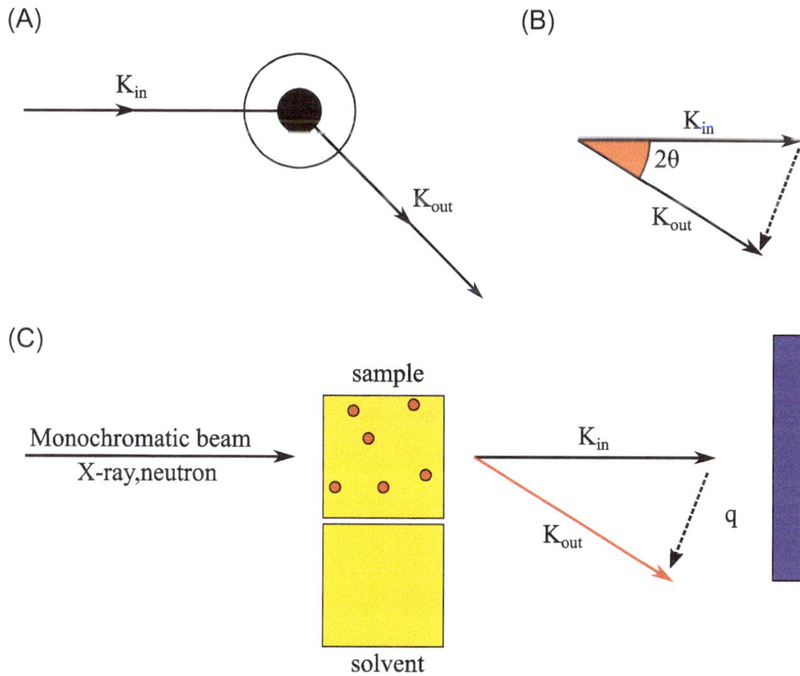

Fig. 1 Schematic overview of a typical BioSAXS experiment. (A) During a SAXS experiment, an incident X-ray beam (K_{in}) is scattered by the sample, and the resulting scattered beam (K_{out}) is recorded by a detector. (B) The scattered beam is measured as a function of the momentum transfer (q), which is, in part defined by the scattering angle (2θ). (C) During a solution SAXS experiment, scattering arises from the sample, the buffer, and the sample holder, typically a capillary. Therefore, it is important to complete an accurate background measurement consisting of both the sample holder and the buffer.

density of the constituent atoms of the scattering molecule. In very simplified terms, the electron density, and hence the SLD, of an atom increases with the atomic number. The main constituent atoms of proteins are carbon, oxygen, hydrogen, and nitrogen, which all have a relatively low electron density and SLD. Therefore, proteins have a low scattering contrast that becomes very small at higher angles, and thus at smaller distances probed. This is a limitation for the study of IDPs due to the dispersion of electron density across larger distances, compared to more globular proteins, where the electron density is concentrated within an overall smaller volume.

The experimentally measured I(q) contains contributions that originate from the protein and the background. In the case of protein solutions, the background consists of the scattering arising from the buffer components, and additionally, the vessel/container for the sample, usually a capillary. To obtain the scattering originating from the protein alone, it is therefore vital that a background measurement is performed. The latter should consist of the buffer that accurately corresponds to the buffer of the protein solution in the same capillary used to measure the sample. To obtain the final protein scattering curve, the background measurement is subtracted from the sample measurement, thereby subtracting contributions from the buffer and capillary from the sample. If this is done accurately, the total scattered intensity measured originates solely from the protein in solution, yielding the final I(q) curve, see Fig. 2.

The final buffer-subtracted I(q) contains information arising from both the structural properties of the protein and, if present, the intermolecular interactions between the proteins. The latter contributions are represented in Eq. (1) by S(q):

$$I(q) = P(q) * S(q). \tag{3}$$

P(q) is the scattered intensity that results from the average shape of the protein in the solution, while S(q) originates from intermolecular interactions between the proteins. Under dilute conditions, where no interactions between the proteins are present, S(q) is equal to 1, and the measured intensity originates purely from P(q). In the case of IDPs, P(q) does not originate from a singular protein conformation, instead, the measured P(q) is a weighted average P(q) of all distinct spatial conformations or "the ensemble" that the protein adopts during the measurement, see Fig. 3. Several different analysis methods exist that can calculate the structural ensemble of IDPs using the SAXS curve under dilute conditions. Determining either the

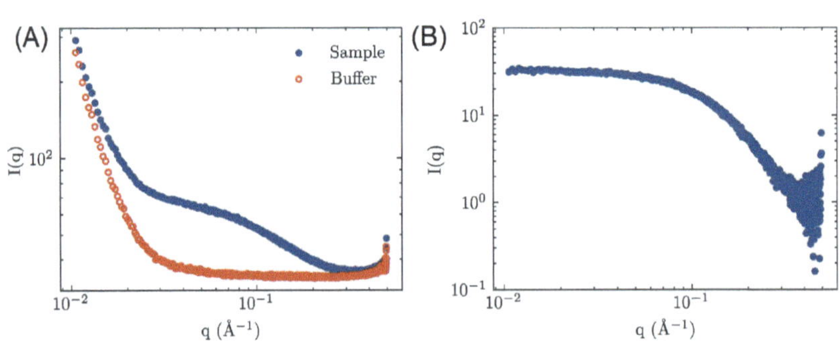

Fig. 2 Examples of experimental scattering profiles where (A) shows both the contributions from a sample and a buffer background, and (B) when the buffer background is subtracted from the sample scattering, resulting in the scattering profile of the protein.

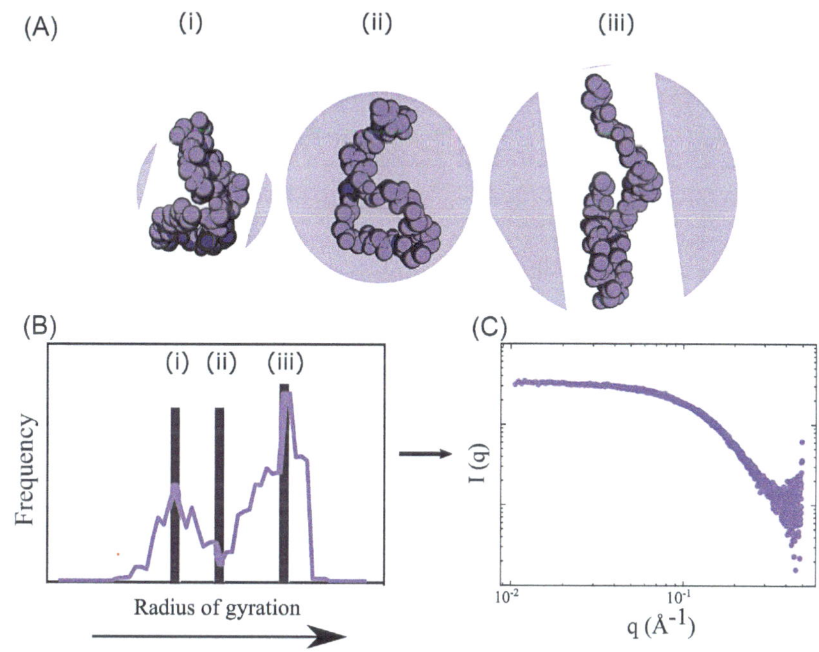

Fig. 3 The scattering curve obtained for disordered proteins arises from weighted contributions of all conformations present in the solution. (A) In solution, IDPs can adopt compact and more extended conformations (top row), which can be characterized by their radius of gyration. (B) The structural properties of IDPs can be summarized in a frequency distribution plot as a function of the radius of gyration. (C) The weighted average distribution gives rise to a single scattering curve.

scattering envelopes for globular proteins or structural ensembles for IDPs from SAXS data are not trivial processes. Less time-consuming and model-independent analysis of the obtained P(q)'s can also be completed, during or after data collection, as will briefly be discussed in the following sections.

P(q) of an IDP can be split into a low-q and a high-q regime. The former contains information about the overall size of the ensemble average of the disordered protein conformations. If present, S(q) also dominates in this region and may distort any further analysis. A simple check that can be performed to determine whether S(q) is contributing to this region is by plotting the low-q region according to the Guinier approximation:

$$\ln(I(q)) = \ln I(0) - \frac{R_g^2}{3} * q^2, \tag{4}$$

thus, a plot of ln(I(q)) against q^2, see Fig. 4. For a homogenous protein solution with no intermolecular interactions, the Guinier plot should be linear. If the Guinier region deviates to higher intensity values at lowest q, *smiling*, this indicates the presence of aggregation or net attractive interactions between proteins, while a decrease in intensity at lowest q, *frowning*, indicates repulsive interactions between proteins within the sample. The presence of S(q) can be analyzed by measuring a concentration series of the protein sample and then comparing the behavior at low-q.

The slope of the Guinier region yields R_g of the protein, which describes the distribution of mass around the center of gravity. For proteins, this is a useful parameter since it gives information about compactness. Globular proteins typically have a more compact conformation and thus a smaller R_g compared to IDPs. Generally, the maximum value of q that should be included in the Guinier plot varies between globular proteins and IDPs. The relationships, $q^*R_g < 1.3$ and $q^*R_g < 1.1$, are commonly used for globular and IDPs (Zheng & Best, 2018), respectively. Extrapolation of the linear fit to the Guinier region to $q = 0$, yields the parameter known as I(0). If no S(q) is present, the measured I(0) depends upon both the molecular mass and the concentration of the measured protein. By calibrating I(0) with known standards it is possible to determine the molecular mass of the protein in solution, and thereby determine if the sample contains monomers or oligomers, see Fig. 4.

The region of q directly following the Guinier region is known as the Porod regime and can provide an overview of the overall shape of the protein measured by plotting the SAXS intensity curve as q^2 *I(q) as a function of q, also denoted as the Kratky plot. The shape of the Kratky transform provides a qualitative assessment of the degree of macromolecular folding or compactness. The Porod exponent (PE) quantifies a power-law relationship describing the degree of foldedness versus flexibility in a sample (Rambo & Tainer, 2011). Complementary power transforms of the scattering curve by q^2, q^3, and q^4 can help detect biomolecular flexibility (Brosey & Tainer, 2019). These further developments of the Porod-Debye interpretation of flexibility and their corresponding quantitation by the PE aid an objective, quantitative assessment of molecular flexibility and compactness important for analyses of IDPs. Furthermore, transforming the scattering data differently allows the calculation of the volume-of-correlation (Vc) parameter, which is a model-independent parameter, or SAXS invariant, like the

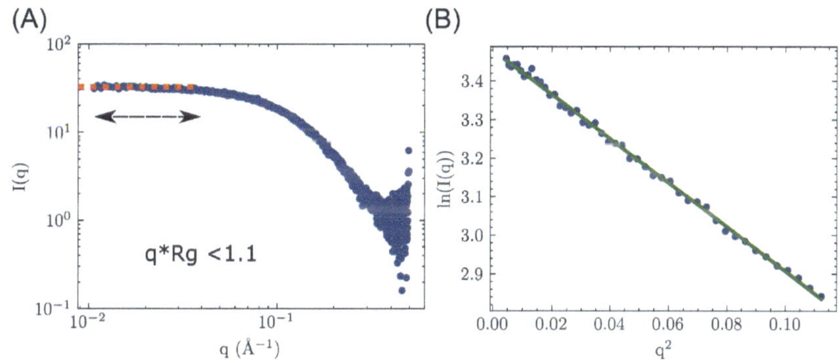

Fig. 4 Rapid analysis of SAXS data that can be completed during data collection. (A) Data at low-q can be used to obtain information about the overall size of the protein measured and determine the presence of intermolecular interactions between the proteins. (B) By plotting the data at low-q as the Guinier plot, ln(I(q)) against q^2, the radius of gyration, R_g, can be determined from the slope. Nonlinearity in the Guinier region is caused by the presence of intermolecular interactions or aggregation.

Porod invariant. Vc is calculated as a scaled ratio of particle volume (Vp) and self-correlation length (lc) and provides complementary monitoring of changes in molecular conformation for flexible systems (Rambo & Tainer, 2013). Increases in Vc reflect decreased compactness and increased flexibility.

In the Kratky plot, globular proteins show a well-defined maximum that decreases towards zero at higher q, while flexible proteins lack a well-defined maximum, and instead, the intensity increases at higher q values, see Fig. 5. For proteins that have a degree of flexibility, there will be an upturn at high-q. Thus, by plotting the Kratky representation, a general overview of the protein shape and flexibility can be obtained.

As previously noted, the measured SAXS intensity is a product of P(q) and S(q), where the former can be used to identify the contribution of the latter. This fact is often exploited for measurements of more concentrated solutions since P(q) cannot be accurately determined due to nonlinear Guinier fits. Thus, the first step in designing SAXS experiments of IDPs, irrespective of dilute or more concentrated protein solutions, should be to accurately measure P(q). Thereafter more complex measurements such as higher protein concentrations, or the presence of crowding molecules, can be performed.

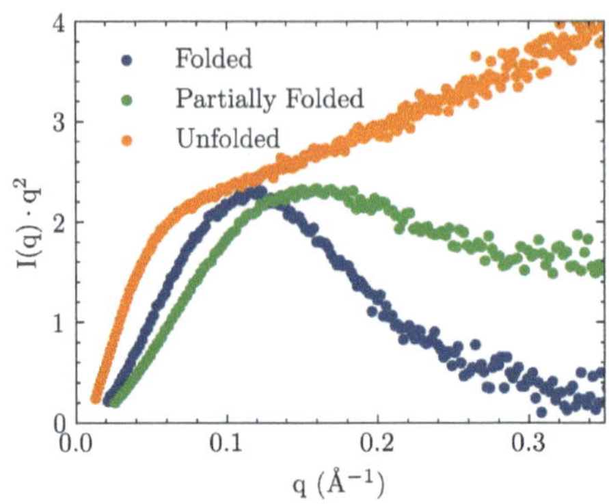

Fig. 5 Plotting the SAXS data in the Kratky format provides a fast assessment of the flexibility of the protein measured. Globular proteins are characterized by a well-defined maximum that decays to zero at higher q, whereas IDPs increase in intensity at higher q.

3. Before a measurement

From a technical point of view, performing protein-solution SAXS measurements is relatively straightforward. This is especially the case when such experiments are performed at synchrotron SAXS beamlines that are optimized for the measurements of protein samples such as BM29 BioSAXS at the European Synchrotron Radiation Facility, ESRF, in Grenoble, France. The major source of problems with performing a successful SAXS measurement is often not directly related to the measurement itself, but rather to the quality and the state of the sample measured, and the accuracy of the buffer background that is required for subtraction. Also, any large aggregates will contribute significantly more to I(q) compared to the protein of interest. When preparing a protein sample, it is important to make sure that there are no contributions from impurities that may be the result of the protein production process, i.e., co-purified proteins or other biological macromolecules. A high-quality sample should be monodispersed and well-characterized. Next follows several tips to help achieve this for dilute and concentrated IDP solutions.

3.1 Dilute conditions

One might assume that measurements at low protein concentration ensure that no intermolecular interactions between the proteins are present.

This is not always the case, even at lower protein concentrations, aggregates or oligomers may form over time. Several controls can be taken during the purification and preparation of protein samples for SAXS, to ensure that the resulting data is informative and of high-quality.

Prior to measurement

1) Buffer matching

   The scattering profile obtained from a SAXS measurement originates from all the components within the experimentally measured system. Poor buffer matching between the background measurement and the sample can lead to either under- or oversubtraction, severely limiting the resolution of the SAXS experiment. The scattering curves might appear reasonable, but analysis of the curves may result in errors when further modeling or interpreting the data. This is particularly problematic in high-q regions, where the contribution from proteins, especially IDPs, is weak. There are several methods that can be used, for example, dialysis. Here, the protein is placed in dialysis tubing, or dialysis cassettes, and is exhaustively dialyzed against the desired buffer. If left for sufficient time, the buffer in- and outside the tubing should equilibrate, where the buffer used for the final step of dialysis can be used for measurement of the background. The precise procedure for dialysis is highly dependent on the specific properties of the protein, buffer, and dialysis tubing/cassette used. In the case of IDPs, extra caution must be taken to choose the correct molecular weight cut-off (MWCO) since they can often pass through smaller pores than indicated due to the non-globular nature. Typically, a pore size equivalent to at least half of the molecular weight of the protein is adequate, however, the lower the better. A second method is to use size-exclusion chromatography (SEC) in combination with SAXS, a now ubiquitous technique on BioSAXS beamlines (Tully, Tarbouriech, Rambo, & Hutin, 2021). Here, the protein of interest is buffer exchanged to the buffer that the SEC column was equilibrated in, directly at the beamline, prior to the measurement. Finally, techniques such as salt-exchange columns, also denoted desalting columns or spin-columns, can be used to exchange the buffer. Be aware that there might be artifacts with buffer components such as glycerol being particularly difficult to match with a spin-column. There are many ways to ensure that the buffer background measured is identical to the buffer, the individual method chosen depends ultimately on the properties of the protein and the buffer.

2) Size exclusion chromatography

During protein purification, a final step including SEC is generally advised. This will remove potential protein containments arising from the protein expression, alongside any larger aggregate species. The presence of any containments should be checked before SEC by using basic molecular biology methods such as sodium dodecyl sulfate–polyacrylamide gel electrophoresis (SDS-PAGE), or more advanced mass-spectrometry experiments. Performing SEC experiments in conjunction with multiangle laser light scattering (SEC-MALLS) can also provide valuable additional information on the presence of oligomeric species i.e., dimers, trimers, etc. formed by the protein, and give an accurate molecular weight measurement.

3) Dynamic light scattering

Dynamic light scattering (DLS) can be used to determine the polydispersity of the sample after purification and monitoring the stability of the protein solution over time. DLS is also non-denaturing and requires a low concentration of protein.

4) Filtration/Centrifugation prior to SAXS

Directly prior to any SAXS measurement, larger protein aggregates that may be present can be easily removed by large MWCO spin filters, which are easy to perform and takes a relatively short time. Note that the compatibility of specific filtration membranes should be tested prior to using the sample for SAXS, for example, by measuring the concentration of the sample before and after filtration. Centrifugation is ubiquitous before any SAXS measurement. Depending on the size of the aggregates, any time between 5 and 30 min may be necessary. Ideally, ultracentrifugation will be used though this is not always available.

3.2 Concentrated conditions

One of the biggest challenges when working with highly concentrated protein solutions is the increase of viscosity, which could be due to reversible self-association driven by charge-charge and hydrophobic interactions between proteins. Since these interactions are protein-dependent, the precise viscosity of a protein solution cannot necessarily be predicted, and the properties of the buffer such as ionic strength, pH, and presence of excipients also play a role. The increased viscosity makes the samples more difficult to handle using traditional laboratory techniques, for example, pipetting and when filling capillaries. Another challenge is the increased density of the solution, which can result in a significant change of volume/concentration as well as cause problems with common characterization techniques, i.e., determination of concentration by UV–Vis (ultraviolet-visible) absorption. For the latter, this can be circumvented by diluting an aliquot of the sample and multiplying the obtained concentration by the dilution factor. A third, less obvious problem, is the formation of foam or bubbles during sample preparation, if these persist then they can disrupt the measurement, here centrifugation is useful.

As highlighted above, the preparation of high concentration protein solutions can be more complex than low concentration solutions. However, the first two stages are the same: ensuring the purity of the sample and obtaining a matching buffer for background measurement. Additional steps are then added to the sample preparation phase to achieve the target high concentration. There are two main methods, the first involves dialyzing the protein against a volatile buffer, or water, and then lyophilizing into a powdered sample. The resulting powder can then be resuspended in a fixed volume of buffer to yield the final concentration. Please notice that local concentration gradients could be induced that potentially promote the formation of aggregates. Furthermore, many proteins do not respond well to lyophilization, and extra care must be taken during resuspension to limit excessive foaming, which as such, can lead to further protein aggregation. The second, and preferred method is to start with a dilute protein solution and increase the concentration using centrifugal filters. This is a time-consuming process and thus should be tested well in advance of any measurements.

4. Performing a measurement

SAXS experiments can be performed either at in-house X-ray sources or at synchrotron facilities. For higher concentrations, quality data can be obtained at modern in-house SAXS setups. Due to the inherently low SLD of protein solutions, it is important to maximize the signal to noise of the samples during measurements. Therefore, performing experiments at a synchrotron beamline specialized in BioSAXS measurements should be considered, such as the BM29 beamline at the ESRF, see Fig. 6. These facilities benefit from the high flux of the synchrotron storage ring resulting in an increased signal-to-noise compared to in-house facilities. Most modern synchrotron BioSAXS beamlines also have specialized sample environments or data collection methods available for the study of

Fig. 6 Overview of the BM29 BioSAXS beamline at the ESRF, in Grenoble, France. The beamline is optimized for the collection of SAXS data from biological samples. Several data-collection methods are available including on-line SEC and the BioSAXS robot.

proteins/IDPs in solution by SAXS. The available hardware differs by beamline, the most frequently used methods for the measurement of protein-solution SAXS are automated BioSAXS robots, SEC-SAXS, and static capillary measurements. The latter method is often used in in-house SAXS setups, however, in recent years advances have resulted in that the former two options are also becoming more common (Bucciarelli et al., 2018; Inoue et al., 2019). Determining the correct data collection method depends upon both the sample and the aim of the experiment. The advantages and disadvantages of the three main data collection methods, most used for protein-solution SAXS experiments, will next be briefly discussed.

4.1 Size exclusion chromatography in combination with SAXS

SEC-SAXS has become an increasingly popular technique to obtain spectra of monodisperse protein samples (Graewert et al., 2020) and to reduce aggregate formation over time (Inoue et al., 2019). Furthermore, SEC-SAXS allows for the investigation of oligomeric species and complexes between proteins, even at low affinities. In principle, every peak apparent on the SEC column can be recorded. Due to the dilution effect, higher concentrations are usually required in comparison to the standard automated batch measurements. Recent advances, for example, the liquid-metal jet, have made these types of experiments feasible to perform on modern in-house SAXS instruments (Bucciarelli et al., 2018) as well.

Here, small volumes of protein (typically 20–100 μL) are loaded onto, and run over, a SEC column, while continuously collecting SAXS data from the elution. During the latter, the sample travels over a small dead volume between the end of the column and the incident X-ray beam, thereby minimizing the time available for any potential aggregate formation. During the whole column run, SAXS and UV–Vis absorption are constantly measured. This data is usually plotted together, resulting in profiles such as those shown in Fig. 7. Typically, the intensity shown for the SAXS experiments corresponds to the sum of the radially averaged intensity per frame. This technique can be somewhat time-consuming as data collection takes the time required for a column run, approximately between 10 and 60 min, depending on the size and type of the column. By measuring the buffer frames prior to the void volume of the respective column, and/or after the solvent front of the column, a well-matching buffer background can be obtained. Prior to running a SEC-SAXS experiment, the column should be well equilibrated, which corresponds to approximately three to five column volumes of buffer unless the buffer contains viscous components such as glycerol; then this should at least be doubled to tripled. Due to the dilution effect, the best quality data is obtained for the higher protein concentrations. Hence, a rule of thumb is to maximize the amount of protein. If the dilution effect is still too large, the protein can be loaded onto the column in a buffer in which the protein is more soluble, for example for IDPs, it is possible to

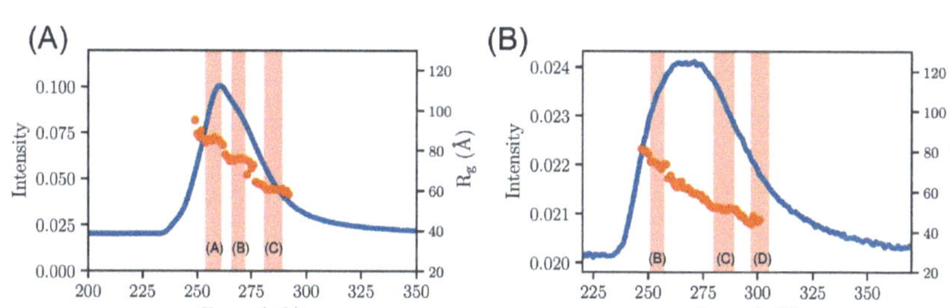

Fig. 7 Example of SEC-SAXS spectra showing the UV-Vis absorption (blue lines) and radius of gyration (R_g) determined from the elution of the column (orange dots). In (A and B) the same samples injected into the column at different protein concentrations are shown. Even though only one peak is apparent by UV–Vis absorption, there are several species present within the peak. Upon diluting the sample, the lowest R_g species becomes apparent, belonging to the monomer of this protein, as shown in (B).

include the addition of a denaturant such as guanidinium (Riback et al., 2017). Please notice, that this is only reasonable if the IDPs lack structural features. When using this strategy care must be taken when selecting the frames for buffer subtraction.

When performing SEC-SAXS experiments, the link between I(q), flow rate, elution volume, and concentration must also be considered. A measurement usually runs at a high flow rate to decrease the time needed per sample. There is a trade-off though, the higher the flow rate, the less time the relevant peak is exposed to X-rays. Hence, by decreasing the flow rate, an increased exposure of the eluted protein can be obtained, which may increase the data quality. However, this may also result in radiation damage if the exposure time is too high. Therefore, the precise protocol and type/volume of column used for a specific SEC-SAXS experiment is dependent on several factors. Ideally, the behavior of the protein on the selected SEC column should be well-characterized in-house prior to performing SEC-SAXS measurements to obtain the highest data quality.

In summary, SEC-SAXS is a useful data collection method to obtain P(q) of protein in solutions and, in some instances, study protein oligomerization and complex formation between proteins. Due to the on-column dilution effect of the loaded sample during the column run, this technique is less useful for the measurement of protein at higher concentrations, or for measuring the effects of concentration through a concentration series.

4.2 Automated BioSAXS robot: Batch measurements

The increased beam intensity at modern synchrotron facilities combined with the increased read-out speed of modern X-ray detectors has massively decreased the exposure time required to obtain high-quality SAXS data from protein solutions. The time for data collection has become so short, that manual sample changing between measurements constitutes a larger amount of time compared to the time required for sample exposure prior to radiation damage. This fact has led to the development of automated BioSAXS sample changer robots, such as those developed by, among others, the European Molecular Biology Laboratory, EMBL (Classen et al., 2013). Hence, it is now possible to collect SAXS data on many protein solutions in a relatively short timeframe. Furthermore, because BioSAXS sample changers automate the whole process of data collection, it is possible to obtain consistent and replicable measurements between individual samples.

Typically, the user loads their samples into a 96-well plate or 8-well PCR strip, which is then placed in the robot. At modern synchrotron SAXS beamlines, the typical volume required for each sample is on the order of 30–50 μL. The BioSAXS robot automatically loads and collects SAXS data of each sample and the corresponding buffer background. The major advantage of using a BioSAXS robot is that the user simply loads the samples into the robot after which no further entry into the experimental hutch is required. This significantly reduces the overhead required for loading samples. An additional benefit is that the sample and the buffer are measured in an identical capillary that is automatically cleaned between each measurement. When determining P(q), it is recommended to perform low concentration measurements at several different concentrations, if no S(q) is present, these curves should overlap. The lack of concentration-dependent change in the measured I(q) is a good indicator that the $S(q) = 1$, and hence, P(q), has been obtained. A downside of using BioSAXS robots and other systems that depend on flowing the sample is that solutions containing a high concentration of either protein or crowding agent have an increased viscosity. Hence, the flow of the sample through the robot is reduced. For these conditions, one should then consider the more traditional, manual, method of loading and measuring capillaries.

4.3 Static capillary measurements

Due to their longer measurement time, and therefore less need for automation, many in-house sources still use static capillaries for the measurement of protein-solution SAXS samples. These capillaries must be manually filled with samples and then placed in the instrument prior to each measurement, removed after each measurement, and cleaned in-between measurements. This is time-consuming and requires significant manual labor. The advantage of the system is that an increased viscosity can be more easily handled. At some synchrotron beamlines, static capillaries, or other devices, are available for viscous measurements (Edwards-Gayle et al., 2021).

Static measurements also pose another problem, especially if carried out at synchrotron beamlines. Radiation damage can cause server denaturation and aggregation in protein samples. The more X-rays or flux density a sample absorbs, the higher is the probability of radiation damage. To mitigate this effect, automated robots and SEC-SAXS, both flow the sample during measurement, reducing the time that a sample is exposed to X-rays.

This is not the case for static measurements. In all cases of SAXS, the addition of excipients may reduce radiation damage, such as ascorbic acid (Jeffries, Graewert, Svergun, & Blanchet, 2015) or sucrose (Kuwamoto, Akiyama, & Fujisawa, 2004), but these should always be tested with the protein prior to the beamtime to see if they react favorably or not.

5. After a measurement

During, or after, a SAXS measurement, the quality of P(q) can be controlled by several different methods. It is prudent to collect SAXS data at different low protein concentrations to rule out the contribution of any S(q). The precise concentration range depends upon the protein of interest and the specifications of the SAXS setup used for the measurement. A general suggestion for SAXS data collection at a modern generation synchrotron beamline is a protein concentration range spanning 1–10 mg/mL. If there is no contribution from S(q), then the resulting experimental scattering curves should superimpose, where special attention should be focused on the low-q scattering. The presence of S(q) would be evident by deviation with increasing concentration in the low-q region, which can easily be observed by plotting the SAXS data as a Guinier plot.

6. Summary and conclusions

An understanding of both the structural and conformational properties as well as the intermolecular interactions between IDPs can provide insight into their biological function. Recent developments for both in-house SAXS sources, and at specialized BioSAXS beamlines, have resulted in a range of SAXS techniques to precisely measure such properties, such as determination of P(q) at low protein concentrations, determination of oligomeric states, monodispersity/polydispersity, or probing the intermolecular interactions at higher protein concentration. The versatility of SAXS measurements provides a method through which these properties can be studied at a range of concentrations. However, obtaining good-quality data that can be accurately interpreted depends upon several factors, as noted in Fig. 8. In the sample preparation phase, the main goal should be to obtain pure and well-characterized samples and the corresponding matching buffer background. One should then carefully consider what type of SAXS

Sample preparation

Step1. Objective
- Ensure sample purity
- Obtain monodisperse samples
- Characterize samples

Step 2. Example methods
- SEC, i.e., separate the sample according to size
- Dialysis i.e., buffer exchange
- DLS, i.e., size and size distribution
- UV-Vis i.e., concentration determination

Step 3. Impact
- Obtain sample with matching buffer
- Where required, sample free from structure factor
- Determine the type of SAXS experiment

Measurement determination

Step 4. Objective
- Determine the goal of the experiment
- Investigate available instruments and methods
- Investigate sample environments

Step 5. Example methods
- BioSAXS robot
- Capillary SAXS
- SEC-SAXS
- More advanced environments

Step 6. Impact
- Determine in-house or synchrotron
- Determine a suitable data collection method
- Determine the quantity of samples required

Fig. 8 Summary of recommended steps to take prior to performing a SAXS experiment. Sample preparation: In this stage, it is vital to understand the physical properties of the protein. For low protein concentration measurements, several experimental methods can be used to ensure that a monodisperse sample is obtained and to characterize the sample. For higher concentration measurements care must be taken when preparing the sample. Furthermore, a key condition for obtaining good SAXS data is the matching of the background measurement with the sample buffer, which can be assured via, e.g., extensive dialysis. Measurement determination: Before collecting SAXS data in-house one should consider the goal of the SAXS experiments. Once this has been determined, it should be decided whether an in-house source is sufficient, or whether the data should be collected at more advanced BioSAXS beamlines. In the latter case, the sample environments at the beamline should be investigated in advance and the most suitable one selected.

measurement best suits the objectives of the experiment. Based on this, a decision can be made to complete the experiments in-house or at a specialized BioSAXS beamline. The latter often has a range of sample environments that can be utilized, the choice of which again depends upon the objective of the experiment. During the measurement, several analyses of the collected data can be completed, yielding information that can influence decisions of ongoing measurements, and provide some initial parameters describing the studied solution.

Acknowledgments

We acknowledge financial support from the Crafoord Foundation, Sweden. The authors would like to thank the European Synchrotron Radiation Facility (ESRF) for the use of beamline BM29 and for the support over the years.

References

Bernado, P., & Svergun, D. I. (2012). Analysis of intrinsically disordered proteins by small-angle X-ray scattering. *Methods in molecular biology (Clifton, N.J.)*, *896*, 107–122. https://doi.org/10.1007/978-1-4614-3704-8_7.

Blake, C. C. F., Koenig, D. F., Mair, G. A., North, A. C. T., Phillips, D. C., & Sarma, V. R. (1965). Structure of hen egg-white lysozyme: A three-dimensional Fourier synthesis at 2 Å resolution. *Nature*, *206*(4986), 757–761. https://doi.org/10.1038/206757a0.

Bonucci, A., Palomino-Schatzlein, M., Malo de Molina, P., Arbe, A., Pierattelli, R., Rizzuti, B., et al. (2021). Crowding effects on the structure and dynamics of the intrinsically disordered nuclear chromatin protein NUPR1. *Frontiers in Molecular Biosciences*, *8*. https://doi.org/10.3389/fmolb.2021.684622.

Brosey, C. A., & Tainer, J. A. (2019). Evolving SAXS versatility: Solution X-ray scattering for macromolecular architecture, functional landscapes, and integrative structural biology. *Current Opinion in Structural Biology*, *58*, 197–213. https://doi.org/10.1016/j.sbi.2019.04.004.

Bucciarelli, S., Midtgaard, S. R., Pedersen, M. N., Skou, S., Arleth, L., & Vestergaard, B. (2018). Size-exclusion chromatography small-angle X-ray scattering of water soluble proteins on a laboratory instrument. *Journal of Applied Crystallography*, *51*, 1623–1632. https://doi.org/10.1107/s1600576718014462.

Classen, S., Hura, G. L., Holton, J. M., Rambo, R. P., Rodic, I., McGuire, P. J., et al. (2013). Implementation and performance of SIBYLS: A dual endstation small-angle X-ray scattering and macromolecular crystallography beamline at the advanced light source. *Journal of Applied Crystallography*, *46*, 1–13. https://doi.org/10.1107/s0021889812048698.

Edwards-Gayle, C. J. C., Khunti, N., Hamley, I. W., Inoue, K., Cowieson, N., & Rambo, R. (2021). Design of a multipurpose sample cell holder for the diamond light source high-throughput SAXS beamline B21. *Journal of Synchrotron Radiation*, *28*, 318–321. https://doi.org/10.1107/s1600577520013831.

Fisher, C. K., & Stultz, C. M. (2011). Constructing ensembles for intrinsically disordered proteins. *Current Opinion in Structural Biology*, *21*(3), 426–431. https://doi.org/10.1016/j.sbi.2011.04.001.

Goldenberg, D. P., & Argyle, B. (2014). Minimal effects of macromolecular crowding on an intrinsically disordered protein: A small-angle neutron scattering study. *Biophysical Journal*, *106*(4), 905–914. https://doi.org/10.1016/j.bpj.2013.12.003.

Gomes, G.-N. W., Krzeminski, M., Namini, A., Martin, E. W., Mittag, T., Head-Gordon, T., et al. (2020). Conformational ensembles of an intrinsically disordered protein consistent with NMR, SAXS, and single-molecule FRET. *Journal of the American Chemical Society*, *142*(37), 15697–15710. https://doi.org/10.1021/jacs.0c02088.

Graewert, M. A., Da Vela, S., Graewert, T. W., Molodenskiy, D. S., Blanchet, C. E., Svergun, D. I., et al. (2020). Adding size exclusion chromatography (SEC) and light scattering (LS) devices to obtain high-quality small angle X-ray scattering (SAXS) data. *Crystals*, *10*(11). https://doi.org/10.3390/cryst10110975.

Grishaev, A. (2012). Sample preparation, data collection, and preliminary data analysis in biomolecular solution X-ray scattering. *Current Protocols in Protein Science*. https://doi.org/10.1002/0471140864.ps1714s70. *Chapter 17*, Unit17.14-Unit17.14.

Inoue, R., Nakagawa, T., Morishima, K., Sato, N., Okuda, A., Urade, R., et al. (2019). Newly developed laboratory-based size exclusion chromatography small-angle x-ray scattering system (La-SSS). *Scientific Reports*, *9*. https://doi.org/10.1038/s41598-019-48911-w.

Jeffries, C. M., Graewert, M. A., Svergun, D. I., & Blanchet, C. E. (2015). Limiting radiation damage for high-brilliance biological solution scattering: Practical experience at the EMBL P12 beamline PETRAIII. *Journal of Synchrotron Radiation*, *22*, 273–279. https://doi.org/10.1107/s1600577515000375.

Johansen, D., Jeffries, C. M. J., Hammouda, B., Trewhella, J., & Goldenberg, D. P. (2011). Effects of macromolecular crowding on an intrinsically disordered protein characterized by small-angle neutron scattering with contrast matching. *Biophysical Journal, 100*(4), 1120–1128. https://doi.org/10.1016/j.bpj.2011.01.020.

Kikhney, A. G., & Svergun, D. I. (2015). A practical guide to small angle X-ray scattering (SAXS) of flexible and intrinsically disordered proteins. *FEBS Letters, 589*(19), 2570–2577. https://doi.org/10.1016/j.febslet.2015.08.027.

Kuwamoto, S., Akiyama, S., & Fujisawa, T. (2004). Radiation damage to a protein solution, detected by synchrotron X-ray small-angle scattering: Dose-related considerations and suppression by cryoprotectants. *Journal of Synchrotron Radiation, 11*, 462–468. https://doi.org/10.1107/s0909049504019272.

Miller, C. M., Kim, Y. C., & Mittal, J. (2016). Protein composition determines the effect of crowding on the properties of disordered proteins. *Biophysical Journal, 111*(1), 28–37. https://doi.org/10.1016/j.bpj.2016.05.033.

Qin, S., & Zhou, H.-X. (2013). Effects of macromolecular crowding on the conformational ensembles of disordered proteins. *Journal of Physical Chemistry Letters, 4*(20), 3429–3434. https://doi.org/10.1021/jz401817x.

Rambo, R. P., & Tainer, J. A. (2011). Characterizing flexible and intrinsically unstructured biological macromolecules by SAS using the Porod-Debye law. *Biopolymers, 95*(8), 559–571. https://doi.org/10.1002/bip.21638.

Rambo, R. P., & Tainer, J. A. (2013). Accurate assessment of mass, models and resolution by small-angle scattering. *Nature, 496*(7446), 477-+. https://doi.org/10.1038/nature12070.

Riback, J. A., Bowman, M. A., Zmyslowski, A. M., Knoverek, C. R., Jumper, J. M., Hinshaw, J. R., et al. (2017). Innovative scattering analysis shows that hydrophobic disordered proteins are expanded in water. *Science, 358*(6360), 238–241. https://doi.org/10.1126/science.aan5774.

Schiro, G., Fichou, Y., Gallat, F.-X., Wood, K., Gabel, F., Moulin, M., et al. (2015). Translational diffusion of hydration water correlates with functional motions in folded and intrinsically disordered proteins. *Nature Communications, 6*. https://doi.org/10.1038/ncomms7490.

Sibille, N., & Bernado, P. (2012). Structural characterization of intrinsically disordered proteins by the combined use of NMR and SAXS. *Biochemical Society Transactions, 40*, 955–962. https://doi.org/10.1042/bst20120149.

Theillet, F.-X., Binolfi, A., Bekei, B., Martorana, A., Rose, H. M., Stuiver, M., et al. (2016). Structural disorder of monomeric alpha-synuclein persists in mammalian cells. *Nature, 530*(7588), 45-+. https://doi.org/10.1038/nature16531.

Theillet, F.-X., Binolfi, A., Frembgen-Kesner, T., Hingorani, K., Sarkar, M., Kyne, C., et al. (2014). Physicochemical properties of cells and their effects on intrinsically disordered proteins (IDPs). *Chemical Reviews, 114*(13), 6661–6714. https://doi.org/10.1021/cr400695p.

Tully, M. D., Tarbouriech, N., Rambo, R. P., & Hutin, S. (2021). Analysis of SEC-SAXS data via EFA deconvolution and scatter. *Jove-Journal of Visualized Experiments, 167*. https://doi.org/10.3791/61578.

Uversky, V. N. (2021). Recent developments in the field of intrinsically disordered proteins: Intrinsic disorder-based emergence in cellular biology in light of the physiological and pathological liquid-liquid phase transitions. In K. A. Dill (Ed.), *50. Annual Review of Biophysics* (pp. 135–156).

Yang, P., Mathieu, C., Kolaitis, R.-M., Zhang, P., Messing, J., Yurtsever, U., et al. (2020). G3BP1 is a tunable switch that triggers phase separation to assemble stress granules. *Cell, 181*(2), 325. https://doi.org/10.1016/j.cell.2020.03.046.

Zheng, W., & Best, R. B. (2018). An extended Guinier analysis for intrinsically disordered proteins. *Journal of Molecular Biology, 430*(16), 2540–2553. https://doi.org/10.1016/j.jmb.2018.03.007.

Zhou, H.-X., & Bilsel, O. (2014). SAXS/SANS probe of intermolecular interactions in concentrated protein solutions. *Biophysical Journal*, *106*(4), 771–773. https://doi.org/10.1016/j.bpj.2014.01.019.

Zimmerman, S. B., & Minton, A. P. (1993). Macromolecular crowding: Biochemical, biophysical, and physiological consequences. *Annual Review of Biophysics and Biomolecular Structure*, *22*, 27–65. https://doi.org/10.1146/annurev.bb.22.060193.000331.

Zimmerman, S. B., & Trach, S. O. (1991). Estimation of macromolecule concentrations and excluded volume effects for the cytoplasm of Escherichia coli. *Journal of Molecular Biology*, *222*(3), 599–620. https://doi.org/10.1016/0022-2836(91)90499-v.

CHAPTER SEVENTEEN

Deriving RNA topological structure from SAXS

Xianyang Fang[a,*], José Gallego[b,*], and Yun-Xing Wang[c,*]

[a]Beijing Advanced Innovation Center for Structural Biology, School of Life Sciences, Tsinghua University, Beijing, China
[b]CITSAM, Universidad Católica de Valencia, C/Quevedo 2, Valencia, Spain
[c]Protein-Nucleic Acid Interaction Section, Center for Structural Biology, National Cancer Institute, National Institutes of Health, Frederick, MD, United States
*Corresponding authors: e-mail address: fangxy@mail.tsinghua.edu.cn; jose.gallego@ucv.es; wangyunx@mail.nih.gov

Contents

Abstract

Structures of well-folded RNA molecules can be determined with atomic resolution by either X-ray crystallography, cryo-EM, or NMR spectroscopy, but those of conformationally-flexible RNAs often are difficult to study with these methods. However, flexible RNAs have biological relevance and likely represent the majority of the RNA conformational space. Due to the high electron density of the phosphate-sugar backbone, RNA is very sensitive to small-angle X-ray scattering (SAXS), and SAXS data can be recorded with sub-μM concentrations and under near-physiological solution conditions without the need for labeling. For these reasons, SAXS has significant advantages

Methods in Enzymology, Volume 677
ISSN 0076-6879
https://doi.org/10.1016/bs.mie.2022.08.037

over other techniques for obtaining global structural information of flexible RNAs in the form of molecular envelopes or low-resolution topological structural models. The SAXS-derived information is extremely valuable for bridging secondary structure data, often determined by other techniques, with a three-dimensional structure description. In this chapter, we present a detailed account of the principle, algorithms, and experimental and computational protocols for topological structure determination of RNA molecules in solution. To illustrate the applications of the methodology, we provide several case studies that cover a broad spectrum of the RNA conformational landscape.

1. Introduction

The diverse roles of non-coding RNAs in biology have led to an increasing interest in the structural basis of these functions (Mattick & Makunin, 2006; Rinn & Chang, 2012). This often requires knowledge of the three-dimensional (3D) arrangement of RNA structural elements or domains. Low-resolution 3D topological structures are frequently sufficient to provide clues and explain RNA functions (Fang, Stagno, Bhandari, Zuo, & Wang, 2015; Lipfert, Ouellet, Norman, Doniach, & Lilley, 2008; Michel & Westhof, 1990; Zhang, Jones, & Ferre-D'Amare, 2014). However, topological structures of highly dynamic RNA molecules are not amenable for dissemination using traditional high-resolution methods like NMR spectroscopy, X-ray crystallography, or cryo-electron microscopy. Thus, different sets of techniques and methods have to be developed and utilized. In this regard, two approaches have been applied: computational prediction, or methods combining computer modeling with experimental information. A pure computational approach has had some success (Lescoute & Westhof, 2006; Parisien & Major, 2008), particularly with recent methodologies based on machine learning and neural networks (Townshend et al., 2021). However, the pure computer-based approach is limited by the fact that the current structural database is very far from complete. More importantly, the structures in the current databases were determined mostly under non-physiologically relevant conditions in order to obtain conformationally homogenous samples amenable to the techniques. This conformational homogeneity is contrary to abundant experimental evidence indicating that RNA often is highly dynamic and conformationally heterogeneous under physiologically relevant solution conditions. The challenges faced in the field of RNA structure prediction are related to the chemical makeup of RNA. It is made of only four different types of residues, sharing the same sugar and phosphate moieties and differing only in base composition. As a result, heterogeneous foldings have high free energy degeneracy and very shallow energy wells. In contrast to folded protein counterparts, folded RNA molecules show very little correlation between possible three-dimensional structures and primary sequence. Consequently, it is extremely challenging to compute RNA structural models with complex folding using a purely computational approach that relies on the limited information available in the current structural databases. Thus, combined approaches using knowledge-based modeling aided by long-range restraints generated from various types of experiments and phylogenetic comparisons might be more appropriate and promising, as shown by several successful studies (Cheng, Chou, & Das, 2015; Harris et al., 1994; Homan et al., 2014; Lehnert, Jaeger, Michel, & Westhof, 1996; Michel & Westhof, 1990; Weinreb et al., 2016; Westhof & Altman, 1994).

It was first demonstrated in the 1980s that small-angle X-ray scattering (SAXS) data contain information about the overall shape of biomacromolecules in solution (Feigin & Svergun, 1987; Glatter & Kratky, 1982). The first application of SAXS for the characterization of an RNA structure was reported in 2000 (Funari et al., 2000). Since then applying SAXS for biomolecule studies exploded after the third generation of synchrotrons became operational and a number of critical computer programs and software tools came to be available (Franke & Svergun, 2009b; Funari et al., 2000; Koch, Vachette, & Svergun, 2003; Konarev, Volkov, Sokolova, Kochb, & Svergun, 2003; Volkov & Svergun, 2003). These programs and software tools are particularly useful for reconstructing low-resolution molecular envelopes or extracting structural parameters for proteins or protein complexes. In almost the same way, SAXS has also been applied to determine structural and physical parameters of RNAs, such as radius of gyration (R_g) (Lake, 1967), maximum particle diameter (D_{max}), molecular weight, Porod volume (V_p) (Rambo & Tainer, 2010), and molecular compactness (Russell, Millett, Doniach, & Herschlag, 2000; Russell et al., 2002).

The sugar-phosphate backbone of RNA is electron-rich and, therefore, makes it a strong scatterer of X-rays than protein. Applying SAXS data for 3D topological RNA structure dissemination is feasible because RNA has the following unique advantages: (a) Secondary structures based on predictions and/or experimental determination are usually known prior to structure determination; (b) RNA structure tends to be relatively modular with a limited variety of secondary structure elements and motifs, and (c) RNA is usually non-spherical in overall shape (Zhang et al., 2014). In essence, RNA tertiary structures can be generally "approximated" with packed A-form duplexes connected by "single-stranded" linkers in junctions (de la Pena, Dufour, & Gallego, 2009; Laing, Wen, Wang, & Schlick, 2012; Lescoute & Westhof, 2006; Wang et al., 2009), interrupted by internal loops or bulges, and "capped" with terminal loops. The junctions themselves are prevalent structural elements, where intricate interactions via base stacking and base pairing among the junction residues as well as remote tertiary contacts dictate the overall RNA folding (de la Pena et al., 2009; Laing et al., 2012; Lescoute & Westhof, 2006). Major structural features such as long duplexes, which have a cross-sectional diameter of ~22 Å, are readily identified in molecular envelopes derived from SAXS data, as demonstrated in many cases, such as the Varkud satellite (*VS*) ribozyme (Hofacker et al., 1994) and the HIV-1 RRE RNA (Fang et al., 2013) or HCV domain 3′X (Cantero-Camacho, Fan, Wang, & Gallego, 2017). Consequently, SAXS has been utilized as a valuable low-resolution technique to determine the 3D topological structures of large RNAs.

2. Outline of methods and experimental strategies

Primarily, there are two basic ways to utilize SAXS for 3D topological structure determination of RNA molecules. First, *de novo* derivation of 3D topological folds simultaneously using SAXS and secondary structure information as input. Additional long-distance restraints generated from other biophysical measurements such as NMR, EPR (Electron paramagnetic resonance), smFRET (Single-molecule Förster resonance energy transfer), or chemical probing (Weeks, 2010) can also be included as input. This approach is discussed in Section 3.2. Second, retrofit RNA atomic models that comply with available secondary structure information to the SAXS-derived molecular envelopes to derive the topological structures. This approach is described in Section 3.3. For multi-domain RNA molecules, this approach requires the use of a "divide-and-conquer" strategy to identify the location of each domain or structural component within the molecular envelope, as well as a verification of the derived structure using other means, such as chemical or biochemical assays.

3. Detailed descriptions of methods and experimental strategies

3.1 RNA sample preparation for SAXS experiments

3.1.1 In vitro *transcription*

RNA can be prepared by various approaches including solid-phase chemical synthesis, enzymatic *in vitro* transcription, or recombinant overexpression in bacteria (Baronti, Karlsson, Marusic, & Petzold, 2018). While each approach has its own set of strengths and limitations, *in vitro* transcription remains the most common approach for RNA synthesis in the laboratory. *In vitro* transcription can be briefly described as the use of a defined DNA template to direct the synthesis of an RNA sequence from ribonucleotide triphosphate (rNTP) mixtures by a polymerase, typically T7 RNA polymerase. As both the rNTP mix and the polymerase can be either commercially purchased or expressed and purified "in-house," only the DNA template needs to be specifically prepared to transcribe each RNA sequence. For short RNAs, two DNA strands can be chemically synthesized, annealed and used for *in vitro* transcription (Lu & Li, 2012). For long RNAs, the cDNA sequence together with the upstream promoter can be either subcloned from an available DNA template into a suitable plasmid or directly custom gene synthesized. Using the plasmid as a template, double-stranded DNA templates containing the promoter can be further prepared by PCR at large scale for *in vitro* transcription. Alternatively, a plasmid template can be simply and cheaply propagated in *E. coli* in large quantities, then extracted and linearized with a restriction enzyme (Nilsen, Rio, & Ares, 2013) prior to transcription. For each RNA to be transcribed, the Mg^{2+} concentration in the transcription buffer usually needs to be optimized prior to large-scale preparation, and DTT needs to be freshly prepared and mixed prior to the reaction.

3.1.2 RNA purification

After *in vitro* transcription, the RNA of interest is usually purified by polyacrylamide gel electrophoresis in denaturing conditions (e.g., in the presence of 8 M urea) and extracted from the gel by electroelution or passive diffusion. The sample is then ethanol precipitated, desalted and transferred to an appropriate buffer by diafiltration or dialysis. Renaturation in the appropriate solution conditions is usually achieved thermally, by annealing or snap-cooling. This procedure usually works well for shorter RNAs for which the unfolding and refolding processes are reversible (Brunelle & Green, 2013). For long RNAs with rugged folding energy landscapes, the denaturing and renaturing processes are irreversible and can sometimes lead to misfolded RNA, so it is recommended to purify these RNA molecules using non-denaturing methods, including native polyacrylamide gel electrophoresis, size exclusion chromatography, or ion-exchange chromatography (Baronti et al., 2018).

3.1.3 *Sample conditioning and buffer matching*

3.1.3.1 Buffer composition

Salts are often useful to suppress long-range electrostatic interactions between solutes (structure factor). They also increase background and decrease solute/solvent contrast, but these effects are often negligible up to ~500 mM salt. High-Z elements should be avoided in the buffer., as they decrease contrast and promote radiation damage by increasing photo-electron production. Free radical scavengers should be included in the buffer when preparing for a synchrotron data collection as they help to minimize radiation damage. Organic buffers, such as TRIS or HEPES, are recommended, which can act as radical scavengers in addition to buffering the pH. Typical buffer concentrations are 20–100 mM. Other common choices are DTT (2–10 mM), TCEP (1–2 mM), or glycerol (~5%). Phosphate buffer should be avoided because it promotes radiation damage. Detergents are best avoided as their signal can be comparable to or exceed the RNA signal, complicating data interpretation.

3.1.3.2 Sample transfer to matching buffer

Preparation of an exactly matching buffer is the most crucial step and this can be achieved by extensive (16–48 h) dialysis or by diafiltration using an appropriate MW-cutoff filter membrane. In this latter case, several cycles of buffer loading are needed to ensure an exact match. Prior to loading the sample for dialysis or filtering, it is recommended to pass it through a 0.22 μm filter.

3.1.3.3 Sample concentration and amount

The typical sample concentration that produces a good scattering signal is 1–10 mg/mL for RNAs, depending on beamlines and molecular weight of RNA. At least three concentrations for each sample, ~1, ~3, and ~5 mg/mL, are recommended if possible, to check particle interactions and extrapolate to an "interaction-free" scattering profile. If there is only enough material for one sample and the sample is free of aggregation, we recommend using a concentration of ~2 mg/mL for RNA molecules.

The minimum volume required for scattering experiments is 160 μL. At least 20 mL of matching buffer are needed for each sample for background measurement and washing of the scattering cell.

3.1.4 Pre-examination before X-ray scattering

Aggregation is the most common problem that can render data uninterpretable. For example, the presence of proportions greater than 5% of dimer relative to monomer may affect the scattering data. Dynamic light scattering, analytical ultracentrifugation, or electrophoresis experiments can be used to detect aggregation/polydispersity before performing X-ray scattering experiments. Solution conditions should be optimized, and methods based on native gel electrophoresis (a single band is required), size exclusion chromatography, or ultracentrifugation through a high-MW cutoff membrane can be applied to obtain homogeneous samples.

3.2 Computing three-dimensional topological structures of RNA molecules based on SAXS and secondary structure

3.2.1 Basic principle and algorithm

The *ab-initio* calculation of a molecular envelope from SAXS uses homogenous dummy atoms (beads) (Svergun, 1999a). The best fit to the SAXS data is achieved by iterative reductions of the number of the dummy atoms to derive an overall shape with low resolution. There is no interaction among the dummy atoms. For rigid RNA molecules, resolution at a residue level can possibly be achieved if the random beads are replaced with globs that represent nucleotide residues that are constrained by covalent linkages between residues (globs) and a secondary structural layout (Bhandari et al., 2017). These two restraints define ranges of relative positions of residues (globs), as well as structural elements in relationship to one another within a given RNA. Such constraints are available *a-priori* (Cruz & Westhof, 2011; Fink, Chen, Noller, & Altman, 1996). Moreover, folded RNA structures are 3D arrangements of common structural motifs such as A-form duplexes connected *via* loops, linkers, and junctions, as well as helical stacking of duplexes involving junctions, many of which are well characterized and can be predicted (Cheng et al., 2015; Cruz & Westhof, 2011; Laing et al., 2012; Ramani, Qiu, & Shendure, 2015).

All of these structural features unique to RNA impose restraints on the conformational space in a *de novo* structure calculation and are the foundation for determining 3D topological structures of RNA with resolution at a residue level from SAXS data. This is achieved by applying the algorithm of

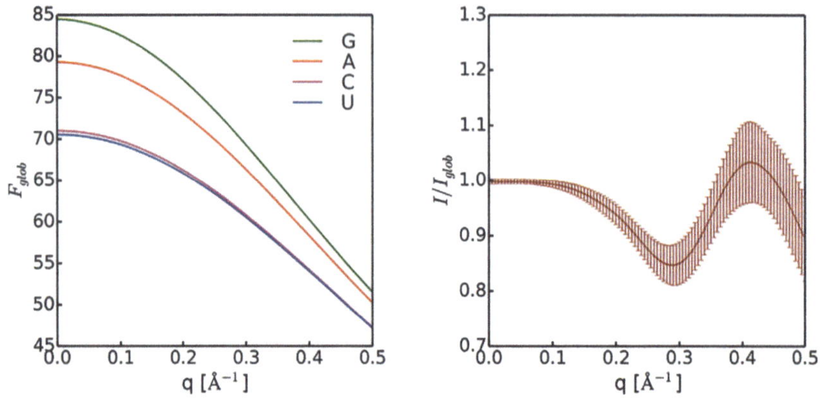

Fig. 1 Glob form factors and intensity correction factors vs momentum transfer. Left: average glob scattering form factors (F_{glob}) as a function of momentum transfer (q) for four types of nucleotides, derived from 32 different RNA structures solved by X-ray crystallography with resolution better than 3 Å. Right: mean glob correction factor (I/I_{glob}) and the associated standard deviation, as a function of q, calculated for 50 different RNA structures from the PDB. *The figure is taken from Bhandari, Y. R., Fan, L. X., Fang, X. Y., Zaki, G. F., Stahlberg, E. A., Jiang, W., et al. (2017). Topological structure determination of RNA using small-angle X-ray scattering.* Journal of Molecular Biology, 429, *3635–3649. https://doi.org/10.1016/j.jmb.2017.09.006.*

natural hierarchical moves (Fig. 1) and simulated annealing for *de novo* topological structure determination (Fig. 2A). This algorithm is implemented in program RS3D (Bhandari et al., 2017) that utilizes RNA secondary and tertiary structure information along with experimental SAXS data simultaneously for obtaining conformational folds with resolution at a nucleotide level, without resorting to any fragment replacement or initial 3D template from the database. To reduce computational loads, each residue is represented with a glob and four different types of globs correspond to A, U, G, and C residues. The form factors of the glob scattering were first generated for each residue type from the structures with a resolution higher than 3 Å in the PDB database (Fig. 1).

One characteristic of RNA structure dynamics is that an RNA secondary structural element moves as a whole (Ehrhardt & Weeks, 2020; Kazantsev et al., 2011). As such, we treat the RNA structure as an assembly of independently moving structural elements. Structural elements include duplexes, loops, and bulges. RS3D uses two types of moves, namely, global/large moves involving both rotation or translation and local/point moves involving only translation (Fig. 2). With the global moves, RS3D randomly picks a duplex or a user-defined helical segment followed by a randomly assigned

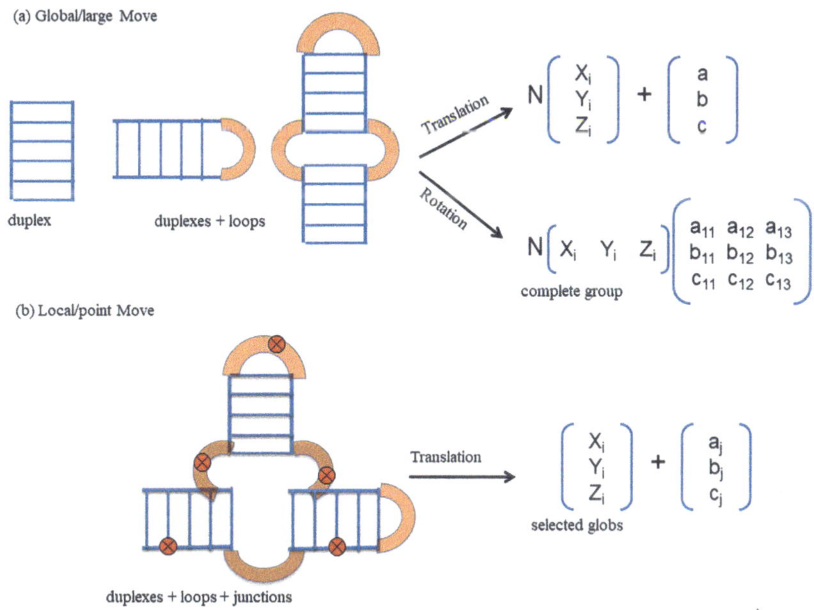

Fig. 2 Global and local moves. A global move (A) carries out a random translation or rotation of a single duplex or helical segment (duplexes and loops), or a complete structural motif such as a subdomain or domain. A local move (B) picks a group of globs (solid red circles with a cross) lying in a duplex, loop, or junction region, and displaces them through a random translation. *The figure is taken from Bhandari, Y. R., Fan, L. X., Fang, X. Y., Zaki, G. F., Stahlberg, E. A., Jiang, W., et al. (2017). Topological structure determination of RNA using small-angle X-ray scattering.* Journal of Molecular Biology, *429, 3635–3649. https://doi.org/10.1016/j.jmb.2017.09.006 with permission.*

translation or rotation. A rotation usually pivots at the first glob position, preferably close to the junction, in the helical segment and around the arbitrarily chosen x-, y-, or z-axis. For a local move, a group of globs is randomly chosen from a structural motif (duplex, loop, or junction), and translated. All rotation and translation moves are carried out in Cartesian space. Overall, the program starts with global moves, along with local moves involving only junctions for the first quart of all moves. In the next step, all global or local moves consisting of any structural motif are carried out simultaneously. All moves are directed towards lowering the value of χ^2 and at the same satisfying the topological constraints imposed by the secondary structure as well as any additional tertiary constraints as described below.

The hierarchical moves are achieved as described in the following. Every new move in the search for the target structure is directed toward the minimization of χ^2 relative to the input SAXS data (Bhandari et al., 2017):

$$\chi^2 = \frac{1}{N_q - 1} \sum_{k=1}^{N_q} \left[\frac{I_{expt}(q_k) - g_k I_{calc}(q_k)}{\sigma(q_k)} \right]^2 \tag{1}$$

where N_q represents the total number of data points, g_k is the glob correction factor, I_{calc} is the back-calculated scattering intensity of the model structure using the Debye equation, and $\sigma(q_k)$ is the experimental noise.

For each attempted move, it must satisfy acceptance criteria, or be rejected, followed by a new move. The χ^2 Metropolis test, together with the secondary and tertiary interaction constraints if present form the basis for the acceptance criteria to avoid any violation of covalent geometry and secondary structures, with the former as the primary and the latter the secondary. The criteria for the acceptance is explicitly expressed as a probability (P_A), expressed as the product of three independent probability terms, where each term must be individually satisfied for the new move to be accepted (Bhandari et al., 2017):

$$P_A = \left[P_{\chi^2}\right]\left[P_{sec-link}\right]\left[P_{ter-link}\right] \tag{2}$$

The first term, $P_{\chi 2}$, in Eq. (2) depends on the value of χ^2 calculated from Eq. (1), and is determined using the standard Metropolis algorithm (Metropolis, Rosenbluth, Rosenbluth, Teller, & Teller, 1953). The χ^2 values for the current configuration (χ_i^2) and the attempted configuration (χ_{i+1}^2) are compared. If $\chi_{i+1}^2 \leq \chi_i^2$, the new move representing a new configuration is accepted. If $\chi_{i+1}^2 > \chi_i^2$, the program selects the direction based on the Boltzmann factor, $\beta = e^{-\Delta\chi 2/\mathrm{T}}$, where $\Delta\chi 2 = \chi_{i+1}^2 - \chi_i^2$ and T is the temperature factor. The Boltzmann factor is then compared to a random number (r) in the interval 0–1, generated by the program. If $\beta > \mathrm{r}$, the move is accepted. If $\beta < \mathrm{r}$, the program rejects the move and attempts a new one. The temperature factor is scaled down by 10% every fixed number of iterations to carry out the simulated annealing. The resulting probability in a Metropolis test is expressed as (Bhandari et al., 2017):

$$P_{\chi^2} = \begin{cases} e^{-\frac{\Delta\chi 2}{\mathrm{T}}} & \textit{if } \Delta\chi 2 > 0 \\ 1 & \textit{otherwise} \end{cases} \tag{3}$$

Consistent with the characteristics of RNA structural dynamics, the amplitude of a hierarchical move heavily depends on the location of the glob. Globs located in junctions or loop regions are relatively more flexible with relatively large motion amplitudes than those in duplexes. Probability $P_{sec\text{-}link}$ is expressed as (Bhandari et al., 2017):

$$P_{sec-link} = \left[\sum_{i=1}^{N}\sum_{j=i+1}^{N} B_{i,j}^{vdw}\right]\left[\sum_{i,\,j=1}^{N_D} P_{i,j}^{duplex}\right]\left[\sum_{i=1}^{N_D}\sum_{j=1}^{N_{ND}} S_{i,j}^{steric}\right]\left[\sum_{i,\,j,\,k=1}^{N_{ND}} M_{i,j,k}^{ang}\right] \tag{4}$$

where N represents the total number of globs, N_D represents the number of globs in duplex regions, and N_{ND} represents the number of globs in non-duplex regions. Each of the four summation terms in square brackets in Eq. (4) represents a penalty score proportional to the number of violations of given secondary structure constraints. Specifically, the first term, $B_{i,j}^{vdw}$, represents the nearest neighbor interactions of any glob, with the van der Waals interaction distances greater than a minimum cutoff distance (to avoid repulsion), while the covalent interaction distances must be shorter than a maximum cutoff distance (to maintain covalent connectivity).The second term, $P_{i,j}^{duplex}$, enforces A-form duplex constraints in the duplex regions of the molecule, involving mainly characteristic distance restraints between covalently connected globs and between the base pairs. The third term $S_{i,j}^{steric}$, parameterizes the steric hindrance between a duplex and the rest of the system. The interaction distance from each glob in the duplex to each glob outside of that duplex is calculated and must be above a specified cutoff distance to return a null penalty. The last term, $M_{i,j,k}^{ang}$, is the angular constraints on the angles formed by each string of three consecutive globs in non-duplex regions. As globs lying in the non-duplex regions, such as the loops and junctions, have a greater range of motion/move, an angle penalty term is implemented to avoid excessive twisting or distortion of the RNA backbone. Every move that decreases individual penalty scores, with zero being the lowest, represented in Eq. (6) is preferred and the new value of the penalty is updated in memory for each round of the calculation. If the penalty score of the new move exceeds that of the current state by a specified threshold, the move is rejected. The implementation of the topological constraints based on secondary structure alone explicitly expressed in Eq. (4), reduces the conformational landscape of RNA to a much narrower subspace than that defined by a SAXS-derived molecular envelope.

The accuracy of such a topological structure is further improved by including approximate long-range restraints, such as semi-quantitative accessible surface probing (Homan et al., 2014) or NMR measurements (Wang et al., 2009). The program has an option that allows user input of any tertiary interaction information available from biochemical/biophysical experiments or bioinformatics studies to further reduce sampling of conformational space. The third term in Eq. (2), $P_{ter-link}$, is the probability of accepting the new conformation based on how well it achieves the tertiary interaction. The frequently encountered tertiary interactions included in the algorithm are interhelical stacking, pseudoknot formation, and other types of long-range tertiary contacts, each of which is represented, respectively, as A_j^{Hstack}, $B_j^{pseudoknot}$, and $D^{long-range}$ in the probability calculation (Bhandari et al., 2017), in the probability calculation,

$$P_{ter-link} = \sum_{i=1}^{n_H} \left[\sum_{j=1}^{N_H} A_j^{Hstack} \right]_{n_i} \sum_{i=1}^{n_{PK}} \left[\sum_{j=1}^{N_{PK}} B_j^{pseudoknot} \right]_{n_i} \left[D^{long-range} \right] \quad (5)$$

where n_H and n_{PK} are the numbers of helical stacking segments and pseudoknots, N_H and N_{PK} are the total numbers of globs in each helical stacking segment and base pairs in each pseudoknot, respectively. The first two terms in Eq. (5) represent the probabilities of achieving the interhelical stacking or pseudoknot formation. As helical stacking across the junctions is an important parameter that determines the tertiary structure (Laing et al., 2012; Lescoute & Westhof, 2006), all different possibilities of coaxial stacking at the junction region are applied exhaustively. To evaluate the contributions from interhelical stacking or pseudoknot formation, a penalty score is assigned for each based on the deviation from the tertiary interaction, which is used to guide the acceptance of the new conformation. The last term in Eq. (5) represents the probability of achieving long-range tertiary interactions between any two regions of the molecule if such a long-range contact is known. A dynamic range for the interaction distance is calculated and updated each time the long-range interaction in the new conformation is closer to the interaction distance in the target structure and improves with each iteration.

3.2.2 CPU and system requirements

Singularity is a container technology that allows the distribution of a program including all dependencies and requirements so that the user can simply use the program out of the box without the need to install or compile it. Instructions on installing singularity on a Linux system can be obtained from http://singularity.lbl.gov/install-linux. The RS3D program is computationally intensive as one of caveats and requires a Linux cluster consisting of several hundred or more CPUs to compute structural models. The resourceful users of the program may make use of computing resources available in various regional and national computing centers. For instance, an RNA of 71-nt long needs approx. 70K core hours to sample 20,000 conformations using a GCC compiler in NIH Biowulf cluster. The use of the intel compiler speeds up the computational efficiency by three times as much on the same HPC platform. It is recommended to generate a minimum pool of 10,000 to 20,000 conformations to explore the folding landscape. The program was on RHEL 6.9 and Ubuntu 14.04, which are common Linux operating systems. Python 2.7 with packages Numpy and Pandas was used for coding.

3.2.3 Procedure

The flowchart illustrating the overall procedure is shown in Fig. 3. To initiate the program, type *RS3D.py input_RNA*. The program takes initial open structures, which are generated based on secondary structures, and converts those open structures into "glob" structures, where each type of residue is represented by a corresponding glob with a defined form factor.

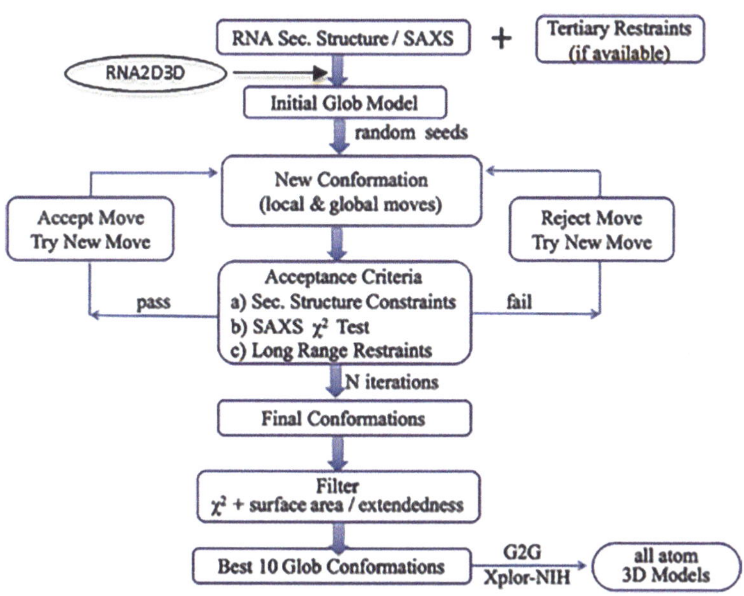

Fig. 3 Flowchart of RS3D depicting various stages of the structure calculations. All steps are carried out within the program, except for the initial open structural model, which is generated using RNA2D3D (Martinez, Maizel Jr., & Shapiro, 2008). *The figure is taken from Bhandari, Y. R., Fan, L. X., Fang, X. Y., Zaki, G. F., Stahlberg, E. A., Jiang, W., et al. (2017). Topological structure determination of RNA using small-angle X-ray scattering.* Journal of Molecular Biology, 429, *3635–3649. https://doi.org/10.1016/j.jmb.2017.09.006 with permission.*

Since the method is aimed at the derivation of low-resolution RNA topological structures, we omit consideration of hydration contribution. The secondary structure is defined in the input file called "*input_RNA*." This file also contains the sequence and tertiary contact information if available. In addition, the user also needs to provide the approximate maximum dimension of the molecule, junction type and SAXS data. The program then performs simulated annealing calculations under the SAXS restraint as well as any other user input restraints such as residue-to-residue distances. The best glob conformations are converted to explicit all-atom models and then undergo an all-atom refinement against the SAXS data. During the whole process of calculation, RS3D makes use of several accessory programs for coordinate conversion, area calculation and refinement. These are RNA 2D3D [2], G2G [3], PyMOL [4] and XPLOR-NIH [5]. Those programs are evoked seamlessly without user intervention.

We have provided example directories corresponding to four RNAs in the folder "RS3D/tests." Each of these examples (except 4MGN-ASA) contains two input files, "*input_RNA*" and "*saxs-file,*" and a python script "*RS3D.py*." In order to set up a new run, specify the path to the job directory (path to the folder where you want your results saved/) in the python script file. Look for the placeholder "self.job_dir" in the *RS3D.py* file. To run the complete workflow, simply run the command "*RS3D.py input_RNA*" in the terminal by sitting on the job directory. The fourth example (4MGN-ASA), aimed at carrying out SAXS as well as solvent accessibility refinement, contains an additional file "asa_data.dat" with HMX scores [6]. In order to set up this run, repeat the above steps but now with the python script "*RS3D-ASA.py*." If you compile RS3D with intel compiler, uncomment the line [self.processing_script = "qsub-run-icpc"] in the "*RS3D.py*" file, and uncomment the lines [self.processing_script = "qsub-run-icpc"], [self.asa_processing_script = "qsub-asa-icpc" in the "*RS3D-ASA.py*" file.

Note: The job submission command in "*RS3D.py/RS3D-ASA.py*" is based on Slurm Job System. If your HPC cluster is based on "SLURM" job submission, it should work mostly as is, but you may still need to modify some platform-dependent variables. Especially, look for the functions "generate_qsub_jobs()" and "generate_qsub_xplor_jobs()" in the python script file. You will need to figure out an equivalent batch job submission command in the line starting with "qsub_template= Template."

3.2.4 Input file format

There is one input configuration file called "*input_RNA*" that contains information on sequence, base pairing, and long-range interactions information such as pseudoknots, helical stack, and long-range contacts. The input configuration file is divided into three sections as described below for the ease of parsing. It is recommended to preserve the order of the sections for the sake of consistency in the file format. Please refer to the sample input file "*input_RNA*.pdf" while reading these instructions. Actual examples of four RNAs are included in the file "*input_examples.pdf*." The numbering of the first residue at the 5′ end starts at 1.

3.2.4.1 Section 1

This section is marked by the term [GENERAL_DESCRIPTION]. This contains the following sub-sections:

(i) SEQUENCE

(ii) RNA

In this subsection, the user needs to specify the name of the RNA. Any space in the RNA name will be stripped and the maximum length of the string is 80 characters.

(iii) NPDBS

In this subsection, the user needs to specify the maximum number of structures (conformers) per helical stack that one wants to calculate. This has to be an integer between 0 and 20000. The number of structures to be computed heavily depends on computation resources (numbers of CPUs) available to users. One can set up a quick test run with a small number of structures, for example, 100, to get a peek into the initial structural model and identify any potential issues in the workflow that can be resolved before running a full-scale calculation. This value can be set between 5000 and 20000 for a production run.

(iv) SAXS_FILE

In this subsection, the user needs to specify the name of SAXS data file. Conventional format with three columns "q_value," "intensity," and "error" is expected. The error column is optional. RS3D makes a default upper cutoff at $q = 0.5\,A^{-1}$.

(v) ASA_FILE

In this subsection, the user needs to specify the name of solvent accessibility file. This file is expected to have two columns, residue number (integer) in the first column and corresponding HMX scores (float) in the second column. You will need to include this subsection only if you are performing the solvent accessibility refinement.

(vi) DMAX

In this subsection, the user needs to specify the maximum particle dimension derived from the SAXS data. This can be an integer or a decimal number.

(vii) JUNCTION_TYPE

In this subsection, the user needs to specify the type of RNA junction as an integer number. In the case of multiple junctions, choose the largest junction type.

(viii) GENERAL_PROPERTIES

In this subsection, the user needs to specify some basic molecule information and RS3D parameters. All the specifier terms should be self-explanatory and are also introduced below.

PDBFILE: name of input pdb file at glob level. This field should not be changed as the input pdb file is generated by the workflow unless you want to provide a custom input file.

NRESIDUES: the size of RNA in terms of total residues.

VDWEXPT: specify the number of bond length exceptions to be ignored [options: integer]. This will allow some unusually long or short bond lengths in the single-stranded region. The default value is 0. It is recommended not to change unless necessary.

VDW: enable optimum distance of separation between nucleotides [options: 1, 0]. Setting this to 0 may cause artificial overlaps.

ASA: enable ASA $\chi 2$ fit [options: 1,0] for conformational sampling. Leave this to the default value [0] as this value is changed on the fly by the python workflow script. You will need to include this line only if you are performing the solvent accessibility refinement.

CHI: enable SAXS $\chi 2$ fit [options: 1,0] for conformational sampling MOVESCALE: scale factor for the size of moves. Can be left to the default value of 1. ITERATIONS: number of steps for each temperature step. A typical value of 20,000. CYCLES: at the end of every cycle, the annealing temperature is reduced by a factor of 0.9. A typical value is 60.

TEMP: annealing temperature (options: 0–1000, float type).

Qmin: minimum value of the momentum transfer vector. Default is the minimum Q-value in data. Qmax: maximum value of the momentum transfer vector. A typical value of 0.3 for glob sampling NGrids: number of grid points for interpolating the scattering intensity based on q-range.

The long-range contact information can be specified under the heading "PROXIMITY_CONSTRAINT." This will calculate the distance of separation between the centers of geometry represented by the two residue groups: GROUPA and GROUPB. Each group may contain a single residue or a group of residues.

The approximate distance (Å) information between the two groups is specified in the field "CONSTRAINT_VALUE," and the program will try to converge to this distance during the conformational sampling. There can be more than one "PROXIMITY_CONSTRAINT" field in the input file (see example RNA 1Y26).

(ix) NONPAIRED

In this section, the user specifies information on the single-stranded region such as terminal loop, internal loop, and junction.

(a) TERMINAL_LOOP and INTERNAL_LOOP

The terms TERMINAL_LOOP and INTERNAL_LOOP are immediately followed by another term "UNIT HUNIT_#," where the user specifies which helical unit the loop is a part of. A helical unit is considered to be a group of duplexes and loops forming a single unit. The number assigned to the helical unit should be an integer and start at 1. The user is expected to enter the same "UNIT HUNIT_#" information while writing the corresponding duplex in the "PAIRED" sub-section. The information on the helical unit is utilized while carrying out the large-scale moves, that transform or rotate a complete helical segment as a rigid body, which consists of internal and terminal loops as well as the duplexes.

Long internal and terminal loops may be given a separate helical unit id (see example RNAs 2M58 and 2N8V). This will help translate and rotate the long loop as a rigid body, resulting in a faster sampling of the conformational space.

(b) JUNCTIONS

Internal and terminal loops are treated equally while carrying out the moves, but junction moves are relatively more frequent as they play an important role in the assembly of the helical segments for achieving the tertiary fold. Junction does not belong to any helical unit.

(c) MY_SEGMENT

Also, users can specify a fragment of nucleotides under the heading MY_SEGMENT. This should be associated with a unique helical unit id specified by UNIT HUNIT_#. This feature is helpful if you are using a fragment template from the database and want to translate or rotate the whole segment as a rigid unit. This feature is also useful in moving a complete domain or subdomain while working with multi-domain RNAs.

(x) PAIRED

In this section, the user needs to specify the base paring information for each duplex separately under the heading "HELICAL_SEGMENT." The helical unit information "UNIT HUNIT_#" should be specified immediately below, which represents the helical unit to which the duplex belongs. The basepair information is represented by writing individual, space-delimited basepair in each line. Several duplexes may belong to the same helical unit represented by "UNIT HUNIT_#." If a short duplex (~3 bp) is connected to a longer duplex through one/two nucleotide internal loops, all of them can be kept under the same helical unit (see example RNA 2M58).

*NOTE: Lines starting with "NAME" keyword are optional and can be used for ease to keep track of the loops and duplexes (see example RNA 2M58).

3.2.4.2 Section 2

This section is marked by the term [PSEUDO_DUP]. The pseudoknot base pair information can be specified in this section under "PSEUDO_DUP#=" heading (see example RNA 2M58). There can be more than one pseudoknot specified, and the program will try to converge the pseudoknot pairs in order to achieve duplex-like base pairing between them. The pseudoknot pairs are expressed like base pairs in the duplex.

3.2.4.3 Section 3

This section is marked by the term [HELICAL_STACKS]. All possible helical stacks across the junctions can be specified under separate "HELICL_STACK_#=" headings. The first helical stack type is numbered at 1. Under each helical stack type, stacking across a junction as well as additional stacking across internal or terminal loop can also be specified, with additional lines of stacking residues (see example RNA 2M58). There will be a separate pool of conformational sampling for each helical stack type. It is recommended that the users be prudent in deciding how many helical stack types they want to run their calculation on. Mainly because helical stacks across junctions play a crucial role in determining long-range interactions and overall topological fold. Additionally, an increase in the number of helical stack types also adds significant computational overload. The best structures will be derived from a combined pool of all different types of helical stacks.

Here is a description of how to write the helical stack information. For example, if the RNA has a three-way junction with three helical segments H1, H2, and H3 around the junctions. There are three unique possibilities of stacking: between H1 and H2 (H1:H2 stack), between H2 and H3 (H2:H3 stack), and between H3 and H1 (H1:H3 stack). This translates to three types of the helical stack, namely three headings HELICAL_STACK_1, HELICAL_STACK_2, and HELICAL_STACK_3 (see example RNA 2N8V). Under each heading, one will need to write all the nucleotides that are supposed to participate in the helical stacking. one may write in an increasing or decreasing order of nucleotides, but they have to be sequential and it is important that the first three nucleotides in the list should belong to a duplex. Helical stacking is implemented in RS3D based on the distance of nucleotides relative to the position of the first three nucleotides lying in a duplex.

Information on long-range interactions available from various biophysical/biochemical data may be helpful in excluding a specific helical stack across the junction. For example, in the case of adenine riboswitch RNA, the kissing loop interaction between the terminal loops of helix H2 and H3 rules out the possibility of H2:H3 stacking.

The following is an example of the input file format.

```
PROPERTY
PDBFILE rna_ini.pdb
NGLOBS 71
VDWEXPT 0
CHI 1
ITERATIONS 20000
CYCLES 60
TEMP 5.0
PROXIMITY_CONSTRAINT
GROUPA 20 21 22 23 24 25 26
GROUPB 48 49 50 51 52 53 54
CONSTRAINT_VALUE 10.0.
PROXIMITY_CONSTRAINT
GROUPA 36 37 38 39
GROUPB 9 63
CONSTRAINT_VALUE 10.0
HELICAL_SEGMENT
NAME HSEGMENT_1
UNIT HUNIT_1
```

```
1 71
2 70
3 69
4 68
5 67
6 66
7 65
8 64
9 63
HELICAL_SEGMENT
NAME HSEGMENT_2
UNIT HUNIT_2
13 33
14 32
15 31
16 30
17 29
18 28
19 27
HELICAL_SEGMENT
NAME HSEGMENT_3
UNIT HUNIT_3
42 60
43 59
44 58
45 57
46 56
47 55
TERMINAL_LOOP
NAME TLOOP1
UNIT HUNIT_2
20 21 22 23 24 25 26
TERMINAL_LOOP.
NAME TLOOP2.
UNIT HUNIT_3.
48 49 50 51 52 53 54
JUNCTION.
10 11 12
JUNCTION.
```

```
34 35 36 37 38 39 40 41
JUNCTION.
61 62
HELICAL_STACK.
58 59 60 61 62 63 64 65 66 67 68 69 70 71
```

3.2.5 Analysis of results

(i) SAXS Refinement

After the calculation is complete, the program creates a number of files and six directories inside the given job directory, namely, /best_pdbs, /best_allatoms, /processing, /rna2d3d, /Xplor- processing, and /xplor_runs. Best 10 glob conformations are present in the folder /best_pdbs. A complete list of all glob conformations properly ranked based on the total score can be found in the file "all_stats1_norm_final.txt," and these conformations are present in the folder "/processing." Top 20 all-atom conformations with actual file path are listed in the file "top20-filter.txt" along with the SAXS χ^2 value.

(ii) Global shape

χ^2 alone generally is not sufficient to identify the best structures because of degeneracy. Consequently, we introduce two additional criteria, extendedness (E) and a minimum solvent accessible surface area (S). A well-folded RNA tends to acquire a minimum surface area. The extendedness relative to a duplex is approximated by the ratio of the maximum length D_{max} of a folded RNA vs the length of a duplex for the same number of residues (Saenger, 1984):

$$E = \frac{D_{max}}{3*(N/2)} \tag{6}$$

where N represents the total number of nucleotides present in the RNA; D_{max} is derived from the SAXS data (Bhandari et al., 2017). This semi-empirical parameter E is useful because it makes use of the information readily available from experimental SAXS and secondary structures to assess the compactness of an unknown structure. For an RNA in a duplex or a single strand, E is expected to be 1 or greater than 1, respectively. Based on a survey of 55 RNA structures in the PDB, RNA with a two-way or three-way/higher junction with $E > 0.7$ or 0.82 respectively is considered to be extended. Thus, a significant deviation in E ($E_{calc} - E_{SAXS}$) is indicative of misfolding during the calculation.

For a folded RNA, an empirical relation between solvent accessible surface area (S) and the number of residues (N) in an RNA (Eq. 7) was formulated based on a survey of the structural database (Bhandari et al., 2017).

$$S = (122^{*}N + 2990) \pm 1675 \quad (7)$$

(iii) SAXS + Solvent Accessibility Refinement

After the calculation is complete, the program should create a number of files and eight directories inside the given job directory, namely, /best_pdbs, /best_pdbs_asa, /best_allatoms, /processing, /asa_processing, /rna2d3d, /Xplor-processing, and /xplor_runs. Best 100 glob conformations based on SAXS refinement are present in the folder /best_pdbs, and these highest-ranking 100 glob conformations are further refined based on the solvent accessibility data to generate the best 10 glob conformations in the folder /best_pdbs_asa. Top 20 all-atom conformations with the actual file path are listed in file "top20-filter.txt" along with the SAXS χ^2 value.

3.3 *Ab initio* calculation of molecular envelopes, generation of atomic models and divide-and-conquer strategy for large multi-domain RNA molecules

The first use of SAXS to outline the overall shape resulted in the description of tRNA as a boomerang-shaped molecule composed of two double-helices connected by a sharp kink in 1964 by J. Witz (Witz, 2003). The rapid development of *ab-initio* 3D-shape reconstruction software, especially DAMMIN (Svergun, 1999b), made it relatively easy to derive molecular envelopes, i.e., shapes, of any monodispersed sample in solution. The first molecular envelope of an RNA generated by *ab-initio* reconstruction was the free *Thermus flavus* 5S ribosomal RNA (rRNA), based on which a putative 5S rRNA atomic model was also proposed (Funari et al., 2000). The molecular envelope derived from the SAXS data was remarkably similar to that in crystal structures. More widespread use of SAXS for RNA shape reconstruction happened 10 years later (Jones, Cantara, Olson, & Musier-Forsyth, 2014; Lipfert, Chu, Bai, Herschlag, & Doniach, 2007; Lipfert et al., 2008; Rambo & Tainer, 2010; Zhang et al., 2014; Zuo et al., 2010). The 3D molecular shapes determined using *ab-initio* reconstruction are highly valuable since they are often the only source of 3D structural information available, which is sufficient to answer some key biological questions. In addition to molecular envelopes, atomic models fitting the dispersion data are selected from ensembles of models complying with base-pair restraints obtained from secondary structure prediction software or experimental methods such as chemical mapping or NMR spectroscopy experiments. For large multi-domain RNA sequences the *ab initio* shape-reconstruction provides the mean for the SAXS-aided divide-and-conquer strategy. RNA structural architectures are highly modular and hierarchical (Cruz & Westhof, 2009). They are dominated by long helices coaxially stacked or packed orthogonally to each other. At the resolution of coarse-grained level, the shapes of isolated individual structural components or domains resemble what they look like in the intact whole RNA molecules. These structural properties provide the foundation of the divide-and-conquer approaches. This approach was first used to determine the 3D topological structure of the *VS* ribozyme, comprising 160 nucleotides (nt) (Lipfert et al., 2008); this structure was found to be consistent with chemical mapping and biological activity data.

4. Case studies

4.1 Computing 3D topological structures from SAXS data and RNA secondary structures

The complete description of the cases illustrated in this section can be found in reference (Bhandari et al., 2017).

MLV-PK RNA. This RNA is a protonation-dependent switch for the translational recoding signal in the murine leukemia virus (Houck-Loomis et al., 2011). The secondary structure of this RNA is shown in Fig. 4A. The calculation generates a pool of 10,000 conformers. This RNA is considered to be well folded based on the E score of less than 0.7. As a result, χ^2 plus the normalized surface area is used to filter the top 10 structures, which fall in the brown rectangle. The RMSDs of those top 10 structures are all within ~7 Å relative to the NMR structure (red) (Houck-Loomis et al., 2011). It is noteworthy to point out that the region with the largest RMSD is the long linker, which was also under-defined in the NMR structure. Excluding this long linker, the best structure in terms of the overall RMSD is ~3 Å.

2′–5′ AG1 lariat-forming ribozyme. 2′–5′ AG1 lariat-forming RNA is an artificial ribozyme that shares the sequence specificity of a lariat formation mimicking the pre-mRNA splicing reaction. This RNA consists of a pseudoknot that is connected to a hairpin *via* a four-residue linker (Fig. 4B). The relative positions between the pseudoknot and the hairpin are not well

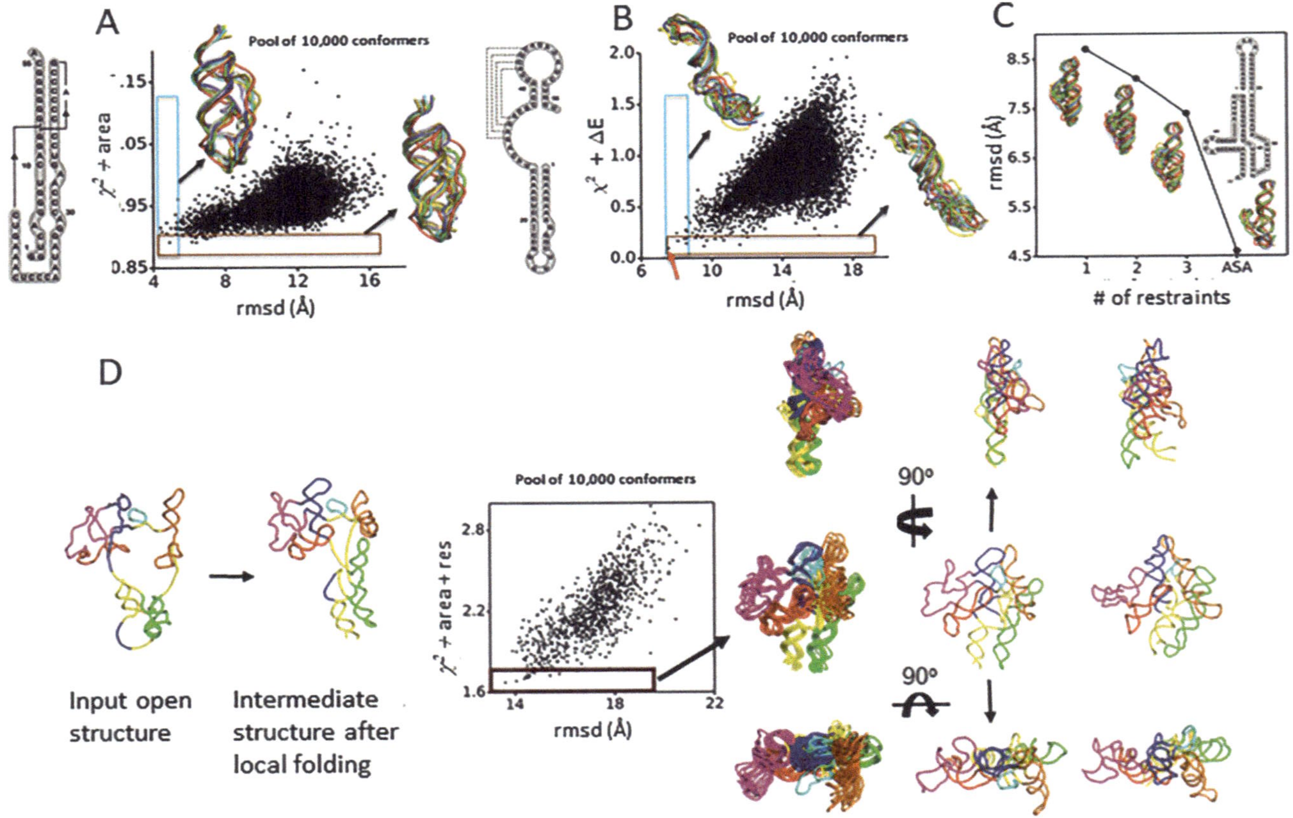

Fig. 4 See figure legend on opposite page.

defined, as in the NMR structure, due to an under-defined link structure (Carlomagno et al., 2013). The overall RMSDs of the top 10 structures relative to the NMR structure (red) are between 8 and 13 Å (brown rectangle). This RNA has an extended structure based on the *E* score. Thus, χ^2 plus ΔE is used to filter the top 10 structures from the pool of 10,000 conformers.

preQ1 RNA. PreQ1 is a riboswitch RNA that regulates genes by binding the pyrrolopyrimidine intermediate preQ1 during the biosynthesis of the essential tRNA base queuosine (Liberman, Salim, Krucinska, & Wedekind, 2013). We use this RNA to demonstrate a dramatic improvement in RMSD by incorporating additional experimental long-range restraints (Fig. 4C). These restraints can be obtained from electron paramagnetic resonance (EPR) (Domnick et al., 2020), Forster resonance energy transfer (FRET) (Vogel, van der Meer, & Blank, 2014), chemical probing (Weeks, 2010), or paramagnetic resonance (PRE) (Orton & Otting, 2018).

RNase P RNA. The bacterial 268-nt RNase P fragment is the RNase P enzymatic core. The structure of this RNA was determined by X-ray crystallography (Kazantsev et al., 2005). We use this molecule as an example to illustrate the utility of the RS3D for folding a large multi-domain RNA. We simulate the hierarchical RNA folding pathway in computing its 3D

Fig. 4 Case studies. (A) The structure of the translational recoding switch in the murine leukemia virus (PDB structure access code: 2LC8). On the left is the secondary structure. The long linker consists of residues 35–47. The χ^2 plus normalized surface area vs the RMSD of a pool of 10,000 conformers is plotted with each dot representing a structure. Superimposed structures are those selected based on the χ^2 plus normalized surface area (brown rectangle), or RMSD (cyan rectangle). (B) Lariat-forming ribozyme (PDB structure access code: 2M58). This RNA consists of a pseudo-knot with a four-residue linker (UACA) to a hairpin. The RMSD of the top structure (indicated by a red arrow) relative to the NMR structure is about 8 Å. (C) Illustration of improvement of RMSDs by incorporating long-range distance restraints or accessible surface area (ASA). Shown in this example is the preQ1 RNA. The crystal structure is in red. (D) Illustration of a multi-domain RNA, the bacterial RNase P RNA (PDB structure access code: 2A64). The folding proceeds in two stages: folding local secondary structural elements in the input open structure, followed by intermediate folding events within individual domains. The final folding is accomplished by incorporating SAXS and all long-range restraints that were derived from phylogenetic comparison. 1000 conformers were generated and the top 10 structures (brown rectangle) were selected based on χ^2, accessible surface area, and tertiary restraint scores. For comparison, the crystal structure in a similar orientation is shown on the far right. *The content of this figure was taken from Bhandari, Y. R., Fan, L. X., Fang, X. Y., Zaki, G. F., Stahlberg, E. A., Jiang, W., et al. (2017). Topological structure determination of RNA using small-angle X-ray scattering.* Journal of Molecular Biology, 429, *3635–3649. https://doi.org/10.1016/j.jmb.2017.09.006 with permission.*

topological structure in two stages. We started with an open structure where local structural elements fold first within each individual domain independently, followed by inter-domain interactions and a higher level of tertiary restraints at the last, to form a compact tertiary structure (Brion & Westhof, 1997) (Fig. 4D). The first stage involves moves of local structural elements, such as duplexes and loops, as well as sub-domains within a domain. In the second stage, the program computes relatively larger moves comprised of a rotation and translation of domains and sub-domains, restrained by SAXS and all reported long-range interactions derived from comparative sequence analysis (Massire & Westhof, 1998).

4.2 The three-dimensional topological structure of the HIV Rev response element (RRE) RNA

4.2.1 Background

The HIV-1 Rev response element (RRE) is a highly structured, ~350 (nt) RNA containing two three-way and four-way junctions. The viral Rev protein recognizes unknown structural features of RRE and forms the RRE-Rev complex that mediates the nucleus-to-cytoplasm export of the viral mRNAs for translation or viral packaging. The detailed local structural contacts between the Rev protein and the high-affinity binding sites of RRE (Battiste et al., 1996; Daugherty, Liu, & Frankel, 2010) have been well-characterized. However, because of the lack of a global structure, the structural basis for the specific mechanism of recognition remained a mystery. Thus, the RRE RNA serves as a good example for functional studies aided by RNA topological structure information (Fang et al., 2013).

The secondary structure of a 233-nt minimal construct of RRE that is fully functional in reporter assays has been mapped by SHAPE (Selective 2′ Hydroxyl Acylation analyzed by Primer Extension) technique. HIV-1 RRE consists of a number of stems, stems I, II, III/IV, and V, which are arranged around a central 4-way junction, with stem-loop II split into a proximal stem (IIA) and two distal stem-loops (IIB and IIC) around a 3-way junction (Legiewicz et al., 2008). Stem IIB and stem IA host two sites that have been identified to bind Rev with high affinity and specificity (Battiste et al., 1996; Daugherty et al., 2010). This 233-nt RNA construct was then chosen for SAXS characterization.

4.2.2 Sample preparation and conditioning

A DNA Ultramer coding for the 233-nt RRE RNA was initially chemically synthesized, then amplified with two primers (5′-CGC<u>GGATCC</u>TAA TACGACTCACTA-3′) and (5′-AGTC<u>GAATTC</u>TGCAGGAGCTGTT.

GATCCTTTAGGTAT-3′) (*Bam*HI and *Eco*RI sites are underlined) by PCR. The PCR product was digested and subcloned into a pUC18 plasmid. The plasmid DNAs were further amplified in *E. coli*, then extracted and linearized as the template for *in vitro* transcription by T7 RNA polymerase. The plasmids for RRE mutants can be generated using QuikChange II Site-Directed Mutagenesis Kit.

The RRE RNA transcript was purified by 10% polyacrylamide non-denaturing gel electrophoresis. In this case, purifying this RNA under non-denaturing conditions is critical. This RNA appears heterogeneous once it is refolded after denaturation regardless of what annealing protocols are used. RNA samples were passively eluted from gel slices into buffer consisting of 0.3 M NaOAc, pH 5.2 and 1 mM EDTA overnight at 4 °C, then buffer exchanged into final buffer conditions containing 10 mM Tris-HCl (pH 8.0), 100 mM KCl and various concentrations of EDTA or $MgCl_2$ by extensive filtration using Amicon Ultracel-10K Centricons (Millipore). Homogeneity of the RNA was confirmed by non-denaturing polyacrylamide gel electrophoresis and dynamic light scattering and in some cases by mass spectroscopy. Before use, all RNA solutions were filtered using 0.2-μm filters and diluted to final concentrations (0.1–2 mg/mL) immediately prior to data collection.

Mg^{2+} ions are known to be important for the structure and stability of RNA molecules (Misra & Draper, 1998). To identify the optimal ionic conditions that promote the homogeneous compaction of RRE, the compactness of RRE was studied as a function of Mg^{2+} concentration (0, 0.2, 0.5, 1.0, 2.0, 4.0 mM) by SAXS. The Guinier analysis, dimensionless Kratky and Porod-Debye plots indicate that the RRE sample is homogeneous, conformationally dynamic in solution, and adopts an extended and open conformation that is not obviously affected by Mg^{2+}. The SAXS scattering profile of RRE obtained at near-physiological Mg^{2+} concentration (0.5 mM) was then used for further analysis.

4.2.3 Deriving the 3D topological structure of the HIV-1 RRE RNA

A two-step shape *ab initio* reconstruction using the program DAMMIN was applied to derive the molecular envelope of RRE (Fig. 5A). In the first step, the search volume was chosen as "sphere" without bias, with a diameter 10–20 Å greater than the D_{max} of the corresponding construct to avoid distortion caused by the possible underestimated D_{max}. The resulting bead model indicates that RRE has a dimension of ~30 × 90 × 180 Å, or is roughly a flat molecule with a thickness of ~30 Å on one of its sides.

The planar feature has been observed in many large RNAs (Holbrook, 2008; Reiter, Chan, & Mondragon, 2010). The thickness of an RNA shape is closely related to the number of duplexes packed in a parallel or antiparallel arrangement (Leontis, Lescoute, & Westhof, 2006), with a dimension of approximately n-fold of 30 Å. Knowing that the overall shape of the

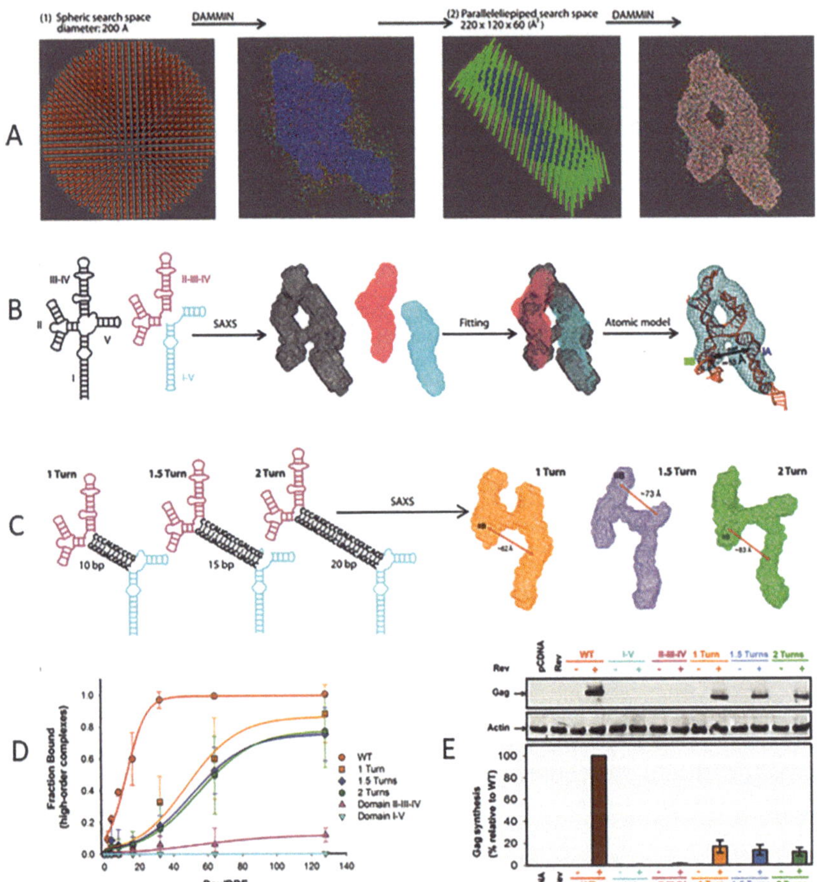

Fig. 5 An illustration of the combined use of SAXS and *in vitro* and *in vivo* assays to study the structure-function relationship of the RRE RNA. (A) the two-step molecular shape reconstruction; (B) divide-and-identify of domain locations in the RRE RNA and final atomic model; (C) extending the central helix to further confirm the global topology and the domain assignments of RRE; (D) & (E) respective *in vitro* and *in vivo* assays to corroborate with the topological structural information. *The content of the figure was taken from Fang, X., Wang, J., O'Carroll, I. P., Mitchell, M., Zuo, X., Wang, Y., et al. (2013). An unusual topological structure of the HIV-1 rev response element.* Cell, 155, *594–605. https://doi.org/10.1016/j.cell.2013.10.008 with permission.*

RRE envelope is flat, to reduce the degeneracy among the bead models and increase the resolution, a second reconstruction step was performed using a confined rectangular search space with a dimension of $60 \times 120 \times 220$ Å, larger than the initial RRE model in all three dimensions. The resulting average envelope of RRE adopts an "A"-like topology, with one leg longer than the other (Fig. 5A).

To localize the individual stems within the RRE bead model, a SAXS-aided divide-and-conquer approach was employed. RRE is comprised of a four-way junction formed among stem I, stem II, stem III/IV and stem V, and a three-way junction formed among stem IIA, IIB and IIC. In a four-way junction (H_1–H_4) with two coaxial stacking, there is a strong preference for stacking between H_1H_4 and H_2H_3 (Laing et al., 2012). Thus, two subconstructs, one composed of stems II, III and IV (II-III-IV) and the other composed of stems I and V (I-V) were constructed. The individual shapes of II-III-IV and I-V constructs match exclusively the specific regions in the "A"-shaped envelope, thus the overall layout of the topological structure and the locations of the domains (Fig. 5B). An analysis of the new RRE mutant constructs that the domains II-III-IV and I-V are separated by 1, 1.5, and 2.0 turns of helixes further confirmed the assignments of the domain layout (Fig. 8C). The derived structural model and the proposed specific mode of binding were validated by both *in vitro* and *in vivo* assays. In conclusion of the study, it was illustrated that the specificity of viral RNA recognition is achieved through a dual "lock-and-key" mechanism between the two binding sites on RRE and the two arginine-rich motifs of a Rev dimer. These combined SAXS envelopes, together with the approximate layout of the domains, make it possible to construct a 3D structural model of the RRE using the software package G2G (Wang et al., 2009). An overlay of the 3D model, drawn in ribbon diagrams, with the 3D RRE SAXS envelope is shown in Fig. 5B.

RRE serves as an excellent case to demonstrate the modularity of RNA structure and the utility of the SAXS-aided divide-and-conquer strategy to dissect an RNA structure by identifying the locations of each structural element in the context of the global molecular envelope. In the case of RRE, to support the coaxial stacking interactions in II-III-IV, two subconstructs, II-III-IV-X and II-III-IV-C, which sequences are identical to that of II-III-IV except for a 7-bp insertion at region IIB in sub-construct II-III-IV-X and an addition of 5 bp at the junction between domain II and domains III-IV in II-III-IV-C, were generated. The subconstructs of II-III-IV-X and II-III-IV-C share a very similar global shape with that of subconstruct II-III-IV, indicating that possible coaxial stacking between domain II and III-IV prevails in all 3 subconstructs. The construct II-III-IV-X with the 7-bp extension of stem-loop IIB was also used to determine whether stem-loop IIB is located at the top or the bottom of the shorter leg of the "A." Comparison of the shape of this construct with that of the RRE envelope confirmed the location of IIB as the lower part of the short leg of the "A." The elongated and bent envelope of the I-V construct suggests that stems I and V may be arranged *via* either a coaxial stacking or helix-loop-helix interaction. The location of domain IA was derived by the following rationale: excluding single bulges, the distance spanned by the 23 bp between the end of Stem I and the beginning of the purine-rich IA internal asymmetrical loop is ~60 Å, assuming a standard A-form duplex. This distance places site IA in the long leg, across from site IIB on the opposing leg of the "A" with a separation of ~55 Å. A summary of the RRE domain locations is illustrated in Fig. 6F.

4.3 Long non-coding subgenomic flavivirus RNAs from dengue virus 2

4.3.1 Background

Long non-coding RNAs (lncRNAs) are generally defined as cellular transcripts ranging from 200 nucleotides (nt) to over 100 kb in length and possessing little or no protein-coding potential (Ma, Bajic, & Zhang, 2013; Zhang et al., 2019). lncRNAs play diverse physiological roles and are implicated in many diseases. Due to their large sizes and inherent flexibility, experimental 3D structural characterization of lncRNAs remains challenging using the conventional high-resolution structural techniques including X-ray crystallography, NMR and cryo-EM. So far, little 3D structural information about lncRNAs is known, which however is often integral to their functions and mechanisms. Like their host, mosquito-borne flaviviruses such as Dengue virus (DENV), Zika virus (ZIKV) and West Nile virus (WNV) produce a large abundance of long non-coding subgenomic flavivirus RNAs (sfRNAs) ranging from 300- to 500-nt in length in infected cells. sfRNA is linked to virus pathogenicity and immune evasion (Clarke, Roby, Slonchak, & Khromykh, 2015). The secondary structures of the complete sfRNAs from DENV2, ZIKV and WNV have been interrogated using SHAPE chemical probing, revealing both similarities and substantial differences among the sfRNAs (Zhang et al., 2019). All sfRNAs exhibit the feature of structural modularity and consist of several conserved and independently folded subdomains (such as the xrRNA1, xrRNA2, DB1, DB2 and 3′SL elements in DENV2) (Fig. 7A).

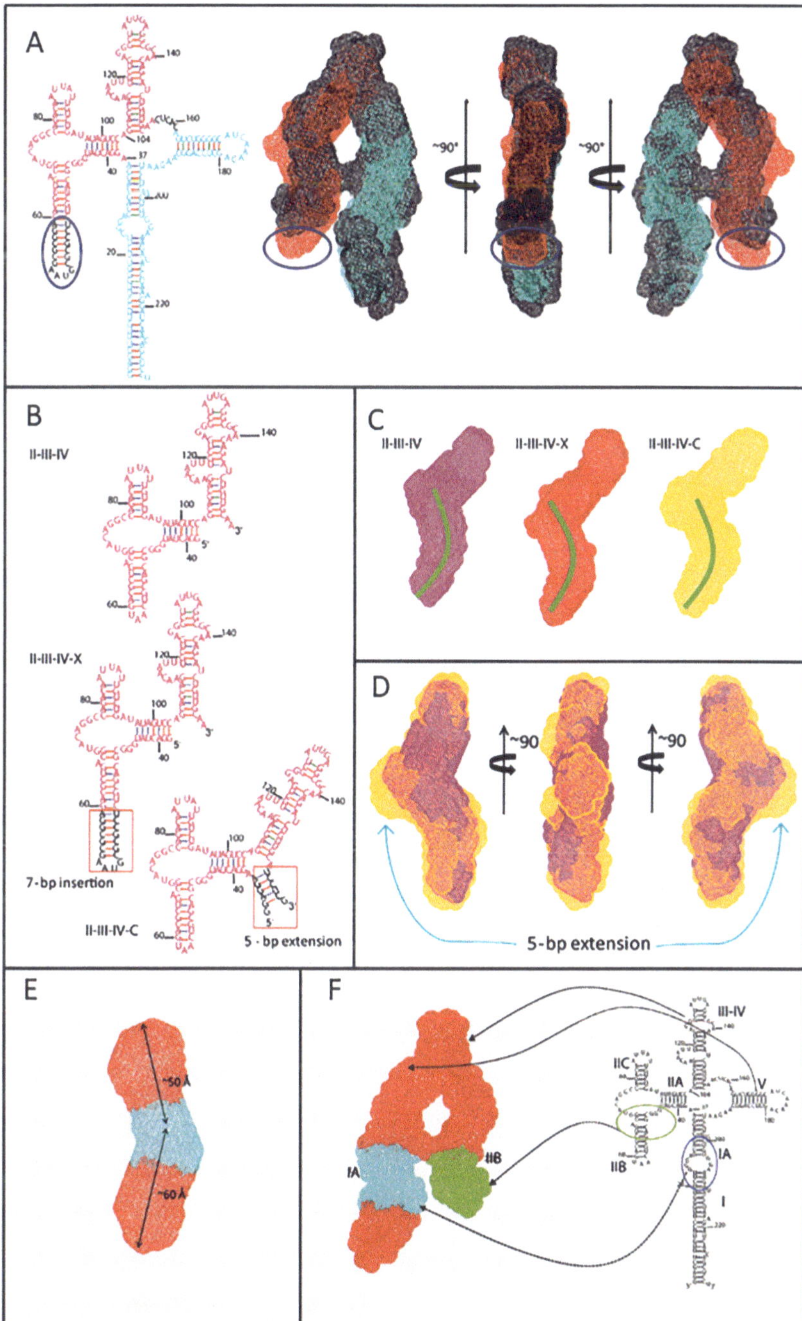

Fig. 6 Illustration of "divide-and-identify" to locate the structural elements in the context of the global envelope. Identifying IIB by extension highlighted by the blue oval (A), the co-axial stacking in stems I-II-IV using two constructions, II-III-IV-X and II-III-IV-C (B–D), or the high-affinity binding site (E). (F) summary of the locations of the key structural elements. *The content of the figure was taken from Fang, X., Wang, J., O'Carroll, I. P., Mitchell, M., Zuo, X., Wang, Y., et al. (2013). An unusual topological structure of the HIV-1 rev response element.* Cell, 155, *594–605. https://doi.org/10.1016/j.cell.2013.10.008 with permission.*

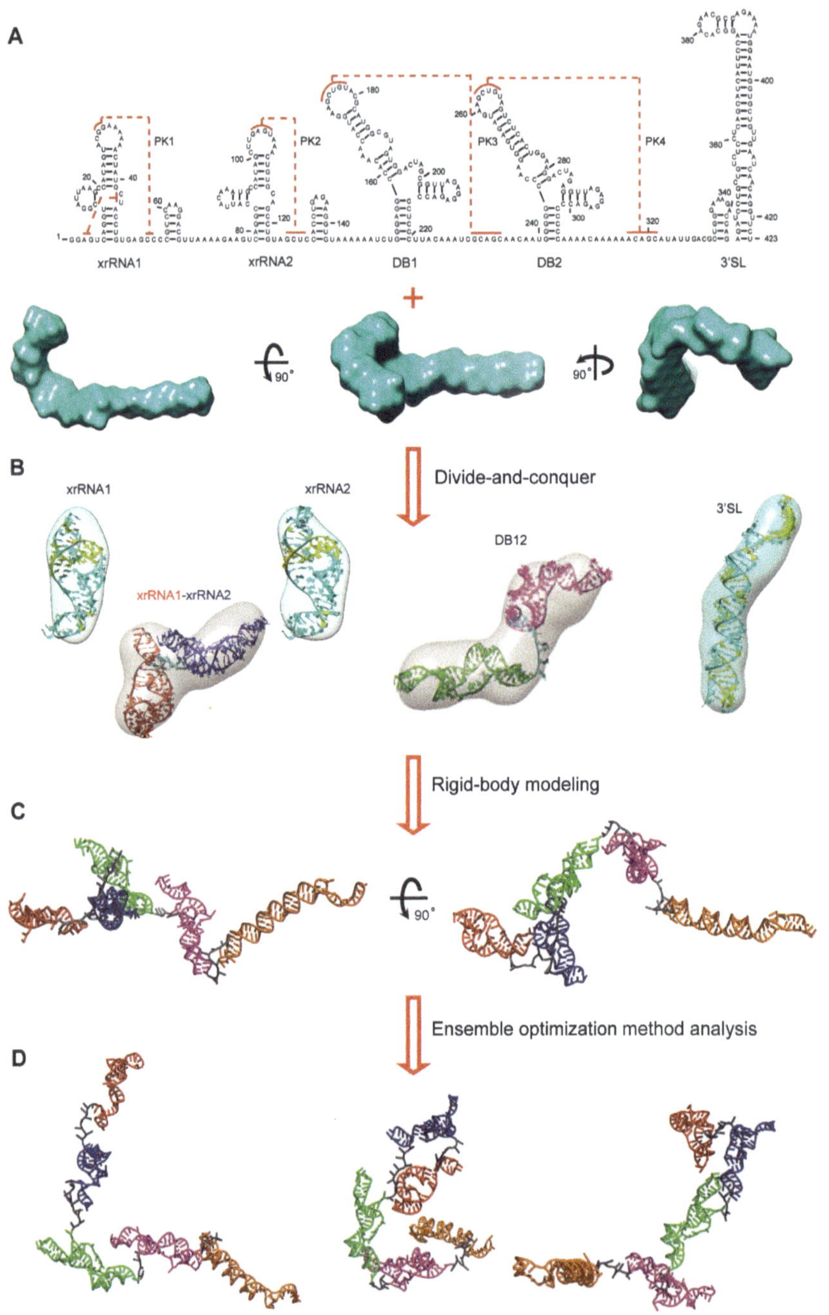

Fig. 7 Probing the 3D topological structure of the DENV2 sfRNA by combined use of SAXS and computational modeling. (A). The secondary structure of DENV2 complete sfRNA reveals five subdomains. (B). *ab initio* shape reconstruction results in an elongated

Although SAXS is limited with a moderate resolution, it remains one of the few techniques that can probe the structure of macromolecules without size limitation under native conditions. The advantages of using SAXS to probe the 3D topological structure of lncRNA are demonstrated as below by using the DENV2 sfRNA (423 nts) as an example.

4.3.2 Sample preparation and conditioning

The plasmid encoding an upstream T7 promoter and the DENV2 sfRNA sequence was synthesized using total gene synthesis and confirmed by DNA sequencing. Using the plasmid as template, plasmids encoding respective subdomains of DENV2 sfRNA were further constructed. The double-stranded DNA fragment templates for *in vitro* RNA production were generated by PCR using an upstream forward primer targeted the plasmids and a downstream reverse primer specific to respective cDNAs. The RNAs were transcribed *in vitro* using T7 RNA polymerase and purified by preparative, non-denaturing polyacrylamide gel electrophoresis. The target RNA bands were cut and passively eluted from gel slides into buffer containing 0.3 M NaOAc, 1 mM EDTA, pH 5.2 overnight at 4 °C. The RNAs were further purified by passing through the size exclusion chromatography column. The homogeneity of the purified sfRNA was characterized by native PAGE gel and dynamic light scattering experiments, indicating high purity and homogeneity of the samples, which was confirmed by further SAXS analysis.

Optimal Mg^{2+} concentration for proper folding of the complete sfRNA was first determined by SAXS, since Mg^{2+} is generally essential to RNA folding and structure. The SAXS-derived parameters, the radius of gyration R_g and the maximum diameters D_{max}, plotted against the concentration of $[Mg^{2+}]$, reach a minimum at 5 mM, indicating that Mg^{2+} promotes the compaction of DENV2 sfRNA at lower concentration but induces attractive intermolecular interactions leading to aggregation at higher

3D structure for DENV2 sfRNA. (C). Atomic models for the individual or combined subdomains of DENV2 sfRNA were fitted into the respective *ab initio* shape envelopes. (D). Starting from the atomic models of individual or combined subdomains, atomic model for DENV2 complete sfRNA is assembled by SAXS-restrained rigid-body modeling using Xplor-NIH. (E). Conformational flexibility of DENV2 sfRNA and the conformational space it samples were analyzed by Ensemble Optimization Method (EOM). *The content of this figure was taken from Zhang, Y., Zhang, Y., Liu, Z.-Y., Cheng, M.-L., Ma, J., Wang, Y., et al. (2019). Long non-coding subgenomic flavivirus RNAs have extended 3D structures and are flexible in solution.* EMBO Reports, 20, *e47016. https://doi.org/10.15252/embr.201847016 with permission.*

concentration. Thus, a concentration of 5 mM Mg^{2+} was chosen as the optimal condition for subsequent SAXS characterization for all subconstructs of DENV2 sfRNA.

4.3.3 Deriving the 3D topological structure

The experimental SAXS curve, the PDDF, the dimensionless Kratky and Porod-Debye plots for DENV2 sfRNA in the presence of 5 mM Mg^{2+} suggest that DENV2 sfRNA is an extended and curved molecule in solution. Its *ab initio* shape envelope built with the program DAMMIN is elongated (Fig. 7B), indicating that DENV2 sfRNA is an open molecules in solution where no long-range interactions between its subdomains exist. However, the nearly featureless envelope prevents direct localization of the individual subdomains. A SAXS-aided divide-and-conquer strategy was then employed to determine how the individual subdomains of DENV2 sfRNA are organized into its 3D structure. The subconstructs for individual (xrRNA1, xrRNA2, 3′SL) and combined subdomains (xrRNA12, DB12) of DENV2 sfRNA were constructed, *in vitro* transcribed, purified under non-denaturing conditions and studied at the optimal Mg^{2+} concentration by SAXS. SAXS data analysis yields overall structural parameters, folding properties and low-resolution shape structures for the individual and combined subdomains. 3D structural models for the individual or combined subdomains were built up from the secondary structure with the aid of SAXS data (Fig. 7C). Currently, high-resolution crystal structures are available for ZIKV xrRNA1 and MVEV (Murray Valley encephalitis virus) xrRNA2, which are homologous to DENV2 xrRNA1 and xrRNA2. Using the crystal structure of ZIKV xrRNA1 as template, the homology modeling program ModeRNA (Rother, Rother, Puton, & Bujnicki, 2011) was used to build up the atomic models for xrRNA1 and xrRNA2 of DENV2. The quality of the homology atomic models was evaluated by comparing the consistency of the respective theoretical and experimental SAXS data and the fitting of the homology models into the *ab initio* shape envelopes. Starting from the atomic models of xrRNA1 and xrRNA2, the atomic model for DENV2 xrRNA12 was built up by SAXS-based rigid-body modeling algorithm using Xplor-NIH (Schwieters, Kuszewski, Tjandra, & Clore, 2003). For DB12 and 3′SL of DENV2, as no homologous high-resolution structures are available, their atomic structures were predicted with Rosetta and then screened against the SAXS data, the best models can be fitted into the respective *ab initio* shape envelopes (Fig. 7C).

With the availability of 3D atomic models for the respective individual or combined subdomains, 3D atomic model of DENV2 complete sfRNA was generated with a SAXS-restrained rigid-body modeling algorithm using Xplor-NIH. The best-fit model of the complete sfRNA is elongated and extended in one dimension consistent with the *ab initio* bead modeling (Fig. 7D). Due to the intrinsic flexibility, large RNAs naturally exist as dynamic structural ensembles in solution. To analyze the conformational flexibility of DENV2 complete sfRNA and the conformational space it samples in solution, the ensemble optimization method (EOM) approach was employed (Bernado, Mylonas, Petoukhov, Blackledge, & Svergun, 2007). In this approach, an initial random pool containing 10,000 coexisting conformers with different conformations that approximate (otherwise infinite) the possible conformational space is generated using the program Xplor-NIH. A genetic algorithm selection process then generates an optimized ensemble containing multiple conformers that best fit the experimental scattering curve. The EOM analysis shows that the complete sfRNA of DENV2 in solution is dynamic and can be best described as a minimal structural ensemble ($N_e = 3$) instead of one single structure (Fig. 7E).

4.4 The 3′X and 5BSL3.2 domains of hepatitis C virus and their distal complexes

4.4.1 Background

The positive-sense RNA genome of hepatitis C virus (HCV) encompasses 9600 nucleotides (nt) and contains a single open reading frame (ORF) flanked by 5′ and 3′ untranslated regions (UTR). The 5′UTR comprises the internal ribosome entry site (IRES), a 340-nt structure that triggers translation of the viral ORF by binding to the small ribosomal subunit (Piñeiro & Martinez-Salas, 2012). On the other side of the genome, the 3′UTR includes a variable segment, a pyrimidine-rich tract and a strongly conserved 98-nt terminal domain called 3′X (Kolykhalov, Feinstone, & Rice, 1996) (Fig. 8A). This domain is essential for replication, as it is recognized by the viral replication complex to start the synthesis of the negative polarity strand (Bartenschlager, 2013). A network of RNA-RNA distal contacts has been shown to regulate the replication and translation stages of the HCV cycle. Subdomain 5BSL3.2 (also named SL9266), located at the 3′ side of the ORF, acts as the main hub of this network by contacting domain 3′X at the 3′UTR (Cantero-Camacho & Gallego, 2018; Castillo-Martínez, Fan, Szewczyk, Wang, & Gallego, 2022; Friebe, Boudet, Simorre, & Bartenschlager, 2005; Shetty, Stefanovic, & Mihailescu, 2013; Tuplin, Struthers, Simmonds, & Evans, 2012; You & Rice, 2008),

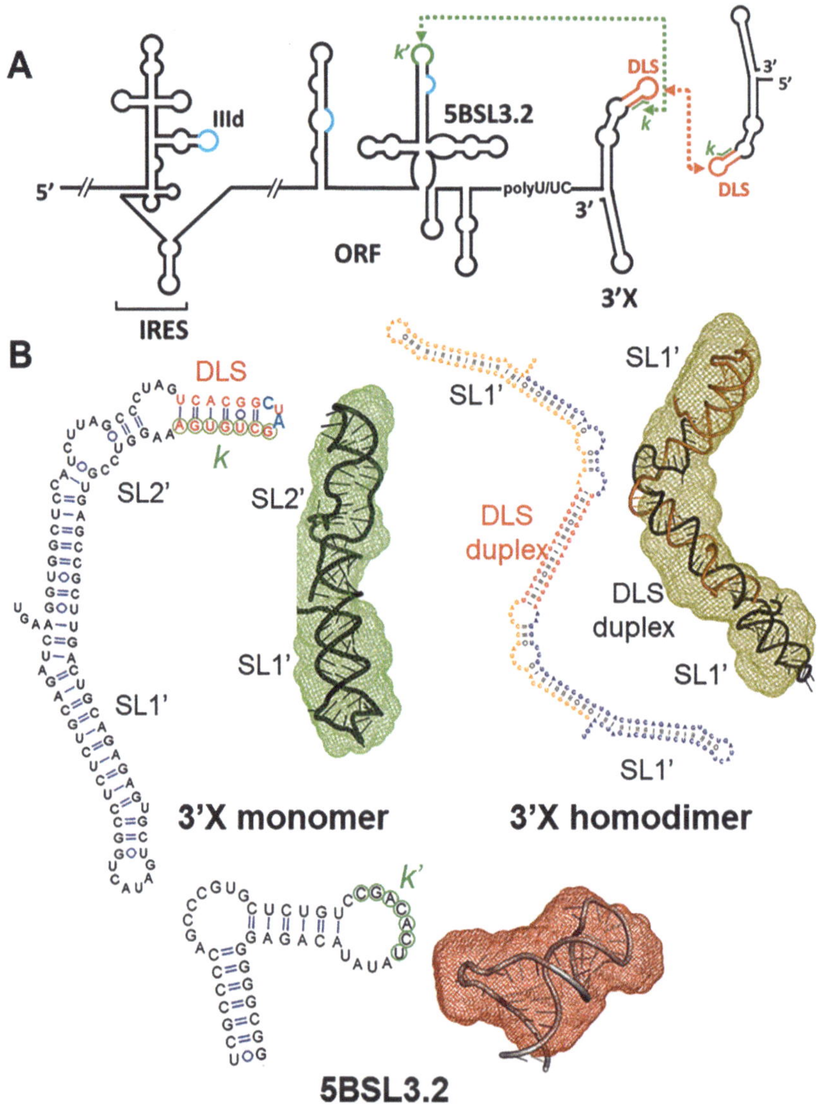

Fig. 8 Structural analysis of the 3′X and 5BSL3.2 domains of HCV and their distal contacts. (A) Schematic representation of the HCV genomic RNA, with the 5BSL3.2-3′X and 3′X-3′X interactions indicated with green and red dotted arrows, respectively. The first interaction involves the *k* and *k*′ sequences and is intramolecular, while the second contact involves the DLS sequence and promotes intermolecular 3′X RNA homodimerization. (B) Solution structure of domains 3′X and 5BSL3.2 determined by NMR spectroscopy and SAXS. The 3′X sequence formed 98-nt monomers at low ionic strength and 196-nt symmetric dimers at higher ionic strength. The image shows dispersion envelopes superposed with the best energy-minimized respective atomic

the IRES at the 5′UTR (Romero-López & Berzal-Herranz, 2009; Shetty et al., 2013), and an additional stem-loop (SL9005) in the ORF (Diviney et al., 2008; Shetty et al., 2013) (Fig. 8A). 5BSL3.2 forms a 46-nt hairpin comprising a 12-nt apical loop and an 8-nt bulge (Friebe et al., 2005; You, Stump, Branch, & Rice, 2004). A 7-nt UCACAGC segment (identified as *k*′) in the apical loop of 5BSL3.2 contacts domain 3′X through a base-pairing interaction involving a complementary GCUGUGA 3′X sequence (*k*) (Cantero-Camacho & Gallego, 2018; Castillo-Martínez et al., 2022; Friebe et al., 2005; Shetty et al., 2013; Tuplin et al., 2012; You & Rice, 2008). This contact is likely required for translation (Castillo-Martínez et al., 2020; Romero-López & Berzal-Herranz, 2012; Tuplin, Struthers, Cook, Bentley, & Evans, 2015). Intriguingly, the *k* sequence of domain 3′X is part of a self-complementary 16-nt segment termed dimer linkage sequence (DLS) that has been shown to promote the dimerization of virus genomes *in vitro* (Cantero-Camacho et al., 2017; Cantero-Camacho & Gallego, 2015; Ivanyi-Nagy et al., 2006; Shetty, Kim, Shimakami, Lemon, & Mihailescu, 2010). On the other hand, the 5BSL3.2 bulge has been proposed to base-pair with a UGGGU sequence contained in the apical loop of IRES subdomain IIId (Romero-López & Berzal-Herranz, 2009; Shetty et al., 2013), or alternatively with a CGGGC segment of SL9005 (Diviney et al., 2008; Shetty et al., 2013). The IRES interaction likely inhibits translation (Romero-López & Berzal-Herranz, 2012), since the GGG residues of IRES subdomain IIId base-pair with residues $C_{1116}CC_{1118}$ of the small ribosomal subunit (Quade, Boehringer, Leibundgut, van den Heuvel, & Ban, 2015) and are essential for IRES activity (Matsuda & Mauro, 2014). All of the components of this HCV RNA-RNA interaction network are evolutionarily conserved within the HCV genome, and in fact, the overlapped DLS and

models, as well as secondary structures determined by NMR. The complementary *k* and *k*′ sequences in domains 3′X and 5BSL3.2 are indicated with green circles; the absolutely conserved 3′X DLS nt are depicted in red. The blue-colored C_{29} and A_{31} nt in the palindromic DLS loop can be mutated to G and U, respectively, to generate a dimerization-defective 3′Xm variant. *This figure was prepared from images contained in Cantero-Camacho, A., Fan, L., Wang, Y.-X., Gallego, J. (2017). Three-dimensional structure of the full-length 3′X-tail of hepatitis C virus RNA in monomeric and dimeric states.* RNA, 23, *1465–1476 and Castillo-Martínez, J., Fan, L., Szewczyk, M.P., Wang, Y. X., Gallego, J. (2022). The low-resolution structural models of hepatitis C virus RNA subdomain 5BSL3.2 and its distal complex with domain 3′X point to conserved regulatory mechanisms within the Flaviviridae family.* Nucleic Acids Research, 50, *2287–2301. https://doi.org/10.1093/nar/gkac061 with permission.*

k sequences of domain 3′X are absolutely conserved (Yamada et al., 1996). This conservation is remarkable in a highly variable virus such as HCV, for which many genotypes and subtypes have been identified around the world (Bartenschlager, 2013). In HCV, where the virions are widely thought to contain a single copy of genomic RNA, the absolutely conserved DLS and *k* sequences are likely involved in controlling the switch between translation and replication by modulating the formation of either 3′X-3′X or 3′X-5BSL3.2 contacts. The NMR and SAXS analyses of these domains and contacts have contributed to explain the molecular mechanism underlaying this switching function, as explained below.

4.4.2 Experimental strategy and ionic conditions

In this case, we used a strategy based on a combination of NMR spectroscopy and SAXS analyses aided by gel electrophoresis experiments (Cantero-Camacho et al., 2017; Castillo-Martínez et al., 2022). Prior to dispersion studies, we determined the secondary structure of each RNA domain by analyzing exchangeable $^{1}H—^{15}N$ HSQC and HNN-COSY (J_{NN} HNN COrrelation SpectroscopY) spectra combined with homonuclear NOESY (Nuclear Overhauser Effect SpectroscopY) and TOCSY (TOtal Correlation SpectroscopY) data. The electrophoresis experiments provided a quick assessment of the stoichiometry of the species in different experimental conditions.

Since both 3′X-3′X homodimerization and 3′X-5BSL3.2 complex formation depend on the presence of salts, we analyzed each domain under two ionic strength conditions: higher ionic strength (2 or 6 mM $MgCl_2$, or 2 mM $MgCl_2$ with 50 mM NaCl), driving homodimer/heterodimer formation, and low ionic strength (no added salts), favoring monomer formation. This strategy allowed us to (i) study RNA-RNA association as a function of solution conditions; (ii) compare monomeric conformations in the absence and presence of ions; (iii) compare the isolated and associated forms of each species. To obtain homogeneous samples, both the 3′X-3′X homodimers and the 3′X-5BSL3.2 heterodimers were purified by size-exclusion chromatography prior to SAXS analyses. The Guinier region of the scattering curves was linear in all cases, indicating that none of the samples were aggregated in any solution condition.

4.4.3 Sequence design

A wild-type 3′X sequence containing an intact DLS was used to study 3′X monomers at low ionic strength and 3′X-3′X homodimers at higher ionic strength. Since magnesium is required for 3′X-5BSL3.2 complex formation and in these conditions domain 3′X also forms homodimers (Cantero-Camacho et al., 2017; Cantero-Camacho & Gallego, 2015, 2018), we used a 3′X sequence containing a double $C_{29}G/A_{31}U$ mutation disrupting the DLS palindrome (hereafter identified as 3′Xm) to study 3′X monomers and 3′X-5BSL3.2 complexes in the presence of magnesium (Figs. 8 and 9). This mutation blocks 3′X homodimerization without

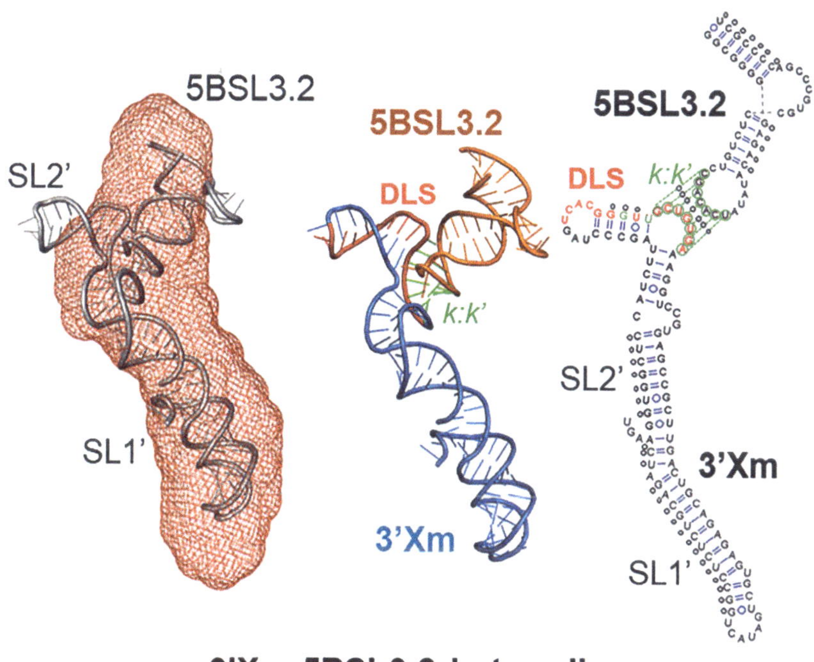

Fig. 9 Solution structure of the 3′X-5BSL3.2 complex of HCV determined by NMR and SAXS. The image shows the three-dimensional dispersion envelope of the complex superposed with the best energy-minimized atomic model, as well as the experimental secondary structure obtained by NMR spectroscopy. In the atomic model shown in the middle, 3′Xm DLS nt, 3′Xm *k* nt, and other 3′Xm nt are respectively indicated in red, green and blue, while 5BSL3.2 *k*′ nt and other 5BSL3.2 nt are respectively depicted in green and orange. In the secondary structure, the $C_{29}G$ and $A_{31}U$ mutations blocking 3′Xm homodimerization are indicated in green, DLS nt are depicted in red, and *k* and *k*′ nt are represented with green circles. The small circles denote NMR-derived base-pairing restraints used for model building. *This figure was prepared from images contained in Castillo-Martínez, J., Fan, L., Szewczyk, M. P., Wang, Y. X., Gallego, J. (2022). The low-resolution structural models of hepatitis C virus RNA subdomain 5BSL3.2 and its distal complex with domain 3′X point to conserved regulatory mechanisms within the Flaviviridae family.* Nucleic Acids Research, 50, *2287–2301. https://doi.org/10.1093/nar/gkac061.*

affecting the structure of the domain or its interaction with 5BSL3.2 (Cantero-Camacho & Gallego, 2018; Castillo-Martínez et al., 2022). In addition to a 46 nt-long molecule comprising the 5BSL3.2 subdomain, the analyses also involved SL1 and SL2′ 3′X subdomains, used for secondary structure NMR analyses and to verify the dispersion results of the full-length domain (Cantero-Camacho et al., 2017; Cantero-Camacho & Gallego, 2018; Castillo-Martínez et al., 2022).

4.4.4 Model building and evaluation

For each RNA domain and ionic condition, we generated a three-dimensional envelope from the scattering data using DAMMIM (Svergun, 1999a) and DAMFILT (Volkov & Svergun, 2003). In parallel, ensembles of all-atom models were generated with either the MC-Sym (Parisien & Major, 2008) or SimRNA (Magnus, Boniecki, Dawson, & Bujnicki, 2016) web servers, using specific base-pair restraints obtained from the NMR analyses. The theoretical scattering profiles of the models were compared with the experimental SAXS data using FoXS (Schneidman-Duhovny, Hammel, & Sali, 2010) to select the 20 best models (with the lowest χ^2 fit values) of each RNA sequence. These models were subsequently refined by progressively minimizing their potential energy with an AMBER force field (Case, 2021) and a generalized Born model for solvent simulation. We evaluated the conformation of the refined models with MolProbity (Williams et al., 2018) and selected the best refined model on the basis of dispersion fit χ^2 values and Molprobity results (Castillo-Martínez et al., 2022). The best models were overlaid with the *ab initio* envelope using SUPCOMB (Kozin & Svergun, 2001). We also calculated the monomer/dimer content of 3′X sequences by comparing the experimental SAXS profiles with the theoretical SAXS curves of monomer, dimer or mixture atomic models, generated using FoXS (Schneidman-Duhovny et al., 2010) for monomer or dimer models, and a combination of MultiFoXS (Schneidman-Duhovny, Hammel, Tainer, & Sali, 2016) and MES (Pelikan, Hura, & Hammel, 2009) for mixture models.

We found it important to use specific base-pairing restraints based on NMR data for generating the ensemble of atomic models. For the monomeric form of the 3′X domain, this information helped us to distinguish between two possible conformations containing either two or three double-helical stems (Cantero-Camacho et al., 2017; Cantero-Camacho & Gallego, 2015). Likewise, the formation of extended antiparallel dimers by DLS-containing 3′X and SL2′ sequences was supported by the detection of diagnostic NMR signals indicating dimerization and by HNN-COSY analyses in the absence and presence of salts (Cantero-Camacho & Gallego, 2015). For the complexes between 3′X sequences and 5BSL3.2 constructs, the NMR analyses indicated that within the 3′X domain, subdomain SL1′ and the lower double-helical stem of subdomain SL2′ remained undisturbed upon complex formation (Cantero-Camacho & Gallego, 2015, 2018; Castillo-Martínez et al., 2020). The changes only affected the upper region of subdomain SL2′ where the *k* nt are base-paired and involved the establishment of new Watson-Crick pairs between these nt and the *k′* nt of the 5BSL3.2 apical loop (Cantero-Camacho & Gallego, 2018; Castillo-Martínez et al., 2020). Additional experiments involving a dimerization-defective SL2′m subdomain with a guanine residue replaced with inosine revealed the establishment of the *k*-*k′* duplex in the SL2′m-5BSL3.2 complex implied the folding of the DLS nt released upon 5BSL3.2 binding into a short hairpin. These experiments allowed us to complete the list of base-pair restrictions for 3′Xm-5BSL3.2 model building (Castillo-Martínez et al., 2022) (Fig. 9).

4.4.5 Topological structure of the 3′X domain in monomeric and dimeric states

The monomeric form of the 3′X domain was studied using the wild-type sequence at low ionic strength (Cantero-Camacho et al., 2017) and the dimerization-defective 3′Xm sequence at both low and higher (6 mM $MgCl_2$) ionic strength (Castillo-Martínez et al., 2022). Both sequences were transcribed from pUC19 vectors containing a *Sca*I restriction site for linearization and run-off transcription, and purified following a procedure based on gel electrophoresis. The Porod-Debye and Kratky plots indicated the presence of stable folds, with slightly less flexibility observed for 3′Xm in the presence of magnesium. The D_{max} values of all structures were similar, but the PDDF curves and R_g values were different in the absence and presence of magnesium, indicating that the overall conformation depended on the presence of the cation. Nevertheless, all of the three-dimensional envelopes obtained from the dispersion data were elongated and the best energy-minimized atomic models comprised two coaxially stacked SL1′ and SL2′ double-helical subdomains, in agreement with the NMR observations (Cantero-Camacho & Gallego, 2015, 2018) (Fig. 8B). In this 3′X monomeric conformations, the DLS sequence is folded in the upper hairpin of the SL2′ subdomain, where the *k* nt are base-paired but the CUAG DLS palindrome is exposed in the apical loop. These exposed palindromes promote dimerization through the formation of a kissing complex (Shetty et al., 2010) involving two 3′X molecules. In spite of being buried in the DLS stem, we showed in subsequent experiments that the *k* nt are capable of pairing with *k′* nt to form the complex with 5BSL3.2 (Cantero-Camacho & Gallego, 2018; Castillo-Martínez et al., 2022) (Fig. 9). On the other hand, the three-terminal nt of the viral genome form an unpaired tail in the 3′X structure. Since the presence of at least three nt after a stable secondary structure has been reported to promote the initiation of RNA synthesis by the viral NS5B polymerase (Kao et al., 2000), the 3′X domain structure is compatible with replication. In fact, we have modeled the domain laterally bound to NS5B with the unpaired 3′-terminal strand entering the template binding channel of the enzyme (Castillo-Martínez et al., 2022).

At higher ionic strength (2 mM $MgCl_2$ and 50 mM NaCl), the D_{max} value of the wild-type 3′X sequence increased by a factor of 1.7 relative to the low ionic strength value (244 vs 148 Å), the three-dimensional envelope almost doubled in size, and molecular weight calculations based on the scattering profiles indicated the presence of homodimers. In these experimental conditions, NMR experiments detected the formation of extended symmetric dimers (Cantero-Camacho & Gallego, 2015). The atomic dimer models generated using NMR base-pair restraints comprised two antiparallel-paired SL2′ strands containing a palindromic DLS duplex in the middle. These strands fold into intramolecular SL1′ hairpins that stack coaxially on each side of the SL2′ duplex (Fig. 8B). These extended dimers have been reported to form in the presence of the core protein of the virus from kissing dimers involving palindromic loops (Ivanyi-Nagy et al., 2006; Shetty et al., 2010).

Confirmatory NMR and SAXS experiments involving SL1 and SL2′ subdomains showed that the envelope of the 3′X domain monomer could be reconstructed with the low-salt envelopes of subdomains SL1 and SL2′, and that the higher-ionic strength shapes of SL1 and SL2′ superposed quite well on the envelope of the 3′X domain dimer, with one SL2′ dimer located in the middle and two SL1 subdomains placed on each side (Cantero-Camacho et al., 2017).

4.4.6 Topological structure of the 5BSL3.2 subdomain and its distal complex with domain 3′X

NMR and SAXS analyses of subdomain 5BSL3.2 in the absence and presence of magnesium indicated that this subdomain adopted a monomeric hairpin structure. The structure was slightly less flexible and more compact in the presence of the cation, but the best energy-minimized atomic models indicated that in both solution conditions 5BSL3.2 formed an "L"-shaped hairpin comprising an extended 12-nt apical loop and two double helical segments separated by an 8-nt bulge. The presence of this bulge induced a strong curvature between the double-helical stems, with the interhelical angle becoming more acute in the presence of magnesium (Fig. 8B). Several reports have shown that subdomain 5BSL3.2 binds specifically to the HCV NS5B polymerase and that this interaction requires the presence of the 8-nt bulge (Kanamori, Yuhashi, Ohnishi, Koike, & Kodama, 2010; Zhang et al., 2005) imposing the "L"-shape to the subdomain. Based on this experimental evidence and on analogies with the SLA stem-loops of the dengue and Zika flaviviruses, which adopt similar "L"-shapes and have been reported to bind to the thumb domain of the NS5 polymerase (Lee et al., 2021), we have built a hypothetical model of the "L"-shaped 5BSL3.2 subdomain bound to the thumb domain of the HCV NS5B polymerase. Formation of this complex would not block the template binding channel of the polymerase nor the double-stranded RNA exit channel and would allow the unpaired 3′-terminal tail of the 3′X domain to enter the active site of the polymerase (Castillo-Martínez et al., 2022), likely favoring replication.

We prepared 3′Xm-5BSL3.2 complexes by mixing 5BSL3.2 and dimerization-defective 3′Xm molecules in the presence of magnesium, followed by purification by size exclusion chromatography. The three-dimensional envelope generated from the dispersion data was elongated like those of the isolated 3′X domain, but exhibited a 42 Å-wide lateral protrusion that was not present in the unbound domain. The best atomic models indicated that the complex adopted a "Y"-shaped structure comprising three double-helical domains and an unusual kissing junction. The lower branch of the "Y" is formed by the coaxially-stacked 3′X subdomain SL1′ and lower helix of subdomain SL2′. The two upper branches, on the other hand, comprise subdomain 5BSL3.2 and a new hairpin formed by DLS nt previously paired to the *k* nt. The kissing junction is formed by an intermolecular helix where the *k′* nt of 5BSL3.2 and *k* nt of 3′X are base-paired; this helix is coaxially stacked with the upper stem of 5BSL3.2 and the lower stem of subdomain SL2′ (Fig. 9). As observed for the isolated molecules, the SAXS data indicated that the complex adopted a well-defined structure with limited flexibility. This means that a hypothetical binding of the 5BSL3.2 subdomain contained in the "Y" to the thumb domain of NS5B would not physically allow the 3′-terminal tail of domain

3′X to reach the template channel of the polymerase and *vice-versa*, binding of the 3′-terminal tail to the template channel would likely be disfavored by the solvent exposure of the bulky "Y." According to this hypothesis, formation of the 5BSL3.2-3′X distal contact would disfavor negative-sense RNA synthesis, indirectly enhancing translation in agreement with previous reports (Castillo-Martínez et al., 2020; Romero-López & Berzal-Herranz, 2012; Tuplin et al., 2015). In addition, 5BSL3.2-3′X complex formation conceals the DLS palindrome (Fig. 9) and inhibits homodimerization. In contrast, the formation of DLS-DLS homodimers inhibits the 5BSL3.2-3′X contact by blocking the *k* sequence of domain 3′X, and would therefore promote 5BSL3.2-NS5B recognition and the concomitant entry of the 3′X terminal tail in the NS5B template channel, stabilizing the replication stage of the HCV cycle (Castillo-Martínez et al., 2022). This structure-based mechanistic model contributes to explain the switching function of the DLS and k/k′ sequences of HCV as well as their unusually high conservation.

5. Summary and conclusions

This chapter describes two basic approaches to the problem of deriving 3D topological structures of RNA molecules. In the first approach, *ab initio* calculations using program RS3D compute 3D topological structures by simultaneously taking into account SAXS and secondary structure information. In this method each residue is represented by a glob that is centered at the center of mass of a residue, and the relative positions of the globs are constrained by covalent linkage and secondary structure, consequently the reduction of the degree of freedom of each residue in the structure (Bhandari et al., 2017). The accuracy of the topological structural models is significantly improved if additional long-range distance restraints are available from various biophysical measurements. The end result is an all-atom 3D structural model, as opposed to the molecular envelope generated with DAMMIN (Franke & Svergun, 2009a). The main disadvantage of this approach is the large demand for CPU hours. The second approach is based on the envelopes generated using programs such as DAMMIN. This approach is widely used in the field, often in combination of a divide-and-conquer strategy. The end result is molecular envelopes as well as all-atom topological structural models that are consistent with the SAXS data. With both approaches, the quality of the topological structural models is significantly improved if secondary structure information and long-range distance restraints are incorporated from other biophysical measurements.

The divide-and-conquer approach works well for multi-domain RNA structures where the layouts of the structural elements are clear. This approach is labor-intensive and requires multiple sample constructs and expert insight. For RNA molecules with complex folds, such as the RNase P RNA, the usefulness of the divide-and-conquer strategy aided with SAXS-derived molecular envelopes is limited.

Acknowledgments

This work is supported by the NIH/NCI Intramural Research Fund to Y.-X. Wang, grants from Universidad Católica de Valencia and Ministerio de Ciencia e Innovación of Spain (BFU2015-65103-R and RTI2018-093935-B-I00) to J. Gallego, and grants from the National Natural Science Foundation of China (No. U1832215) and the Beijing Advanced Innovation Center for Structural Biology to X. Fang. The authors thank Y.R. Bhandari, Y. Zhang, A. Cantero-Camacho, J. Castillo-Martínez and L. Fan for their contributions in the studies illustrated in this chapter.

References

Baronti, L., Karlsson, H., Marusic, M., & Petzold, K. (2018). A guide to large-scale RNA sample preparation. *Analytical and Bioanalytical Chemistry, 410*, 3239–3252. https://doi.org/10.1007/s00216-018-0943-8.

Bartenschlager, R. (Ed.). (2013). *Hepatitis C virus: From molecular virology to antiviral therapy* Springer-Verlag.

Battiste, J. L., Mao, H., Rao, N. S., Tan, R., Muhandiram, D. R., Kay, L. E., et al. (1996). Alpha helix-RNA major groove recognition in an HIV-1 rev peptide-RRE RNA complex. *Science, 273*, 1547–1551.

Bernado, P., Mylonas, E., Petoukhov, M. V., Blackledge, M., & Svergun, D. I. (2007). Structural characterization of flexible proteins using small-angle X-ray scattering. *Journal of the American Chemical Society, 129*, 5656–5664. https://doi.org/10.1021/ja069124n.

Bhandari, Y. R., Fan, L. X., Fang, X. Y., Zaki, G. F., Stahlberg, E. A., Jiang, W., et al. (2017). Topological structure determination of RNA using small-angle X-ray scattering. *Journal of Molecular Biology, 429*, 3635–3649. https://doi.org/10.1016/j.jmb.2017.09.006.

Brion, P., & Westhof, E. (1997). Hierarchy and dynamics of RNA folding. *Annual Review of Biophysics and Biomolecular Structure, 26*, 113–137. https://doi.org/10.1146/annurev.biophys.26.1.113.

Brunelle, J. L., & Green, R. (2013). In vitro transcription from plasmid or PCR-amplified DNA. *Methods in Enzymology, 530*, 101–114. https://doi.org/10.1016/B978-0-12-420037-1.00005-1.

Cantero-Camacho, A., Fan, L., Wang, Y.-X., & Gallego, J. (2017). Three-dimensional structure of the full-length 3'X-tail of hepatitis C virus RNA in monomeric and dimeric states. *RNA, 23*, 1465–1476.

Cantero-Camacho, Á., & Gallego, J. (2015). The conserved 3'X terminal domain of hepatitis C virus genomic RNA forms a two-stem structure that promotes viral RNA dimerization. *Nucleic Acids Research, 43*, 8529–8539. https://doi.org/10.1093/nar/gkv786.

Cantero-Camacho, Á., & Gallego, J. (2018). An unexpected RNA distal interaction mode found in an essential region of the hepatitis C virus genome. *Nucleic Acids Research, 46*, 4200–4212. https://doi.org/10.1093/nar/gky074.

Carlomagno, T., Amata, I., Codutti, L., Falb, M., Fohrer, J., Masiewicz, P., et al. (2013). Structural principles of RNA catalysis in a 2′-5′ lariat-forming ribozyme. *Journal of the American Chemical Society, 135*, 4403–4411. https://doi.org/10.1021/ja311868t.

Case, D.A.e.a. (2021). *Amber21.*

Castillo-Martínez, J., Fan, L., Szewczyk, M. P., Wang, Y. X., & Gallego, J. (2022). The low-resolution structural models of hepatitis C virus RNA subdomain 5BSL3.2 and its distal complex with domain 3'X point to conserved regulatory mechanisms within the Flaviviridae family. *Nucleic Acids Research, 50*, 2287–2301. https://doi.org/10.1093/nar/gkac061.

Castillo-Martínez, J., Ovejero, T., Romero-López, C., Sanmartín, I., Berzal-Herranz, B., Oltra, E., et al. (2020). Structure and function analysis of the essential 3'X domain of hepatitis C virus. *RNA, 26*, 186–198. https://doi.org/10.1261/rna.073189.119.

Cheng, C. Y., Chou, F. C., & Das, R. (2015). Modeling complex RNA tertiary folds with Rosetta. *Methods in Enzymology, 553*, 35–64. https://doi.org/10.1016/bs.mie.2014.10.051.

Clarke, B. D., Roby, J. A., Slonchak, A., & Khromykh, A. A. (2015). Functional non-coding RNAs derived from the flavivirus 3′ untranslated region. *Virus Research, 206*, 53–61. https://doi.org/10.1016/j.virusres.2015.01.026.

Cruz, J. A., & Westhof, E. (2009). The dynamic landscapes of RNA architecture. *Cell, 136*, 604–609. https://doi.org/10.1016/j.cell.2009.02.003.

Cruz, J. A., & Westhof, E. (2011). Sequence-based identification of 3D structural modules in RNA with RMDetect. *Nature Methods, 8*, 513–521. https://doi.org/10.1038/nmeth.1603.

Daugherty, M. D., Liu, B., & Frankel, A. D. (2010). Structural basis for cooperative RNA binding and export complex assembly by HIV rev. *Nature Structural & Molecular Biology, 17*, 1337–1342.

de la Pena, M., Dufour, D., & Gallego, J. (2009). Three-way RNA junctions with remote tertiary contacts: A recurrent and highly versatile fold. *RNA, 15*, 1949–1964. https://doi.org/10.1261/rna.1889509.

Diviney, S., Tuplin, A., Struthers, M., Armstrong, V., Elliott, R. M., Simmonds, P., et al. (2008). A hepatitis C virus cis-acting replication element forms a long-range RNA-RNA interaction with upstream RNA sequences in NS5B. *Journal of Virology, 82*, 9008–9022. https://doi.org/10.1128/JVI.02326-07.

Domnick, C., Eggert, F., Wuebben, C., Bornewasser, L., Hagelueken, G., Schiemann, O., et al. (2020). EPR distance measurements on long non-coding RNAs empowered by genetic alphabet expansion transcription. *Angewandte Chemie (International Ed. in English), 59*, 7891–7896. https://doi.org/10.1002/anie.201916447.

Ehrhardt, J. E., & Weeks, K. M. (2020). Time-resolved, single-molecule, correlated chemical probing of RNA. *Journal of the American Chemical Society, 142*, 18735–18740. https://doi.org/10.1021/jacs.0c06221.

Fang, X., Stagno, J. R., Bhandari, Y. R., Zuo, X., & Wang, Y. X. (2015). Small-angle X-ray scattering: A bridge between RNA secondary structures and three-dimensional topological structures. *Current Opinion in Structural Biology, 30*, 147–160. https://doi.org/10.1016/j.sbi.2015.02.010.

Fang, X., Wang, J., O'Carroll, I. P., Mitchell, M., Zuo, X., Wang, Y., et al. (2013). An unusual topological structure of the HIV-1 rev response element. *Cell, 155*, 594–605. https://doi.org/10.1016/j.cell.2013.10.008.

Feigin, L. A., & Svergun, D. I. (1987). *Structure analysis by small angle X-ray.* and Neutron Scattering (Plenum Press).

Fink, D. L., Chen, R. O., Noller, H. F., & Altman, R. B. (1996). Computational methods for defining the allowed conformational space of 16S rRNA based on chemical footprinting data. *RNA, 2*, 851–866.

Franke, D., & Svergun, D. I. (2009a). DAMMIF, a program for rapid ab-initio shape determination in small-angle scattering. *Journal of Applied Crystallography*, *42*(Pt 2), 342–346. https://doi.org/10.1107/S0021889809000338.

Franke, D., & Svergun, D. I. (2009b). Rapid ab initio shape determination by simulated annealing using a single phase dummy atom model. *Journal of Applied Crystallography*, *42*, 342–346.

Friebe, P., Boudet, J., Simorre, J. P., & Bartenschlager, R. (2005). Kissing-loop interaction in the 3′ end of the hepatitis C virus genome essential for RNA replication. *Journal of Virology*, *79*, 380–392. https://doi.org/10.1128/JVI.79.1.380-392.2005.

Funari, S. S., Rapp, G., Perbandt, M., Dierks, K., Vallazza, M., Betzel, C., et al. (2000). Structure of free Thermus flavus 5 S rRNA at 1.3 nm resolution from synchrotron X-ray solution scattering. *The Journal of Biological Chemistry*, *275*, 31283–31288. https://doi.org/10.1074/jbc.M004974200.

Glatter, O., & Kratky, O. (1982). *Small angle X-ray scattering*. Academic Press.

Harris, M. E., Nolan, J. M., Malhotra, A., Brown, J. W., Harvey, S. C., & Pace, N. R. (1994). Use of photoaffinity crosslinking and molecular modeling to analyze the global architecture of ribonuclease P RNA. *The EMBO Journal*, *13*, 3953–3963.

Hofacker, I. L., Fontana, W., Stadler, P. F., Bonhoeffer, S., Tacker, M., & Schuster, P. (1994). Fast folding and comparison of RNA secondary structures. *Monatshefte fur Chemie*, *125*, 167–188.

Holbrook, S. R. (2008). Structural principles from large RNAs. *Annual Review of Biophysics*, *37*, 445–464. https://doi.org/10.1146/annurev.biophys.36.040306.132755.

Homan, P. J., Tandon, A., Rice, G. M., Ding, F., Dokholyan, N. V., & Weeks, K. M. (2014). RNA tertiary structure analysis by 2′-hydroxyl molecular interference. *Biochemistry*, *53*, 6825–6833. https://doi.org/10.1021/bi501218g.

Houck-Loomis, B., Durney, M. A., Salguero, C., Shankar, N., Nagle, J. M., Goff, S. P., et al. (2011). An equilibrium-dependent retroviral mRNA switch regulates translational recoding. *Nature*, *480*, 561–564. https://doi.org/10.1038/nature10657.

Ivanyi-Nagy, R., Kanevsky, I., Gabus, C., Lavergne, J. P., Ficheux, D., Penin, F., et al. (2006). Analysis of hepatitis C virus RNA dimerization and core-RNA interactions. *Nucleic Acids Research*, *34*, 2618–2633. https://doi.org/10.1093/nar/gkl240.

Jones, C. P., Cantara, W. A., Olson, E. D., & Musier-Forsyth, K. (2014). Small-angle X-ray scattering-derived structure of the HIV-1 5' UTR reveals 3D tRNA mimicry. *Proceedings of the National Academy of Sciences of the United States of America*, *111*, 3395–3400. https://doi.org/10.1073/pnas.1319658111.

Kanamori, H., Yuhashi, K., Ohnishi, S., Koike, K., & Kodama, T. (2010). RNA-dependent RNA polymerase of hepatitis C virus binds to its coding region RNA stem-loop structure, 5BSL3.2, and its negative strand. *The Journal of General Virology*, *91*, 1207–1212. https://doi.org/10.1099/vir.0.016907-0.

Kao, C. C., Yang, X., Kline, A., Wang, Q. M., Barket, D., & Heinz, B. A. (2000). Template requirements for RNA synthesis by a recombinant hepatitis C virus RNA-dependent RNA polymerase. *Journal of Virology*, *74*, 11121–11128.

Kazantsev, A. V., Krivenko, A. A., Harrington, D. J., Holbrook, S. R., Adams, P. D., & Pace, N. R. (2005). Crystal structure of a bacterial ribonuclease P RNA. *Proceedings of the National Academy of Sciences of the United States of America*, *102*, 13392–13397. https://doi.org/10.1073/pnas.0506662102.

Kazantsev, A. V., Rambo, R. P., Karimpour, S., Santalucia, J., Jr., Tainer, J. A., & Pace, N. R. (2011). Solution structure of RNase P RNA. *RNA*, *17*, 1159–1171. https://doi.org/10.1261/rna.2563511.

Koch, M. H., Vachette, P., & Svergun, D. I. (2003). Small-angle scattering: A view on the properties, structures and structural changes of biological macromolecules in solution. *Quarterly Reviews of Biophysics*, *36*, 147–227.

Kolykhalov, A. A., Feinstone, S. M., & Rice, C. M. (1996). Identification of a highly conserved sequence element at the 3′ terminus of hepatitis C virus genome RNA. *Journal of Virology*, *70*, 3363–3371.

Konarev, P. V., Volkov, V. V., Sokolova, A. V., Kochb, M. H. J., & Svergun, D. I. (2003). PRIMUS: A windows PC-based system for small-angle scattering data analysis. *Journal of Applied Crystallography*, *36*, 1277–1282.

Kozin, M. B., & Svergun, D. I. (2001). Automated matching of high- and low-resolution structural models. *Journal of Applied Crystallography*, *34*, 33–41. https://doi.org/10.1107/S0021889800014126.

Laing, C., Wen, D., Wang, J. T., & Schlick, T. (2012). Predicting coaxial helical stacking in RNA junctions. *Nucleic Acids Research*, *40*, 487–498. https://doi.org/10.1093/nar/gkr629.

Lake, J. A. (1967). Yeast transfer RNA: A small-angle x-ray study. *Science*, *156*, 1371–1373.

Lee, E., Bujalowski, P. J., Teramoto, T., Gottipati, K., Scott, S. D., Padmanabhan, R., et al. (2021). Structures of flavivirus RNA promoters suggest two binding modes with NS5 polymerase. *Nature Communications*, *12*, 2530. https://doi.org/10.1038/s41467-021-22846-1.

Legiewicz, M., Badorrek, C. S., Turner, K. B., Fabris, D., Hamm, T. E., Rekosh, D., et al. (2008). Resistance to RevM10 inhibition reflects a conformational switch in the HIV-1 rev response element. *Proceedings of the National Academy of Sciences of the United States of America*, *105*, 14365–14370.

Lehnert, V., Jaeger, L., Michel, F., & Westhof, E. (1996). New loop-loop tertiary interactions in self-splicing introns of subgroup IC and ID: A complete 3D model of the Tetrahymena thermophila ribozyme. *Chemistry & Biology*, *3*, 993–1009.

Leontis, N. B., Lescoute, A., & Westhof, E. (2006). The building blocks and motifs of RNA architecture. *Current Opinion in Structural Biology*, *16*, 279–287.

Lescoute, A., & Westhof, E. (2006). Topology of three-way junctions in folded RNAs. *RNA*, *12*, 83–93. https://doi.org/10.1261/rna.2208106.

Liberman, J. A., Salim, M., Krucinska, J., & Wedekind, J. E. (2013). Structure of a class II preQ1 riboswitch reveals ligand recognition by a new fold. *Nature Chemical Biology*, *9*, 353–355. https://doi.org/10.1038/nchembio.1231.

Lipfert, J., Chu, V. B., Bai, Y., Herschlag, D., & Doniach, S. (2007). Low-resolution models for nucleic acids from small-angle X-ray scattering with applications to electrostatic modeling. *Journal of Applied Crystallography*, *40*, S229–S234. https://doi.org/10.1107/S0021889807001707.

Lipfert, J., Ouellet, J., Norman, D. G., Doniach, S., & Lilley, D. M. (2008). The complete VS ribozyme in solution studied by small-angle X-ray scattering. *Structure*, *16*, 1357–1367. https://doi.org/10.1016/j.str.2008.07.007.

Lu, C., & Li, P. (2012). Preparation of short RNA by in vitro transcription. *Methods in Molecular Biology*, *941*, 59–68. https://doi.org/10.1007/978-1-62703-113-4_5.

Ma, L., Bajic, V. B., & Zhang, Z. (2013). On the classification of long non-coding RNAs. *RNA Biology*, *10*, 925–933. https://doi.org/10.4161/rna.24604.

Magnus, M., Boniecki, M. J., Dawson, W., & Bujnicki, J. M. (2016). SimRNAweb: A web server for RNA 3D structure modeling with optional restraints. *Nucleic Acids Research*, *44*, W315–W319. https://doi.org/10.1093/nar/gkw279.

Martinez, H. M., Maizel, J. V., Jr., & Shapiro, B. A. (2008). RNA2D3D: A program for generating, viewing, and comparing 3-dimensional models of RNA. *Journal of Biomolecular Structure & Dynamics*, *25*, 669–683. https://doi.org/10.1080/07391102.2008.10531240.

Massire, C., & Westhof, E. (1998). MANIP: An interactive tool for modelling RNA. *Journal of Molecular Graphics & Modelling*, *16*, 197–205. 255-197.

Matsuda, D., & Mauro, V. P. (2014). Base pairing between hepatitis C virus RNA and 18S rRNA is required for IRES-dependent translation initiation in vivo. *Proceedings of the National Academy of Sciences of the United States of America, 111*, 15385–15389. https://doi.org/10.1073/pnas.1413472111.

Mattick, J. S., & Makunin, I. V. (2006). Non-coding RNA. *Human Molecular Genetics, 15Spec No 1*, R17–R29. https://doi.org/10.1093/hmg/ddl046.

Metropolis, N., Rosenbluth, A. W., Rosenbluth, M. N., Teller, A. H., & Teller, E. (1953). Equation of state calculations by fast computing machines. *The Journal of Chemical Physics, 21*, 1087–1092.

Michel, F., & Westhof, E. (1990). Modelling of the three-dimensional architecture of group I catalytic introns based on comparative sequence analysis. *Journal of Molecular Biology, 216*, 585–610. https://doi.org/10.1016/0022-2836(90)90386-Z.

Misra, V. K., & Draper, D. E. (1998). On the role of magnesium ions in RNA stability. *Biopolymers, 48*, 113–135. https://doi.org/10.1002/(SICI)1097-0282(1998)48:2<113::AID-BIP3>3.0.CO;2-Y.

Nilsen, T. W., Rio, D. C., & Ares, M., Jr. (2013). High-yield synthesis of RNA using T7 RNA polymerase and plasmid DNA or oligonucleotide templates. *Cold Spring Harbor Protocols, 1*, 1091–1093. https://doi.org/10.1101/pdb.prot078535.

Orton, H. W., & Otting, G. (2018). Accurate electron-nucleus distances from paramagnetic relaxation enhancements. *Journal of the American Chemical Society, 140*, 7688–7697. https://doi.org/10.1021/jacs.8b03858.

Parisien, M., & Major, F. (2008). The MC-fold and MC-Sym pipeline infers RNA structure from sequence data. *Nature, 452*, 51–55. https://doi.org/10.1038/nature06684.

Pelikan, M., Hura, G. L., & Hammel, M. (2009). Structure and flexibility within proteins as identified through small angle X-ray scattering. *General Physiology and Biophysics, 28*, 174–189. https://doi.org/10.4149/gpb_2009_02_174.

Piñeiro, D., & Martinez-Salas, E. (2012). RNA structural elements of hepatitis C virus controlling viral RNA translation and the implications for viral pathogenesis. *Viruses, 4*, 2233–2250. https://doi.org/10.3390/v4102233.

Quade, N., Boehringer, D., Leibundgut, M., van den Heuvel, J., & Ban, N. (2015). Cryo-EM structure of hepatitis C virus IRES bound to the human ribosome at 3.9-Å resolution. *Nature Communications, 6*, 7646. https://doi.org/10.1038/ncomms8646.

Ramani, V., Qiu, R., & Shendure, J. (2015). High-throughput determination of RNA structure by proximity ligation. *Nature Biotechnology, 33*, 980–984. https://doi.org/10.1038/nbt.3289.

Rambo, R. P., & Tainer, J. A. (2010). Improving small-angle X-ray scattering data for structural analyses of the RNA world. *RNA, 16*, 638–646. https://doi.org/10.1261/rna.1946310.

Reiter, N. J., Chan, C. W., & Mondragon, A. (2010). Emerging structural themes in large RNA molecules. *Current Opinion in Structural Biology, 21*, 319–326.

Rinn, J. L., & Chang, H. Y. (2012). Genome regulation by long noncoding RNAs. *Annual Review of Biochemistry, 81*, 145–166. https://doi.org/10.1146/annurev-biochem-051410-092902.

Romero-López, C., & Berzal-Herranz, A. (2009). A long-range RNA-RNA interaction between the 5′ and 3′ ends of the HCV genome. *RNA, 15*, 1740–1752. https://doi.org/10.1261/rna.1680809.

Romero-López, C., & Berzal-Herranz, A. (2012). The functional RNA domain 5BSL3.2 within the NS5B coding sequence influences hepatitis C virus IRES-mediated translation. *Cellular and Molecular Life Sciences, 69*, 103–113. https://doi.org/10.1007/s00018-011-0729-z.

Rother, M., Rother, K., Puton, T., & Bujnicki, J. M. (2011). ModeRNA: A tool for comparative modeling of RNA 3D structure. *Nucleic Acids Research*, *39*, 4007–4022. https://doi.org/10.1093/nar/gkq1320.

Russell, R., Millett, I. S., Doniach, S., & Herschlag, D. (2000). Small angle X-ray scattering reveals a compact intermediate in RNA folding. *Nature Structural Biology*, *7*, 367–370. https://doi.org/10.1038/75132.

Russell, R., Millett, I. S., Tate, M. W., Kwok, L. W., Nakatani, B., Gruner, S. M., et al. (2002). Rapid compaction during RNA folding. *Proceedings of the National Academy of Sciences of the United States of America*, *99*, 4266–4271. https://doi.org/10.1073/pnas.072589599.

Saenger, W. (1984). *Principles of nucleic acid structure*. NY: Springer-Verlag.

Schneidman-Duhovny, D., Hammel, M., & Sali, A. (2010). FoXS: A web server for rapid computation and fitting of SAXS profiles. *Nucleic Acids Research*, *38*, W540–W544. https://doi.org/10.1093/nar/gkq461.

Schneidman-Duhovny, D., Hammel, M., Tainer, J. A., & Sali, A. (2016). FoXS, FoXSDock and MultiFoXS: Single-state and multi-state structural modeling of proteins and their complexes based on SAXS profiles. *Nucleic Acids Research*, *44*, W424–W429. https://doi.org/10.1093/nar/gkw389.

Schwieters, C. D., Kuszewski, J. J., Tjandra, N., & Clore, G. M. (2003). The Xplor-NIH NMR molecular structure determination package. *Journal of Magnetic Resonance*, *160*, 65–73. https://doi.org/10.1016/s1090-7807(02)00014-9.

Shetty, S., Kim, S., Shimakami, T., Lemon, S. M., & Mihailescu, M. R. (2010). Hepatitis C virus genomic RNA dimerization is mediated via a kissing complex intermediate. *RNA*, *16*, 913–925. https://doi.org/10.1261/rna.1960410.

Shetty, S., Stefanovic, S., & Mihailescu, M. R. (2013). Hepatitis C virus RNA: Molecular switches mediated by long-range RNA-RNA interactions? *Nucleic Acids Research*, *41*, 2526–2540. https://doi.org/10.1093/nar/gks1318.

Svergun, D. I. (1999a). Restoring low resolution structure of biological macromolecules from solution scattering using simulated annealing. *Biophysical Journal*, *76*, 2879–2886. https://doi.org/10.1016/s0006-3495(99)77443-6.

Svergun, D. I. (1999b). Restoring low resolution structure of biological macromolecules from solution scattering using simulated annealing. *Biophysical Journal*, 77, 2896.

Townshend, R. J. L., Eismann, S., Walkins, A. M., Rangan, R., Karelina, M., Das, R., et al. (2021). Geometric deep learning of RNA structure. *Science*, *373*, 4. https://doi.org/10.1126/science.abe5650.

Tuplin, A., Struthers, M., Cook, J., Bentley, K., & Evans, D. J. (2015). Inhibition of HCV translation by disrupting the structure and interactions of the viral CRE and 3′ X-tail. *Nucleic Acids Research*, *43*, 2914–2926. https://doi.org/10.1093/nar/gkv142.

Tuplin, A., Struthers, M., Simmonds, P., & Evans, D. J. (2012). A twist in the tail: SHAPE mapping of long-range interactions and structural rearrangements of RNA elements involved in HCV replication. *Nucleic Acids Research*, *40*, 6908–6921. https://doi.org/10.1093/nar/gks370.

Vogel, S. S., van der Meer, B. W., & Blank, P. S. (2014). Estimating the distance separating fluorescent protein FRET pairs. *Methods*, *66*, 131–138. https://doi.org/10.1016/j.ymeth.2013.06.021.

Volkov, V. V., & Svergun, D. I. (2003). Uniqueness of ab initio shape determination in small-angle scattering. *Journal of Applied Crystallography*, *36*, 860–864. https://doi.org/10.1107/S0021889803000268.

Wang, J., Zuo, X., Yu, P., Xu, H., Starich, M. R., Tiede, D. M., et al. (2009). A method for helical RNA global structure determination in solution using small-angle x-ray scattering and NMR measurements. *Journal of Molecular Biology*, *393*, 717–734. https://doi.org/10.1016/j.jmb.2009.08.001.

Weeks, K. M. (2010). Advances in RNA structure analysis by chemical probing. *Current Opinion in Structural Biology*, *20*, 295–304. https://doi.org/10.1016/j.sbi.2010.04.001.

Weinreb, C., Riesselman, A. J., Ingraham, J. B., Gross, T., Sander, C., & Marks, D. S. (2016). 3D RNA and functional interactions from evolutionary couplings. *Cell*, *165*, 963–975. https://doi.org/10.1016/j.cell.2016.03.030.

Westhof, E., & Altman, S. (1994). 3-dimensional working model of M1 RNA, the catalytic RNA subunit of ribonuclease-P from Escherichia coli. *Proceedings of the National Academy of Sciences of the United States of America*, *91*, 5133–5137. https://doi.org/10.1073/pnas.91.11.5133.

Williams, C. J., Headd, J. J., Moriarty, N. W., Prisant, M. G., Videau, L. L., Deis, L. N., et al. (2018). MolProbity: More and better reference data for improved all-atom structure validation. *Protein Science*, *27*, 293–315. https://doi.org/10.1002/pro.3330.

Witz, J. (2003). 1964: The first model for the shape of a transfer RNA molecule. An account of an unpublished small-angle X-ray scattering study. *Biochimie*, *85*, 1265–1268.

Yamada, N., Tanihara, K., Takada, A., Yorihuzi, T., Tsutsumi, M., Shimomura, H., et al. (1996). Genetic organization and diversity of the 3′ noncoding region of the hepatitis C virus genome. *Virology*, *223*, 255–261. https://doi.org/10.1006/viro.1996.0476.

You, S., & Rice, C. M. (2008). 3' RNA elements in hepatitis C virus replication: Kissing partners and long poly(U). *Journal of Virology*, *82*, 184–195. https://doi.org/10.1128/JVI.01796-07.

You, S., Stump, D. D., Branch, A. D., & Rice, C. M. (2004). A cis-acting replication element in the sequence encoding the NS5B RNA-dependent RNA polymerase is required for hepatitis C virus RNA replication. *Journal of Virology*, *78*, 1352–1366.

Zhang, J., Jones, C. P., & Ferre-D'Amare, A. R. (2014). Global analysis of riboswitches by small-angle X-ray scattering and calorimetry. *Biochimica et Biophysica Acta*, *1839*, 1020–1029. https://doi.org/10.1016/j.bbagrm.2014.04.014.

Zhang, J., Yamada, O., Sakamoto, T., Yoshida, H., Araki, H., Murata, T., et al. (2005). Inhibition of hepatitis C virus replication by pol III-directed overexpression of RNA decoys corresponding to stem-loop structures in the NS5B coding region. *Virology*, *342*, 276–285. https://doi.org/10.1016/j.virol.2005.08.003.

Zhang, Y., Zhang, Y., Liu, Z.-Y., Cheng, M.-L., Ma, J., Wang, Y., et al. (2019). Long non-coding subgenomic flavivirus RNAs have extended 3D structures and are flexible in solution. *EMBO Reports*, *20*, e47016. https://doi.org/10.15252/embr.201847016.

Zuo, X. B., Wang, J. B., Yu, P., Eyler, D., Xu, H., Starich, M. R., et al. (2010). Solution structure of the cap-independent translational enhancer and ribosome-binding element in the 3 ' UTR of turnip crinkle virus. *Proceedings of the National Academy of Sciences of the United States of America*, *107*, 1385–1390. https://doi.org/10.1073/pnas.0908140107.

CHAPTER EIGHTEEN

Disentangling polydisperse biomolecular systems by Chemometrics decomposition of SAS data

Amin Sagar* and Pau Bernadó*
Centre de Biologie Structurale (CBS), Université de Montpellier, INSERM, CNRS, Montpellier, France
*Corresponding authors: e-mail address: amin.sagar@cbs.cnrs.fr; pau.bernado@cbs.cnrs.fr

Contents

Abstract

The structural characterization of polydisperse systems consisting of multiple coexisting species or conformations is very challenging or impossible with classical approaches. As a consequence, the structural bases of relevant questions related to protein folding, transient partner recognition, conformational transitions or fibrillation remain poorly understood. Small-Angle Scattering (SAS) techniques structurally probe species present in solution in a population-weighted manner, enabling the inspection of polydisperse systems. However, decomposition of these data to derive the contribution of individual components is not straightforward and requires the acquisition of large SAS datasets and adapted mathematical tools. Here, we present a detailed procedure for the usage of the program COSMiCS for the decomposition of SAS datasets. COSMiCS adapts the popular MCR-ALS chemometrics routine to the specificities of scattering data. Through the use of multiple SAS representations, the appropriate scaling of the data and the

Methods in Enzymology, Volume 677
ISSN 0076-6879
https://doi.org/10.1016/bs.mie.2022.08.038

possibility to simultaneously decompose multiple orthogonal datasets, COSMiCS efficiently disentangles mixtures and provides species-specific structural and thermodynamic/kinetic information of the process under investigation. Although exemplified for a transient biomolecular interaction, our chemometrics strategy can be applied to many other biological processes that can be straightforwardly probed in last generation SAS beamlines. Indeed, recent experimental setups, including microfluidics and stop-flow devices, coupled to fast-reading detectors can yield large concentration or time-dependent datasets that can be decomposed with COSMiCS. Importantly, as an open-source code, previously known features of the system of interest can be introduced as constraints in the optimization, producing robust solutions for biological systems of increasing complexity.

1. Introduction

Proteins orchestrate the vast majority of biological processes either in isolation or by forming active complexes with other biomolecules (Petsko & Ringe, 2004). Often, protein activity is regulated by external stimuli that trigger a conformational change or the recognition of physiological partners to form the active complex. Photons, ions, small molecules or post-translational modifications are among the most common external stimuli regulating protein function. Understanding biological processes requires a detailed three-dimensional description of molecules involved and the conformational transitions facilitating the function (Bernadó & Blackledge, 2010). Structural genomics initiatives and technological developments in the last decades have enormously boosted the range of biological systems amenable to structural biology, as well as the quality and resolution of the resulting models (Chandonia & Brenner, 2006; Kuhlbrandt, 2014). Moreover, recent algorithmic developments, based on the accumulated knowledge on protein structure, enable the prediction of accurate models from amino acid sequence (Baek et al., 2021; Jumper et al., 2021). As a consequence, our knowledge of biomolecular structure has been enormously enriched enabling a molecular understanding of key biological processes.

Despite these advances, several biological processes remain challenging for traditional structural biology approaches. For instance, conformational rearrangements occurring upon ligand binding and low affinity, transient biomolecular complexes are not always captured by high-resolution methods. The inherent conformational and/or compositional polydispersity of these systems is at the origin of the difficulties experienced by structural biology methods for their study. Small-angle X-ray scattering (SAXS), despite its low resolution, is a very well adapted technique to study these challenging systems in near physiological conditions (Bernadó, Shimizu,

Zaccai, Kamikubo, & Sugiyama, 2018; Feigin & Svergun, 1987; Jacques & Trewhella, 2010; Putnam, Hammel, Hura, & Tainer, 2007). SAXS is sensitive to the overall size and shape of biomolecules in solution and, in the presence of polydispersity, measured scattering profiles report on the different species present in the solution in a population-weighted manner (Bernadó & Svergun, 2012; Doniach, 2001). Importantly, the versatility of present dedicated biological SAXS beamlines in brilliant synchrotron X-ray sources also allows fast and precise measurements of large SAXS datasets in which the experimental conditions (concentration, ligand, pH, ionic strength, etc.) can be modified in a controlled manner (Hura et al., 2013). Alternatively, by triggering a specific biological process using rapid mixers, microfluidic chips or optical devices, time-resolved SAXS (TR-SAXS) data can be recorded, probing kinetics in a broad range of timescales (Akiyama et al., 2002; Heyes et al., 2019; Josts et al., 2018; Josts et al., 2020). Using these strategies, it is straightforward to gradually modify the relative populations of the distinct conformational or compositional states of the protein of interest. However, these redundant data must be disentangled to derive valuable structural and thermodynamic/kinetic information of the process. SAXS data decomposition is straightforward if structural models of the involved species are known or can be reasonably predicted (Bernadó et al., 2009; Konarev, Volkov, Sokolova, Koch, & Svergun, 2003; Petoukhov et al., 2012). However, this premise is not always fulfilled, and more general approaches must be applied.

Objective decomposition of large and complex experimental datasets using chemometrics is routinely used in many scientific disciplines spanning from analytical and organic chemistry to metabolomics (de Juan & Tauler, 2003; Trygg, Holmes, & Lundstedt, 2007). One of the most widely applied chemometric approaches is the Multivariate Curve Resolution using Alternate Least Squares (MCR-ALS) (Jaumot, Gargallo, De Juan, & Tauler, 2005; Tauler, 1995). The basic assumption of MCR is that the measured data, **D**, can be decomposed as the sum of individual contributions of the coexistent components, each one of them defined by a spectroscopic profile, composing the matrix $\mathbf{S}^{\mathbf{T}}$, and weighted according to their respective concentrations, **C**. The bilinear model is described by:

$$\mathbf{D} = \mathbf{C}\mathbf{S}^{\mathbf{T}} + \mathbf{R} \tag{1}$$

where **R** is the matrix of residuals. Eq. (1) can be resolved using different approaches, including the ALS algorithm. However, MCR-ALS solutions are very often ambiguous (Rajkó, 2010; Tauler, 2001). This problem can be partially alleviated by applying mathematical restraints along the optimization process, for example, forcing non-negative values for the spectra and concentrations or preserving the mass balance (see below). A more efficient strategy to reduce ambiguity is the simultaneous optimization of multiple experimental datasets that can be achieved by either monitoring the same samples using different techniques, *multiway*, using the same technique but modifying the experimental conditions, *multiset*, or both simultaneously, *multiway/multiset* (de Juan & Tauler, 2003).

Chemometrics in general, and MCR-ALS in particular, have been used to decompose SAXS data in order to investigate transient biomolecular interactions (Chandola, Williamson, Craig, Friedman, & Bailey-Kellogg, 2014; Lipfert, Coulmbus, Chu, & Doniach, 2007; Minh & Makowski, 2013; Sagar, Herranz-Trillo, Langkilde, Vestergaard, & Bernadó, 2021; Williamson, Craig, Kondrashkina, Bailey-Kellogg, & Friedman, 2008), protein oligomerization (Blobel, Bernadó, Svergun, Tauler, & Pons, 2009; Taube et al., 2019), conformational transitions upon ligand binding (Meisburger, Xu, & Ando, 2021), protein and RNA folding (Akiyama et al., 2002; Ayuso-Tejedor, García-Fandiño, Orozco, Sancho, & Bernadó, 2011; Chen, Hodgson, & Doniach, 1996; Meisburger et al., 2021; Plumridge et al., 2018; Segel et al., 1999), viral capsid assembly (Chevreuil et al., 2020; Le Vay et al., 2020; Oliver et al., 2020), and amyloidogenic processes (Corsini et al., 2022; Herranz-Trillo et al., 2017). Moreover, chemometrics approaches have also been found useful in interpreting size-exclusion (SEC) and ion-exchange (IEX) chromatographies coupled to SAXS where they are applied to decompose partially overlapped chromatographic peaks (Chevreuil et al., 2020; Hopkins, Gillilan, & Skou, 2017; Konarev et al., 2022; Meisburger et al., 2016; Tully, Tarbouriech, Rambo, & Hutin, 2021). In addition to the shape of the peaks, which represent their population, these strategies provide the SAXS profiles of the eluting species. In this chapter, SEC-SAXS and IEX-SAXS will not be described, and the reader is referred to other dedicated chapters in this book.

Here, we describe the use of the program COSMiCS (Complex Objective Structural analysis of Multi-Component Systems) to decompose SAS datasets derived from titration or time-resolved experiments. Note that although the procedure is applied to the analysis of SAXS datasets, Small-Angle Neutron Scattering (SANS) data can be analyzed in the same way. COSMiCS adapts MCR-ALS to the specific characteristics of SAXS data, i.e., the disparity of intensity values within and between experimental curves and the limited presence of features in SAXS curves (Herranz-Trillo et al., 2017; Sagar et al., 2021). Concretely, COSMiCS uses multiple scattering data representations in a *multiway* manner to emphasize the signal at different resolutions and scales the curves to minimize the intensity differences between the curves of the dataset. Moreover, the program can handle multiple SAXS datasets measured in different conditions, enabling the *multiset/multiway* mode, which notably improves the accuracy of the solutions. By addressing these issues, COSMiCS has become a general tool to address challenging biological questions involving conformational transitions and transient protein complexes exploiting SAXS data that can be measured in present synchrotrons. COSMiCS has already been used to study amyloid formation by Insulin and α-synuclein (Herranz-Trillo et al., 2017) and biofilm formation by bacterial protein Sse1 (Corsini et al., 2022).

2. General description of COSMiCS

In the following section, we will describe COSMiCS including the mathematical bases of MCR-ALS, the overall description of COSMiCS architecture, and the description of the restraints that can be implemented for data analyses.

2.1 Mathematical considerations of MCR-ALS

MCR is a powerful tool for advanced multivariate data analysis used in a broad range of disciplines (Jaumot et al., 2005; Tauler, 1995). Note that the following mathematical description of MCR-ALS is adapted to SAXS experiments and COSMiCS usage. The basic assumption of MCR is that measured data variance can be decomposed as the weighted sum of individual contributions from the coexisting components, each one of them defined by a set of profiles corresponding to each data representation applied and weighted according to their composition in the analyzed mixture. The bilinear MCR model is described in Eq. (1), where the dimensions of the matrices are: $\mathbf{D}(I,J)$, $\mathbf{C}(I,N)$, $\mathbf{S}^T$ (N,J) and $\mathbf{R}(I,J)$, where I the number of rows in data matrix $\mathbf{D}$ (i.e., the number of profiles measured varying, for instance, protein concentrations or time), J the number of columns in data matrix $\mathbf{D}$ (the number of momentum transfer, q, points), and N the number of components (chemical species contributing to the scattering signal). $\mathbf{C}$ describes the composition contributions of the N components for each of the experimental conditions tested. $\mathbf{S}^T$ is the matrix containing the pure SAXS profiles of the N components. Due to unavoidable experimental and modeling uncertainties, including experimental noise, this matrix decomposition is not perfect, and the differences between the measured data and their decomposition are collected in **R,** the residual matrix.

The problem to solve in MCR can be mathematically stated in the following way: given the data matrix **D** find: (i) N, the number of species causing the observed data variance, (ii) find the concentration profiles of these components in matrix **C**, and (iii) find the pure scattering profiles of these components in matrix $\mathbf{S}^{\mathrm{T}}$. Stated in this way, and without any further constraint, there is not a unique set of matrices **C** and $\mathbf{S}^{\mathrm{T}}$ that solve Eq. (1). Indeed, there is an infinite number of mathematically equivalent solutions of **C** and $\mathbf{S}^{\mathrm{T}}$ that, when multiplied by each other, give the same result due to the rotational and scaling degrees of freedom (Malinowski, 2002). Fortunately, it is usually possible to considerably reduce the number of potential solutions by introducing prior knowledge of the problem under study and/or constraints derived from the physical nature of the system.

2.2 Constraints used in COSMiCS

A constraint is defined as a particular characteristic of chemical or mathematical nature that the SAXS curves of the pure components or their concentration profiles must obey. The following constraints are implemented in COSMiCS:

- *Non-negativity:* This constraint forces the decomposed matrices to only adopt positive or zero values. It can always be applied to the concentration profiles and also to several types of spectra, such as SAXS curves. The application of the *non-negativity* constraint can be carried out according to different least squares approaches, the classical non-negative least squares (nnls) (Lawson & Hanson, 1974) and the fast non-negative least squares (fnnls) (Bro & de Jong, 1997). An additional option is replacing negative values by zeros, which is useful when the other algorithms fail or take too long. Although the three algorithms are implemented, fnnls is used by default in COSMiCS
- *Unimodality:* This constraint forces the presence of a single maximum per profile, which can be potentially applied to curves and concentrations. MCR-ALS can apply this constraint using three different options. "Vertical" and "horizontal" options cut the secondary maxima vertically or horizontally, respectively (de Juan, van der Heyden, Tauler, & Massart, 1997). A third implementation, "average," which is used by default in COSMiCS, applies the constraint in a smoother way, taking averages as used in unimodal least squares algorithms (Bro & Sidiropoulos, 1998).

- *Selectivity*: This constraint is applied when one or several species are known to be absent in a particular window of the dataset. The most restrictive use of the *selectivity* constraint is when only a single species is present in a region of the dataset, significantly reducing the ambiguity problem. The SAXS curves of the individual partners in titration experiments are incorporated in COSMiCS analyses by applying the *selectivity* constraint (see below).
- *Equality:* This constraint is similar to *selectivity* except it specifies fixed SAXS profiles rather than concentrations. This ensures that the specified initial estimates are not changed during the optimization process. In some circumstances this constraint can be useful. For example, if a reaction intermediate with known structure/SAXS profile is present in small concentrations, its SAXS profile can be included as one of the initial estimates and kept unchanged during the optimization. However, imposing *selectivity* should be preferred over *equality* as it profits from the information coded in the whole dataset. Also, note that *selectivity* and *equality* have different meanings for the MCR-ALS program. The users are advised to read the MCR-ALS manual if they want to alter the constraints inside MCR-ALS
- *Closure:* This constraint, which is related to mass balance, is introduced for the concentration matrix **C**. In COSMiCS, the total concentration of the system can be fixed to a single value, usually equal to 1, (i.e., folding and conformational transitions) or a variable value if the specific concentrations of the partners along the experiment are known (i.e., titration experiments with known dilutions). In addition, the stoichiometry of the interaction can be defined. For example, for a monomer-dimer system, $[M]+2[D]$ is constant along a concentration series
- *Custom restraints:* It is possible to introduce custom restraints adapted to the system under study to improve decomposition. For example, to improve the performance for monomer-dimer equilibrium of albumin, we implemented a custom restraint that does not allow the concentrations of either monomer or dimer to decrease when increasing the total concentration of the protein (Sagar et al., 2021).

2.3 Practical aspects of COSMiCS

The software has been implemented under MATLAB© and does not need any toolbox apart from the MATLAB standard core program. It has been tested on MATLAB© R2016a, R2020a and R2021a. It needs MCR-ALS to be in the path of MATLAB©. The command line version of

MCR-ALS can be downloaded from http://www.mcrals.info/. The software has been tested on computers under Linux Ubuntu 12, Mac OS 10.10 "Yosemite" and Windows 10 with no need of any additional resources. COSMiCS uses the MATLAB interface to introduce the input data and the parameters of the optimization. COSMiCS can be downloaded from: https://github.com/BernadoLab. Theoretical SAXS datasets generated by simulating titrations for 18 transient protein–protein complexes can be accessed from https://github.com/BernadoLab/SimulatedSAXSdata-Structure2021. The complexes included in this dataset have diverse dissociation constants, shapes (aspect ratio/asphericity) and sizes making it a useful benchmark for assessing the performance of other decomposition methods.

3. Considerations for SAXS measurements

The successful application of COSMiCS requires relatively large SAXS datasets (>10 curves/dataset) appropriately sampling the compositional or conformational space of interest. The experimental conditions modified along the experiment must induce systematic changes in the composition/conformation of the protein. For instance, to study transient biomolecular interactions, titration experiments, in which the relative concentrations of the interacting proteins are gradually changed, are necessary. Equivalently, to study an oligomerization process, one must systematically modify the concentration of the protein. Note that the species exhibiting the same population profiles will not be decomposed by chemometrics approaches and will provide a mixed scattering curve. Indeed, enhancing the differences between the population profiles increases the decomposition capacity of chemometrics approaches. This highlights the relevance of designing an appropriate measurement strategy.

COSMiCS allows the simultaneous decomposition of multiple SAXS datasets. Datasets measured using distinct experimental conditions (*multiset*) increase the disparity of concentration profiles and facilitate decomposition. However, these multiple datasets must contain (at least) one common species, which will evolve differently in each of the datasets. For instance, when studying a transient biomolecular complex ($A + B \rightarrow AB$), one can imagine titrating a given amount of A with increasing quantities of B, and the reverse experiment. The two resulting datasets will contain information about the same three species, but they will display completely different concentration profiles in both datasets. In other types of scenarios, such as folding or ligand titrations, measuring complementary datasets is possible, although the equivalence of the contributing species is not always guaranteed.

Although the preparation of tens of samples by hand is feasible, it is prone to pipetting errors, which could challenge subsequent analyses. Note that inaccurate concentration estimation can lead to erroneous solutions when applying the *closure* restraint. Microfluidics devices are a very well-suited alternative to obtain large SAXS datasets with precise modification of the solutions composing the mixture using a limited amount of material (Silva, 2017). Multiple chip architectures adapted to SAXS beamlines have been described enabling all types of mixtures (Lafleur et al., 2011; Rodríguez-Ruiz et al., 2017; Schwemmer et al., 2016; Toft et al., 2008).

Experimental datasets can also originate from time-resolved SAXS experiments (TR-SAXS) (Cho, Schotte, Stadnytskyi, & Anfinrud, 2021). Biomolecular folding, conformational changes and aggregation are examples of biological processes, with timescales spanning from microseconds to hours, that can benefit from chemometrics decomposition (Giehm, Svergun, Otzen, & Vestergaard, 2011; Graceffa et al., 2013; Josts et al., 2020; Plumridge et al., 2018). A well-defined event triggering the process is mandatory to initiate the reaction that will be monitored in a time-dependent manner. This can be achieved by temperature (Thompson et al., 2019) or pH jumps (Rimmerman et al., 2019), rapid mixing of components (Plumridge et al., 2018), the use of photoinduced reactions (Berntsson et al., 2019; Heyes et al., 2019), or by initiating a defined aggregation protocol (Langkilde, Herranz-Trillo, Bernadó, & Vestergaard, 2018).

4. Procedure for COSMiCS decomposition

4.1 Decomposition steps

This section describes in a step-by-step manner the decomposition of SAXS datasets using COSMiCS, which is graphically depicted in Fig. 1. We will use the 1AK4 dataset available from the GitHub repository as an example. The statements printed on the MATLAB command window are shown in blue font while the input introduced by the user are shown in black bold font.

1. *Initialization*: Open MATLAB, navigate to the COSMiCS directory and open the script COSMiCS_multi.m. Also add COSMiCS2021-main directory to the MATLAB's path by simply right-clicking on the directory name followed by Add to path/Selected folders and subfolders. Also add the MCR-ALS directory to the path using the same method. Now, all the COSMiCS and MCR-ALS scripts should be accessible to MATLAB. Execute the COSMiCS_multi.m script by clicking on "Run" in the Editor tab

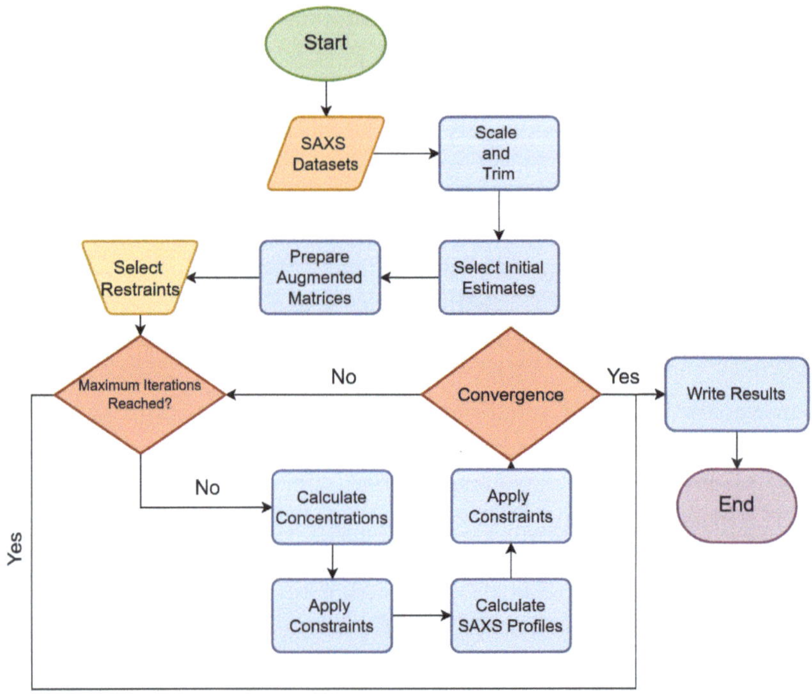

Fig. 1 A flowchart showing the steps involved in SAS data decomposition with COSMiCS.

2. *Input/Output definition*: As COSMiCS allows for simultaneous decomposition of multiple SAXS datasets, the user is first asked to define the number of datasets to be decomposed. As in this example we will decompose two datasets (one titration per partner), enter 2 and press "Enter."

Please enter the number of experiments performed

Number of experiments[1] **2**

This is followed by specifying the locations (paths) of the datasets. All the files with names following a particular pattern or having the same extension, specified by the user, will be then loaded from the defined locations.

Select the folder containing the data from experiment number 1 (press any key to continue)

Press 'Enter' and browse to the folder /path-to-data/1ak4/50-curves/TitrAwithB0p2to5 and click on 'Select Folder'.

What is the pattern of the files (e.g. curve*.dat) ?? ***.dat**

As all the files with extension ".dat" are SAXS profiles, we enter "*.dat" when the program asks for the pattern of the files. Similarly, introduce the information for the second dataset.

Select the folder containing the data from experiment number 2 (press any key to continue)

Press 'Enter' and browse to the folder /path-to-data/1ak4/50-curves/TitrBwithA0p2to5 and click on 'Select Folder'.

What is the pattern of the files (e.g. curve*.dat) ?? ***.dat**

The selected datasets will be concatenated in a *multi-set* manner (one under the other).

An output folder is then defined for saving the results.

Where do you want to save the output? (press any key to continue)

Press 'Enter' and browse to the folder /path-to-data/output-folder and click on 'Select Folder'.

The program then asks for other information to properly load the SAXS profiles including the number of header lines, which are not read by the program, and the number of columns in the data.

Number of header lines [3]: **0**

Number of columns [3]: **3**

COSMiCS then performs a sanity check on the profiles to make sure that the profiles are of equal length and are sampled at the same q-values (see note 1). In general, COSMiCS accepts files containing at least three columns containing, from left to right, q-values, scattering intensities and experimental errors.

3. *Preprocessing of data*: The raw SAXS datasets need to be processed before starting the decomposition procedure. The first step consists of downscaling the curves to reduce the intensity differences between profiles. First, COSMiCS scales the profile with the lowest intensity in such a way that the smallest q-value for this curve adopts a value of 100 in arbitrary units (see note 2). Then, the rest of the curves are individually scaled with respect to the curve with the lowest intensity to minimize the intensity differences between them. The program can display the plots of the scaled SAXS curves to make sure the data is correctly read and loaded (Fig. 2A).

Do you want to see the curves (semi-logarithmic scale)?? [Y]/N: **Y**

The user is then prompted to specify the angular units of the input dataset with an option of converting the units from $Å^{-1}$ to nm^{-1} and vice-versa.

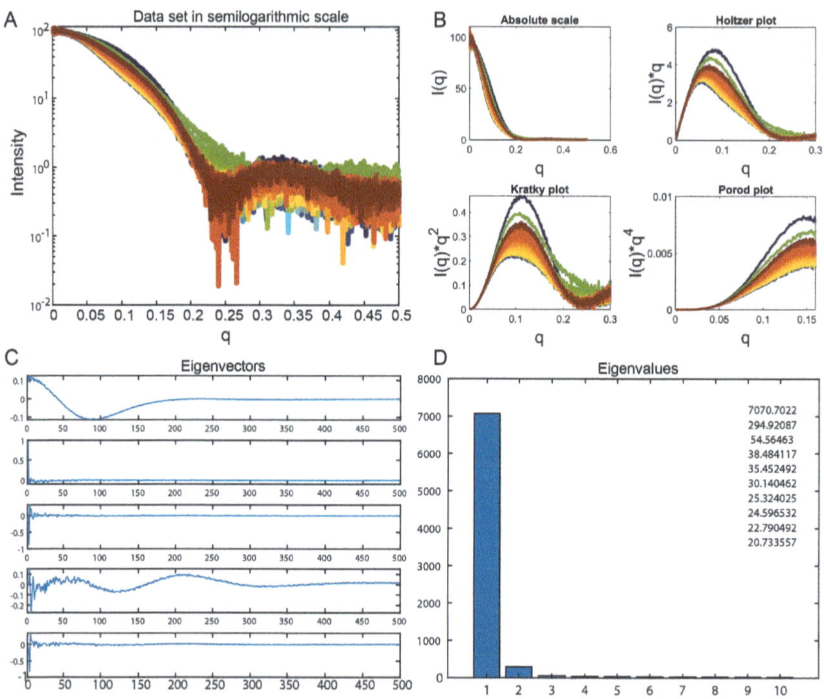

Fig. 2 (A) The plot of the curves loaded into COSMiCS after rescaling. (B) The four different representations generated by COSMiCS, i.e., Absolute (I(*q*) vs *q*), Holtzer (I(*q*).*q* vs *q*), Kratky (I(*q*).q^2 vs *q*) and Porod (*I*(*q*).q^4 vs *q*) after specifying the ranges for each representation. (C and D) The plots of Eigenvectors and Eigenvalues, respectively, obtained after principal component analysis.

Units of your curves: 1/Angstrom [A] or 1/nanometers (N): **A**

Do you want to change the units?? Y/[N]: **N**

The curves are displayed again for the user to make sure the input units are correct. The user is then prompted to choose the number of points to be removed from the beginning of the curves (see note 3). In this specific case, we do not need to remove any points.

Do you want to eliminate some data points at the beginning?? [Y]/N: **N**

4. *Generating multiple representations of SAXS datasets*: Our previous studies showed that simultaneously fitting multiple representations of SAXS datasets leads to better decomposition and reduces the chances of getting unrealistic SAXS profiles, which nevertheless fit the data (Herranz-Trillo et al., 2017; Sagar et al., 2021). Using multiple representations of the SAXS data emphasizes different regions of the *q*-range to improve fitting

throughout the angular range. COSMiCS allows the users to generate Holtzer ($I(q)^*q$ vs q) (Holtzer, 1955; Rambo & Tainer, 2013), Kratky ($I(q)^*q^2$ vs q) (Glatter & Kratky, 1982) and Porod ($I(q)^*q^4$ vs q) (Porod, 1951) representations of the input datasets (Fig. 2B) and build concatenated matrices in *multiway* mode (one besides the other). Then, the program allows to iteratively modify the q-ranges of different representations, by defining the largest q-value for each representation, until the appropriate ranges are found (see note 4).

```
Do you want to cut the matrices?? Y/N [Y]: Y
Default (1/???):
Cut for Absolute scale  = 0.3
Cut for Holtzer scale  = 0.16
Cut for Kratky scale = 0.16
Cut for Porod scale  = 0.07
Do you want default values?? [Y]/N: Y
```

In this case, the default values are correct.

5. *Determination of the number of significant components*: An essential step in decomposition is to determine the number of significant components (N) present in the matrix **D**. Note that this information might be known to the user from previous experiments. After loading and scaling the datasets, COSMiCS automatically runs Principal Component Analysis (PCA), which can be used to determine the number of components. The first 10 eigenvalues are displayed on the console and the plots showing the size of the eigenvalues and shape of the eigenvectors derived from the data matrix **D** analysis can be displayed (Fig. 2C and D) to allow the user to identify the number of relevant components (see note 5).

```
Do you want to see the results of PCA?? Y/N [Y]: Y
```

6. *Generation of initial estimates of SAXS profiles*: The next step is to select initial estimates of the scattering curves of the pure species in order to initiate the optimization process. Normally, these initial estimates are curves selected from the experimental dataset, **D**, after rescaling

In the first step, the number of components needs to be defined. We know from our knowledge of the systems as well as from the PCA done in the previous step that this example is a three-component system.

```
Number of species [3]: 3
```

Now the program will ask for choosing a method for selecting the initial estimates. If previous information is available, the curves of the dataset representing estimates can be introduced manually. For example, in case of a concentration series experiments monitoring a monomer-dimer equilibrium, the curves at the lowest and the highest concentrations can be selected as estimates of monomer and dimer, respectively. In case of a transient complex, the SAXS profiles of the individual components, which can be easily obtained and included in the dataset for decomposition, would serve as two of the initial estimates for the unbound partners. However, in the majority of cases it is not possible to provide initial estimates for all species and numerical strategies are necessary to identify them.

The methods presently available in COSMiCS are "Pure" from MCR-ALS and a χ^2-based method specifically developed for the analysis of SAXS data of transient complexes. The "Pure" option selects the purest curves in the experimental data matrix **D** based on the SIMPLISMA approach (Windig & Guilment, 1991; Windig & Stephenson, 1992). The reader is referred to the original publications for a detailed explanation of the method. For transient complexes, COSMiCS offers a novel strategy for choosing the initial estimate of the profile of the complex by comparing all the remaining curves of the dataset to those of the unbound partners. This analysis identifies the curve most different from the two corresponding to the unbound partners as the initial estimate of the complex (Fig. 3).

Selection Method for Initial Estimates

For Pure based selection enter 1

For Chi based selection enter 2 (Currently only available for 3 species)

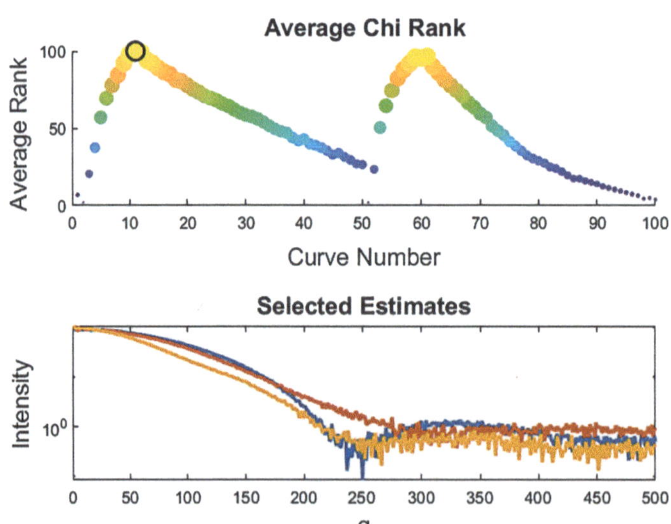

Fig. 3 (Top) Plot of average rank resulting from the χ^2-based approach for all the input SAXS curves. Curve 11 is the top-ranked one and it is selected to be the initial estimate for the complex. (Bottom) the three curves selected as initial estimates.

If the 'Pure' approach is used to analyze our test dataset,

How would you like to generate the initial estimates: **1**

The program finds the two curves belonging to the pure partners without any additional information (1 and 51), while curve 11 will be the starting profile for the complex.

If the χ^2 approach is used, the two curves for the unbound partners need to be defined. In this case, they are curves numbered 1 and 51 (First curves of each titration dataset). This approach also identifies curve 11 as the starting profile for the complex.

How would you like to generate the initial estimates: **2**

Enter two of the initial estimates (For example [1,51]) : **[1,51]**

Note that "Pure" based estimates are expected to be acceptable for relatively high-affinity complexes like "1AK4". However, χ^2-based approach is better adapted to decompose lower affinity complexes.

It is important to note that the identity of the initial estimates needs to be understood before applying *selectivity*. Sorting the curves in numerical order helps in this regard, so COSMiCS offers to sort the initial estimates.

Do you want to sort the curves?? [Y]/N: **Y**

7. *Definition of constraints*: The next step is to define the constraints to be imposed on the decomposition procedure. COSMiCS offers to use some predefined combinations of constraints suitable for certain kinds of experiments including fibrillation or folding, titration and SEC-SAXS. Otherwise, the user can input any combination in a vector form. The codes to apply various constraints are printed by COSMiCS in option 4 (see below). For example, to apply *non-negativity* and *closure*, the user can enter "[1,3]". In our example, we can use the "Titration" mode that applies *non-negativity*, *closure* and *selectivity*

Type of experiment
(1) Fibrillation or folding (default) (Non-negativity for conc and spectra, closure=1)
(2) Titration (Non-negativity for conc and spectra, closure, selectivity)
(3) SEC-SAXS experiment (Non-negativity for conc and spectra)
(4) User Defined Constraints (Non-negativity=1, Unimodality=2, Closure=3, Selectivity=4, Equality=5)
Select type of experiment: **2**

For almost all types of SAXS datasets, *non-negativity* can be imposed on both SAXS profiles and concentrations. Note that the fnnls algorithm is used by default.

Closure can be applied whenever a mass-balance is expected to be observed. *Closure* allows for reliably retrieving the concentrations in addition to the SAXS profiles but requires information about the stoichiometry of interaction (see note 6). In our example, *closure* can be applied as the initial concentrations are known and the stoichiometry is 1:1:1. The initial concentrations can be supplied as a text file in each of the specified input folders having a single line specifying the input concentration of the component kept at constant concentration through the titrations and setting the concentrations for the absent components as zero, in the correct order (see note 7). For example, in this case, the two input files should have "500 0 0" and "0 500 0" as they correspond to titrations with the first and second component kept constant, respectively. The zeros are used to ascertain which components are absent in the curves specified when implementing *selectivity*.

What is the pattern of the files (e.g. *.txt) ?? ***.txt**
What is the stoichiometry of the system (e.g. [1,1,1])??: **[1,1,1]**

The *selectivity* constraint also significantly improves decomposition performance of COSMiCS and should be used whenever possible. *Selectivity* or local rank implies the absence of certain components in defined regions of the dataset. For example, in the case of a transient complex, the SAXS curves of the individual partners can be defined to have "zero" concentration of the complex and of the other partner (see note 8). In our example, curves 1 and 51 correspond to the unbound partners and *selectivity* constraint will be applied for them.

Enter the curve numbers to apply selectivity restraint :**[1,51]**

If none of the predefined options are suitable, the user can input any combination in a vector form by selecting **4** as a type of experiment.

8. *Alternating least squares*: The final step is defining the parameters for running the ALS algorithm including the maximum number of steps (default = 1000) and the convergence criteria (default = 0.01% change in standard deviation of residuals). Moreover, a graphical display of the optimization process can be enabled

Convergence criterion (0.01% by default): **Enter**

Maximum number of iterations (1000 by default): **Enter**

Do you want graphic output during the ALS optimization [Y]/N)? **Y**

9. *Saved Results*: After successful completion of decomposition, nine subfolders are created in the specified output directory. Eight of these folders have the results for the eight combinations of the representations of the SAXS data. The names of the folders have the information about the combination of the representations used. For example, Test01_A and Test05_AHK have the results obtained by using just the Absolute representation and the combination of Absolute, Holtzer and Kratky representations, respectively (see note 9). These folders contain the decomposed SAXS profiles of the three (or the specified number) of species named "SpeciesX.dat" (see Note 10). The values of χ^2 between the experimental and fitted curves are given in a file called "chiSquareCurves.txt." The concentrations derived directly from decomposition (i.e., scaled by the same factor as the SAXS profiles) and rescaled to reflect the actual experiment are saved in files named "concentration.txt" and "ScaledConc.txt," respectively. The values used for scaling each of the curves are also saved in a file called "Scale.txt" in the output directory. Each of these folders contains a folder named "reconstruction" that has all the scaled input curves named "curveExp_xx.dat" and the fitted curves named "curvesMCRALS_xx.dat" from the decomposition. The output directory also has files with the eigenvectors and eigenvalues obtained in the PCA analysis. Additionally, the output directory has a subdirectory called "Report" which contains the plots of the various representations of the SAXS data used for decomposition, the eigenvalues and eigenvectors from PCA analysis, χ^2 values (Fig. 4A), the SAXS profiles and concentrations (Fig. 4B) obtained when using the different combinations of representations. An html report named 'Report.html' is also saved, which can be used for convenient viewing of all the results in a browser

4.2 Notes

1. All the SAXS profiles need to have exactly the same number of points at the same q-values. If some of the SAXS curves are sampled at different q-values, because they have been measured at different synchrotrons or different modes (batch vs SEC-SAXS), they should be re-binned using, for example, the program DATREGRID from the ATSAS package (Manalastas-Cantos et al., 2021).
2. The rescaling is done to improve the performance of fnnls algorithm when imposing *non-negativity* and to reduce the difference between the SAXS profiles due to the vertical translation

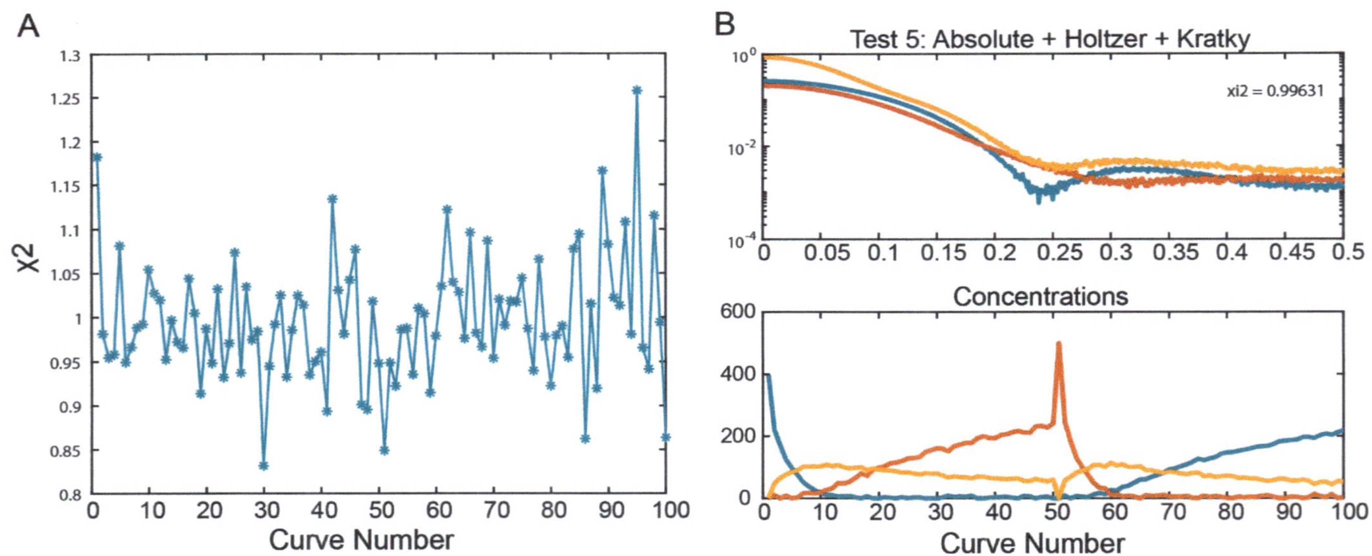

Fig. 4 (A) χ^2 values for all the input SAXS profiles and corresponding fitted curves. (B) The decomposed SAXS profiles for the three components (top) and their concentrations at each point of the titrations (bottom). Such plots are saved for all the combinations of SAXS data representations used for decomposition.

3. It is possible to observe deviations in the low-q data especially at high concentrations due to inter-particle interference effects or slight aggregation. In such cases, the same number of points should be removed from all the curves as explained in note 1
4. The q-ranges for the Holtzer and Kratky representations should be chosen to include the peak but exclude the high angle noisy data in case of folded proteins. For disordered proteins, the q-ranges can be chosen using the exclusion of the noisy high q-region as the only criteria
5. In this context, the number of relevant components refers to the number of independently evolving components, i.e., whose concentrations change differently through the dataset. For example, SAXS data collected for the disorder-to-order transition of a protein or by the addition of a ligand can be considered to be a two-component system, although physically, the number of conformations present in the system can be huge. For flexible systems, the curve derived from the decomposition corresponds to the average of all coexisting conformations in a given state
6. In cases where the stoichiometry is not known beforehand, a first COSMiCS run can be executed without using the *closure* constraint. The decomposed SAXS profiles can then be used to estimate the molecular weight and determine stoichiometry. The determined stoichiometry can be used to apply *closure* in a subsequent COSMiCS run to retrieve concentrations
7. In the current implementation of *closure*, only the concentration of the component kept fixed is used and the sum of the concentrations of the fixed component and the complex is constrained. This is done to avoid pipetting errors to negatively affect the decomposition. Therefore, the text file used as input for setting up the *closure* constraint can have all just one line specifying the used concentration of the fixed component and the rest as zero. However, if the users are confident about the precision of the input concentrations, the *closure* constraint can be modified to constrain the sum of all the components equal to the sum of input concentrations
8. *Selectivity* constraint can be applied by either defining SAXS profiles or concentrations. However, imposing *selectivity* on the concentrations is preferred because it allows for the SAXS profiles to be refined during the decomposition procedure. This generally increases the quality of the resulting profiles as they incorporate information from the whole dataset

9. COSMiCS uses all combinations of SAS data representations (8 in total), always including the Absolute values. In principle, all solutions should be equivalent. In practice, we have observed that in the vast majority of cases, the AHK combination (Absolute, Holtzer and Kratky) offers the highest quality solution (Sagar et al., 2021). We have also observed that combinations containing Porod's representation tend to provide the lowest quality solutions
10. The error bars in the output decomposed SAXS profiles are error bars of the experimental SAXS profiles used as corresponding initial estimates and do not represent the uncertainty in decomposition

5. Indications of unreliable/incorrect decomposition and possible solutions

In general, the performance of COSMiCS can be negatively affected if the dataset has too few curves, the SAXS profiles are excessively noisy, the number of components used for decomposition is incorrect or the initial estimates are incorrectly identified. There are two possible scenarios in cases where COSMiCS or any decomposition procedure gives incorrect results. The first scenario, which is straightforward to identify, is when the experimental data is not fitted by decomposition. This is clearly detectable by large values of χ^2 between the experimental and fitted curves and systematic deviations in the residual plots. In such cases, the user can try increasing the number of components, removing the low-q data if inter-particle interference signatures are identified, or choosing a different method for selecting the initial estimates.

The second scenario, which is harder to identify, is when data are fitted correctly but the solutions are incorrect. This occurs when some ambiguity in the solutions exists even after imposing constraints. Still, there are a few indications of potential problems in decomposition and possible measures to deal with them.

(1) One indication of a problem in decomposition is manifested when two or more of the decomposed components have very similar SAXS profiles. This typically occurs due to overestimation of the number of components. This overestimation could be either because some of the expected components simply do not exist, they are sampled in very small fractions with the performed experiments, or they evolve simultaneously (see note 5). In such cases, the number of components should be reduced till all the decomposed SAXS profiles are dissimilar.

The analysis of decomposed SAXS profiles should provide some indication about whether some of the expected components are missing or the procedure is unable to separate their contributions. The components with small contributions can be extracted by either imposing additional constraints on decomposition or modifying the experimental conditions to increase their contribution to the scattering signal in some of the SAXS profiles

(2) Another indication of inaccurate decomposition is the presence of artifacts in the decomposed SAXS profiles, i.e., sharp peaks and dips which are not expected for typical biomolecular SAXS profiles. The use of multiple representations of SAXS data drastically reduces the chances of getting such solutions. In cases where such artifacts are observed, even when using multiple representations, the q-ranges used for the alternative representations should be checked to ensure that the complete peak is considered for Kratky and Holtzer representations and the plateau is reached in Porod representation

6. Summary and conclusions

Scattering techniques are well-suited for the structural investigation of polydisperse systems in nearly physiological conditions. However, despite their sensitivity to the presence of multiple species, decomposing averaged curves in the different contributions is extremely challenging if no additional structural information is available. Here, we have presented a detailed procedure to run a program, COSMiCS, to decompose SAXS and SANS datasets sampling complex mixtures. The example chosen to exemplify its usage corresponds to a transient biomolecular complex, but small changes in COSMiCS parameters enable its usage to oligomerization processes, folding reactions or time-dependent experiments. Its modular and open-source code enables the user to adapt COSMiCS to a broad range of systems and to integrate other biophysical and spectroscopic measurements. The output provided by the program allows the evaluation of its performance and the quality of the resulting curves. In this way, SAS-driven modeling tools could be applied to derive structural models of molecular species that are scarcely populated in solution. The continuous improvement of Bio-SAXS beamlines with increasing brilliance and automatized protocols allow the fast and robust measurements of hundreds of samples in a reasonable time. Moreover, the amount of material required for these experiments is continuously reduced by implementing microfluidic devices in the sample environment. In the context of this continuous technical evolution, COSMiCS can be a highly valuable tool to address biological questions that are out of the reach of traditional structural biology methods, including encounter complexes, amyloidogenic soluble oligomers, kinetics of structural changes using TR-SAXS, capsid assembly mechanisms, etc.

Acknowledgments

This work was supported by the Labex EpiGenMed, an "Investissements d'avenir" program (ANR-10-LABX-12-01). The CBS is a member of France-BioImaging (FBI) and the French Infrastructure for Integrated Structural Biology (FRISBI), 2 national infrastructures supported by the French National Research Agency (ANR-10-INBS-04-01 and ANR-10-INBS-05, respectively).

References

Akiyama, S., Takahashi, S., Kimura, T., Ishimori, K., Morishima, I., Nishikawa, Y., et al. (2002). Conformational landscape of cytochrome c folding studied by microsecond-resolved small-angle x-ray scattering. *Proceedings of the National Academy of Sciences of the United States of America*, *99*, 1329–1334.

Ayuso-Tejedor, S., García-Fandiño, R., Orozco, M., Sancho, J., & Bernadó, P. (2011). Structural analysis of an equilibrium folding intermediate in the apoflavodoxin native ensemble by small-angle X-ray scattering. *Journal of Molecular Biology*, *406*, 604–619.

Baek, M., DiMaio, F., Anishchenko, I., Dauparas, J., Ovchinnikov, S., Lee, G. R., et al. (2021). Accurate prediction of protein structures and interactions using a three-track neural network. *Science*, *373*(6557), 871–876.

Bernadó, P., & Blackledge, M. (2010). Structural biology: Proteins in dynamic equilibrium. *Nature*, *468*(7327), 1046–1048.

Bernadó, P., Pérez, Y., Blobel, J., Fernández-Recio, J., Svergun, D. I., & Pons, M. (2009). Structural characterization of unphosphorylated STAT5a oligomerization equilibrium in solution by small-angle X-ray scattering. *Protein Science*, *18*(4), 716–726.

Bernadó, P., Shimizu, N., Zaccai, G., Kamikubo, H., & Sugiyama, M. (2018). Solution scattering approaches to dynamical ordering in biomolecular systems. *Biochimica et Biophysica Acta-General Subjects*, *1862*(2), 253–274.

Bernadó, P., & Svergun, D. I. (2012). Structural analysis of intrinsically disordered proteins by small-angle X-ray scattering. *Molecular BioSystems*, *8*, 151–167.

Berntsson, O., Rodriguez, R., Henry, L., Panman, M. R., Hughes, A. J., Einholz, C., et al. (2019). Photoactivation of Drosophila melanogaster cryptochrome through sequential conformational transitions. *Science. Advances*, *5*(7), eaaw1531.

Blobel, J., Bernadó, P., Svergun, D. I., Tauler, R., & Pons, M. (2009). Low-resolution structures of transient protein-protein complexes using small-angle X-ray scattering. *Journal of the American Chemical Society*, *131*, 4378–4386.

Bro, R., & de Jong, S. (1997). A fast non-negativity-constrained least squares algorithm. *Journal of Chemometrics*, *11*(5), 393–401.

Bro, R., & Sidiropoulos, N. D. (1998). Least squares algorithms under unimodality and non-negativity constraints. *Journal of Chemometrics*, *12*, 223–247.

Chandola, H., Williamson, T. E., Craig, B. A., Friedman, A. M., & Bailey-Kellogg, C. (2014). Stoichiometries and affinities of interacting proteins from concentration series of solution scattering data: Decomposition by least squares and quadratic optimization. *Journal of Applied Crystallography*, *47*, 899–914.

Chandonia, J. M., & Brenner, S. E. (2006). The impact of structural genomics: Expectations and outcomes. *Science*, *311*(5759), 347–351.
Chen, L., Hodgson, K. O., & Doniach, S. (1996). A lysozyme folding intermediate revealed by solution X-ray scattering. *Journal of Molecular Biology*, *261*(5), 658–671.
Chevreuil, M., Lecoq, L., Wang, S., Gargowitsch, L., Nhiri, N., Jacquet, E., et al. (2020). Nonsymmetrical dynamics of the HBV capsid assembly and disassembly evidenced by their transient species. *Journal of Physical Chemistry B*, *124*(45), 9987–9995.
Cho, H. S., Schotte, F., Stadnytskyi, V., & Anfinrud, P. (2021). Time-resolved X-ray scattering studies of proteins. *Current Opinion in Structural Biology*, *70*, 99–107.
Corsini, P. M., Wang, S., Rehman, S., Fenn, K., Sagar, A., Sirovica, S., et al. (2022). Molecular and cellular insight into *Escherichia coli* SslE and its role during biofilm maturation. *NPJ Biofilms and Microbiomes*, *8*(1), 9.
de Juan, A., & Tauler, R. (2003). Chemometrics applied to unravel multicomponent processes and mixtures: Revisiting latest trends in multivariate resolution. *Analytica Chimica Acta*, *500*, 195–210.
de Juan, A., van der Heyden, Y., Tauler, R., & Massart, D. L. (1997). Assessment of new constraints applied to the alternating least squares method. *Analytica Chimica Acta*, *346*, 307–318.
Doniach, S. (2001). Changes in biomolecular conformation seen by small angle X-ray scattering. *Chemical Reviews*, *101*, 1763–1778.
Feigin, L. A., & Svergun, D. I. (1987). *Structure analysis by small-angle X-ray and neutron scattering*. New York: Plenum Press.
Giehm, L., Svergun, D. I., Otzen, D. E., & Vestergaard, B. (2011). Low-resolution structure of a vesicle disrupting α-synuclein oligomer that accumulates during fibrillation. *Proceedings of the National Academy of Sciences of the United States of America*, *108*(8), 3246–3251.
Glatter, O., & Kratky, O. (1982). *Small angle X-ray scattering*. Academic press.
Graceffa, R., Nobrega, R. P., Barrea, R. A., Kathuria, S. V., Chakravarthy, S., Bilsel, O., et al. (2013). Sub-millisecond time-resolved SAXS using a continuous-flow mixer and X-ray microbeam. *Journal of Synchrotron Radiation*, *20*(6), 820–825.
Herranz-Trillo, F., Groenning, M., van Maarschalkerweerd, A., Tauler, R., Vestergaard, B., & Bernadó, P. (2017). Structural analysis of multi-component amyloid systems by chemometric SAXS data decomposition. *Structure*, *25*, 5–15.
Heyes, D. J., Hardman, S. J. O., Pedersen, M. N., Woodhouse, J., De La Mora, E., Wulff, M., et al. (2019). Light-induced structural changes in a full-length cyanobacterial phytochrome probed by time-resolved X-ray scattering. *Communications in Biology*, *2*, 1.
Holtzer, A. (1955). Interpretation of the angular distribution of the light scattered by a polydisperse system of rods. *Journal of Polymer Science*, *17*(85), 432–434.
Hopkins, J. B., Gillilan, R. E., & Skou, S. (2017). BioXTAS RAW: Improvements to a free open-source program for small-angle X-ray scattering data reduction and analysis. *Journal of Applied Crystallography*, *50*, 1545–1553.
Hura, G. L., Budworth, H., Dyer, K. N., Rambo, R. P., Hammel, M., McMurray, C. T., et al. (2013). Comprehensive macromolecular conformations mapped by quantitative SAXS analyses. *Nature Methods*, *10*(6), 453–454.
Jacques, D. A., & Trewhella, J. (2010). Small-angle scattering for structural for structural biology-expanding the frontier while avoiding the pitfalls. *Protein Science*, *2010*(19), 642–657.
Jaumot, J., Gargallo, R., De Juan, A., & Tauler, R. (2005). A graphical user-friendly interface for MCR-ALS: A new tool for multivariate curve resolution in MATLAB. *Chemometrics and Intelligent Laboratory Systems*, *76*(1), 101–110.
Josts, I., Gao, Y., Monteiro, D. C. F., Niebling, S., Nitsche, J., Veith, K., et al. (2020). Structural kinetics of MsbA investigated by stopped-flow time-resolved small-angle X-ray scattering. *Structure*, *28*(3), 348–354.

Josts, I., Niebling, S., Gao, Y., Levantino, M., Tidow, H., & Monteiro, D. (2018). Photocage-initiated time-resolved solution X-ray scattering investigation of protein dimerization. *IUCrJ*, *5*(6), 667–672.
Jumper, J., Evans, R., Pritzel, A., Green, T., Figurnov, M., Ronneberger, O., et al. (2021). Highly accurate protein structure prediction with AlphaFold. *Nature*, *596*(7873), 583–589.
Konarev, P. V., Graewert, M. A., Jeffries, C. M., Fukuda, M., Cheremnykh, T. A., Volkov, V. V., et al. (2022). EFAMIX, a tool to decompose inline chromatography SAXS data from partially overlapping components. *Protein Science*, *31*(1), 269–282.
Konarev, P. V., Volkov, V. V., Sokolova, A. V., Koch, M. H. J., & Svergun, D. I. (2003). PRIMUS: A windows PC-based system for small-angle scattering data analysis. *Journal of Applied Crystallography*, *36*, 1277–1282.
Kuhlbrandt, W. (2014). The resolution revolution. *Science*, *2014*(343), 1443–1444.
Lafleur, J. P., Snakenborg, D., Nielsen, S. S., Møller, M., Toft, K. N., Menzel, A., et al. (2011). *Journal of Applied Crystallography*, *44*, 1090–1099.
Langkilde, A. E., Herranz-Trillo, F., Bernadó, P., & Vestergaard, B. (2018). Noninvasive structural analysis of intermediate species during fibrillation: An application of small-angle X-ray scattering. *Methods in Molecular Biology*, *1779*, 209–239.
Lawson, C. L., & Hanson, R. J. (1974). *Solving least squares problems*. New York: Prentice-Hall.
Le Vay, K., Carter, B. M., Watkins, D. W., Dora Tang, T. Y., Ting, V. P., Cölfen, H., et al. (2020). Controlling protein Nanocage assembly with hydrostatic pressure. *Journal of the American Chemical Society*, *142*, 20640–20650.
Lipfert, J., Coulmbus, L., Chu, V. B., & Doniach, S. (2007). Analysis of small-angle X-ray scattering data of protein-detergent complexes by singular value decomposition. *Journal of Applied Crystallography*, *40*, s235–s239.
Malinowski, E. (2002). *Factor analysis in chemistry ed*. New York, NY: Wiley-Interscience.
Manalastas-Cantos, K., Konarev, P. V., Hajizadeh, N. R., Kikhney, A. G., Petoukhov, M. V., Molodenskiy, D. S., et al. (2021). ATSAS 3.0: Expanded functionality and new tools for small-angle scattering data analysis. *Journal of Applied Crystallography*, *54*(1), 343–355.
Meisburger, S. P., Taylor, A. B., Khan, C. A., Zhang, S., Fitzpatrick, P. F., & Ando, N. (2016). Domain movements upon activation of phenylalaninehydroxylase characterized by crystallography and chromatography-coupled small-angle X-ray scattering. *Journal of the American Chemical Society*, *138*, 6506–6516.
Meisburger, S. P., Xu, D., & Ando, N. (2021). REGALS: A general method to deconvolve X-ray scattering data from evolving mixtures. *IUCrJ*, *8*(2), 225–237.
Minh, D. D. L., & Makowski, L. (2013). Wide-angle X-ray solution scattering for protein-ligand binding: Multivariate curve resolution with Bayesian confidence intervals. *Biophysical Journal*, *104*(4), 873–883.
Oliver, R. C., Potrzebowski, W., Najibi, S. M., Pedersen, M. N., Arleth, L., Mahmoudi, N., et al. (2020). Assembly of capsids from hepatitis B virus Core protein progresses through highly populated intermediates in the presence and absence of RNA. *ACS Nano*, *14*(8), 10226–10238.
Petoukhov, M. V., Franke, D., Shkumatov, A. V., Tria, G., Kikhney, A. G., Gajda, M., et al. (2012). New developments in the ATSAS program package for small-angle scattering data analysis. *Journal of Applied Crystallography*, *45*, 342–350.
Petsko, G., & Ringe, D. (2004). *Protein structure and function*. News Science Press.
Plumridge, A., Katz, A. M., Calvey, G. D., Elber, R., Kirmizialtin, S., & Pollack, L. (2018). Revealing the distinct folding phases of an RNA three-helix junction. *Nucleic Acids Research*, *46*(14), 7354–7365.
Porod, G. (1951). Die Röntgenkleinwinkelstreuung von dichtgepackten kolloiden Systemen. *Colloid and Polymer Science*, *124*, 83–114.

Putnam, C. D., Hammel, M., Hura, G. L., & Tainer, J. A. (2007). X-ray solution scattering (SAXS) combined with crystallography and computation: Defining accurate macromolecular structures, conformations and assemblies in solution. *Quarterly Reviews of Biophysics*, *40*, 191–285.

Rajkó, R. (2010). Additional knowledge for determining and interpreting feasible band boundaries in self-modeling/multivariate curve resolution of two-component systems. *Analytica Chimica Acta*, *661*(2), 129–132.

Rambo, R. P., & Tainer, J. A. (2013). Accurate assessment of mass, models and resolution by small-angle scattering. *Nature*, *496*, 477–481.

Rimmerman, D., Leshchev, D., Hsu, D. J., Hong, J., Abraham, B., Henning, R., et al. (2019). Revealing fast structural dynamics in pH-responsive peptides with time-resolved X-ray scattering. *Journal of Physical Chemistry B*, *123*(9), 2016–2021.

Rodríguez-Ruiz, I., Radajewski, D., Charton, S., Phamvan, N., Brennich, M., Pernot, P., et al. (2017). Innovative high-throughput SAXS methodologies based on photonic lab-on-a-Chip sensors: Application to macromolecular studies. *Sensors (Basel)*, *17*(6), 1266.

Sagar, A., Herranz-Trillo, F., Langkilde, A. E., Vestergaard, B., & Bernadó, P. (2021). Structure and thermodynamics of transient protein-protein complexes by chemometric decomposition of SAXS datasets. *Structure*, *29*, 1074–1090.

Schwemmer, F., Blanchet, C. E., Spilotros, A., Kosse, D., Zehnle, S., Mertens, H. D., et al. (2016). LabDisk for SAXS: A centrifugal microfluidic sample preparation platform for small-angle X-ray scattering. *Lab on a Chip*, *16*(7), 1161–1170.

Segel, D. J., Eliezer, D., Uversky, V., Fink, A. L., Hodgson, K. O., & Doniach, S. (1999). Transient dimer in the refolding kinetics of cytochrome c characterized by small-angle X-ray scattering. *Biochemistry*, *38*, 15352–15359.

Silva, B. F. B. (2017). SAXS on a chip: From dynamics of phase transitions to alignment phenomena at interfaces studied with microfluidic devices. *Physical Chemistry Chemical Physics*, *19*(35), 23690–23703.

Taube, M., Pietralik, Z., Szymanska, A., Szutkowski, K., Clemens, D., Grubb, A., et al. (2019). The domain swapping of human cystatin C induced by synchrotron radiation. *Scientific Reports*, *9*(1), 8548.

Tauler, R. (1995). Multivariate curve resolution applied to second order data. *Chemometrics and Intelligent Laboratory Systems*, *30*, 133–146.

Tauler, R. (2001). Calculation of maximum and minimum band boundaries of feasible solutions for species profiles obtained by multivariate curve resolution. *Journal of Chemometrics*, *15*(8), 627–646.

Thompson, M. C., Barad, B. A., Wolff, A. M., Sun Cho, H., Schotte, F., Schwarz, D. M. C., et al. (2019). Temperature-jump solution X-ray scattering reveals distinct motions in a dynamic enzyme. *Nature Chemistry*, *11*(11), 1058–1066.

Toft, K. N., Vestergaard, B., Nielsen, S. S., Snakenborg, D., Jeppesen, M. G., Jacobsen, J. K., et al. (2008). High-throughput small angle X-ray scattering from proteins in solution using a microfluidic front-end. *Analytical Chemistry*, *80*(10), 3648–3654.

Trygg, J., Holmes, E., & Lundstedt, T. (2007). Chemometrics in metabonomics. *Journal of Proteome Research*, *6*, 469–479.

Tully, M. D., Tarbouriech, N., Rambo, R. P., & Hutin, S. (2021). Analysis of SEC-SAXS data via EFA deconvolution and scatter. *Journal of Visualized Experiments*, *167*, e61578.

Williamson, T. E., Craig, B. A., Kondrashkina, E., Bailey-Kellogg, C., & Friedman, A. M. (2008). Analysis of self-associating proteins by singular value decomposition of solution scattering data. *Biophysical Journal*, *94*, 4906–4923.

Windig, W., & Guilment, J. (1991). Interactive self-modeling mixture analysis. *Analytical Chemistry*, *63*(14), 1425–1432.

Windig, W., & Stephenson, D. A. (1992). Self-modeling mixture analysis of second-derivative near-infrared spectral data using the simplisma approach. *Analytical Chemistry*, *64*, 2735–2742.

FSC
www.fsc.org
MIX
Papier | Fördert gute Waldnutzung
FSC® C083411

Kolibri 360 GmbH
Ferdinand-Jühlke-Straße 7
99095 Erfurt, Deutschland
produktsicherheit@kolibri360.de